**ADVANCES IN
CELLULAR NEUROBIOLOGY
Volume 1**

ADVANCES IN CELLULAR NEUROBIOLOGY

Volume 1

EDITED BY

SERGEY FEDOROFF

LEIF HERTZ

College of Medicine
University of Saskatchewan
Saskatoon, Canada

1980

ACADEMIC PRESS
A Subsidiary of Harcourt Brace Jovanovich, Publishers

New York   London   Toronto   Sydney   San Francisco

ACADEMIC PRESS, INC.
111 Fifth Avenue, New York, New York 10003

United Kingdom Edition published by
ACADEMIC PRESS, INC. (LONDON) LTD.
24/28 Oval Road, London NW1 7DX

ISSN 0270–0794

ISBN 0–12–008301–9

PRINTED IN THE UNITED STATES OF AMERICA

80 81 82 83   9 8 7 6 5 4 3 2 1

CONTENTS

BIOCHEMICAL CHARACTERISTICS OF INDIVIDUAL NEURONS

Takahiko Kato

SECTION 2. AGING AND PATHOLOGY

CEREBELLAR GRANULE CELLS IN NORMAL AND NEUROLOGICAL MUTANTS OF MICE

Anne Messer

CELL GENERATION AND AGING OF NONTRANSFORMED GLIAL CELLS FROM ADULT HUMANS

Jan Pontén and *Bengt Westermark*

AGE-RELATED CHANGES IN NEURONAL AND GLIAL ENZYME ACTIVITIES

Antonia Vernadakis and *Ellen Bragg Arnold*

GLIAL FIBRILLARY ACIDIC (GFA) PROTEIN IN NORMAL NEURAL CELLS AND IN PATHOLOGICAL CONDITIONS

Amico Bignami, Doris Dahl, and *David C. Rueger*

SECTION 3. METHODOLOGIES

IN VITRO BEHAVIOR OF ISOLATED OLIGODENDROCYTES

Sara Szuchet and *Kari Stefansson*

BIOCHEMICAL MAPPING OF SPECIFIC NEURONAL PATHWAYS

E. G. McGeer and *P. L. McGeer*

SEPARATION OF NEURONAL AND GLIAL CELLS AND SUBCELLULAR CONSTITUENTS

Fritz A. Henn

SEPARATION OF NEURONS AND GLIAL CELLS BY AFFINITY METHODS

Silvio Varon and *Marston Manthorpe*

LIST OF CONTRIBUTORS

Numbers in parentheses indicate the pages on which the authors' contributions begin.

J.J. Anders (3), Laboratory of Neuropathology and Neuroanatomical Sciences, National Institute of Neurological, Communicative Disease and Stroke, National Institutes of Health, Bethesda, Maryland 20205

Ellen Bragg Arnold (229), Department of Psychiatry, University of Colorado, School of Medicine, Denver, Colorado 80262

Amico Bignami (285), Spinal Cord Injury Research Laboratory, West Roxbury Veterans Administration Medical Center, and Department of Neuropathology, Harvard Medical School, Boston, Massachusetts 02132

M.A. Bisby (69), Division of Medical Physiology, University of Calgary, Calgary, Alberta T2N 1N4, Canada

M.W. Brightman (3), Laboratory of Neuropathology and Neuroanatomical Sciences, National Institute of Neurological, Communicative Disease and Stroke, National Institutes of Health, Bethesda, Maryland 20205

Doris Dahl (285), Spinal Cord Injury Research Laboratory, West Roxbury Veterans Administration Medical Center, and Department of Neuropathology, Harvard Medical School, Boston, Massachusetts 02132

Bernd Hamprecht (31), Physiologisch–Chemisches Institut, Universität Würzburg, Würzburg 8700, Federal Republic of Germany

Fritz A. Henn (373), Neurochemical Research Laboratories, Department of Psychiatry, University of Iowa, Iowa City, Iowa 52242

Takahiko Kato (119), Department of Biochemistry, Institute of Brain Research, University of Tokyo, Faculty of Medicine, Bunkyoku, Tokyo, Japan

ix

E.G. McGeer (347), Kinsmen Laboratory of Neurological Research, Department of Psychiatry, University of British Columbia, Vancouver, British Columbia V6T 1W5, Canada

P.L. McGeer (347), Kinsmen Laboratory of Neurological Research, Department of Psychiatry, University of British Columbia, Vancouver, British Columbia V6T 1W5, Canada

Marston Manthorpe (405), Department of Biology, University of California at San Diego, La Jolla, California 92037

Anne Messer (179), Division of Laboratories and Research, New York State Department of Health, Albany, New York 12201

Jan Pontén (209), Department of Tumor Biology, Wallenberg Laboratory, S-751-22 Uppsala, Sweden

J.M. Rosenstein (3), Laboratory of Neuropathology and Neuroanatomical Sciences, National Institute of Neurological, Communicative Disease and Stroke, National Institutes of Health, Bethesda, Maryland 20205

David C. Rueger (285), Spinal Cord Injury Research Laboratory, West Roxbury Veterans Administration Medical Center, and Department of Neuropathology, Harvard Medical School, Boston, Massachusetts 02132

Sara Szuchet (313), Department of Neurology, Pritzker School of Medicine, University of Chicago, Chicago, Illinois 60637

Kari Stefansson (313), Department of Neurology, Pritzker School of Medicine, University of Chicago, Chicago Illinois 60637

Dietrich van Calker (31), Max-Planck-Institut für Biochemie, Martinsried, Federal Republic of Germany

Silvio Varon (405), Department of Biology and School of Medicine, University of California at San Diego, La Jolla, California 92037

Antonia Vernadakis (229), Departments of Psychiatry and Pharmacology, University of Colorado School of Medicine, Denver, Colorado 80262

Bengt Westermark (209), Department of Tumor Biology, Wallenberg Laboratory, S-751-22 Uppsala, Sweden

PREFACE

Despite tremendous advances in biomedical research in the last 20 years, understanding of the nervous system has lagged behind. The topographical anatomy and functional relation of the brain and spinal cord were already established at the turn of the century but the structure and function at the cellular level and higher brain functions such as learning, memory, and intelligence are still poorly understood.

The central nervous system is a heterogeneous system composed of many small populations of cells, each differing from the other in cell numbers, cell size, cell types, and density, and each occupying a distinct topographical area of the brain. At the same time, these distinct cell populations (nuclei) are developmentally and functionally integrated into one nervous system. Each of these areas of the central nervous system is composed of several types of neuronal and glial cells with different biochemical, biophysical, and pharmacological characteristics. Apart from electrophysiological recording, it is only very recently that techniques have become available to study the characteristics of the individual cell types, establishing a new area of neurobiology which could appropriately be named "cellular neurobiology" and which draws on some aspects of morphology, biochemistry, pharmacology, physiology, endocrinology, embryology, and genetics.

Discussion with many neurobiologists has convinced us that it would be very useful, at a time when neurobiology is expanding exponentially, to have a publication that would regularly review new information in the area of cellular neurobiology. As a result *Advances in Cellular Neurobiology* was undertaken. It will differ from existing publications in the area of neurobiology in that attention will be focused on the cellular basis of neurobiological events rather than molecular or organismic organizations and functions of the nervous systems.

Each volume will have three subdivisions: Cell differentiation and interaction, aging and pathology, and methodology. We intend that the series will provide a compendium of the rapidly increasing body of information about neural cells and as an interface between normal and abnormal (pathological) reactions of the cells, that it will stimulate new ideas and research. At the same time the transfer of basic information to clinical situations and application may be facilitated. Advancement of research in any field is dependent on the methodologies available, and such has certainly been the case in the development of cellular neurobiology. For this reason we have included a section on methods. The intention is to review both those methods currently used and also methods from other fields which might possibly be applied in neurobiology.

SERGEY FEDOROFF
LEIF HERTZ

ADVANCES IN
CELLULAR NEUROBIOLOGY
Volume 1

Section 1

CELL DIFFERENTIATION AND INTERACTION

SPECIALIZATIONS OF NONNEURONAL CELL MEMBRANES IN THE VERTEBRATE NERVOUS SYSTEM

M. W. BRIGHTMAN, J. J. ANDERS, AND J. M. ROSENSTEIN

Laboratory of Neuropathology and Neuroanatomical Sciences,
National Institute of Neurological, Communicative Disease and Stroke,
National Institutes of Health, Bethesda, Maryland

I. Introduction

The plasma membranes of nonneuronal cells within the central nervous system (CNS), being derived from ectoderm, have in common with the

ISBN 0-12-008301-9

epithelia of other organs, a number of morphological properties. These properties include, first, the sharing of membrane material between contiguous cells to form intercellular junctions and, second, an intrinsic membrane specialization—orthogonal arrays of particles within the cell membranes of astrocytes and ependymal cells. These arrays are characteristic of the plasma membrane of only six other organs so far described.

The intercellular junctions that tether nonneuronal cells of the brain and spinal cord are of four main types: gap junctions, tight junctions, hemidesmosomes, and adhering junctions. We shall restrict our comments to gap junctions and tight junctions, whose functions in neural tissue are partly inferred from their roles in other organs. Although these junctions are involved in fastening cells to each other, the extent of adhesion at tight junctions and the cell-to-cell communications at gap junctions set them apart from purely adhesive contacts. These observations will be summarized here to help us guess how the junctions may influence the activities of astrocytes and choroid plexus epithelium in particular.

An account of the intramembranous specialization of astrocytes and ependymal cells, the orthogonal congeries of particles or "assemblies," is also set forth in some detail. Our intention is to describe them qualitatively and quantitatively and to present a comparison with those of other organs. This information should enable us to discern any changes that would result from experimental manipulation of the assemblies, the functions of which are unknown.

II. Gap Junctions

A. Structure

The nonneuronal cells of the nervous system that form gap junctions are astrocytes and ependymal cells (Brightman and Reese, 1969) and the cells of the arachnoid and pia mater (Nabeshima et al., 1975). Consequently, within and around the CNS this type of intercellular contact is ubiquitous.

In electron micrographs of tissue that has been block-mordanted in uranyl acetate solution before plastic embedment, the junctions appear septalaminar: The unit membrane of each contiguous cell contributes three laminae on either side of a seventh, median layer (Fig. 1) (Revel and Karnovsky, 1967). The median layer is a cleft about 2–3 nm wide and confluent with the extracellular space. These junctions never form continuous belts or zonules around cell bodies but are discontinuous plaques which can be penetrated or circumvented by molecules moving through the extracellular

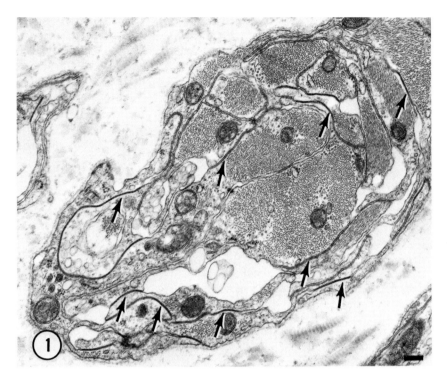

FIG. 1. The interior of this Schwann cell tunnel, once occupied by neurites, is now filled with reactive Schwann cell processes identified by their numerous cytoplasmic filaments and many gap junctions (arrows). The specimen is from a fragment of rat superior cervical ganglion transplanted to the cerebellar surface 14 months prior to fixation. Bar = 225 nm.

space. Since the median gap is so narrow, most molecules move primarily around the junctions which thus act as impediments but not complete barriers to free, extracellular flow.

When outlined with colloidal lanthanum hydroxide and viewed *en face,* the gap junction appears as an aggregate of closely packed polygons with a center-to-center spacing of 8–9 nm. Each polygon is the subunit or "connexon," a designation suggested by Goodenough (1975), of the gap junction. Although the connexon is impermeable to extracellular solutes, it is purported to enclose a core that is permeable, as we shall see, to certain intracellular ions and molecules, thus permitting the migration of material from cell to cell.

In replicas of freeze-fractured gap junctions, the connexons appear in bold relief as large, round particles about 8–9 nm in diameter, which form aggregates on the inner P face of the membrane, with complementary pits on the outer E face. The center-to-center spacing of adjacent connexons varies with the packing density, the determinants of which will be considered in some detail, as will the association of the subunits with those of tight junctions. An extensive, detailed account of the structural, biochemical, and functional aspects of gap junctions has recently been presented by Bennett and Goodenough (1978). Variations in the form of gap junctions have been summarized by Larsen (1977), and for a discussion of other types of junctions that occur between nonneuronal cells, such as adhering junctions, the reader is also referred to the review by Staehelin (1974).

B. Functions Attributed to Gap Junctions

1. ELECTROTONIC COUPLING

A wide variety of epithelial cell types, including nonexcitable ones such as liver (Penn, 1966), salivary gland, and excretory tubule epithelium (Loewenstein and Kanno, 1964; Loewenstein, 1975), are electrotonically coupled. Electrical communication between cells has been determined by injecting electric current into one cell of a cluster of cells in visible contact with each other and then measuring the resulting voltage across the injected cell and one of its neighbors. A critical appraisal of such techniques and the concept of electrotonic interaction is presented by Bennett (1977).

How specific is this intercellular communication? When genetically different cells are cocultured, electrical communication can be established between some but not all of them. For example, one type of hamster fibroblast can communicate with rabbit lens epithelial cells which can also interact with rat liver cells (Michalke and Loewenstein, 1971). Similarly, myocardial and 3T3 cells from the mouse, when cocultured, become ionically coupled (Epstein and Gilula, 1977). Thus, genetically disparate cells can, *in vitro,* communicate with each other. The establishment of electrical interaction is nonspecific.

Not only ions and electric current, but relatively large molecules, including fluorescein (MW 332), can pass between cells in contact with each other (Loewenstein and Kanno, 1964). The sites of passage were assumed to be intercellular junctions. Junctions between normal cells *in situ,* however, are almost always composite, with tight junctions, adhering junctions, and gap junctions arranged in series (Farquhar and Palade, 1963). The gap junction has, because of its substructure, become the preferred candidate for the site of electrical and molecular exchange between conjoined cells. The structural attribute that was suggestive of a communicating pore appeared subtly in *en face* views as a faint dot stained by lanthanum (Revel and Karnovsky, 1967). This core, about 1.5–2 nm wide, was situated in the center of the connexon and has since been clearly depicted in gap junctions isolated from rat liver and negatively stained with uranyl formate by Gilula (1977). The central core is regarded as the hydrated channel through which electric current and molecules in the cytoplasm are exchanged between coupled cells (Bennett, 1973; McNutt and Weinstein, 1970, 1973; Gilula *et al.,* 1972). Each channel is insulated from extracellular fluid by the surrounding wall of the subunit or connexon.

In addition to its substructure, there are other indications that the gap junction, rather than the other members of junctional complexes, mediates electrotonic coupling. Vertebrate cells *in vitro* that become ionically connected are joined, according to the descriptions, by gap junctions only (Azarnia *et al.,* 1974). A greater sampling of such junctions is achievable by freeze-fracture, and here too the only junction mentioned is of the gap type between vertebrate cells (Epstein and Gilula, 1977). To establish the specificity of the gap junction as the route for cytoplasmic interchange *in situ,* an epithelium must be tethered by gap junctions only, an exclusiveness so far undisclosed. Conversely, an epithelium with junctional complexes lacking gap junctions might not be electrically coupled. Such an epithelium is the choroid plexus of the rat where, if present at all, gap junctions are very few and small (Brightman *et al.,* 1975b); more likely, they are entirely absent (Van Deurs and Koehler, 1979). Only tight junctions and adhering junctions unite these cells. The cells are accessible, lying within the fourth ventricle, and large enough, about 15 μm wide and 30 μm deep, to be impaled by electrodes. If these cells are ionically coupled, then the interaction might be mediated by the other types of junctions or by the nonjunctional portion of the lateral cell membrane (Bennett, 1977).

2. METABOLIC COOPERATION

The hydrated channel of the connexon is purported to subserve not only electrotonic interaction but metabolic cooperation as well. As first enunciated by Pitts (1971), the concept of metabolic cooperation is exemplified by cocultivating one cell type that is incapable of incorporating a metabolite from the medium with a cell type that can. The incompetent cell can receive the metabolite from the cytoplasm of the competent one after the two have made physical contact. That this contact is intimate and involves gap junctions has been convincingly demonstrated by Gilula *et al.* (1972). Hypoxanthine, a purine which cannot be incorporated from the fluid medium by one type of fibroblast, is received by this fibroblast from a competent type when gap junctions have been established between the two. Although these junctions are believed to be the route of transfer, these authors state that "structures resembling focal tight junctions occur frequently" between coupled cells. Significantly, however, cells that were united only by tight junctions were neither ionically nor metabolically coupled. The coincidence between the advent of coupling and the formation of gap junctions provides strong, though circumstantial, reinforcement of the hypothesis that the gap junction is the type of junction responsible for both electrical and metabolic coupling.

The cell-to-cell transfer of nucleotides also takes place *in vivo*. In the ovary of the rat and other mammals, granulosa cells and oocytes of the follicles are interconnected by numerous small gap junctions and desmosomes (Anderson and Albertini, 1976). Electrotonic coupling and the exchange of fluorescein were demonstrable at the stage of oocyte differentiation (Gilula *et al.,* 1978). In freeze-fracture replicas, the aggregates of connexons were either pleiomorphic strands or polygonal clumps on the P face with corresponding pits on the E face. Gilula *et al.* (1978) suggest that, since much more uridine is incorporated into oocytes while they are still in contact with cumulus cells (Wasserman and Letourneau, 1976), these cells may be vital for oocyte metabolism. Other "specific factors" may also be transferred from cumulus to oocyte, since oocytes continue to enlarge only so long as they are in contact with the cumulus (see Gilula *et al.,* 1978, for references).

On the basis of such evidence, Pitts and Simms (1977) make a trenchant qualification to the cell theory: The cell is not a metabolically independent unit of tissue structure. Although a cell maintains a degree of individuality by retaining macromolecules within its borders, its functional identity is shared by the transfer, to its neighbors, of the smaller metabolic products of its enzyme–protein activities. All cells that are appropriately connected can respond in concert to fluctuating demands on intermediate metabolism. Stated more tersely, the gap junction is the route by which conjoined cells can act as a metabolic syncytium.

3. ROLES OF GAP JUNCTIONS *in Situ*

In these terms, reactive Schwann cells (Fig. 2) and both normal and reactive astrocytes may also act *in situ* as metabolic syncytia. A substantial proportion of their cell membrane area, by being junctional and therefore permeable, would provide an extended surface for the interchange of ions and metabolites. A critical survey of ionic and metabolic interactions between glial cells and neurons has been presented by Kuffler and Nicholls (1976). The size of the molecules that can be exchanged by way of gap junc-

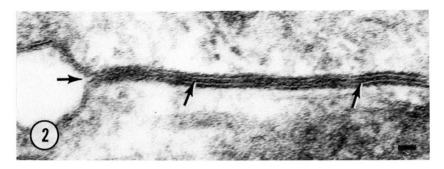

FIG. 2. This gap junction, like all others in thin, plastic sections, is septalaminar and has a median cleft (arrows) that communicates with the extracellular space, which is distended (at left). Rat. Bar = 25 nm.

tions is, however, limited. Although nucleotides can be transferred, larger molecules such as nucleic acids and proteins cannot (Pitts and Simms, 1977). The ranges of molecular size and junction permeation are nicely tabulated by Bennett (1973).

Gap junctions may thus serve to integrate the metabolic activities of many cell types *in situ*. In an organized bulk, only some cells may receive innervation and, in order for all the cells to respond in unison, the stimulus must be transferred. It has been suggested that gap junctions provide the transcellular route (Sheridan, 1971) into the noninnervated but conjoined cells. Thus, in the rat salivary gland, some acinar cells receive dual innervation (both adrenergic and cholinergic), some receive a single innervation, and some are not innervated. In order for the cells to act in concert to secrete enzymes in appropriate bulk and ratios, some means of integrating their activity seems necessary. Gap junctions unite acinar cells of the parotid gland of the rat (Hand, 1972), and these cells are electrically coupled to at least one of their neighbors.

Dyes such as fluorescein and Procion yellow, when injected into one acinar cell, appear within the cytoplasm of the adjacent cell to which it is electrically coupled in over 50% of trials in isolated parotid glands, according to Hammer and Sheridan (1978). These authors summarize the evidence that the ratio of water to proteins in saliva may be indirectly influenced by the levels of cyclic nucleotide, such as adenosine monophosphate, and make the large but reasonable inferential jump that the transfer of ions and cyclic nucleotide between cells with disparate innervation permits integration of their activity to produce a particular type of secretory product.

The involvement of gap junctions in cell-to-cell coupling is also indicated by structural changes that follow uncoupling. When contiguous cells become physiologically uncoupled, the connexons of their gap junctions become more tightly and regularly packed. Thus, when mouse livers were perfused with hyperosmotic sucrose, fixed with aldehydes, freeze-fractured, and replicated, the gap junctions appeared to be split so that the extracellular space intruded between the P and E faces of the cleaved membranes. This "unzippering" was accompanied by a marked increase in the regularity of the connexon lattice (Goodenough and Gilula, 1972).

Comparable changes have been brought about by other uncoupling procedures. When the mucosa of rat stomach was exposed to 2,4-dinitrophenol for 1 hr or rendered hypoxic, the center-to-center spacing of the connexons decreased from an average, in control tissue, of about 10 nm to about 8 nm. A closer packing also accompanied hypoxia of liver cells, and it was concluded that these conformational changes—condensed and regular packing—resulted in junctional uncoupling (Peracchia, 1977).

The uncoupling maneuvers just described must be drastic, because they are detectable beyond the level of uncoupling produced by chemical fixation alone. Fixation greatly increases the electrical resistance between cells, a change that signifies uncoupling (Bennett, 1973) as manifested by the crystalline array of connexons. Chemical fixation may be avoided by freezing tissue very rapidly; the tissue is dropped onto a highly polished metallic surface maintained at liquid helium temperature, about $4°K$ (Heuser et al., 1976). In ciliary epithelium so frozen, the connexons of gap junctions were displayed as patches of randomly dispersed particles on the P face of fractured cells. The complementary pits on the E face were likewise scattered. When the epithelium was made anoxic before rapid freezing, the connexons aggregated into rows of hexagonally packed particles, the pattern so characteristic of chemically fixed specimens (Raviola et al., 1978). It is likely that the dispersed pattern rather than the crystalline one more closely represents the native state of junction subunits in other epithelia.

III. Tight Junctions

A. Structure and Distribution

A tight or occluding junction is formed where the outer leaflets of the apposed cell membranes make such close contact that the extracellular space between the adjacent cells is occluded (Farquhar and Palade, 1963). In thin sections of plastic-embedded tissue examined electron microscopically, the tight junction is always pentalaminar: two electron-dense lines representing the cytoplasmic leaflet of either cell membrane followed by two electron-lucent middle leaflets and the shared median lamina derived from the outer leaflet of each membrane. Instead of a median extracellular slit as in the gap junction, there is the median lamina. The functional consequence of this thin, 2- to 3-nm-wide lamina is enormous; it prevents or impedes the passive, extracellular flow of substances between cells and so is responsible for establishing gradients between fluid compartments within the CNS and other organs.

Within the CNS, the tight junctions are located between the endothelial cells of blood vessels (Reese and Karnovsky, 1967), between specialized ependymal cells or tanycytes overlying presumptive neuroendocrine sites around the cerebral ventricles (Brightman *et al.,* 1975a), between the epithelial cells of the choroid plexus (Brightman and Reese, 1969), between the cells of the arachnoid mater around the CNS (Nabeshima *et al.,* 1975), between the cells of the perineurium around the peripheral nervous system (Andres, 1967; Olsson and Reese, 1971), and within the myelin sheath (Brightman and Reese, 1969; Mugnaini and Schnapp, 1974).

B. Extensiveness

The feature of the tight junction that is functionally significant for the CNS is that it is zonular; where these zones are incomplete, the junction consists of overlapping belts. In some other organs, e.g., the endothelium of skeletal and cardiac muscle (Karnovsky, 1967), the tight junction does not form a continuous belt around connected cells but is instead interrupted to form fasciae. Although the basic structure in tight junctions is the same irrespective of the organ, it is the *extent* of the junction that sets it apart from comparable junctions outside the CNS. The extent of the junc tion is best appreciated with the use of electron-dense tracers. Because of the junction's completeness, macromolecules such as the protein horseradish peroxidase (HRP, MW 40,000) cannot pass from blood to extracellular fluid (Reese and Karnovsky, 1967) or in the opposite direction (Bodenheimer and Brightman, 1968). The zonular tight junction is not peculiar to the CNS but also acts to isolate fluid compartments in endocrine organs such as the thyroid gland (Tice *et al.,* 1975).

C. Degree of Tightness

A second feature of the tight junction, in addition to its extensiveness, is the degree of tightness. Despite attempts to define this property morphologically, the only certain means of doing so at this time is by means of tracers of different sizes. An example of how the tightness can be revealed is provided by the junctions in choroidal epithelium and cerebral endothelium. In thin sections, the tight junctions between epithelium and those between endothelium appear to be identical: a series of pentalaminar configurations arranged in tandem. The junctions at both sites prevent the extracellular movement of the protein HRP or a much smaller heme peptide (MW 1900; Feder *et al.*, 1969) and the yet smaller colloidal suspension of lanthanum hydroxide (Brightman and Reese, 1969). When ionic lanthanum, the smallest electron-dense tracer so far discovered, is used, however, the tight junctions of the endothelium cannot be penetrated (Bouldin and Krigman, 1975; L. Zis, M. Sato, S. Rapoport, G. Goping, and M. W. Brightman, unpublished observation), but the identically appearing junctions of the choroidal epithelium can (Castel *et al.*, 1974).

An analog of norepinephrine, 5-hydroxydopamine (MW 256; diameter, 0.5–0.7 nm) has been recently introduced as an extracellular tracer (Richards, 1978). This molecule, larger than ionic lanthanum but comparable in size to another favorite extracellular tracer, sucrose (MW 342), was infused into the cerebral ventricles of rats. The amine could not penetrate the choroidal tight junction but did pass through the tight junctions between tanycytes of the median eminence. Richards (1978) concluded that, for molecules of this size, such as neurotransmitters, the zonular tight junction acted as a barrier to their exchange between blood and cerebrospinal fluid (CSF), but that the tanycyte junctions were perhaps fascial rather than zonular and therefore permeable to this tracer.

The tightness of the occlusion, in addition to its linear extent at tight junctions, must also be defined in terms of the structure of the junction. Although tight junctions from different tissues all look alike in thin sections, consistent differences are revealed by freeze-fracturing the junctions and replicating the cleaved faces with platinum and carbon. In the electron microscope, such junctions appear as strands or ridges on the inner P face and as complementary grooves on the E face of the junction membrane (Kreutziger, 1968; Wade and Karnovsky, 1974a). The strands of some tight junctions anastamose with each other in patterns characteristic of the cell type being joined. Thus, the epithelial cells of the gastrointestinal mucosa are linked by tight junctions which, because of numerous anastomotic loops, have a netlike appearance, whereas the parallel strands comprising the junctions between the epithelial cells of the choroid plexus (Brightman *et al.*, 1975b) and of the arachnoid mater (Figs. 3 and 4) have very few interconnections.

Another difference between these two types is that the junctions of the digestive epithelium usually appear as smooth, continuous ridges, whereas the ridges in the choroid plexus are composed of large individual or

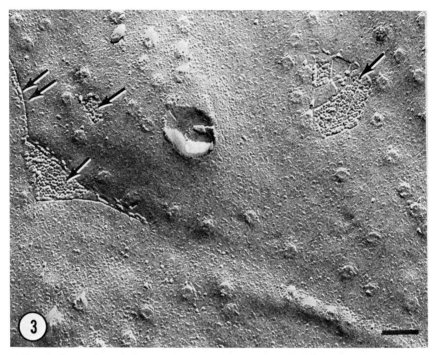

FIG. 3. Freeze-cleaved pia-arachnoid mater of rat. The perimeter of gap junctions (arrows) is marked by a strand or ridge (lines). Many typical, round ostia of caveolae indent this P fracture face. Adult mouse. Bar = 72 nm.

"fused" particles with particle-free intervals between them. Some of these particle-free areas may be due to plastic deformation: The particles are pulled out of the P face during cleavage and stick to the complementary groove on the E face. Such dislocations are not common in the fractured epithelium of the intestine, and this difference may reflect a physical difference in the state of the junction particles within the two types of epithelium.

The systematic comparison of tight-junction structure, as revealed by replicas of freeze-cleaved tissues, appeared to offer a means of predicting, on morphological grounds, the degree of tightness. Claude and Goodenough (1973) compared the appearance of fractured tight junctions in different epithelia of known physiological permeability, electrical resistance, and conductance. By counting the number of parallel strands on the cleaved P face and by considering the frequency of anastomoses, they hypothesized that, in junctions that were physiologically tight, the number of strands was greater than 5 or 6 and that their interconnections were more

FIG. 4. Freeze-cleaved pia-arachnoid mater of rat. Strands of particles (lines) not only demarcate the perimeter of gap junctions on the E fracture face but actually penetrate the junction. Ostia of caveolae abut the strands. Bar = 94 nm.

frequent. There were fewer strands in the "leakier" type of tight junction, such as that of rabbit gallbladder and intestine, where ionic lanthanum penetrates the tight junction (Machen *et al.,* 1972).

Several contradictory observations have invalidated the hypothesis. The tight junctions between epithelial cells of the mucosa in the rabbit ileum are permeable to ionic lanthanum, yet the junctions are composed of many rows of highly interconnected junction strands, according to Martinez-Palomo and Erlij (1975). The complexity of the junction fibrils is similar to that of the epithelium of the toad urinary bladder where the tight junctions are highly impermeable to ions. These authors also noted that the junctions of the toad urinary bladder, which has a high electrical resistance of 12,000 Ω/cm^2 across its epithelium (Reuss and Finn, 1974), is normally impermeable to ionic lanthanum but can be rendered permeable when bathed on the luminal side of the epithelium by hyperosmotic urea (Erlij and Martinex-Palomo, 1972) and lysine. The junctions then let lanthanum pass through. When freeze-cleaved after hyperosmotic lysine exposure, the constituent strands of the junctions are not noticeably different from those of junctions bathed by isosmotic solutions. It is conceivable that, if these specimens had been fixed in hyperosmotic buffer (i.e., greater than the 0.1 M of sodium cocodylate used) or quick-frozen without chemical fixation, some differences may have been discerned. However, in urinary bladders exposed to hyperosmotic urea on the mucosal surface and fixed in aldehyde with 0.1 M cacodylate, the junctional strands are bowed away from each other to form "blisters," a structural modulation that accompanies increased permeability of the junction (Wade and Karnovsky, 1974b).

The lack of correlation between the number and complexity of junctional strands and the degree of tightness of an occluding junction led Martinez-Palomo and Erlij (1975) to suggest that there were other determinants not yet revealed by electron microscopy. In the nervous system, the choroid plexus is considered a leaky epithelium with a high conductance of, for example, 6.3 mmho/cm^2 in the cat (Welch and Araki, 1975). In freeze-fractured choroid plexus and tanycytes of the rat, it has been suggested that the nature rather than the number of constituent strands may influence junctional permeability. Here the strands are often discontinuous, and these interruptions might provide channels through which ions can move (Brightman et al., 1975a). These particle-free discontinuities could arise from some physical peculiarity of this tight junction such that the particles of the strands are readily torn from the P face during cleavage. This artefact is probably not produced because, in complementary replicas, there are fewer particles in the corresponding grooves on the E face than could be accounted for by plastic dislocation, as reported by Van Deurs and Koehler (1979). The analysis of these last two authors lends further credence to the possibility that the continuity of an individual junctional strand may affect the permeability of the junction and that the particle-free discontinuities may be equivalent to the "pores" deduced on physiological grounds by Wright (1972).

A comparable variant of leaky tight junctions may be formed where single strands are isolated from their neighbors by considerable intervals in the pia–arachnoid mater (Fig. 3). The absence of tight-junction particles from these intervals is not due to plastic deformation; background particles and the ostia of caveolae may occupy the intervals (Fig. 3). The isolated strands are characteristically associated with leptomeningeal cells where the constituent particles lie upon the E face (Figs. 3 and 4) rather than the P face (Dermietzel, 1975). So intimate are the tight and gap junctions in these cells that a strand not only forms a border in contact with the outermost connexons (Dermietzel, 1975) but actually intrudes into the domain of the gap junction (Fig. 4). The bordering strand may, in turn, be in contact with fusion spots marking the ostia of numerous caveolae (Fig. 4).

The significance of the single strand that so faithfully outlines the perimeter of gap junctions (Fig. 4) is unclear. This admixture of different subunits of the membrane at some leptomeningeal junctions may signify a developmental peculiarity of the gap junctions. Decker and Friend (1974) have demonstrated, in a thorough study of alterations in junctions within the developing neural tube of the frog, that junctions that were once zonular fragment into single isolated strands. Such strands appear to act as foci at which gap junctions are assembled. The consistent remnants of tight-junction strands within the plaque of connexons in the pia–arachnoid may signify that this developmental relationship is even closer in some cells of mature leptomeninges and that connexons can aggregate completely around a ridge and not just along one of its sides.

IV. Assemblies

A. Morphology and Distribution

An intramembranous specialization, revealed only by freeze-fracture, has so far been found within plasma membranes but not within cytoplasmic membranes. This specialization consists of an orthogonal array of small, 6- to 7-nm-wide particles tightly packed so as to form rectangles or squares on the P face of the fractured cell membrane. Correspondingly small pits, imprinted by each subunit and arrayed in perfect linearity, form reticules in the complementary E face (Fig. 5, inset). Unlike tight or gap junctions, the subunits are confined to the thickness of the cell membrane and do not bridge the extracellular cleft to insert into the plasma membrane of an adjacent cell; the assemblies do not form junctions. Assemblies occur independently of junctions but can also be found in the vicinity of tight junctions (Elfvin and Forsman, 1978) and intimately associated with gap junctions (Fig. 5).

Within the CNS, these arrays or assemblies occupy the membranes of astrocytes (Dermietzel, 1974), being most abundant in astrocytes lining fluid compartments such as subarachnoid CSF and perivascular fluid and less numerous in those that are satellite to neurons (Landis and Reese, 1974). Assemblies also reside within the plasma membranes of Müller cells at the inner limiting membrane of the retina (Reale and Luciano, 1974) and the ependyma (Brightman et al., 1975b; Privat, 1977), but not within those of the choroid plexus epithelium (Brightman et al., 1975a), oligodendrocytes, or endothelium. As in satellite astrocytes, assemblies are also part of the cell membrane belonging to satellite cells in sympathetic ganglia

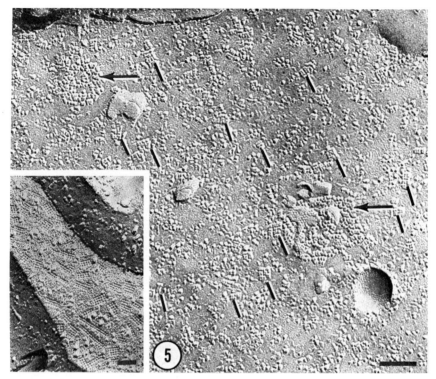

FIG. 5. P face of cleaved membrane of astrocytic process belonging to glial scar. As in normal membrane, assemblies (lines) are randomly scattered; some lie next to gap junctions (arrows). Rat. Bar = 120 nm. Inset: E face of reactive astrocyte with reticules formed by aligned pits complementary to assembly subunits. Bar = 25 nm.

(Elfvin and Forsman, 1978) and in dorsal root ganglia (Pannese et al., 1977).

Outside the nervous system, assemblies occur within the plasma membranes of hepatocytes (Kreutziger, 1968), epithelial cells of the intestine (Staehelin, 1974) and trachea (Inoue and Hogg, 1977), light cells of the renal collecting tubules (Humbert et al., 1975), skeletal muscle (Heuser et al., 1976; Smith et al., 1975; Rash et al., 1974; Ellisman and Rash, 1977), and parietal cells of the stomach (Bordi and Perrelet, 1978). Because the number of assemblies appears to be far greater in the mammalian astrocyte than in most other cell types, the astrocyte appears to be best suited for ascertaining the functions of this membrane specialization.

In order to determine the functions of assemblies, some preliminary information is needed concerning (1) their morphology and distribution (see

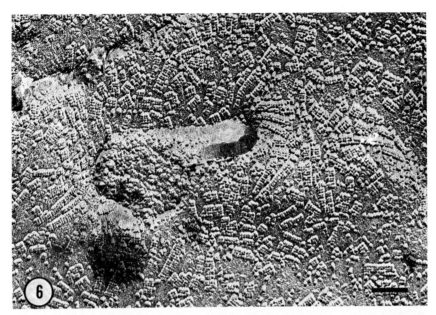

FIG. 6. Trains of subunits form linear assemblies which do not appear to be randomly dispersed. In the center is probably the rim of a branch emanating from this astrocytic process. Assemblies are aligned in parallel around the rim. Normal human frontal lobe, marginal astrocyte. Bar = 90 nm.

the preceding discussion), (2) the nature of the assembly subunit, (3) the distribution, number, and area of assemblies at different stages of maturation of normal astrocytes, (4) what happens to the assembly in reactive astrocytes, (5) what, if anything, can be gleaned about their possible function from their distribution in other organs.

B. Nature of Assemblies

It is currently held that the particles on the faces of freeze-cleaved membranes are protein that has been inserted into the surrounding particle-free lipid matrix (Branton, 1971). In order to see whether this assumption is valid for the subunit particles of the assembly, primary cultures of astrocytes were made from the cerebral cortex of 6-day-old rats. It had been determined (Anders and Brightman, 1978, 1979) that, at early stages of development, assemblies were still being added to the membrane and were fewer than in adult animals (Fig. 7). When $10^{-6}M$ cycloheximide, an inhibitor of protein synthesis, was added to the culture medium, the

FIG. 7. Developing astrocyte. P face of cleaved membrane from marginal astrocyte with relatively few assemblies (lines) dispersed at random among background particles. Rat, 20-day-old fetus. Bar = 110 nm.

assemblies disappeared after 3 hr (Anders and Pagnanelli, 1979). Background particles and those belonging to gap junctions persisted on the same fracture face. In paired cultures not exposed to the inhibitor, assemblies were present.

Our results also imply that the turnover rate of the gap junction subunit or connexon is much slower than that of the assembly. This contention is supported by the observations of Epstein *et al.* (1977): The inhibition of protein synthesis does not prevent the formation of gap junctions between hepatoma cells *in vitro*. When these cells were dissociated and reaggregated in the presence of cycloheximide at concentrations (100 μg/ml) that almost completely halted protein synthesis, there was no significant decrease in the percentage of ionically coupled cells. They took their results to support the view (Decker and Friend, 1974) that gap junctions are formed by a spontaneous assembly or aggregation of formed, intramembranous subunits. In the immature astrocyte, assembly subunits are still being formed and, presumably, degraded, so that a cessation of protein synthesis would lead to their disappearance. The connexons of gap junctions already inserted into the membrane of these young cells would and do persist.

It appears that, at least under these *in vitro* conditions, and compared with gap junctions and background particles, assemblies turn over rapidly. A less likely alternative is that a hypothetical "polymerase" enzyme protein responsible for aggregating the subunits is being rapidly synthesized. A further indication that the assembly subunit contains protein is the disappearance of assemblies when the cultures, though viable, are not supplied with amino acids or protein.

These experiments have further import because not all intramembranous particles are necessarily protein in nature. They may instead consist of lipid or lipopolysaccharide. Monolayer artificial membranes have been prepared from equimolar mixtures of the phospholipid, egg lecithin, diphosphatidylglycerol, and cardiolipin. In the presence of calcium ion, the cardiolipin changes from a bilayer phase to a micellar one in the form of particles (Verkleij *et al.*, 1979). These lipid particles, revealed by freeze-fracture, are about 10 nm in diameter, are uniform in size, and form chains or strands with complementary pits on the E face. The pits are about 7 nm wide. Thus, lipid alone can form intramembranous particles having complementary pits. Our experiments with cycloheximide indicate, however, that the assembly subunit contains protein.

C. Number, Area, and Distribution in Brain

The subpial astrocytes comprise the outermost margin of the CNS and can consequently be readily located and are accessible to agents infused into the CSF. The glia limitans is thus useful for establishing a baseline whereby induced changes can be recognized. Although the smallest assembly in the adult marginal glia could contain as few as four subunits, most consisted of a greater number. Morphometry demonstrated that the average diameter of each subunit was 6.7 nm (5.6–7.8 nm range) and did not alter with the age of the rat or the region of the brain sampled. However, the number of assemblies in the adult glia limitans of the medulla (469–595 μm^2) is consistently greater than that of the cerebral cortex (185–495 μm^2). Among brain regions in the adult, there is no significant difference in the total area of membrane occupied by the assemblies, but the small size of the sample may not have revealed such a difference. The assemblies in unfixed, rapidly frozen astrocytes are similar to those in chemically fixed cells (Reese and Landis, 1974).

D. Development

During development, assemblies and background particles are constantly being added to the astrocytic membrane. Assemblies first appeared in the astrocytes of rat fetuses between 19 and 20 days and reached adult levels in number and in the area of membrane they occupied after 40 days postpartum (Table I). Thus, the area of fractured membrane occupied by assemblies when they first appeared in the 20-day-old fetus was 1%, in the newborn about 14%, and in the young adult about 36%. Irrespective of the age of the rat, the average area of the assemblies remained the same (0.0010 μm^2).

There is no synchrony in the ontogenetic appearance of assemblies. In the 20-day-old rat, one subpial foot process of an astrocyte may have more

TABLE I [a]

DEVELOPMENTAL CHANGES IN NUMBER AND AREA OF ASSEMBLIES

Age of rat	Mean number of assemblies per square micrometer of membrane	Total area of assemblies per square micrometer of membrane (μm^2)
20-day fetus (4)[b]	7–10	1
Postpartum		
3 days (3)	160–190	14
9 days (4)	150–276	18
Over 41 days (3)	303–439	36

[a] Adapted from Anders and Brightman (1979).
[b] Number of rats.

or fewer assemblies than its immediate neighbor, a difference that could reflect different rates of maturation. Neighbors without assemblies could be considered as still in the astroblastic stage, whereas acquisition of the first assemblies indicates differentiation into an astrocyte. This interpretation concurs with that of Pannese *et al.* (1977) who conclude that satellite cells of the dorsal root ganglia in chick embryos acquire assemblies at the time of differentiation, when they have become satellite to neuroblasts. The advent of assemblies may thus be an indicator of the maturational state of a particular type of cell. Whether the assemblies are intercalated into the membrane as singlet or doublet units or whether they are inserted *de novo* with the full complement of subunits is unknown; single 6- to 7-nm-wide particles are observed occasionally in the astrocyte membrane of fetal rats (Anders and Brightman, 1979).

E. Glial Scar

A subtle gliosis, one confined to a discrete region of the CNS, can be elicited by transplanting a fragment of autonomic ganglion to the surface of the brain in young (6- to 10-day-old) rats (Rosenstein and Brightman, 1978). In response to the close proximity of the transplant, the underlying brain surface emits excrescences from its marginal astrocytic processes (Fig. 8). There may be many such excrescences, and their plasma membranes contain numerous assemblies that differ in shape and position from the usual pattern. The assemblies of the reactive astrocytes consist of subunit particles that become attached linearly to form long trains that are orderly arranged in parallel (Fig. 8). A comparable pattern occurs in normal astrocytes where they emit branches (Fig. 6), but it is not known whether the assemblies are merely conforming to the decreased radius of curvature of the cell branch. If the radius were the determinant, then one would expect side-to-side packing of some subunits to form square assemblies in addition to linear ones. The exclusively end-to-end arrangement suggests that this configuration might be related to a functional rather than a mechanical property. A mechanical alignment of subunits has not, however, been ruled out.

The gliotic reaction is not confined to the marginal subpial process but involves the formation of parallel sheets of reactive astrocytic processes beneath it. Another departure from the normal pattern is seen within these laminae. Normally, the number of assemblies is greatest in the marginal process and diminishes with successively deeper layers. In the scar, all astrocytic layers have appreciable numbers of astrocytes. It is as though the function of the assemblies, whatever that may be, is now shared by all the laminae in a gliotic scar. However, in the reactive marginal astrocyte that did not emit excrescences, the total area of cell membrane occupied by assemblies was the same as that in normal rats.

F. Distribution in Other Organs

The modicum of information available on the distribution of assemblies in organs other than the nervous system is still insufficient for drawing any well-founded inference as to the assemblies' functions. If only to stress the amount of information yet to be gathered, a brief summary of where assemblies occur in different cell types may be useful. The list is a short one.

In the sartorius muscle of the frog, the assemblies within the sarcolemma often form short "trains" or linear assemblies, as they do at branch points and in plicae of astrocytic processes. According to Smith et al. (1975), the linear assemblies of the frog muscle are nonrandomly dispersed, lying parallel to one another but relatively widely separated. These authors also have noted that the assemblies of rabbit sarcolemma are square or rectilinear in shape. The assemblies in amphibian and mammalian forms are dispersed at random with respect to all levels of the sarcomere. The configuration and degree of randomness signify, in the absence of other information, nothing about function. The average number of assemblies in the sarcolemma of skeletal muscle from different vertebrate species is far less than that in the plasma membrane of adult marginal astrocytes (Table II). Moreover, the changes in distribution of assemblies at spatial increments away from the myoneural junction may have functional significance.

In the light cells of the renal collecting tubule, assemblies occur only within the basal and lateral portions of the cell membrane but not above the level of the tight junctions or at the apical surface. The general conclusion is that they may have "a possible role in the permeability of the membrane" (Humbert et al., 1975).

In the parietal cells of the gastric mucosa in rats, assemblies are situated predominantly at the basal pole and less so on the lateral aspect, as reported by Bordi and Perrelet (1978). In contrast to the plicated surface of the reactive astrocyte, the frequency of assemblies is greater in the flatter portions of the basal plasmalemma than in the folded portions of the parietal cell. Regrettably, the assemblies were not counted but, in comparing them with other cell types, these authors had the impression that the parietal cells were "particularly rich" in assemblies at the surface facing connective and vascular tissue. Thus, both parietal cells and astrocytes are similar in having numerous assemblies in a membrane that borders a

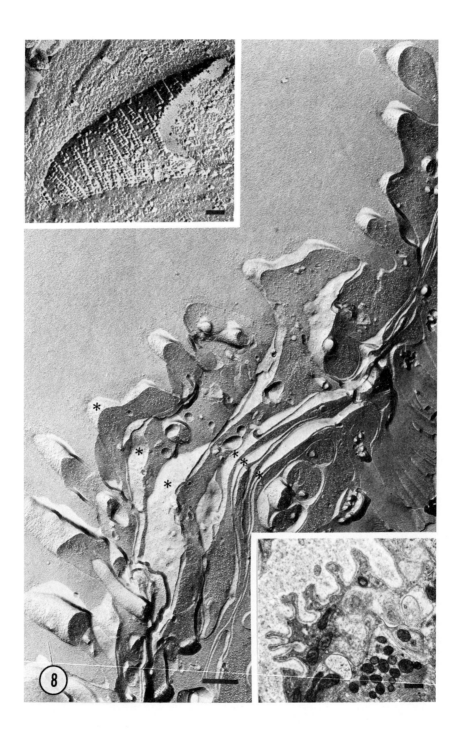

TABLE II

COMPARISONS OF ASSEMBLY NUMBERS IN DIFFERENT CELL TYPES

Specimen	Average number of assemblies per square micrometer of membrane
Subpial astrocyte	
Rat[a]	303–429
Satellite cell	
Guinea pig[b]	50–175
Skeletal muscle	
Frog[c]	15–30
Rabbit[c]	50
Rat[d]	30–100 about 0.5 mm from myoneural junction; < 1 to 5–10 about 0.2 mm from myoneural junction

[a] Anders and Brightman (1979).
[b] Elfvin and Forsman (1978).
[c] Smith et al. (1975). Values based on their estimate of 10 subunits per assembly (p. 372).
[d] Rash et al. (1974).

stromal compartment, and it is conceivable that the assemblies in these two cell types have a similar function.

Although the size and shape of assemblies are remarkably constant throughout the vertebrate species examined to date, the degree of randomness and their distribution along the cell membrane do vary. The assemblies, which all look alike except for the trains or linear ones, might have quite different functions in various cell types. If, on the other hand, all assemblies have the same function, then their preferential concentration at certain poles in a cell type implies that different portions of the cell membrane may carry out the same activity.

In order to gain some insight into this activity, it is necessary to see whether the number and configuration of assemblies can be manipulated by changing the cells' metabolic activity. Such experiments are now underway.

FIG. 8. Astrocytic scar. Marginal astrocytes fronting CSF form excrescences displayed in a thin section (lower right inset) and in a replica of a freeze-fractured surface (center). Linear assemblies of subunit trains, arrayed nonrandomly in parallel, lie within the cell membrane of the excrescences (upper left inset). Unlike the successively deep astrocytic sheets of the normal glia limitans, the lamellae (asterisks) of the scar all contain assemblies. Rat. Upper left inset: bar = 25 nm; lower right inset: bar = 100 nm; center: bar = 400 nm. (From Anders and Brightman, 1979).

References

Anders, J. J., and Brightman, M. W. (1978). Developmental and regional variations in astrocytic assemblies. *Anat. Rec.* **190**, 323–324 (abstr.).

Anders, J. J., and Brightman, M. W. (1979). Assemblies of particles in the cell membranes of developing, mature and reactive astrocytes. *J. Neurocytol.* (in press).

Anders, J. J., and Pagnanelli, D. M. (1979). The protein nature and arrangement of intramembranous particle assemblies in normal and reactive astrocytes. *Anat. Rec.* **193**, 470 (abstr.).

Anderson, R., and Albertini, D. F. (1976). Gap junctions between the oocyte and companion follicle cells in the mammalian ovary. *J. Cell Biol.* **71**, 680–686.

Andres, K. (1967). Über die fine struktur der Arachnoidea und Dura Mater von Mammalia. *Z. Zellforsch. Mikrosk. Anat.* **79**, 292–295.

Azarnia, R., Larsen, W. J., and Loewenstein, W. R. (1974). The membrane junctions in communicating and noncommunicating cells, their hybrids and segregants. *Proc. Natl. Acad. Sci. U.S.A.* **71**, 880–884.

Bennett, M. V. L. (1973). Permeability and structure of electronic junctions and intercellular movement of tracers. *In* "Intracellular Staining Techniques in Neurobiology" (S. D. Kater and C. Nicholson, eds.), pp. 115–133. Am. Elsevier, New York.

Bennett, M. V. L. (1977). Electrical transmission: A functional analysis and comparison to chemical transmission. *In* "Handbook of Physiology" (J. M. Brookhart *et al.,* eds.), Sect. 1, Vol. I, Part I, pp. 357–416. Am. Physiol. Soc., Bethesda, Maryland.

Bennett, M. V. L., and Goodenough, D. A. (1978). Gap junctions: Electrotonic coupling and intercellular communication. *Neurosci. Res. Program, Bull.* **16**, 373–486.

Bodenheimer, T., and Brightman, M. W. (1968). A blood–brain barrier to peroxidase in capillaries surrounded by perivascular spaces. *Am. J. Anat.* **122**, 249–268.

Bordi, C., and Perrelet, A. (1978). Orthogonal arrays of particles in plasma membranes of the gastric parietal cell. *Anat. Rec.* **192**, 297–304.

Bouldin, T. W., and Krigman, M. R. (1975). Differential permeability of cerebral capillary and choroid plexus to lanthanum ion. *Brain Res.* **99**, 444–448.

Branton, D. (1971). Freeze-etching studies of membrane structure. *Philos. Trans. R. Soc. London, Ser. B* **261**, 133–138.

Brightman, M. W., and Reese, T. S. (1969). Junctions between intimately apposed cell membranes in the vertebrate brain. *J. Cell Biol.* **40**, 648–677.

Brightman, M. W., Prescott, L., and Reese, T. S. (1975a). Intercellular junctions of special ependyma. *In* "Brain-Endocrine Interaction" (K. M. Knigge, D. E. Scott, M. Kobayashi, and S. Ishii, eds.), Vol. 2, pp. 146–165. Karger, Basel.

Brightman, M. W., Shivers, R., and Prescott, L. (1975b). Morphology of the walls around fluid compartments in nervous tissue. *In* "Fluid Environment of the Brain" (H. Cserr, J. Fenstermacher, and V. Fencl, eds.), pp. 3–24. Academic Press, New York.

Castel, M., Sahar, A., and Erlij, D. (1974). The movement of lanthanum across diffusion barriers in the choroid plexus of the cat. *Brain Res.* **67**, 178–184.

Claude, P., and Goodenough, D. A. (1973). Fracture faces of zonulae occludentes from "tight" and "leaky" epithelia. *J. Cell Biol.* **58**, 390–400.

Decker, R. S., and Friend, D. S. (1974). Assembly of gap junctions during amphibian neurulation. *J. Cell Biol.* **62**, 32–47.

Dermietzel, R. (1974). Junctions in the central nervous system of the cat. III. Gap junctions and membrane-associated orthogonal particle complexes (MOPC) in astrocytic membranes. *Cell Tissue Res.* **149**, 121–135.

Dermietzel, R. (1975). Junctions in the central nervous system of the cat. V. The junctional complex of the pia-arachnoid membrane. *Cell Tissue Res.* **164,** 309–329.

Elfvin, L. G., and Forsman, C. (1978). The ultrastructure of junctions between satellite cells in mammalian sympathetic ganglia as revealed by freeze-etching. *J. Ultrastruct. Res.* **63,** 261–274.

Ellisman, M. H., and Rash, J. E. (1977). Studies of excitable membranes. III. Freeze-fracture examination of the membrane specializations at the neuromuscular junction and in the non-junctional sarcolemma after denervation. *Brain Res.* **137,** 197–206.

Epstein, M. L., and Gilula, N. B. (1977). A study of communication specificity between cells in culture. *J. Cell Biol.* **75,** 769–787.

Epstein, M. L., Sheridan, J. D., and Johnson, R. G. (1977). Formation of low-resistance junctions *in vitro* in the absence of protein synthesis and ATP production. *Exp. Cell Res.* **104,** 25–30.

Erlij, D., and Martinez-Paloma, A. (1972). Opening of tight junctions in frog skin by hypertonic urea solutions. *J. Membr. Biol.* **9,** 229–240.

Farquhar, M. G., and Palade, G. E. (1963). Junctional complexes in various epithelia. *J. Cell Biol.* **17,** 375–412.

Feder, N., Reese, T. S., and Brightman, M. W. (1969). Microperoxidase, a new tracer of low molecular weight. A study of the interstitial compartments of the mouse brain. *J. Cell Biol.* **43,** 35A–36A (Abstract).

Gilula, N. B. (1977). Gap junctions and cell communication. *In* "International Cell Biology. 1976–1977" (B. R. Brinkley and K. R. Porter, eds.), pp. 61–69. Rockefeller Univ. Press, New York.

Gilula, N. B., Reeves, O. R., and Steinbach, A. (1972). Metabolic coupling and cell contacts. *Nature (London)* **235,** 262–265.

Gilula, N., Epstein, M. L., and Beers, W. H. (1978). Cell-to-cell communication and ovulation. *J. Cell Biol.* **78,** 58–75.

Goodenough, D. A. (1975). Methods for the isolation and structural characterization of hepatocyte gap junctions. *Methods Membr. Biol.* **3,** 51–80.

Goodenough, D. A., and Gilula, N. B. (1972). Cell junctions and intercellular communication. *In* "Membranes and Viruses in Immunopathology" (S. B. Day and R. A. Good, eds.), pp. 155–168. Academic Press, New York.

Hammer, M. G., and Sheridan, J. D. (1978). Electrical coupling and dye transfer between acinar cells in rat salivary glands. *J. Physiol. (London)* **275,** 495–505.

Hand, A. H. (1972). Adrenergic and cholinergic nerve terminals in the rat parotid gland. Electron microscopic observations on permanganate-fixed glands. *Anat. Rec.* **173,** 131–139.

Heuser, J. E., Reese, T. S., and Landis, D. M. D. (1976). Preservation of synaptic structure by rapid freezing. *Cold Spring Harbor Symp. Quant. Biol.* **40,** 17–24.

Humbert, F., Pricam, C., Perrelet, A., and Orce, L. (1975). Specific plasma membrane differentiations in the cells of the kidney collecting tubule. *J. Ultrastruct. Res.* **52,** 13–20.

Inoue, S., and Hogg, J. C. (1977). Freeze-etch study of the tracheal epithelium of normal guinea pigs with particular reference to intercellular junctions. *J. Ultrastruct. Res.* **61,** 89–99.

Karnovsky, M. J. (1967). The ultrastructural basis of capillary permeability studies with peroxidase as a tracer. *J. Cell Biol.* **35,** 213–236.

Kreutziger, G. O. (1968). Freeze-etching of intercellular junctions of mouse liver. *Proc., Electron Microsc. Soc. Am.* **26,** 234.

Kuffler, S. W., and Nicholls, J. G. (1976). "From Neuron to Brain." Sinauer Assoc., Sunderland, Massachusetts.

Landis, D. M. D., and Reese, T. S. (1974). Arrays of particles in freeze-fractured astrocytic membranes. *J. Cell Biol.* **60**, 316–320.

Larsen, W. J. (1977). Structural diversity of gap junctions. A review. *Tissue & Cell* **9**, 373–394.

Loewenstein, W. R. (1975). Permeability of membrane junctions. *Ann. N. Y. Acad. Sci.* **137**, 441–472.

Loewenstein, W. R., and Kanno, Y. (1964). Studies on an epithelial (gland) cell junction. I. Modifications of surface membrane permeability. *J. Cell Biol.* **22**, 565–586.

Machen, T. E., Erlij, D., and Wooding, F. B. P. (1972). Permeable junctional complexes. The movement of lanthanum across rabbit gallbladder and intestine. *J. Cell Biol.* **54**, 302–312.

McNutt, N. S., and Weinstein, R. S. (1970). The ultrastructure of the nexus. A correlated thin-section and freeze-cleave study. *J. Cell Biol.* **47**, 666–688.

McNutt, N. S., and Weinstein, R. S. (1973). Membrane ultrastructure at mammalian intercellular junctions. *Prog. Biophys. Mol. Biol.* **26**, 45–101.

Martinez-Palomo, A., and Erlij, D. (1975). Structure of tight junctions in epithelia with different permeability. *Proc. Natl. Acad. Sci. U.S.A.* **72**, 4487–4491.

Michalke, W., and Loewenstein, W. R. (1971). Communication between cells of different types. *Nature (London)* **232**, 121–122.

Mugnaini, E., and Schnapp, B. (1974). Possible role of the zonula occludens of the myelin sheath in demyelinating conditions. *Nature (London)* **251**, 725–727.

Nabeshima, S., Reese, T. S., Landis, D. M. D., and Brightman, M. W. (1975). Junctions in the meninges and marginal glia. *J. Comp. Neurol.* **164**, 127–170.

Olsson, Y., and Reese, T. S. (1971). Permeability of vasa nervorum and perineurium in mouse sciatic nerve studied by fluorescence and electron microscopy. *J. Neuropathol. Exp. Neurol.* **30**, 105–119.

Pannese, E., Luciano, L., Iurato, S., and Reale, E. (1977). Intercellular junctions and other membrane specializations in developing spinal ganglia: A freeze-fracture study. *J. Ultrastruct. Res.* **60**, 169–180.

Penn, R. D. (1966). Ionic communication between liver cells. *J. Cell Biol.* **29**, 171–174.

Peracchia, C. (1977). Gap junctions. Structural changes after uncoupling procedures. *J. Cell Biol.* **72**, 628–641.

Pitts, J. D. (1971). Molecular exchange and growth control in tissue culture. *In* "Growth Control in Cell Cultures" (G. E. W. Wolstenholme and J. Knight, eds.), pp. 89–105. Churchill-Livingstone, Edinburgh and London.

Pitts, J. D., and Simms, J. W. (1977). Permeability of junctions between animal cells. *Exp. Cell Res.* **104**, 153–163.

Privat, A. (1977). The ependyma and subependymal layer of the young rat: A new contribution with freeze-fracture. *Neurosci. Symp.* **2**, 447–457.

Rash, J. E., Ellisman, M. H., Staehelin, L. A., and Porter, K. R. (1974). Molecular specializations of excitable membranes in normal, chronically denervated, and dystrophic muscle fibers. *In* "Exploratory Concepts in Muscular Dystrophy II" (A. T. Milhorat, ed.), pp. 271–289. Exerpta Med. Found., Amsterdam.

Raviola, E., Goodenough, D., and Raviola, G. (1978). The native structure of gap junctions rapidly frozen at 4°K. *J. Cell Biol.* **79**, 229a.

Reale, E., and Luciano, L. (1974). Introduction to freeze-fracture method in retinal research. *Albrecht von Graefes Arch. Klin. Exp. Ophthalmol.* **192**, 73–87.

Reese, T. S., and Karnovsky, M. J. (1967). Fine structural localization of a blood-brain barrier to exogenous peroxidase. *J. Cell Biol.* **34**, 207–217.

Reese, T. S., and Landis, D. (1974). Structure of cerebellar cortex prepared for freeze-fracturing by rapid freezing. *Neurosci. Abstr.* p. 388.

Reuss, L., and Finn, A. (1974). Passive electrical properties of toad urinary bladder. *J. Gen. Physiol.* **64**, 1-25.

Revel, J. P., and Karnovsky, M. J. (1967). Hexagonal arrays of subunits in intercellular junctions of the mouse heart and liver. *J. Cell Biol.* **33**, C7-C12.

Richards, J. G. (1978). Permeability of intercellular junctions in brain epithelia and endothelia to exogenious amine: Cytochemical localization of extracellular 5-hydroxydopamine. *J. Neurocytol.* **7**, 61-70.

Rosenstein, J. M., and Brightman, M. W. (1978). Intact cerebral ventricle as a site for tissue transplantation. *Nature (London)* **275**, 83-85.

Sheridan, J. D. (1971). Electrical coupling between fat cells in newt fat body and mouse brown fat. *J. Cell Biol.* **50**, 795-803.

Smith, D. S., Baerwald, R. J., and Hart, M. A. (1975). The distribution of orthogonal assemblies and other intercalated particles in frog sartorius and rabbit sacrospinalis muscle. *Tissue & Cell* **7**, 369-382.

Staehelin, L. A. (1974). Structure and function of intercellular junctions. *Int. Rev. Cytol.* **39**, 191-283.

Tice, L., Wollman, S. H., and Carter, R. C. (1975). Changes in tight junctions of thyroid epithelium with changes in thyroid activity. *J. Cell Biol.* **66**, 657-663.

Van Deurs, B., and Koehler, J. K. (1979). Tight junctions in the choroid plexus epithelium. *J. Cell Biol.* **80**, 662-673.

Verkleij, A. J., Mombers, C., Leunissen-Bijvelt, J., and Ververgaert, P. H. (1979). Lipidic intramembranous particles. *Nature (London)* **279**, 162-163.

Wade, J. B., and Karnovsky, M. J. (1974a). The structure of the zonula occludens. *J. Cell Biol.* **60**, 168-180.

Wade, J. B., and Karnovsky, M. J. (1974b). Fracture faces of osmotically disrupted zonulae occludentes. *J. Cell Biol.* **62**, 344-350.

Wasserman, P. M., and Letourneau, G. E. (1976). RNA synthesis in fully grown mouse oocytes. *Nature (London)* **261**, 73-74.

Welch, K., and Araki, H. (1975). Features of the choroid plexus of the cat, studies *in vitro*. *In* "Fluid Environment of the Brain" (H. Cserr, J. Fenstermacher, and V. Fencl, eds.), pp. 157-165. Academic Press, New York.

Wright, E. M. (1972). Mechanisms of ion transport across the choroid plexus. *J. Physiol. (London)* **226**, 545-571.

ADVANCES IN CELLULAR NEUROBIOLOGY, VOLUME 1

EFFECTS OF NEUROHORMONES ON GLIAL CELLS[1]

DIETRICH VAN CALKER
Max-Planck-Institut für Biochemie
Martinsried, Federal Republic of Germany
AND
BERND HAMPRECHT
Physiologisch-Chemisches Institut
Universität Wüzburg
Würzburg, Federal Republic of Germany

[1] Dedicated to the memory of our friend Philippe Benda.

I. Introduction

The enormous effort being made to understand the functions of neuroglia may be judged from the increasing number of monographs, anthologies, and reviews dealing with the subject (Glees, 1955; Windle, 1958; Nakai, 1963; Galambos, 1964; De Robertis and Carrea, 1965; Kuffler and Nicholls, 1966, 1977; Bunge, 1968; Lasansky, 1971; Fleischhauer, 1972; Johnston and Roots, 1972; de Vellis and Kukes, 1973; Watson, 1974; Privat, 1975; Somjen, 1975; Fedoroff and Hertz, 1978; Schoffeniels et al., 1978; Varon and Somjen, 1979). Nevertheless, our knowledge of what might be the biochemical function of glial cells is still very limited. The two most enigmatic fields of neuroscience, glial function and the problem of learning and memory, have been combined in the speculation that interactions between neurons and glial cells may be the basis of the "higher" adaptive functions of the brain (Galambos, 1961, 1964; Svaetichin et al., 1965; Roitbak, 1970). A prerequisite for this idea is that glial cells and neurons not only exchange trophic or mitotic factors (for a review of this subject, see Varon and Bunge, 1978) but also integrate their information-processing capacity by the exchange of hormonal signals. The question as to whether glial cells possess receptors for such signals is the subject of this review. Such signaling compounds neither fit into the definition of a "hormone" (which is released into the circulation) nor represent a classic "neurotransmitter" (which is released by the presynaptic terminal and acts at the membranes facing the synaptic cleft, the pre- and postsynaptic membranes). Indeed, both anatomical and physiological evidence suggests that, in addition to their role as neurotransmitters, biogenic amines might also act as "neuromodulators" or "neurohormones," "a mode of operation intermediate between the private addressing of classical synaptic messengers and the broadcasting of neuroendocrine secretion" (Dismukes, 1977; see also Henn, 1978). A similar role may be assessed to the newly discovered "peptidergic" systems in the brain (Scharrer, 1978). In this article the term "neurohormone" is used to designate compounds that have been shown to or are suspected of, transmitting information across the extracellular space, classic hormones and neurotransmitters as well as agents not yet classified. However, we shall not discuss factors, the only known functions of which are to supply growth-promoting, trophic, mitotic, or "differentiating" influences. These have been comprehensively reviewed by others (Westermark and Wasteson, 1975; Lim et al., 1978; Varon and Bunge, 1978).

II. Models for Glial Cells

The problem of obtaining mature glial cells of sufficient purity and integrity from adult brain is still unsolved, although some progress has been made recently (see Henn, this volume). Most studies, therefore, were performed using cultured glial cells, either permanent cell lines derived from tumors or primary cultures from perinatal rodent brain enriched in glia-like cells. This section briefly discusses the question of the reliability of these cells as models for neuroglia.

A. Clonal Glial Cell Lines

Since glial cell lines have recently been reviewed (Pfeiffer et al., 1978), only some brief remarks are given to provide background. By far the most widely studied glial cell line is the rat glioma line C6, which was cloned from a N-nitrosomethylurea-induced brain tumor (Benda et al., 1968, 1971) of Wistar–Furth strain rats (P. Benda, personal communication). Although often referred to as "astrocytoma" cells, C6 cells display markers of oligodendroglia such as 2′, 3′-cyclic-AMP phosphohydrolase (Zanetta et al., 1972; Volpe et al., 1975) and inducibility by hydrocortisone or glycerolphosphate dehydrogenase (GPDH) (Davidson and Benda, 1970; de Vellis et al., 1971, 1977; de Vellis and Brooker, 1973), as well as markers of astrocytes such as the glial fibrillary acidic (GFA) protein (Bissell et al., 1975). C6 cells also contain S-100 protein, another putative glial marker (Benda et al., 1968, 1971). For recent reviews on glial markers see Varon (1978), Varon and Somjen (1979), and Laerum et al. (1978). Of the numerous other glial lines that have been developed (Pfeiffer et al., 1978) only human astrocytoma cells were investigated in some detail as far as receptors for neurohormones were concerned. 138MG cells (Pontén and Macintyre, 1968) were derived from a grade III astrocytoma–glioblastoma. These cells contain S-100 and GFA proteins (Edström et al., 1973; Walum, 1975). 1181N1 cells (Perkins et al., 1971) and 1321N1 cells (a subclone of 1181N1, see Clark et al., 1974) are clonal cell lines derived from 118MG cells which originated from human glioblastoma multiforme (Pontén and Macintyre, 1968). EH-118MG cells were developed through the action of the Engelbreth–Holm strain of Rous sarcoma virus on 118MG cells (Pontén and Macintyre, 1968; Macintyre et al., 1969). It is possible that the combinations of markers or functions expressed in the various clonal glial strains reflect the potentialities of glial cells, but that they are not identical with those of any kind of glial cell in the nervous system.

B. Primary Cultures from Perinatal Brain Tissue

Primary cultures prepared from dissociated late fetal or neonatal brain tissue mainly consist of flat epithelioid cells (Shein, 1965; Varon and Raiborn, 1969; Booher and Sensenbrenner, 1972). When cultured in the presence of monobutyryl cyclic AMP, dibutyryl cyclic AMP, or brain extract, these cells differentiate and assume the morphology of astrocytes (Sensenbrenner et al., 1972; Lim et al., 1973; Shapiro, 1973; Moonen et al., 1975). A glial nature of the cells in these cultures is suggested by the expression of a number of putative glial markers (Laerum et al., 1978; Varon, 1978; Varon and Somjen, 1979). These include GFA protein (Bock et al., 1975, 1977), S-100 protein (Lim et al., 1977a), carbonic anhydrase activity (Kimelberg et al., 1978), and glial filaments as seen by electron microscopy (Moonen et al., 1976; Lim et al., 1977b; Kimelberg et al., 1978). The flat epithelioid cells have been identified as immature astrocytes using specific staining methods (Shein, 1965; Cummins and Glover, 1978). Scattered on this bottom layer of astroblasts are "phase dark cells" (Schrier, 1973). According to their staining properties and morphology they were classified as "migratory spongioblasts" (Shein, 1965) which appeared to develop into oligodendroblasts (Shein, 1965). The presence of oligodendroglia in the cultures is further suggested by the induction by hydrocortisone of GPDH activity (Breen and de Vellis, 1974), a presumed marker of oligodendroglia (de Vellis et al., 1978). In some regions of the cultures one finds groups of cells with beating cilia, presumably ependymal cells (Schrier, 1973). The advantage of these cultures is that the majority of the cells represent normal, untransformed glial cells, although they are in a state of immaturity. However, since several different glial cell types are present in these cultures, a certain property, e.g., receptors for neurohormones, can only with difficulty be allocated to a specific type of glial cell. These drawbacks may be overcome by the development of techniques for the fractionation and selective purification of cells from such heterogeneous cultures.

III. Receptors for Putative Neurohormones

A. Norepinephrine

1. ADRENERGIC β RECEPTORS

A variety of different glial cell preparations respond to norepinephrine or isoproterenol with an increase in the intracellular level of cyclic AMP.

After the original description of this phenomenon for rat glioma C6 cells (Gilman and Nirenberg, 1971a; Schimmer, 1971; Schultz et al., 1972) and for human astrocytoma cells 1181N1 (Clark and Perkins, 1971), similar effects were described for several other clonal cell lines of glial origin (Schubert et al., 1974, 1976; Edström et al., 1975; Sundarraj et al., 1975). Furthermore, norepinephrine and isoproterenol stimulate the accumulation of cyclic AMP in primary cultures of perinatal mouse or rat brain, a model for untransformed glial cells (Gilman and Schrier, 1972; van Calker, 1977; van Calker et al., 1977, 1978a; Hauser, 1977; McCarthy and de Vellis, 1978). All these effects are inhibited by antagonists of adrenergic β receptors but not by α blockers, suggesting their mediation by β receptors. Cholera toxin increases the potency of isoproterenol in stimulating the accumulation of cyclic AMP in human astrocytoma cells 1321N1 and EH118MG (Su et al., 1975; Johnson et al., 1978a). With avian erythrocytes as a model, this may be explained in the following way. The effectomer subunit A of the toxin enters the cell, where it acts as an ADP ribosyltransferase (Cassel and Pfeuffer, 1978; Moss and Vaughan, 1978). ADP ribosylation inhibits a membrane-bound GTP hydrolase (Cassel and Selinger, 1977) that is somehow associated with adenylate cyclase (Pfeuffer and Helmreich, 1975). Inhibition of the GTP hydrolase causes a permanent activation of each adenylate cyclase molecule, which is switched on by the interaction (Tolkovsky and Levitzki, 1978) with an adrenergic β-receptor-agonist complex. Provided sufficient time is allowed, probably extremely low concentrations of β agonist would permanently activate all the adenylate cyclase molecules (Tolkovsky and Levitzki, 1978). Unconvincingly low activations (15%) by norepinephrine of adenylate cyclase in bulk-prepared glial fractions from various brain areas have been reported (Palmer, 1973). The maximal stimulation by catecholamines of both intracellular cyclic AMP accumulation and adenylate cyclase activity is enhanced when C6 cells reach confluency (Morris and Makman, 1976b). Lithium and, less pronouncedly, sodium and potassium ions, were shown to potentiate epinephrine-stimulated adenylate cyclase activity in C6 cells (Schimmer, 1971, 1973). Catecholamines have been reported also to increase the level of cyclic GMP in C6 cells. Furthermore, this effect appears to be mediated by an adrenergic β receptor (Hsu and Brooker, 1976; Schwartz, 1976; Bottenstein and de Vellis, 1978).

Early attempts to characterize the β receptors on C6 cells by the binding of [^{3}H]norepinephrine were unsuccessful (Premont et al., 1975), probably because of the large number of catechol-binding sites known to be present on cell membranes (e.g., Cuatrecasas et al., 1974). Reliable binding studies have recently been performed using the newly developed β-receptor antagonists $(-)-[^3$H]dihydroalprenolol and [^{125}I]iodohydroxybenzylpindolol

(Maguire *et al.*, 1976; Lucas and Bockaert, 1977; Schmitt and Pochet, 1977; Terasaki and Brooker, 1978). The figures given for the number of β receptors per C6 cell are 4000 for C6TG1A (Maguire *et al.*, 1976), 4400 for C62B (Terasaki and Brooker, 1978), and 10,000 (Lucas and Bockaert, 1977) and 7000 (Schmitt and Pochet, 1977) for C6. Also, the β receptors expressed by human astrocytoma cells EH118MG have been characterized by binding studies using [^{125}I]iodohydroxybenzylpindolol (Johnson *et al.*, 1978a).

2. ADRENERGIC α RECEPTORS

The maximal stimulation by norepinephrine, an α- and β-adrenergic agonist, of cyclic AMP accumulation in primary cultures from perinatal mouse brain is much less pronounced than the maximal stimulation by isoproterenol, a relatively specific β-adrenergic agonist. However, in the presence of an α-adrenergic antagonist the full β-adrenergic potency of norepinephrine is revealed. Adrenergic α agonists decrease the effect of isoproterenol. This decrease is prevented by α blockers. From these results it has been concluded that, in addition to the adrenergic β receptors, a substantial number of cells in the cultures also express adrenergic α receptors which mediate an inhibition of cyclic AMP accumulation (van Calker, 1977; van Calker *et al.*, 1977, 1978a). Similar results have been reported for cultured rat brain cells (Hauser, 1977; McCarthy and de Vellis, 1978) and for C62B cells, a subclone of rat glioma C6 cells (Bottenstein and de Vellis, 1978). It has been argued that at least one cell type in the primary cultures must express both adrenergic receptors (van Calker, 1977; van Calker *et al.*, 1978a). There is evidence that neurons also may coexpress two functionally opposite receptors for the same neurotransmitter (Szabadi, 1977, 1978). The precise biological role of this arrangement is not known. One possibility is that it enables the cell to preset precisely its maximal response by way of varying the densities of the two species of receptors and/or modifying their coupling to second messenger systems.

B. Dopamine

Dopamine increases the level of cyclic AMP in primary cultures of perinatal mouse brain (van Calker, 1977; van Calker *et al.*, 1979a), in C62B cells (Bottenstein and de Vellis, 1978), in human astrocytoma cells 1321N1 (Clark *et al.*, 1975), and in a number of other clonal glial cell lines (Schubert *et al.*, 1976). However, these effects appear to be due to a weak β-adrenergic activity of dopamine, since they are antagonized by propranolol, a β-adrenergic blocking agent, but only to a lesser degree by dopamine antagonists (Schubert *et al.,* 1976; Bottenstein and de Vellis, 1978). Furthermore, isoproterenol and norepinephrine are much more potent than dopamine in stimulating the cyclic AMP accumulation in these cells.

However, Henn and co-workers (1977, 1978; Henn, 1978) reported that bulk-prepared glial cells from the caudate nucleus of bovine brain expressed dopamine receptors (assayed by specific binding of [³H]dopamine and [³H]haloperidol) and contained a dopamine-sensitive adenylate cyclase. The stimulation by dopamine of enzyme activity appears to be mediated by dopamine receptors rather than adrenergic β receptors, since (1) dopamine is more potent than norepinephrine and the β agonist isoproterenol is without effect, and (2) neuroleptics such as chlorpromazine and haloperidol, known antagonists of dopamine receptors, but not the β-adrenergic antagonist propranolol block the stimulation by dopamine (Henn *et al.,* 1978).

These results seem to indicate the presence on certain glial cells of functional dopamine receptors. Yet the question remains why kainic acid, an analog of glutamic acid that selectively destroys neurons, should almost completely abolish dopamine-stimulated adenylate cyclase in the striatum (McGeer *et al.,* 1976; Di Chiara *et al.,* 1977; Schwarcz and Coyle, 1977) if a substantial proportion of the dopamine receptors are localized on glial cells.

The presence of dopamine-sensitive adenylate cyclase in bulk-prepared neuronal and glial fractions from various areas of the brain was also reported by another group (Palmer, 1973; Palmer and Manian, 1976). The slight stimulation by dopamine of enzyme activity was inhibited by phenothiazines (Palmer and Manian, 1976). No attempt was made in these studies to determine whether the effect was mediated by dopamine receptors or adrenergic receptors.

C. Histamine

Histamine increases the level of cyclic AMP in human astrocytoma cells 1181N1 (Clark and Perkins, 1971) and in primary cultures of perinatal mouse brain (van Calker, 1977; van Calker *et al.,* 1977, 1979a), but not in rat glioma C6 cells (Gilman and Nirenberg, 1971a; Schultz *et al.,* 1972). The effect on astrocytoma cells is lost after repeated subculturing of the cells (Clark *et al.,* 1975). The stimulation by histamine of cyclic AMP accumulation in cultured mouse brain cells is very small and only significant in the presence of inhibitors of phosphodiesterase (PDE). The effect appears to be mediated by histamine H_2 receptors, since it can be blocked by the H_2 antagonist metiamide (Black *et al.*, 1974) but is not influenced by α- or β-adrenergic antagonists, and is mimicked by 4-methylhistamine, a specific H_2 agonist (Black *et al.*, 1972), but not by the H_1 agonists (Durant *et al.*, 1975) 2-(2-aminoethyl)thiazole and 2-(2-aminoethyl)pyridine. For unknown reasons, the extent of stimulation by histamine varies greatly among different preparations, and several cultures do not respond at all (D. van Calker, unpublished observations; see also Gilman and Schrier, 1972). Thus, only a small proportion of the different cell types present in the cultures may respond to histamine, and the survival and/or differentiation of these cells could depend on conditions that are not yet controlled.

D. Prostaglandins

Stimulation of cyclic AMP accumulation by prostaglandin E_1 (PGE_1) has been shown for rat glioma cells (clone C6-BU-1, Hamprecht and Schultz, 1973), for human astrocytoma cells 138MG (Edström *et al.*, 1975) and 1321N1 (Clark *et al.*, 1975; Ortmann and Perkins, 1977), and for cultured rat (Gilman and Schrier, 1972; McCarthy and de Vellis, 1978) and mouse (van Calker, 1977; van Calker *et al.*, 1977, 1979a) brain cells. In primary cultures from perinatal mouse brain PGE_1 and PGE_2 were equally potent as stimulators of cyclic AMP accumulation (van Calker, 1977; van Calker *et al.*, 1979a). The dose–response curves for PGE_1 and PGE_2 extend over a concentration range of two orders of magnitude (0.1–10 μM) without reaching a plateau. Prostaglandins A_1, A_2, B_1, $F_{1\alpha}$, and $F_{2\alpha}$ caused only minor effects and only at concentrations above 1 μM.

Similar results were reported for human astrocytoma cells 1321N1 (Ortmann and Perkins, 1977). PGE_1 and PGE_2 have the same potency, while the maximal effect of PGE_2 is smaller than that of PGE_1. The dose–response curves resemble those obtained in the system of cultured mouse brain cells, and a plateau is reached at about 300 μM, a concentration that was not tried in the other system. Also, in the human astrocytoma cells the prostaglandins of the A, B, and F series are less potent than PGE_1 and PGE_2. The effects of combinations of the various prostaglandins indicate that all efficacious prostaglandins interact with a common receptor. 7-Oxa-13-prostynoic acid and indomethacin were shown to be competitive inhibitors of the effect of PGE_1, while polyphloretin phosphate and the mefenamate class of nonsteroidal antiinflammatory agents caused a complex pattern of inhibition. At high concentrations prostacyclin (PGI_2), its stable derivative $6\beta H$-5, 6α-dihydro-PGI_2, and a stable analog of prostaglandin endoperoxides also stimulate the accumulation of cyclic AMP in the cells (Ortmann, 1978). Interestingly, cholera toxin increases both the potency and the maximal effect of PGE_1 on human astrocytoma cells 1321N1 and EH118MG (Su et al., 1975; Johnson et al., 1978a). The PGE concentrations needed to elicit an increase in the level of cyclic AMP in the different glial preparations are about two orders of magnitude higher than those that are effective with neuroblastoma (Gilman and Nirenberg, 1971b; Ortmann, 1978) or neuroblastoma × glioma hybrid cells (Traber et al., 1974; Hamprecht, 1976, 1977; Traber, 1976). A similarly low PGE potency has been observed in brain slices (for review, see Daly, 1976). Thus far, no explanation for this intriguing phenomenon has been offered. However, the increase in the potency of PGE_1 caused by cholera toxin may be a key to its understanding (see Section III,A,1).

C6 glioma cells stimulated by β-adrenergic agonists release cyclic AMP into the culture medium (Penit et al., 1974; Doore et al., 1975). This release appears to be an energy-dependent secretion process that can be inhibited by PGA_1 and, less potently, by PGE_1 (Doore et al., 1975). The biological function of the various actions of prostaglandins mentioned is completely unclear. Studies on the regulation of prostaglandin synthesis in cells of the nervous system should help to elucidate this problem.

E. Acetylcholine

Rat glioma cells (clone C6-BU-1) grown in the presence of $N^6,O^{2'}$-dibutyryl cyclic AMP (DBCA) respond to the iontophoretic application of acetylcholine by slow hyperpolarization (Hamprecht et al., 1976). This effect is inhibited by atropine and α-bungarotoxin but not by d-tubocurarine. Similar results were recently obtained in our laboratory (G. Reiser, R. Heumann, and B. Hamprecht, unpublished) using a clonal hybrid line prepared by the fusion of C6 cells with each other (Heumann, 1978; R. Heumann, M. Öcalan, and B. Hamprecht, unpublished). Because of the enlarged size of these glioma × glioma hybrid cells, electrophysiological measurements can also be performed with cells grown in the absence of DBCA (G. Reiser, R. Heumann, and B. Hamprecht, unpublished).

Carbamylcholine, a stable analog of acetylcholine, has been reported to decrease the level of cyclic AMP and cyclic GMP in C62B cells (Bottenstein and de Vellis, 1978). The nicotinic blocker hexamethonium, but not the muscarinic antagonist atropine, prevented the decrease in the level of cyclic GMP but not that in the level of cyclic AMP. Also, the increase in the accumulation of cyclic AMP induced by norepinephrine was inhibited by carbamycholine, while that of cyclic GMP remained unaltered. The interpretation of these findings seems difficult (Bottenstein and de Vellis, 1978). Carbamylcholine also inhibits the increase in the level of cyclic AMP caused in human astrocytoma cells (clone 1321N1) by isoproterenol, adenosine, and PGE_1 (Gross and Clark, 1977). In contrast to the findings with the C6 cells mentioned above this effect seems to be due to the activation of a muscarinic receptor, since it is antagonized by atropine but not by nicotinic or α-adrenergic blocking agents. The effect is dependent on the presence of Ca^{2+} ions in the incubation medium. Carbamylcholine did not influence the basal level of cyclic AMP.

Acetylcholine receptors were identified in the Schwann cell membranes of the squid giant nerve fiber (Villegas, 1973, 1974, 1975, 1978; Rawlins and Villegas, 1978). External application of acetylcholine or carbamylcholine to the resting nerve fiber causes long-lasting hyperpolarization of the Schwann cells, which can be blocked by d-tubocurarine. Carbamylcholine increases the relative permeability of the Schwann cell membrane to potassium ions (Villegas, 1974), and α-bungarotoxin irreversibly blocks the hyperpolarization caused by carbamylcholine (Villegas, 1975). Nicotine, but not muscarine, mimicks the action of acetylcholine and carbamylcholine (Villegas, 1974, 1975). These results suggest the presence of nicotinic acetylcholine receptors in the Schwann cell membrane. The location of these receptors was identified by the binding of $[^{125}I]\alpha$-bungarotoxin and electron microscope autoradiography (Rawlins and Villegas, 1978). Radioactive label was mainly located over the cell surface facing the neighboring axon and the adjacent Schwann cells. Both the binding of $[^{125}I]\alpha$-bungarotoxin and the irreversibility of the blockage by the toxin of the effect of carbamylcholine were prevented by d-tubocurarine (Villegas, 1975; Rawlins and Villegas, 1978).

F. Adenosine

Adenosine stimulates the accumulation of cyclic AMP in human astrocytoma cells 1321N1 (Clark et al., 1974) and in primary cultures from perinatal mouse (van Calker, 1977; van Calker et al., 1977, 1979a,b) and rat (Gilman and Schrier, 1972: Schrier and Gilman, 1973; McCarthy and de Vellis, 1978) brain. No effect of adenosine on cyclic AMP levels in rat glioma C6 cells was observed (Gilman and Nirenberg, 1971a; Schultz et al., 1972; van Calker, 1977). However, hybrid cells (C6-4) prepared by the fusion of C6 cells with each other (Heumann, 1978; R. Heumann, M. Öcalan, and B. Hamprecht, unpublished) respond to adenosine with an increase in the intracellular level of cyclic AMP (R. Heumann, D. van Calker, and B. Hamprecht, unpublished). Adenosine not only stimulates the accumulation of cyclic AMP in cultured mouse glioblasts but also (at much lower concentrations) strongly inhibits the increase in the level of cyclic AMP evoked by β-adrenergic agonists. These two effects are mediated by two different types of receptors for the nucleoside, as concluded from the following findings (van Calker, 1977; van Calker *et al.*, 1978b, 1979a,b):

1. Both effects are mediated by extracellular receptors, since they are not inhibited by blockage of the uptake of adenosine or inhibition of PDE activity but are antagonized by methylxanthines, known antagonists of adenosine receptors.
2. The receptors are specific for adenosine or a closely related compound, since the effects are not blocked by antagonists of other receptors known to be present on the cells.
3. Both receptors require for activity the integrity of the ribose moiety of the nucleoside, in accordance with the stereochemical requirements of the "R" site of adenylate cyclase (which is identical with the adenosine receptor), as recently defined by Londos and Wolff (1977).
4. The receptors mediating, respectively, the stimulatory and the inhibitory action of adenosine, are different from each other, since the order of potency of derivatives of adenosine (altered at the purine ring) is different for the two effects.

The names A1 receptor for the receptor that mediates the inhibition and A2 for the receptor that mediates the stimulation of cyclic AMP accumulation have been proposed (van Calker *et al.*, 1978b, 1979a,b). At the present time only speculations on the function of these receptors are possible.

G. Peptides

Glioblasts cultured from dissociated neonatal mouse brain respond to secretin with an increase in the intracellular level of cyclic AMP (Propst *et al.*, 1979b). This effect is especially pronounced in the presence of a PDE inhibitor. It is not antagonized by blockers of adrenergic α or β receptors or by the adenosine antagonist isobutylmethylxanthine. However, secretin-(5–27), a synthetic fragment of secretin, known to act as a secretin antagonist (Robberecht *et al.*, 1976; Propst *et al.*, 1979a), induces a parallel shift to the right of the dose–response curve for secretin. Thus, the effect of secretin appears to be mediated by specific receptors for this peptide. The effect of secretin is inhibited by agonists of adrenergic α receptors and adenosine A1 receptors (D. van Calker, M. Müller, and B. Hamprecht, unpublished; see also Sections III,A and F) and by somatostatin (unpublished work; see below). This case of secretin action provides another example of a gastrointestinal hormone that may serve a function in the central nervous system (for review, see Pearse, 1976).

In the same cultures somatostatin evoked a strong inhibition of the increase in the level of cyclic AMP caused by isoproterenol. (Propst *et al.*, 1979b). This effect is probably specific for this peptide, since it is observed at very low concentrations (half-maximal inhibition occurred at about 3 nM) and is not antagonized by blockers of α receptors or A1 receptors. Although somatostatin shows some affinity for opiate receptors (Terenius, 1976), its action on cultured brain cells cannot be mediated by opiate receptors, since the cultures do not respond to opioids (Probst *et al.*, 1979b), and the opiate antagonist naloxone does not block the action of somatostatin (D. van Calker, M. Müller, and B. Hamprecht, unpublished).

Specific binding of [5-L-*Ile*-^{125}I]angiotensin II to monolayer cultures of fetal rat brain cells has been described (Raizada *et al.*, 1978). Since the nature of the cells in these cultures was not reported, it is unclear whether or not they represent mainly glioblasts (which would be expected for cultures from fetal rats at 20 days of gestation). Nevertheless, this result is highly noteworthy in view of the presence of the renin–angiotensin system in the brain (for review, see Severs and Daniels-Severs, 1973).

H. Serotonin

An effect of serotonin on the pulsatile movements of oligodendroglia *in vitro* has been noticed by Murray (1958). Recently, serotonin was reported to increase slightly the level of cyclic GMP and to decrease the basal level of cyclic AMP in C62B cells. The increase evoked by norepinephrine in the levels of both cyclic nucleotides was inhibited by serotonin (Bottenstein and de Vellis, 1978). Since these effects were observed at very high serotonin concentrations (300 μM) and no pharmacological characterizations of the receptors were attempted, the significance of the findings remains unclear.

I. Benzodiazepine

The newly discovered receptors for benzodiazepines (for review, see Iversen, 1977, 1978) may at least in part be located on glial cells. These receptors are likely to be the binding sites for endogenous factor(s) that are mimicked by benzodiazepines. It is reported that the destruction of neuronal elements in the corpus striatum or the cerebellum of rat brain does not alter the density of benzodiazepine-binding sites in these areas.

High-affinity benzodiazepine-binding sites were detected in membranes prepared from rat C6 glioma cells and were enriched in subcellular fractions from the brain containing glial cells (Chang *et al.*, 1978).

IV. Events Secondary to Receptor Activation

The expression by different glial preparations of receptors for various putative neurotransmitters or neuromodulators lends support to the hypothesis that mature glial cells *in situ* also carry these receptors. It would, therefore, be highly desirable to know which physiological functions of glial cells are subject to this kind of hormonal control. Present knowledge regarding this question is examined in this section.

A. Enzyme Activities

The majority of receptors present on glia-like cells has been detected by monitoring the effect of the hormones on the accumulation of intracellular cyclic AMP or on the activity of adenylate cyclase in cell homogenates. All enzymes known to be necessary for a control by cyclic AMP of cellular metabolism, i.e., adenylate cyclase and cyclic nucleotide-dependent protein kinase, have been shown to be present in rat C6 and human 1181N1 astrocytoma cells (Perkins *et al.*, 1971; Jard *et al.*, 1972; Opler and Makman, 1972; Salem and de Vellis, 1976; Schlaeger and Köhler, 1976). Stimulation by norepinephrine changes the pattern of protein phosphorylation and induces a subcellular redistribution of protein kinase activity in C6 cells (Salem and de Vellis, 1976; Groppi and Browning, 1977). The elevation by norepinephrine of the level of cyclic AMP in C6 glioma cells is followed by an increased degradation of cyclic AMP (Schultz *et al.*, 1972; Uzunov *et al.*, 1973; Weiss *et al.*, 1973; Browning, 1974; Browning *et al.*, 1974a,b, 1976; Schwartz and Passoneau, 1974, 1975). This increased degradation of cyclic AMP is due to induction of both the low- and the high-K_m forms of cyclic AMP–PDE in C6 cells by activation of β-adrenergic receptors. The induction could possibly be mediated by cyclic AMP, since DBCA exerts the same effect. An increase in PDE activity may be one of the reasons for "refractoriness" toward repeated challenge with a hormone (see Section IV, D). Induction of PDE activity by norepinephrine or DBCA is inhibited by actinomycin D or cycloheximide, indicating the requirement for *de novo* RNA and protein synthesis. Hormonal regulation of cyclic nucleotide PDE is common to a wide variety of cell types. For a recent review see Thompson and Strada (1978).

De novo RNA and protein synthesis is also required for the induction by norepinephrine of lactate dehydrogenase (LDH) in C6 glioma cells. The effect is mediated by β-adrenergic receptors, as shown by blockage with the β-adrenergic antagonist propranolol. This effect can also be mimicked by exogenously applied DBCA (de Vellis and Brooker, 1973). Dopamine (30 μM), acting apparently as a weak β-adrenergic agonist, also fully induces LDH activity but evokes only a small increase in the level of cyclic AMP. This raise in the cyclic AMP level is observed only during a short period after application of the drug, reaching basal levels again at 5 min (Bottenstein and de Vellis, 1978).

However, in a previous report from the same laboratory (de Vellis and Brooker, 1973), evidence was given that a maximal elevation of cyclic AMP was necessary during the initial period of RNA synthesis (up to 2.5 hr) to achieve LDH induction. Furthermore, depletion of intracellular Ca^{2+} abolishes the induction of LDH evoked by norepinephrine without affecting the increase in the level of cyclic AMP (Bottenstein and de Vellis, 1978). Thus, the mediation by cyclic AMP of LDH induction in C6 glioma cells does not seem to be proven beyond doubt.

Similarly, the role of cyclic AMP in the induction by norepinephrine of glycogenolysis in C6 cells is not fully understood. Norepinephrine and epinephrine, acting via β-adrenergic receptors and DBCA reduce the glycogen content of C6 cells (Opler and Makman, 1972). However, histamine, which in contrast to the catecholamines does not elevate the level of cyclic AMP in the cells, also was found to stimulate glycogenolysis. Furthermore, pretreatment of the cells with norepinephrine for 3 hr reduces the increase in the accumulation of cyclic AMP evoked by a subsequent incubation with norepinephrine (see Section IV, D for a discussion of this phenomenon). Development of this status of refractoriness does not significantly alter the increase in glycogenolysis evoked by norepinephrine (Browning et al., 1974a). Norepinephrine inhibits the uptake of isotopically labeled glucose and increases the release of label from C6 cells preloaded with radioactive glucose (Newburgh and Rosenberg, 1972). However, an increased uptake into C6 cells of glucose in response to norepinephrine and histamine has also been reported (Opler and Makman, 1973). Cyclic AMP is known from other systems to mediate the conversion of phosphorylase b to the active form phosphorylase a via a cascade-like chain of enzyme inductions (Jost and Rickenberg, 1971; Robison et al., 1971). Agents that increase the level of cyclic AMP in C6 cells (norepinephrine, PGE_1, the PDE inhibitor isobutylmethylxanthine) and DBCA induce this conversion to a variable degree (Browning *et al.,* 1974a; Passoneau and Crites, 1976; Drummond *et al.,* 1977). Recent evidence (Drummond *et al.,* 1977) suggests that even a minute increase in the level of cyclic AMP might be sufficient to induce phosphorylase conversion. This finding might explain some of the discrepancies mentioned above. However, other mechanisms beyond cyclic AMP also might regulate glycogen metabolism. For instance, the glucose content in the medium regulates the activity of glycogen synthetase and phosphorylase in cultured glioma and neuroblastoma cells (Passoneau and Crites, 1976).

The activity of ornithine decarboxylase of C6-Bu-1 glioma cells increases after incubation of the cells with DBCA and theophylline or agents that stimulate intracellular accumulation of cyclic AMP (norepinephrine, isoproterenol) (Bachrach, 1975). A similar effect in glioma-like hamster brain tumor cells has been reported to be mediated by adrenergic β-2 receptors (Hsu *et al.,* 1977). Norepinephrine and DBCA increase the activity of Na^+, K^+-ATPase in primary cultures from rat brain consisting predominantly of glioblasts (Kimelberg *et al.,* 1978; Narumi *et al.,* 1978). Effects of catecholamines on ATPase activity are known from other systems (Herd *et al.,* 1970; Schaefer *et al.,* 1972, 1974; Tria *et al.,* 1974; Hexum, 1977).

These enzymes, the activities of which are enhanced by agents that increase the cyclic AMP content, are common to many cell types. Thus, they do not seem to reflect a function specific for glial cells. However, two recent papers report the regulation by norepinephrine of enzymes presumed to be glial markers. Norepinephrine acting via an adrenergic β receptor and analogs of cyclic AMP cause an increase in the activity of $2'$, $3'$-cyclic-nucleotide $3'$-phosphohydrolase in several clonal glioma cells lines (McMorris, 1977). This enzyme is believed to be a specific biochemical marker for myelin and oligodendroglia (Norton, 1976; Varon, 1978). Furthermore, norepinephrine and DBCA increase the activity of carbonic anhydrase in primary glioblast cultures (Hertz and Sapirstein, 1978; Kimelberg *et al.,* 1978; Narumi *et al.,* 1978). This enzyme was reported to be localized in glia (Giacobini, 1962) and is considered a glial marker (Laerum *et al.,* 1978; Varon, 1978; Roussel *et al.,* 1979). However, C6 glioma cells do not express carbonic anhydrase activity (de Vellis and Brooker, 1973; Kimelberg *et al.,* 1978). Carbonic anhydrase from bovine erythrocytes is activated by cyclic AMP-dependent protein kinase, and this activation is correlated with a cyclic AMP-dependent phosphorylation of the enzyme protein (Narumi and Miyamoto, 1974). The activation by norepinephrine of the enzyme in cultured glioblasts may, therefore, also be mediated by cyclic AMP.

B. *Interaction with Glucocorticoids*

The interaction of glucocorticoids with agents that increase the intracellular accumulation of cyclic AMP has been documented in many systems (for review, see Jost and Rickenberg, 1971; Thompson and Lippman, 1974; Wicks, 1974; for recent work, see references in Breen *et al.*, 1978). GPDH is induced by hydrocortisone in C6 glioma cells (de Vellis and Brooker, 1973) and in primary cultures from rat brain (Breen and de Vellis, 1974, 1975). This induction is specific for brain tissue (de Vellis and Inglish, 1968) and was recently reported to be confined to oligodendroglia (de Vellis *et al.*, 1978). It is apparently not mediated by cyclic AMP, since norepinephrine and DBCA do not induce the enzyme (de Vellis and Brooker, 1973). However, norepinephrine and DBCA in the presence or absence of the PDE inhibitor isobutylmethylxanthine potentiate the induction by hydrocortisone of GPDH (Breen *et al.*, 1978). This effect is blocked by actinomycin D, α-amanitin, and cycloheximide, but not by cytosine arabinoside, indicating the requirement for *de novo* RNA and protein synthesis. On the other hand, glucocorticoids increase basal and norepinephrine-evoked accumulation of cyclic AMP as well as adenylate cyclase activity (Brostrom *et al.*, 1974). Only the maximal effect of norepinephrine is increased, and the concentration needed for half-maximal stimulation and the time course remain unaltered. In contrast, in human astrocytoma cells 1321N1 glucocorticoids not only increase the maximal effect of PGE_1 on cyclic AMP accumulation but are also claimed to increase its potency fivefold. No change in the response to isoproterenol was observed (Foster and Perkins, 1977). At a concentration that blocks protein synthesis in the cells cycloheximide totally prevents the effect of dexamethasone. Therefore, the increased response to PGE_1 was suggested to be due to the synthesis of a protein that modifies the sensitivity of adenylate cyclase (Foster and Perkins, 1977). Interestingly, the phenomenon is dependent on the presence of serum, indicating that factors other than glucocorticoids may also be required for the increase in sensitivity to PGE_1. Brostrom *et al.* (1974) found that the glucocorticoid-induced increased response to norepinephrine of C6 cells was correlated with an increased activity of adenylate cyclase. However, Foster and Perkins (1977) reported that glucocorticoids did not cause a higher sensitivity to PGE_1 of C6RC16 cells, thioguanine-resistant mutants of C6. Thus it appears that, unless the two clones of C6 behave very differently, glucocorticoids could specifically regulate the sensitivity of adenylate cyclase toward one out of several hormones. Further work is clearly needed to clarify this point.

C. Morphological Effects

Agents that stimulate the accumulation of cyclic AMP induce a morphological transformation in certain glial lines (Kanje *et al.*, 1973; Edström *et al.*, 1974, 1975; Oey, 1975, 1976; Schlegel and Oey, 1975) and in primary cultures of neonatal rat brain consisting mainly of glioblasts (Narumi *et al.*, 1978). Cells treated with catecholamines (Edström *et al.*, 1975; Oey, 1975; Schlegel and Oey, 1975; Narumi *et al.*, 1978) or PGE$_1$ (Kanje *et al.*, 1973; Edström *et al.*, 1974, 1975) change from a flattened irregular shape to a bipolar or astrocyte-like morphology with extensive process formation. The effect of the catecholamines is inhibited by antagonists of adrenergic β but not α receptors (Edström *et al.*, 1975; Oey, 1975; Schlegel and Oey, 1975). The induced morphological alterations disappear during prolonged incubation (Edström *et al.*, 1975; Oey, 1976), as does the elevation of the level of cyclic AMP (see Section IV, D). Similar morphological changes are observed when clonal glial cells (Mcintyre *et al.*, 1972; Edström *et al.*, 1973, 1974, 1975; Hamprecht *et al.*, 1973; Kanje *et al.*, 1973; Daniels and Hamprecht, 1974; Klier *et al.*, 1975; Oey, 1975; Sato *et al.*, 1975; Schlegel and Oey, 1975; Steinbach and Schubert, 1975) or primary glioblast cultures (Lim *et al.*, 1973; Shapiro, 1973; Moonen *et al.*, 1975) are treated with analogs of cyclic AMP. Agents that interfere with microtubule assembly, such as colchicine (Oey, 1975) and vinblastine (Kanje *et al.*, 1973; Edström *et al.*, 1974; Sheppard *et al.*, 1975), prevent the changes in morphology or abolish the morphological differentiation, although colchicine and vinblastine act by different mechanisms (Reiser *et al.*, 1975). Accordingly, cells morphologically differentiated by DBCA show a large number of microfilaments and microtubules, while "undifferentiated" cells are relatively devoid of these structures (Klier *et al.*, 1975). This is in contrast to the reports of other workers (Daniels and Hamprecht, 1974) who did not observe such ultrastructural changes in a mutant of C6 cells. Actinomycin D and cycloheximide do not prevent the morphological changes (Kanje *et al.*, 1973; Edström *et al.*, 1974; Sheppard *et al.*, 1975).

The evidence summarized above supports the hypothesis of a functional role of cyclic AMP in the process of "morphological differentiation" of glial cells. However, a number of substances unrelated to cyclic AMP, such as bromodeoxyuridine (Silbert and Goldstein, 1972; Schwartz *et al.*, 1973), amethopterin (Silbert and Goldstein, 1972), glucocorticoids (Brostrom *et al.*, 1974), dibutyryl cyclic GMP, and sodium butyrate (Sato *et al.*, 1975) also change the morphology of the cells. Furthermore, morphological differentiation by DBCA is accompanied by the ablation of intracellular membrane systems of human astrocytoma cells (Macintyre *et al.,* 1972), and DBCA in combination with serum withdrawal suppresses the increase in the rate of oxygen uptake caused by high concentrations of potassium in primary glioblast cultures (Hertz *et al.,* 1976). Thus, cell morphology is not always a reliable indicator of the differentiated state of the cell.

D. Refractoriness

When exposed repeatedly or for a prolonged period to β-adrenergic agonists or PGE_1, human and rat glioma cells lose their responsiveness to the hormone (Schultz *et al.,* 1972; Browning *et al.,* 1974a, 1976; de Vellis and Brooker, 1974; Schwartz and Passoneau, 1974; Johnson and Perkins, 1975, 1976; Perkins *et al.,* 1975, 1978; Franklin and Twose, 1976; Leichtling *et al.,* 1976; Morris and Makman, 1976a; Oey, 1976; Su *et al.,* 1976a,b; Johnson *et al.,* 1978b; Nickols and Brooker, 1978; Terasaki *et al.,* 1978). Similar effects were also observed with primary cultures from perinatal mouse brain (D. van Calker, M. Müller, and B. Hamprecht, unpublished work). This "desensitization" or refractoriness is not restricted to glia-like cells but is a common phenomenon elicited in a wide variety of tissues and cell types by a number of different hormones (Kakiuchi and Rall, 1968; Makman, 1971; Franklin and Foster, 1973; Remold-O'Donnell, 1974; Franklin *et al.,* 1975; Newcombe *et al.,* 1975; Lesniak and Roth, 1976; Raff, 1976; Shear *et al.,* 1976; Lefkowitz, 1978; Lefkowitz and Williams, 1978; Hunzicker-Dunn *et al.,* 1978). At least three processes seem to be responsible for the desensitization, the contribution of each varying among the different cell types and hormones studied: (1) loss or inactivation of hormone receptors, (2) increased degradation of cyclic AMP, (3) inhibition of the adenylate cyclase system distal to the hormone–receptor interaction. For a detailed discussion of the phenomenon the reader is referred to a number of recent reviews (Hunzicker-Dunn *et al.,* 1978; Lefkowitz, 1978; Lefkowitz and Williams, 1978; Perkins *et al.,* 1978; Terasaki *et al.,* 1978).

E. Factors Produced by Glial Cells

A possible interaction between glial cells and neurons implies an exchange of signals between the two cell types. While the presence on glia-like cells of receptors for putative neurotransmitters or neuromodulators (see Section III) provides a possible mechanism for a signal transfer from neurons to glial cells, much less is known about the signaling in the reverse direction. The finding that rat glioma tumors contain a protein that is im-

munologically, chemically, and biologically similar to nerve growth factor
(NGF) (Longo and Penhoet, 1974) led to a controversy as to whether synthesis and secretion of NGF might be "a new role for the glial cell" (Anonymous, 1974; Levi-Montalcini, 1975). More recently, Murphy *et al.* (1977)
and Schwartz *et al.* (1977) have shown that this NGF-like material is synthesized and secreted into the medium by cultured clonal rat glioma C6
cells, thereby excluding the possibility of its origin from fibroblasts contaminating a solid tumor (Levi-Montalcini, 1975). Furthermore, Schwartz
and collaborators (1977; Schwartz and Costa, 1977) found that stimulation
of the β receptors present on the cells resulted in an increased amount of
NGF-like material both in the cells and in the surrounding medium. In the
same cells 17β-estradiol stimulates the secretion of NGF into the surrounding medium (Perez-Polo *et al.*, 1977).

In contrast to these studies, Monard *et al.* (1975) did not find NGF-like
material in medium conditioned by C6 cells. However, they found (Monard
et al., 1973, 1975) a factor ("glial factor") released by C6 cells, which was
not related to NGF and induced morphological differentiation of neuroblastoma cells. This factor has recently been partially purified (Lindsay and
Monard, 1976). A similar activity is also present in the media of rat brain
primary cultures, the amount of activity being dependent on the developmental age of the animal the cultures were derived from (Schürch-Rathgeb
and Monard, 1978). Recently, Monard and co-workers (Barde *et al.*, 1978)
reported the presence in glia-conditioned medium of a factor that supports
the survival and growth of sensory neurons but is unrelated to NGF or glial
factor. Thus, a whole family of compounds, acting on different types of
neuronal cells, may be produced by glial cells. This could explain also the
results of Varon and co-workers, who were among the first to postulate a
glial origin of NGF-like molecules (Burnham *et al.*, 1972). The supportive
role of NGF in fiber outgrowth and survival of ganglionic neurons can be
mimicked, in the absence of NGF, by an increase in the nonneuronal (glial)
cells in the cultures (Burnham *et al.*, 1972). In the presence of antibodies to
NGF or, if the nonneuronal cells are incubated with anti-NGF and then
washed prior to their presentation to the neurons, their supporting activity
is impaired (Varon *et al.*, 1974a–d). Even more interesting, this NGF-like
activity of the nonneuronal ganglionic cells appears to be very specific.
NGF-like competence could be observed only when the nonneuronal cells
and the neurons were derived from the same ganglia. In contrast, the NGF
isolated from mouse submaxillary gland, mouse sarcomas, or snake venom
acts on a variety of sympathetic neurons and on dorsal root ganglia (for a
recent review, see Bradshaw, 1978). Not only the survival and fiber
outgrowth but also the type of neurotransmitter synthesized by sympathetic
neurons is influenced by nonneuronal cells. Conditioned medium from

nonneuronal cells reduces the adrenergic properties of pure sympathetic neurons and induces choline acetyltransferase and thus the production of acetylcholine (for review, see Patterson, 1978). Similarly, medium conditioned by cell line C6-4 obtained by fusion of C6 glioma cells with themselves (Heumann, 1978; R. Heumann, M. Öcalan, and B. Hamprecht, unpublished) induces choline acetyltransferase activity in a simultaneously cholinergic and adrenergic hybrid cell line (Heumann et al., 1979). Unfortunately, it is not known whether the production by glial cells of these factors different from NGF is also modulated via receptors present on the surface of the cells. In addition, it is unclear whether the factors described above play an important role in the communication between mature glial and neuronal cells or may rather be necessary during the development and differentiation of these cells.

V. Conclusions

Thus far, there are no cell preparations available that could convincingly be called clear-cut models of mature astrocytes and oligodendrocytes. For example, the most widely used cell line, C6 glioma (Benda et al., 1968), displays properties of both kinds of glial cells and, therefore, may be a model for a rather immature, poorly differentiated glial precursor cell. One problem is that too few markers of glial cells are known. The situation is gradually improving as new putative markers are being discovered. Nevertheless, we should be prepared for the possibility that even in the intact nervous system several combinations of markers out of an extended list will be found associated with various kinds of glial cells, depending on the location and the stage of ontogenic development. Such a situation would be analogous to that of neurons; a cell can still be classified as a neuron even if it lacks cell processes or electrical excitability, as long as certain other criteria are fulfilled. Classifications are helpful guidelines, not more. It is of little use to try to "press" the cells of the nervous system into preconceived idealized categories. The objective should be to describe the properties of the existing cells and understand their function in the network of the nervous system.

The results reviewed above indicate that receptors for several putative neurohormones are expressed by transformed glial cells and by immature glial cells from rodent brain. They are summarized in Tables I and II. Human astrocytoma cells express receptors for adenosine, norepinephrine (β), acetylcholine (muscarinic), and prostaglandins (Table II). One cell type present in primary cultures of glioblasts appears to carry both adrenergic α

TABLE I

EFFECTS OF ADRENERGIC AGONISTS ON GLIA-LIKE CELLS

Receptor	Cells studied	Effects
β	Rat glioma C6	Cyclic AMP increased [a]
		Cyclic GMP increased [b]
		Morphological differentiation [c]
		Increased degradation of cyclic AMP [d]
		Increased LDH activity [e]
		Increased glycogenolysis [f]
		Increased activity of ornithine decarboxylase [g]
		Increased activity of CNPase [h]
		Increased NGF content [i]
		Potentiation of GPDH induction by hydrocortisone [j]
	Human astrocytoma 1181N1 and EH118MG	Cyclic AMP increased [k]
	Human glioma 138MG	Cyclic AMP increased [l]
	Several clonal rat glioma lines	Cyclic AMP increased [m]
	Several clonal lines from mouse glioma G26	Cyclic AMP increased [n]
	Primary cultures of astroblasts	Cyclic AMP increased [o]
		Morphological differentiation [p]
		Increase in Na^+, K^+-ATPase activity [p]
		Increase in activity of carbonic anhydrase [p]
$\beta(?)$	Bulk-prepared glial cells	Increase in adenylate cyclase activity [q]
α	Rat glioma C6	Inhibition of cyclic AMP increase [r]
	Primary cultures of glioblasts	Inhibition of cyclic AMP increase [s]

[a] Gilman and Nirenberg, 1971a; Schultz *et al.*, 1972; Bottenstein and de Vellis, 1978; [b] Schwartz, 1976; Bottenstein and de Vellis, 1978; Hsu and Brooker, 1976; [c] Oey, 1975, 1976; Schlegel and Oey, 1975; [d] Schultz *et al.*, 1972; Uzunov *et al.*, 1973; Weiss *et al.*, 1973; Browning, 1974; Browning *et al.*, 1974a,b, 1976; Schwartz and Passoneau, 1974, 1975; [e] de Vellis and Brooker, 1973; Bottenstein and de Vellis, 1978; [f] Opler and Makman, 1972; Browning *et al.*, 1974a; Passoneau and Crites, 1976; Drummond *et al.*, 1977; [g] Bachrach, 1975; [h] McMorris, 1977; [i] Schwartz and Costa, 1977; Schwartz *et al.*, 1977; [j] Breen *et al.*, 1978; [k] Clark and Perkins, 1971; Su *et al.*, 1975; Johnson *et al.*, 1978a; [l] Edström *et al.*, 1975; [m] Schubert *et al.*, 1976; [n] Sundarraj *et al.*, 1975; [o] Gilman and Schrier, 1972; van Calker, 1977; van Calker *et al.*, 1977, 1978a; McCarthy and de Vellis, 1978; [p] Narumi *et al.*, 1978; [q] Palmer, 1973; [r] Bottenstein and de Vellis, 1978; [s] Hauser, 1977; van Calker, 1977; van Calker *et al.*, 1977, 1978a; McCarthy and de Vellis, 1978.

TABLE II

EFFECTS OF PUTATIVE NEUROHORMONES OR THEIR DERIVATIVES ON GLIA-LIKE CELLS

Agent and receptor	Cells studied	Effects
Dopamine/dopamine?	Bulk-prepared glial cells from caudate nucleus	Activation of adenylate cyclase [a]
Dopamine/?	Bulk-prepared glial cells (various brain areas)	Activation of adenylate cyclase [b]
Histamine/?	Human astrocytoma 1181N1	Increase in cyclic AMP [c]
Histamine/H_2?	Primary cultures of glioblasts	Increase in cyclic AMP [d]
PGE_1/prostaglandin?	Rat glioma C6	Increase in cyclic AMP [e] Increased glycogenolysis [f]
	Human astrocytoma 138MG	Increase in cyclic AMP [g] Morphological differentiation [h]
PGE_1, PGE_2, PGI_2/ prostaglandin?	Human astrocytoma cells 1321N1 and EH118MG	Increase in cyclic AMP [i]
PGE_1, PGE_2/ prostaglandin?	Primary cultures of glioblasts	Increase in cyclic AMP [d]
PGA_1, PGE_1/?	Rat glioma C6	Inhibition of the release of cyclic AMP [j]
Acetylcholine/ muscarinic?	Rat glioma C6-BU-1 (grown in the presence of DBCA)	Hyperpolarization [k]
	Rat glioma hybrid C6-4	Hyperpolarization [l]
Carbamylcholine/ receptor?	Rat glioma C62B	Decrease in cyclic AMP and cyclic GMP Inhibition of increase in cyclic AMP [m]
Carbamylcholine/ muscarinic	Human astrocytoma 1321N1	Inhibition of increase in cyclic AMP [n]
Acetylcholine, carbamylcholine, nicotine/ nicotinic	Schwann cells of squid giant nerve fiber	Hyperpolarization, increased permeability for K^+ [o]
Adenosine/A2?	Rat glioma hybrid C6-4	Increase in cyclic AMP [p]
	Human astrocytoma 1321N1	Increase in cyclic AMP [q]
Adenosine/A2	Primary cultures of glioblasts	Increase in cyclic AMP [r]
Adenosine/A1	Primary cultures of glioblasts	Inhibition of increase in cyclic AMP [s]

TABLE II (continued)

EFFECTS OF PUTATIVE NEUROHORMONES OR THEIR DERIVATIVES ON GLIA-LIKE CELLS

Agent and receptor	Cells studied	Effects
Secretin / secretin	Primary cultures of glioblasts	Increase in cyclic AMP [t]
Somatostatin / somatostatin?	Primary cultures of glioblasts	Inhibition of increase in cyclic AMP [t]
Serotonin / ?	Oligodendrocytes in culture	Contractions [u]
	Rat glioma C62B	Increase in cyclic GMP, decrease in cyclic AMP, inhibition of increase in cyclic AMP and cyclic GMP [m]

[a] Henn et al., 1977, 1978; Henn, 1978; [b] Palmer, 1973; Palmer and Manian, 1976; [c] Clark and Perkins, 1971; [d] van Calker, 1977; van Calker et al., 1977, 1979a; [e] Hamprecht and Schultz, 1973; [f] Drummond et al., 1977; [g] Edström et al., 1975; [h] Kanje et al., 1973; Edström et al., 1974, 1975; [i] Su et al., 1975; Ortmann and Perkins, 1977; Johnson et al., 1978a; Ortmann, 1978; [j] Doore et al., 1975; [k] Hamprecht et al., 1976; [l] G. Reiser, R. Heumann, and B. Hamprecht, unpublished; [m] Bottenstein and de Vellis, 1978; [n] Gross and Clark, 1977; [o] Villegas, 1973, 1974, 1975, 1978; [p] R. Heumann, D. van Calker and B. Hamprecht, unpublished; [q] Clark et al., 1974; [r] Gilman and Schrier, 1972; Schrier and Gilman, 1973; van Calker, 1977; van Calker et al., 1977, 1979a,b; McCarthy and de Vellis, 1978; [s] van Calker, 1977; van Calker et al., 1977, 1978b, 1979a,b; [t] Probst et al., 1979b; [u] Murray, 1958.

and β receptors (van Calker, 1977; van Calker et al., 1978a). In addition, as the response of the cells to β-adrenergic agonists is almost completely inhibited by A1 agonists, at least one cell type must exist that carries α, β, and A1 receptors (van Calker et al., 1978b, 1979a,b). Thus, a set of different receptors can be allocated to a single type of cell (of unknown nature) in the culture. If such a combination of receptors were specific for a given cell type, it could provide an identity label allowing the identification of a given cell type during its isolation from heterogeneous cultures.

It also appears likely that mature glial cells receive signals via such receptors. However, the evidence for this is scanty. The few studies that have been performed using glia-enriched fractions from adult brain suffer from such drawbacks as questionable purity, lack of viability of the cells, and minuteness of the response. Nevertheless, slices from rodent brains respond to norepinephrine with a remarkably high increase in the level of cyclic AMP (for review, see Daly, 1976). The results obtained with the brain cell culture system suggest a strong participation of glial cells in this response. Certain receptors carried by immature glial cells might be important only during the process of migration and differentiation. Therefore, in mature glial cells they would no longer be expressed or would be replaced by other receptors serving vital functions only in this final stage of development. The presence of (yet hypothetical) hormone receptors on mature glial cells provokes several important questions: (1) How many different types of receptors are expressed by different classes of glial cells (astrocytes, ependymal cells, oligodendrocytes)? (2) Do the different classes of glial cells differ in the sets of receptors they express? (3) Does the receptor makeup of a given class of glial cells vary with the brain region? (4) Does it change with the animal species or even with the strain? (5) Which cells (adjacent neurons, secretory neurons in other parts of the brain that release the neurohormone(s) into the cerebrospinal fluid, or other glial cells) are the source of the neurohormones to which a given glial cell is susceptible? (6) Which processes in the glial cell are regulated by means of these receptors?

It is very striking that almost every glial cell system investigated so far expresses adrenergic β receptors that can elicit the formation of cyclic AMP in the cells. In contrast, only a very small number of neurons in the brain synthesize norepinephrine. Nevertheless, they may distribute this compound throughout various regions of the brain with the aid of their enormously arborized axons and thus reach a large number of neurons (for discussion, see Dismukes, 1977) and glial cells. Only a small fraction of the many axonal varicosities of the adrenergic projections in rat cerebral cortex appear to participate in classic synapses (Chan-Palay, 1977; Descarries *et al.,* 1977). Thus, norepinephrine released from these nonsynaptic boutons could diffuse away and reach receptors on large populations of surrounding cells (Dismukes, 1977), including glial cells. Even if neurohormones were released by producer cells far away from their target cells, they could be conveyed there by the cerebrospinal fluid.

Such hormone-like signals could preset a "tonus" of the glial cells (and, most likely, the neurons) and modify their response to the more "private" message of the transmitter released by the adjacent synapses. What then could be the response of the glial cell to such messages? Glial cells are known to produce factors which have trophic and differentiating influences on nuerons and can also regulate the synthesis by neurons of neurotransmitters (see Section IV,E). The production of at least one of these factors (NGF-like material) is regulated via adrenergic β receptors (Schwartz and Costa, 1977; Schwartz *et al.,* 1977). It is tempting to speculate that other, perhaps yet unknown factors are also produced by glial cells, the synthesis of which could be regulated by neurohormones. The release of such factors could then be induced by the increase in the K^+ concentration around the glial cell, which is caused by the firing of adjacent neurons (Kuffler and Nicholls, 1966, 1977). The ability of the glial cells to release certain substances on depolarization by high concentrations of extracellular K^+ has been shown in the case of γ-aminobutyric acid (Minchin and Iversen, 1974; Iversen and Kelly, 1975; Bowery *et al.,* 1976). If the function of this hypothetical factor were to regulate the efficiency of neuronal transmission at the synapses in the close environment of the glial cell, then this sequence of events would provide a molecular mechanism for Hebb's postulate of learning (Hebb, 1949; see also Marr, 1969; Stent, 1973). Synaptic connections made by neuron A with neuron B would only be strengthened if the adjacent glial cell(s) coincidently received two signals: (1) the neurotransmitter released at the synapses, indicating firing of neuron A, and (2) increased K^+ concentration, indicating a high level of impulse traffic in the environment. If one of these signals were missing, the hypothetical factor would, respectively, not be made or released, provided its turnover were rapid enough.

Alternatively, or in addition, the β-adrenergic stimulus could influence processes not unique to the brain, such as intermediary metabolism. Availability of glucose and oxygen on the one hand and removal of catabolites such as carbon dioxide, NH_4^+, and lactic acid on the other hand are of utmost importance for brain function. Thus the homeostatic control of the composition of the extracellular space around the neurons (e.g., ion composition, pH value, content of glucose, amino acids, and their metabolites) might well be among the most important functions of glial cells. In fact, astrocytes *in situ* (Oksche, 1958; Caley and Maxwell, 1968), as well as C6 glioma cells (Opler and Makman, 1972), contain glycogen. In the latter case an influence of norepinephrine on glycogenolysis has been clearly demonstrated.

In spite of the relatively recent dates of the discoveries of these details, the reader should be aware of the fact that behind this discussion are ideas of glial function almost 100 years old: Golgi's view of a nutritive neuron-supporting function, Lugaro's hypothesis of a secretory function, and Cajal's concept of neuron–glia interaction (for review, see Kuffler and Nicholls, 1966). It appears that present-day scientists are just fleshing out the skeleton of ideas put forward by these pioneers, and it becomes evident that these ideas are not mutually exclusive. Rather, they are likely to turn out to be complementary.

References

Anonymous. (1974). A new role of the glial cell? *Nature (London)* **251,** 100–101.
Bachrach, U. (1975). Cyclic AMP-mediated induction of ornithine decarboxylase of glioma and neuroblastoma cells. *Proc. Natl. Acad. Sci. U.S.A.* **72,** 3087–3091.
Barde, Y. A., Lindsay, R. M., Monard, D., and Thoenen, H. (1978). New factor released by cultured glioma cells supporting survival and growth of sensory neurons. *Nature (London)* **274,** 818.

Benda, P., Lightbody, J., Sato, G., Levine, L., and Sweet, W. (1968). Differentiated rat glial cell strain in tissue culture. *Science* **161**, 370–371.

Benda, P., Someda, K., Messer, J., and Sweet, W. (1971). Morphological and immunochemical studies of rat glial tumors and clonal strains propagates in culture. *J. Neurosurg.* **34**, 310–323.

Bissell, M. G., Eng, L. F., Herman, M. M., Bensch, K. G., and Miles, L. E. M. (1975). Quantitative increase of neuroglia-specific GFA-protein in rat C-6 glioma cells *in vitro. Nature (London)* **255**, 633–634.

Black, J. W., Duncan, W. A. M., Durant, C. J., Ganellin, C. R., and Parsons, E. M. (1972). Definition and antagonism of histamine H_2-receptors. *Nature (London)* **236**, 385–390.

Black, J. W., Durant, G. J., Emmett, J. C., and Ganellin, C. R. (1974). Sulphur-methylene isosterism in the development of metiamide, a new histamine H_2-receptor antagonist. *Nature (London)* **248**, 65–67.

Bock, E., Jorgensen, O. S., Dittmann, L., and Eng, L. F. (1975). Determination of brain specific antigens in short-term cultivated rat astroglial cells and in rat synaptosomes. *J. Neurochem.* **25**, 867–870.

Bock, E., Møller, M., Nissen, C., and Sensenbrenner, M. (1977). Glial fibrillary acidic protein in primary astroglial cell cultures derived from newborn rat brain. *FEBS Lett.* **83**, 207–211.

Booher, J., and Sensenbrenner, M. (1972). Growth and cultivation of dissociated neurons and glial cells from embryonic chick, rat and human brain in flask cultures. *Neurobiology* **2**, 97–105.

Bottenstein, J. E., and de Vellis, J. (1978). Regulation of cyclic GMP, cyclic AMP and lactate dehydrogenase by putative neurotransmitters in the C6 rat glioma cell line. *Life Sci.* **23**, 821–834.

Bowery, N. G., Brown, D. A., Collins, G. G. S., Galvan, M., Marsh, S., and Yamini, G. (1976). Indirect effects of amino acids on sympathetic ganglion cells mediated through the release of γ-aminobutyric acid from glial cells. *Br. J. Pharmacol.* **57**, 73–91.

Bradshaw, R. A. (1978). Nerve growth factor. *Annu. Rev. Biochem.* **47**, 191–216.

Breen, G. A. M., and de Vellis, J. (1974). Regulation of glycerol phosphate dehydrogenase by hydrocortisone in dissociated rat cerebral cell cultures. *Dev. Biol.* **41**, 255–266.

Breen, G. A. M., and de Vellis, J. (1975). Regulation of glycerol phosphate dehydrogenase by hydrocortisone in rat brain explants. *Exp. Cell Res.* **91**, 159–169.

Breen, G. A. M., McGinnis, J. F., and de Vellis, J. (1978). Modulation of the hydrocortisone induction of glycerol phosphate dehydrogenase by N^6, $O^{2'}$-dibutyryl cyclic AMP, norepinephrine, and isobutylmethylxanthine in rat brain cell cultures. *J. Biol. Chem.* **253**, 2554–2562.

Brostrom, M. A., Kon, C., Olson, D. R., and Breckenridge, B. McL. (1974). Adenosine 3′,5′-monophosphate in glial tumor cells treated with glucocorticoids. *Mol. Pharmacol.* **10**, 711–720.

Browning, E. T. (1974). Suppression of the norepinephrine (NE) effect on cAMP content of C6 astrocytoma cells: Mode of expression. *Fed. Proc., Fed. Am. Soc. Exp. Biol.* **33**, 507.

Browning, E. T., Schwartz, J. P., and Breckenridge, B. McL. (1974a). Norepinephrine-sensitive properties of C-6 astrocytoma cells. *Mol. Pharmacol.* **10**, 162–174.

Browning, E. T., Groppi, V. E., and Kon, C. (1974b). Papaverine, a potent inhibitor of respiration in C-6 astrocytoma cells. *Mol. Pharmacol.* **10**, 175–181.

Browning, E. T., Brostrom, C. O., and Groppi, V. E. (1976). Altered adenosine cyclic 3′,5′-monophosphate synthesis and degradation by C-6 astrocytoma cells following prolonged exposure to norepinephrine. *Mol. Pharmacol.* **12**, 32–40.

Bunge, R. P. (1968). Glial cells and the central myelin sheath. *Physiol. Rev.* **48**, 197–251.

Burnham, P., Raiborn, C., and Varon, S. (1972). Replacement of nerve growth factor by ganglionic non-neuronal cells for the survival *in vitro* of dissociated ganglionic neurons. *Proc. Natl. Acad. Sci. U.S.A.* **69**, 3556–3560.

Caley, D. W., and Maxwell, D. S. (1968). An electron microscopic study of the neuroglia during postnatal development of the rat cerebrum. *J. Comp. Neurol.* **133**, 45–69.

Cassel, D., and Pfeuffer, T. (1978). Mechanism of cholera toxin action: Covalent modification of the guanyl nucleotide-binding protein of the adenylate cyclase system. *Proc. Natl. Acad. Sci. U.S.A.* **75**, 2661–2673.

Cassel, D., and Selinger, Z. (1977). Mechanism of adenylate cyclase activation by cholera toxin inhibition of GTP hydrolysis at the regulatory site. *Proc. Natl. Acad. Sci. U.S.A.* **74**, 3307–3311.

Chang, R. S. L., Tran, V. T., Poduslo, S. E., and Snyder, S. H. (1978). Glial localization of benzodiazepine receptors in the mammalian brain. *Abstr., Soc. Neurosci.* **4**, 511.

Chan-Palay, V. (1977). "Cerebellar Dentate Nucleus." Springer-Verlag, Berlin and New York.

Clark, R. B., and Perkins, J. P. (1971). Regulation of adenosine $3':5'$-cyclic monophosphate concentration in cultured human astrocytoma cells by catecholamines and histamine. *Proc. Natl. Acad. Sci. U.S.A.* **68**, 2757–2760.

Clark, R. B., Gross, R., Su, Y.-F., and Perkins, J. P. (1974). Regulation of adenosine $3':5'$-monophosphate content in human astrocytoma cells by adenosine and the adenine nucleotides. *J. Biol. Chem.* **249**, 5296–5303.

Clark, R. B., Su, Y.-F., Ortmann, R., Cubeddu, X. L., Johnson, G. L., and Perkins, J. P. (1975). Factors influencing the effect of hormones on the accumulation of cyclic AMP in cultured human astrocytoma cells. *Metab., Clin. Exp.* **24**, 343–358.

Cuatrecasas, P., Tell, G. P. E., Sica, V., Parikh, I., and Chang, K.-J. (1974). Noradrenaline binding and the search for catecholamine receptors. *Nature (London)* **247**, 92–97.

Cummins, C. J., and Glover, R. A. (1978). Propagation and histological characterization of a homotypic population of astrocytes derived from neonatal rat brain. *J. Anat.* **125**, 117–125.

Daly, J. W. (1976). The nature of receptors regulating the formation of cyclic AMP in brain tissue. *Life Sci.* **18**, 1349–1358.

Daniels, M. P., and Hamprecht, B. (1974). The ultrastructure of neuroblastoma × glioma somatic cell hybrids. Expression of neuronal characteristics stimulated by dibutyryl adenosine $3',5'$-cyclic monophosphate. *J. Cell Biol.* **63**, 691–699.

Davidson, R. L., and Benda, P. (1970). Regulation of specific functions of glial cells in somatic hybrids. II. Control of inducibility of glycerol-3-phosphate dehydrogenase. *Proc. Natl. Acad. Sci. U.S.A.* **67**, 1870–1877.

De Robertis, E. D. P., and Carrea, R., eds. (1965). "Biology of Neuroglia." Am. Elsevier, New York.

Descarries, L., Watkins, K. C., and Lapierre, Y. (1977). Noradrenergic axon terminals in the cerebral cortex of rat. III. Topometric ultrastructural analysis. *Brain Res.* **133**, 197–222.

de Vellis, J., and Brooker, G. (1973). Induction of enzymes by glucocorticoids and catecholamines in a rat glial cell line. *In* "Tissue Culture of the Nervous System" (G. Sato, ed.), pp. 231–246. Plenum, New York.

de Vellis, J., and Brooker, G. (1974). Reversal of catecholamine refractoriness by inhibitors of RNA and protein synthesis. *Science* **186**, 1221–1223.

de Vellis, J., and Inglish, D. (1968). Hormonal control of glycerolphosphate dehydrogenase in the rat brain. *J. Neurochem.* **15**, 1061–1070.

de Vellis, J., and Kukes, G. (1973). Regulation of glial cell functions by hormones and ions: A review. *Tex. Rep. Biol. Med.* **31**, 271–293.

de Vellis, J., Inglish, D., and Galey, F. (1971). Effects of cortisol and epinephrine on glial cells in culture. *Cell. Aspects Neural Growth Differ., Proc. Conf., 1969* UCLA Forum Med. Sci., No. 14.

de Vellis, J., McGinnis, J. F., Breen, G. A. M., Leveille, P., Bennett, K., and McCarthy, K. (1978). Hormonal effects on differentiation in neural cultures. *In* "Cell, Tissue and Organ Cultures in Neurobiology" (S. Fedoroff and L. Hertz, eds.), pp. 485–511. Academic Press, New York.

Di Chiara, G., Porceddu, M. L., Fratta, W., and Gessa, G. L. (1977). Postsynaptic receptors are not essential for dopaminergic feedback regulation. *Nature (London)* **267**, 270–272.

Dismukes, K. (1977). New look at the aminergic nervous system. *Nature (London)* **269**, 557–558.

Doore, B. J., Bashor, M. M., Spitzer, N., Mawe, R. C., and Saier, M. H., Jr. (1975). Regulation of adenosine 3′:5′-monophosphate efflux from rat glioma cells in culture. *J. Biol. Chem.* **250**, 4371–4372.

Drummond, A. H., Baguley, B. C., and Staehelin, M. (1977). Beta-adrenergic regulation of glycogen phosphorylase activity and adenosine cyclic 3′,5′-monophosphate accumulation in control and desensitized C-6 astrocytoma cells. *Mol. Pharmacol.* **13**, 1159–1169.

Durant, G. J., Ganellin, C. R., and Parsons, M. E. (1975). Chemical differentiation of histamine H_1- and H_2-receptor agonists. *J. Med. Chem.* **18**, 905–909.

Edström, A., Haglid, K. G., Kanje, M., Rönnbäck, L., and Walum, E. (1973). Morphological alterations and increase of S-100 protein in cultured human glioma cells deprived of serum. *Exp. Cell Res.* **83**, 426–429.

Edström, A., Kanje, M., and Walum, E. (1974). Effects of dibutyryl cyclic AMP and prostaglandin E_1 on cultured human glioma cells. *Exp. Cell Res.* **85**, 217–223.

Edström, A., Kanje, M., Löfgren, P., and Walum, E. (1975). Drug-induced alterations in morphology and level of cAMP in cultured human glioma cells. *Exp. Cell Res.* **95**, 359–364.

Fedoroff, S., and Hertz, L., eds. (1978). "Cell, Tissue, and Organ Cultures in Neurobiology." Academic Press, New York.

Fleischhauer, K. (1972). Ependyma and subependymal layer. In "The Structure and Function of Nervous Tissue" (G. H. Bourne, ed.), Vol. 6, pp. 1–46. Academic Press, New York.

Foster, S. J., and Perkins, J. P. (1977). Glucocorticoids increase the responsiveness of cells in culture to prostaglandin E_1. *Proc. Natl. Acad. Sci. U.S.A.* **74**, 4816–4820.

Franklin, T. J., and Foster, S. J. (1973). Hormone-induced desensitisation of hormonal control of cyclic AMP levels in human diploid fibroblasts. *Nature (London), New Biol.* **246**, 146–148.

Franklin, T. J., and Twose, P. A. (1976). Desensitization of beta-adrenergic receptors of glioma cells: Studies with intact and broken cell preparations. *FEBS Lett.* **66**, 225–229.

Franklin, T. J., Morris, W. P., and Twose, P. A. (1975). Desensitization of beta-adrenergic receptors in human fibroblasts in tissue culture. *Mol. Pharmacol.* **11**, 485–491.

Galambos, R. (1961). A glial-neural theory of brain function. *Proc. Natl. Acad. Sci. U.S.A.* **47**, 129–136.

Galambos, R. (1964). Glial cells. *Neurosci. Res. Program, Bull.* **2**, Part 6.

Giacobini, E. (1962). A cytochemical study of the localization of carbonic anhydrase in the nervous system. *J. Neurochem.* **9**, 169–177.

Gilman, A., and Nirenberg, M. (1971a). Effect of catecholamines on the adenosine 3′:5′-cyclic monophosphate concentrations of clonal satellite cells of neurons. *Proc. Natl. Acad. Sci. U.S.A.* **68**, 2165–2168.

Gilman, A., and Nirenberg, M. (1971b). Regulation of adenosine 3′,5′-cyclic monophosphate metabolism in cultured neuroblastoma cells. *Nature (London)* **234**, 356–357.

Gilman, A. G., and Schrier, B. K. (1972). Adenosine cyclic 3′,5′-monophosphate in fetal rat brain cell cultures. I. Effect of catecholamines. *Mol. Pharmacol.* **8**, 410–416.

Glees, P. (1955). "Neuroglia, Morphology and Function." Thomas, Springfield, Illinois.

Groppi, V. E., and Browning, E. T. (1977). Norepinephrine-induced protein phosphorylation and dephosphorylation in intact C-6 glioma cells: Analysis by two-dimensional gel electrophoresis. *Fed. Proc., Fed. Am. Soc. Exp. Biol.* **36**, 773.

Gross, R. A., and Clark, R. B. (1977). Regulation of adenosine 3′,5′-monophosphate content in human astrocytoma cells by isoproterenol and carbachol. *Mol. Pharmacol.* **13**, 242–250.

Hamprecht, B. (1976). Neuron models. *Angew, Chem., Int. Ed. Engl.* **15**, 194–206.

Hamprecht, B. (1977). Structural, electrophysiological, biochemical and pharmacological properties of neuroblastoma × glioma cell hybrids in cell culture. *Int. Rev. Cytol.* **49**, 99–170.

Hamprecht, B., and Schultz, J. (1973). Influence of noradrenaline, prostaglandin E₁ and inhibitors of phosphodiesterase activity on levels of the cyclic adenosine 3′,5′-monophosphate in somatic cell hybrids. *Hoppe-Seyler's Z. Physiol. Chem.* **354**, 1633–1641.

Hamprecht, B., Jaffe, B. M., and Philpott, G. W. (1973). Prostaglandin production by neuroblastoma, glioma and fibroblast cell lines: Stimulation by $N^6,O^{2′}$-dibutyryladenosine 3′:5′-cyclic monophosphate. *FEBS Lett.* **36**, 193–198.

Hamprecht, B., Kemper, W., and Amano, T. (1976). Electrical response of glioma cells to acetylcholine. *Brain Res.* **101**, 129–135.

Hauser, K. L. (1977). Adrenergic receptors in glioblast cultures from newborn rat cortex. *Experientia* **33**, 805.

Hebb, D. O. (1949). "Organization of Behavior." Wiley, New York.

Henn, F. A. (1978). Dopamine and schizophrenia. *Lancet* **2**, 293–295.

Henn, F. A., Anderson, D. J., and Sellström, Å. (1977). Possible relationship between glial cells, dopamine and the effects of antipsychotic drugs. *Nature (London)* **266**, 637–638.

Henn, F. A., Anderson, D. J., and Sellström, Å. (1978). The role of neuroleptic drug receptors on astroglial cells. In "Dynamic Properties of Glial Cells" (E. Schoffeniels, G. Frank, L. Hertz, and D. B. Tower, eds.), pp. 435–441. Pergamon, Oxford.

Herd, P. A., Horwitz, B. A., and Smith, R. E. (1970). Norepinephrine-sensitive Na⁺/K⁺ ATPase activity in brown adipose tissue. *Experientia* **26**, 825–826.

Hertz, L., and Sapirstein, V. (1978). Carbonic anhydrase activity in astrocytes in primary cultures grown in the presence of or absence of dibutyryl cyclic AMP. *In Vitro* **14**, 376.

Hertz, L., Oteruelo, F., and Fedoroff, S. (1976). Dissociation between effects of cyclic AMP on morphology and on metabolism in primary culture of glial cells. *J. Cell Biol.* **70**, 135a.

Heumann, R. (1978). Herstellung und Charakterisierung von polyploiden Gliomazellen und von zugleich cholinerg-adrenergen Hybridzellen. Ph.D. Thesis, Technical University of Munich.

Heumann, R., Öcalan, M., Kachel, V., and Hamprecht, B. (1979). Clonal hybrid cell lines expressing cholinergic and adrenergic properties. *Proc. Natl. Acad. Sci. U.S.A.* **76**, 4674–4677.

Hexum, T. D. (1977). The effect of catecholamines on transport (Na, K) adenosine triphosphatase. *Biochem. Pharmacol.* **26**, 1221–1227.

Hsu, C.-Y., and Brooker, G. (1976). Adrenergic agonists increase cyclic GMP in a rat glial cell line (C6-2B). *Fed. Proc., Fed. Am. Soc. Exp. Biol.* **35**, 295.

Hsu, W. H., Coppoc, G. L., and Burnstein, T. (1977). Effects of adrenergic agents on ornithine decarboxylase activity in hamster brain tumor cells. *Fed. Proc., Fed. Am. Soc. Exp. Biol.* **36**, 950.

Hunzicker-Dunn, M., Bockaert, J., and Birnbaumer, L. (1978). Physiological aspects of appearance and desensitization of gonadotropin-sensitive adenylyl cyclase in ovarian tissues and membranes of rabbits, rats, and pigs. *Recept. Horm. Action* **3**, 393–433.

Iversen, L. (1977). Anti-anxiety receptors in the brain? *Nature (London)* **266**, 678.

Iversen, L. (1978). GABA and benzodiazepine receptors. *Nature (London)* **275**, 477.

Iversen, L. L., and Kelly, J. S. (1975). Uptake and metabolism of γ-aminobutyric acid by neurones and glial cells. *Biochem. Pharmacol.* **24**, 933–938.

Jard, S., Premont, J., and Benda, P. (1972). Adenylate cyclase, phosphodiesterases and protein kinase of rat glial cells in culture. *FEBS Lett.* **26**, 344–348.

Johnson, G. L., and Perkins, J. P. (1975). Desensitization of adenylate cyclase (AC) in human astrocytoma cells. *Pharmacologist* **17**, 233.

Johnson, G. L., and Perkins, J. P. (1976). Effect of cholera toxin on desensitization of adenylate cyclase of human astrocytoma cells. *Fed. Proc., Fed. Am. Soc. Exp. Biol.* **35**, 1394.

Johnson, G. L., Harden, T. K., and Perkins, J. P. (1978a). Regulation of adenosine 3':5'-monophosphate content of Rous sarcoma virus-transformed human astrocytoma cells. *J. Biol. Chem.* **253**, 1465–1471.

Johnson, G. L., Wolfe, B. B., Harden, T. K., Molinoff, P. B., and Perkins, J. P. (1978b). Role of β-adrenergic receptors in catecholamine-induced desensitization of adenylate cyclase in human astrocytoma cells. *J. Biol. Chem.* **253**, 1472–1480.

Johnston, P. V., and Roots, B. J. (1972) "Nerve Membranes. A Study of the Biological and Chemical Aspects of Neuron–Glia Relationship." Pergamon, Oxford.

Jost, J.-P., and Rickenberg, H. V. (1971). Cyclic AMP. *Annu. Rev. Biochem.* **40**, 741–774.

Kakiuchi, S., and Rall, T. W. (1968). The influence of chemical agents on the accumulation of adenosine 3',5'-phosphate in slices of rabbit cerebellum. *Mol. Pharmacol.* **4**, 367–378.

Kanje, M., Walum, E., and Edström, A. (1973). Dibutyryl cyclic AMP and prostaglandin E_1 induce morphological alterations in cultured human glioma cells. *Acta Physiol. Scand., Suppl.* p. 107.

Kimelberg, H. K., Narumi, S., and Bourke, R. S. (1978). Enzymatic and morphological properties of primary rat brain astrocyte cultures, and enzyme development *in vivo*. *Brain Res.* **153**, 55–77.

Klier, G., Schubert, D., and Heinemann, S. (1975). Ultrastructural changes accompanying the induced differentiation of clonal rat nerve and glia. *Neurobiology* **5**, 1–7.

Kuffler, S. W., and Nicholls, J. G. (1966). The physiology of neuroglial cells. *Ergeb. Physiol., Biol. Chem. Exp. Pharmakol.* **57**, 1–90.

Kuffler, S. W., and Nicholls, J. G. (1977). "From Neuron to Brain. A Cellular Approach to the Function of the Nervous System." Sinauer Assoc., Sunderland, Massachusetts.

Laerum, O. D., Bigner, D. D., and Rajewski, M. F., eds. (1978). Biology of brain tumors. *Int. Union Cancer, Tech. Rep. Ser.* **30**, 143–157.

Lasansky, A. (1971). Nervous function at the cellular level: Glia. *Annu. Rev. Physiol.* **33**, 241–256.

Lefkowitz, R. J. (1978). Regulation of β-adrenergic receptors by β-adrenergic agonists. *Recept. Horm. Action* **3**, 179–194.

Lefkowitz, R. J., and Williams, L. T. (1978). Molecular mechanisms of activation and desensitization of adenylate cyclase coupled beta-adrenergic receptors. *Adv. Cyclic Nucleotide Res.* **9**, 1–17.

Leichtling, B. H., Drotar, A. M., Ortmann, R., and Perkins, J. P. (1976). Growth of astrocytoma cells in the presence of prostaglandin E_1: Effect on the regulation of cyclic AMP metabolism. *J. Cyclic Nucleotide Res.* **2**, 89–98.

Lesniak, M. A., and Roth, J. (1976). Regulation of receptor concentration by homologous hormone. *J. Biol. Chem.* **251**, 3720-3729.

Levi-Montalcini, R. (1975). A new role for the glial cell? *Nature (London)* **253**, 687.

Lim, R., Mitsunobu, K., and Li, W. K. P. (1973). Maturation-stimulating effect of brain extract and dibutyryl cyclic AMP on dissociated embryonic brain cells in culture. *Exp. Cell Res.* **79**, 243-246.

Lim, R., Turriff, D. E., Troy, S. S., Moore, B. W., and Eng, L. F. (1977a). Glia maturation factor: Effect on chemical differentiation of glioblasts in culture. *Science* **195**, 195-196.

Lim, R., Troy, S. S., and Turriff, D. E. (1977b). Fine structure of cultured glioblasts before and after stimulation by a glia maturation factor. *Exp. Cell Res.* **106**, 357-372.

Lim, R., Turriff, D. E., Troy, S. S., and Kato, T. (1978). Differentiation of glioblasts under the influence of glial maturation factor. *In* "Cell, Tissue, and Organ Cultures in Neurobiology" (S. Fedoroff and L. Hertz, eds.), pp. 223-235. Academic Press, New York.

Lindsay, R. M., and Monard, D. (1976). Isolation and partial purification of glial factor. *Experientia* **32**, 801-802.

Londos, C., and Wolff, J. (1977). Two distinct adenosine sensitive sites on adenylate cyclase. *Proc. Natl. Acad. Sci. U.S.A.* **74**, 5482—5486.

Longo, A. M., and Penhoet, E. E. (1974). Nerve growth factor in rat glioma cells. *Proc. Natl. Acad. Sci. U.S.A.* **71**, 2347-2349.

Lucas, M., and Bockaert, J. (1977). Use of $(-)-[^3H]$dihydroalprenolol to study beta-adrenergic receptor-adenylate cyclase coupling in C6 glioma cells: Role of $5'$-guanylylimidodiphosphate. *Mol. Pharmacol.* **13**, 314-329.

McCarthy, K. D., and de Vellis, J. (1978). Alpha-adrenergic receptor modulation of beta-adrenergic, adenosine and prostaglandin E_1 increased adenosine $3':5'$-cyclic monophosphate levels in primary cultures of glia. *J. Cyclic Nucleotide Res.* **4**, 15-26.

McGeer, E. G., Innanen, V. T., and McGeer, P. L. (1976). Evidence on the cellular localization of adenyl cyclase in the neostriatum. *Brain Res.* **118**, 356-358.

Macintyre, E. H., Grimes, R. A., and Vatter, A. E. (1969). Cytology and growth characteristics of human tumour astrocytes transformed by Rous sarcoma virus. *J. Cell Sci.* **5**, 583-602.

Macintyre, E. H., Wintersgill, C. J., Perkins, J. P., and Vatter, A. E. (1972). The responses in culture of human tumour astrocytes and neuroblasts to $N^6,O^{2'}$-dibutyryl adenosine $3',5'$-monophophoric acid. *J. Cell Sci.* **11**, 639-667.

McMorris, F. A. (1977). Norepinephrine induces glial-specific enzyme activity in cultured glioma cells. *Proc. Natl. Acad. Sci. U.S.A.* **74**, 4501-4504.

Maguire, M. E., Wiklund, R. A., Anderson, H. J., and Gilman, A. G. (1976). Binding of $[^{125}I]$iodohydroxybenzylpindolol to putative β-adrenergic receptors of rat glioma cells and other cell clones. *J. Biol. Chem.* **251**, 1221-1231.

Makman, M. H. (1971). Properties of adenylate cyclase of lymphoid cells. *Proc. Natl. Acad. Sci. U.S.A.* **68**, 885-889.

Marr, D. (1969). A theory of cerebellar cortex. *J. Physiol. (London)* **202**, 437-470.

Minchin, M. C. W., and Iversen, L. L. (1974). Release of $[^3H]$gamma-aminobutyric acid from glial cells in rat dorsal root ganglia. *J. Neurochem.* **23**, 533-540.

Monard, D., Solomon, F., Rentsch, M., and Gysin, R. (1973). Glia-induced morphological differentiation in neuroblastoma cells. *Proc. Natl. Acad. Sci. U.S.A.* **70**, 1894-1897.

Monard, D., Stockel, K., Goodman, R., and Thoenen, H. (1975). Distinction between nerve growth factor and glial factor. *Nature (London)* **258**, 444-445.

Moonen, G., Cam, Y., Sensenbrenner, M., and Mandel, P. (1975). Variability of the effects of serum-free medium, dibutyryl-cyclic AMP or theophylline on the morphology of cultured new-born rat astroblasts. *Cell Tissue Res.* **163**, 365-372.

Moonen, G., Heinen, E., Goessens, G. (1976). Comparative ultrastructural study of the effects of serum-free medium and dibutyryl-cyclic AMP on newborn rat astroblasts. *Cell Tissue Res.* **167,** 221–227.

Morris, S. A., and Makman, M. H. (1976a). On the mechanism of receptor inactivation in the rat astrocytoma cell. *Pharmacologist* **18,** 186.

Morris, S. A., and Makman, M. H. (1976b). Cell density and receptor-adenylate cyclase relationships in the C-6 astrocytoma cell. *Mol. Pharmacol.* **12,** 362–372.

Moss, J., and Vaughan, M. (1978). Isolation of an avian erythrocyte protein possessing ADP-ribosyl-transferase activity and capable of activating adenylate cyclase. *Proc. Natl. Acad. Sci. U.S.A.* **75,** 3621–3624.

Murphy, R. A., Oger, J., Saide, J. D., Blanchard, M. H., Arnason, B. G. W., Hogan, C., Pantazis, N. J., and Young, M. (1977). Secretion of nerve growth factor by central nervous system glioma cells in culture. *J. Cell Biol.* **72,** 769–773.

Murray, M. R. (1958). Response of oligodendrocytes to serotonin. *In* "Biology of Neuroglia" (W. F. Windle, ed.), pp. 176–180. Thomas, Springfield, Illinois.

Nakai, J. E. (1963). "Morphology of Neuroglia." Ikagu Shoin Ltd., Tokyo.

Narumi, S., and Miyamoto, E. (1974). Activation and phosphorylation of carbonic anhydrase by adenosine $3',5'$-monophosphate dependent protein kinases. *Biochim. Biophys. Acta* **350,** 215–224.

Narumi, S., Kimelberg, H. K., and Bourke, R. S. (1978). Effects of norepinephrine on the morphology and some enzyme activities of primary monolayer cultures from rat brain. *J. Neurochem.* **31,** 1479–1490.

Newburgh, R. W., and Rosenberg, R. N. (1972). Effect of norepinephrine on glucose metabolism in glioblastoma and neuroblastoma cells in cell culture. *Proc. Natl. Acad. Sci. U.S.A.* **69,** 1677–1680.

Newcombe, D. S., Ciosek, C. P., Jr., Ishikawa, Y., and Fahey, J. V. (1975). Human synoviocytes: Activation and desensitization by prostaglandins and l-epinephrine. *Proc. Natl. Acad. Sci. U.S.A.* **72,** 3124–3128.

Nickols, G. A., and Brooker, G. (1978). Temperature sensitivity of cyclic AMP production and catecholamine-induced refractoriness in a rat astrocytoma cell line. *Proc. Natl. Acad. Sci. U.S.A.* **75,** 5520–5524.

Norton, W. T. (1976). Formation, structure and biochemistry of myelin. *In* "Basic Neurochemistry" (G. J. Siegel, R. W. Albers, R. Katzman, and B. W. Agranoff, eds.), pp. 74–99. Little, Brown, Boston, Massachusetts.

Oey, J. (1975). Noradrenaline induces morphological alterations in nucleated and enucleated rat C6 glioma cells. *Nature (London)* **257,** 317–319.

Oey, J. (1976). Noradrenaline refractoriness in cultured glioma cells. *Exp. Brain Res.* **25,** 359–368.

Oksche, A. (1958). Histologische Untersuchungen über die Bedeutung des Ependyms der Glia und der Plexus chorioidei für den Kohlenhydratstoffwechsel des ZNS. *Z. Zellforsch. Mikrosk. Anat.* **48,** 74–129.

Opler, L. A., and Makman, M. H. (1972). Mediation by cyclic AMP of hormone-stimulated glycogenolysis in cultured rat astrocytoma cells. *Biochem. Biophys. Res. Commun.* **46,** 1140–1145.

Opler, L. A., and Makman, M. H. (1973). Mediation by cyclic AMP of hormone-stimulated glucose uptake in cultured rat astrocytoma cells. *Fed. Proc., Fed. Am. Soc. Exp. Biol.* **32,** 610.

Ortmann, R. (1978). Effect of PGI_2 and stable endoperoxide analogues on cyclic nucleotide levels in clonal cell lines of CNS origin. *FEBS Lett.* **90,** 348–352.

Ortmann, R., and Perkins, J. P. (1977). Stimulation of adenosine $3':5'$-monophosphate formation by prostaglandins in human astrocytoma cells. *J. Biol. Chem.* **252,** 6018–6025.

Palmer, G. C. (1973). Adenyl cyclase in neuronal and glial-enriched fractions from rat and rabbit brain. *Res. Commun. Chem. Pathol. Pharmacol.* **5**, 603-613.

Palmer, G. C., and Manian, A. A. (1976). Actions of phenothiazine analogues on dopamine-sensitive adenylate-cyclase in neuronal and glia-enriched fractions from rat brain. *Biochem. Pharmacol.* **25**, 63-71.

Passonneau, J. V., and Crites, S. K. (1976). Regulation of glycogen metabolism in astrocytoma and neuroblastoma cells in culture. *J. Biol. Chem.* **251**, 2015-2022.

Patterson, P. H. (1978). Environmental determination of autonomic neurotransmitter functions. *Annu. Rev. Neurosci.* **1**, 1-17.

Pearse, A. G. E. (1976). Peptides in brain and intestine. *Nature (London)* **262**, 92-94.

Penit, J., Jard, S., and Benda, P. (1974). Probenecide sensitive 3′,5′-cyclic AMP secretion by isoproterenol-stimulated glial cells in culture. *FEBS Lett.* **41**, 156-160.

Perez-Polo, J. R., Hall, K., Livingston, K., and Westlund, K. (1977). Steroid induction of nerve growth factor synthesis in cell culture. *Life Sci.* **21**, 1535-1544.

Perkins, J. P., Macintyre, E. H., Riley, W. D., and Clark, R. B. (1971). Adenyl cyclase phosphodiesterase and cyclic AMP dependent protein kinase of malignant glial cells in culture. *Life Sci.* **10**, 1069-1080.

Perkins, J. P., Moore, M. M., Kalisker, A., and Su, Y.-F. (1975). Regulation of cyclic AMP content in normal and malignant brain cells. *Adv. Cyclic Nucleotide Res.* **5**, 641-660.

Perkins, J. P., Johnson, G. L., and Harden, T. K. (1978). Drug-induced modification of the responsiveness of adenylate cyclase to hormones. *Adv. Cyclic Nucleotide Res.* **9**, 19-32.

Pfeiffer, S. E., Betschart, B., Cook, J., Mancini, P., and Morris, R. (1978). Glial cell lines. *In* "Cell, Tissue and Organ Cultures in Neurobiology" (S. Fedoroff and L. Hertz, eds.), pp. 287-346. Academic Press, New York.

Pfeuffer, T., and Helmreich, E. J. M. (1975). Activation of pigeon erythrocyte membrane adenylate cyclase by guanyl nucleotide analogues and separation of a nucleotide binding protein. *J. Biol. Chem.* **250**, 867-876.

Pontén, J., and Macintyre, E. H. (1968). Long term culture of normal and neoplastic human glia. *Acta Pathol. Microbiol. Scand.* **74**, 465-486.

Premont, J., Benda, P., and Jard, S. (1975). [³H]Norepinephrine binding by rat glial cells in culture. Lack of correlation between binding and adenylate cyclase activation. *Biochim. Biophys. Acta* **381**, 368-376.

Privat, A. (1975). Postnatal gliogenesis in the mammalian brain. *Int. Rev. Cytol.* **40**, 281-323.

Propst, F., Moroder, L., Wünsch, E., and Hamprecht, B. (1979a). The influence of secretin, glucagon and other peptides, and of amino acids, prostaglandin endoperoxide analogues and diazepam on the level of adenosine 3′,5′-cyclic monophosphate in neuroblastoma × glioma hybrid cells. *J. Neurochem.* **32**, (1459-1500).

Propst, F., van Calker, D., Moroder, L., Wünsch, E., and Hamprecht, B. (1979b). The influence of gastrointestinal hormones on the level of cyclic AMP in neuroblastoma × glioma hybrid cells and in cells of primary cultures of perinatal mouse brain. *In* "Hormonal Receptors in Digestive Tract Physiology" (G. Rosselin, ed.). Elsevier, Amsterdam (in press).

Raff, M. (1976). Self-regulation of membrane receptors. *Nature (London)* **259**, 265-266.

Raizada, M. K., Yang, J. W., Phillips, M. I., and Fellows, R. E. (1978). Angiotensin II receptors of rat brain cells in culture. *In Vitro* **14**, 376-377.

Rawlins, F. A., and Villegas, J. (1978). Autoradiographic localization of acetylcholine receptors in the Schwann cell membrane of the squid nerve fiber. *J. Cell Biol.* **77**, 371-376.

Reiser, G., Lautenschlager, E., and Hamprecht, B. (1975). Effects of Colcemid and lithium ions on processes of cultured cells derived from the nervous system. *In* "Microtubules and Microtubule Inhibitors" (M. Borger and M. de Brabander, eds.), pp. 259–268. North-Holland Publ., Amsterdam.

Remold-O'Donnell, E. (1974). Stimulation and desensitization of macrophage adenylate cyclase by prostaglandins and catecholamines. *J. Biol. Chem.* **249**, 3615–3621.

Robberecht, P., Conlon, T. P., and Gardner, J. D. (1976). Interaction of porcine vasoactive intestinal peptide with disposed pancreatic acinar cells from the guinea pig. *J. Biol. Chem.* **251**, 4635–4639.

Robison, G. A., Butcher, R. W., and Sutherland, E. W. (1971). "Cyclic AMP." Academic Press, New York.

Roitbak, A. J. (1970). A new hypothesis concerning the mechanism of formation of the conditioned reflex. *Acta Neurobiol. Exp.* **30**, 81–94.

Roussel, G., Delaunoy, J.-P., Nussbaum, J.-L., and Mandel, P. (1979). Demonstration of a specific localization of carbonic anhydrase C in the glial cells of rat CNS by an immunohistochemical method. *Brain Res.* **160**, 47–55.

Salem, R., and de Vellis, J. (1976). Protein-kinase (PK) activity and cAMP-dependent protein phosphorylation in subcellular fractions after norepinephrine (NE) treatment of glial cells. *Fed. Proc., Fed. Am. Soc. Exp. Biol.* **35**, 296.

Sato, S., Sugimura, T., Yoda, K., and Fujimura, S., (1975). Morphological differentiation of cultured mouse glioblastoma cells induced by dibutyryl cyclic adenosine monophosphate. *Cancer Res.* **35**, 2494–2499.

Schaefer, A., Unyi, G., and Pfeifer, A. K. (1972). The effects of a soluble factor and of catecholamines on the activity of adenosine triphosphatase in subcellular fractions of rat brain. *Biochem. Pharmacol.* **21**, 2289–2294.

Schaefer, A., Seregi, A., and Komlós, M. (1974). Ascorbic acid like effect of the soluble fraction of rat brain on adenosine triphosphatase and its relation to catecholamines and chelating agents. *Biochem. Pharmacol.* **23**, 2257–2271.

Scharrer, B. (1978). Peptidergic neurons: Facts and trends. *Gen. Comp. Endocrinol.* **34**, 50–62.

Schimmer, B. P. (1971). Effects of catecholamines and monovalent cations on adenylate cyclase activity in cultured glial tumor cells. *Biochim. Biophys. Acta* **252**, 567–573.

Schimmer, B. P. (1973). Influence of lithium ions on epinephrine-stimulated adenylate cyclase activity in cultured glial tumor cells. *Biochim. Biophys. Acta* **327**, 186–192.

Schlaeger, E. J., and Köhler, G. (1976). External cyclic AMP-dependent protein kinase activity in rat C6 glioma cells. *Nature (London)* **260**, 705–707.

Schlegel, B., and Oey, J. (1975). Catecholamine-induced cyclic AMP and morphological alterations in glioma cells. *Naturwissenschaften* **62**, 534–535.

Schmitt, H., and Pochet, R. (1977). *In vivo* labelling of β-adrenergic receptors on rat glioma cells. *FEBS Lett.* **76**, 302–305.

Schoffeniels, E., Franck, G., Hertz, L., and Tower, D. B., eds. (1978). "Dynamic Properties of Glia Cells." Pergamon, Oxford.

Schrier, B. K. (1973). Surface cultures of fetal mammalian brain cells: Effect of subculture on morphology and choline acetyltransferase activity. *J. Neurobiol.* **4**, 117–124.

Schrier, B. K., and Gilman, A. G. (1973). Elevation of cyclic AMP of rat brain cell cultures by adenosine. *Fed. Proc., Fed. Am. Soc. Exp. Biol.* **32**, 680.

Schubert, D., Heinemann, S., Carlisle, W., Tarikas, H., Kimes, B., Patrick, J., Steinbach, J. H., Culp, W., and Brandt, B. L. (1974). Clonal lines from the rat central nervous system. *Nature (London)* **249**, 224–227.

Schubert, D., Tarikas, H., and LaCorbiere, M. (1976). Neurotransmitter regulation of adenosine 3′:5′-monophosphate in clonal nerve, glia, and muscle cell lines. *Science* **192**, 471–472.

Schultz, J., Hamprecht, B., and Daly, J. W. (1972). Accumulation of adenosine 3':5'-cyclic monophosphate in clonal glial cells: Labeling of intracellular adenine nucleotides with radioactive adenine. *Proc. Natl. Acad. Sci. U.S.A.* **69**, 1266–1270.

Schürch-Ratgeb, Y., and Monard, D. (1978). Brain development influences the appearance of glial factor-like activity in rat brain primary cultures. *Nature (London)* **273**, 308–309.

Schwarcz, R., and Coyle, J. T. (1977). Neurochemical sequelae of kainate injections in corpus striatum and substantia nigra of the rat. *Life Sci.* **20**, 431–436.

Schwartz, J. P. (1976). Catecholamine-mediated elevation of cyclic GMP in the rat C-6 glioma cell line. *J. Cyclic Nucleotide Res.* **2**, 287–296.

Schwartz, J. P., and Costa, E. (1977). Regulation of nerve growth factor content in C-6 glioma cells by β-adrenergic receptor stimulation. *Naunyn-Schmiedeberg's Arch. Pharmacol.* **300**, 123–129.

Schwartz, J. P., and Passonneau, J. V. (1974). Cyclic AMP-mediated induction of the cyclic AMP phosphodiesterase of C-6 glioma cells. *Proc. Natl. Acad. Sci. U.S.A.* **71**, 3844–3848.

Schwartz, J. P., and Passonneau, J. V. (1975). Correlation between intracellular cyclic neucleotide levels and cyclic nucleotide phosphodiesterase activity in C-6 glioma and C-1300 neuroblastoma cells. *Fed. Proc., Fed. Am. Soc. Exp. Biol.* **34**, 694.

Schwartz, J. P., Morris, N. R., and Breckenridge, B. McL. (1973). Adenosine 3',5'-monophosphate in glial tumor cells. *J. Biol. Chem.* **218**, 2699–2704.

Schwartz, J. P., Chuang, D.-M., and Costa, E. (1977). Increase in nerve growth factor content of C6 glioma cells by the activation of a β-adrenergic receptor. *Brain Res.* **137**, 369–375.

Sensenbrenner, M., Springer, N., Booher, J., and Mandel, P. (1972). Histochemical studies during the differentiation of dissociated nerve cells cultivated in the presence of brain extracts. *Neurobiology* **2**, 49–60.

Severs, W. B., and Daniels-Severs, A. E. (1973). Effects of angiotensin on the central nervous system. *Pharmacol. Rev.* **25**, 415–449.

Shapiro, D. L. (1973). Morphological and biochemical alterations in foetal rat brain cells cultured in the presence of monobutyryl cyclic AMP. *Nature (London)* **241**, 203–204.

Shear, M., Insel, P. A., Melmon, K. L., and Coffino, P. (1976). Agonist-specific refractoriness induced by isoproterenol. *J. Biol. Chem.* **251**, 7572–7576.

Shein, H. M. (1965). Propagation of human fetal spongioblasts and astrocytes in dispersed cell cultures. *Exp. Cell Res.* **40**, 554–569.

Sheppard, J. R., Hudson, T. H., and Larson, J. R. (1975). Adenosine 3',5'-monophosphate analogs promote a circular morphology of cultured schwannoma cells. *Science* **187**, 179–181.

Silbert, S. W., and Goldstein, M. N. (1972). Drug-induced differentiation of a rat glioma *in vitro*. *Cancer Res.* **32**, 1422–1427.

Somjen, G. G. (1975). Electrophysiology of neuroglia. *Annu. Rev. Physiol.* **37**, 163–190.

Steinbach, J. H., and Schubert, D. (1975). Multiple modes of dibutyryl cyclic AMP-induced process formation by clonal nerve and glial cells. *Exp. Cell Res.* **91**, 449–453.

Stent, G. S. (1973). A physiological mechanism for Hebb's postulate of learning. *Proc. Natl. Acad. Sci. U.S.A.* **70**, 997–1001.

Su, Y.-F., Johnson, G. L., and Perkins, J. P. (1975). Influence of cholera toxin on the responsiveness of adenylate cyclase to catecholamines and prostaglandins. *Pharmacologist* **17**, 233.

Su, Y.-F., Cubeddu, X. L., and Perkins, J. P. (1976a). Regulation of adenosine 3':5'-monophosphate content of human astrocytoma cells: Desensitization to catecholamines and prostaglandins. *J. Cyclic Nucleotide Res.* **2**, 257–270.

Su, Y.-F., Johnson, G. L., Cubeddu, X. L., Leichtling, B. H., Ortmann, R., and Perkins, J. P. (1976b). Regulation of adenosine 3':5'-monophosphate content of human astrocytoma cells: Mechanism of agonist specific desensitization. *J. Cyclic Nucleotide Res.* **2**, 271–285.

Sundarraj, N., Schachner, M., and Pfeiffer, S. E. (1975). Biochemically differentiated mouse glial cell lines carrying a nervous system specific cell surface antigen (NS-1). *Proc. Natl. Acad. Sci. U.S.A.* **72**, 1927–1931.

Svaetichin, G., Negishi, K., Fatehchand, R., Drujan, B. D., and Selvin de Testa, A. (1965). Nervous function based on interactions between neuronal and non-neuronal elements. *Prog. Brain Res.* **15**, 243–266.

Szabadi, E. (1977). A model of two functionally antagonistic receptor populations activated by the same agonist. *J. Theor. Biol.* **69**, 101–112.

Szabadi, E. (1978). Minireview. Functionally opposite receptors on neurones. *Life Sci.* **23**, 1889–1898.

Terasaki, W. L., and Brooker, G. (1978). [^{125}I]iodohydroxybenzylpindolol binding sites on intact rat glioma cells. *J. Biol. Chem.* **253**, 5418–5425

Terasaki, W. L., Brooker, G., de Vellis, J., Inglish, D., Hsu, C.-Y., and Moylan, R. D. (1978). Involvement of cyclic AMP and protein synthesis in catecholamine refractoriness. *Adv. Cyclic Nucleotide Res.* **9**, 33–52.

Terenius, L. (1976). Somatostatin and ACTH are peptides with partial antagonist-like selectivity for opiate receptors. *Eur. J. Pharmacol.* **38**, 211–213.

Thompson, E. B., and Lippman, M. E. (1974). Mechanism of action of glucocorticoids. *Metab., Clin. Exp.* **23**, 159–202.

Thompson, W. J., and Strada, S. J. (1978). Hormonal regulation of cyclic nucleotide phosphodiesterases. *Recept. Horm. Action* **3**, 553–577.

Tolkovsky, A. M., and Levitzki, A. (1978). Model of coupling between the β-adrenergic receptor and adenylate cyclase in turkey erythrocytes. *Biochemistry* **17**, 3795–3810.

Traber, J. (1976). Untersuchungen zur Wirkung von Opiaten und Neurohormonen an klonalen Zellkulturen des Nervensystems. Ph.D. Thesis, University of Munich.

Traber, J., Fischer, K., Latzin, S., and Hamprecht, B. (1974). Cultures of cells derived from the nervous system: Synthesis and action of prostaglandin E. *Proc. Int. Congr. Collegium Int. Neuropsychopharmac., 9th,* 1974 pp. 956–969.

Tria, E., Luly, P., Tomasi, V., Trevisani, A., and Barnabei, O. (1974). Modulation by cyclic AMP *in vitro* of liver plasma membrane (Na$^+$-K$^+$)-ATPase and protein kinases. *Biochim. Biophys. Acta* **343**, 297–306.

Uzunov, P., Shein, H. M., and Weiss, B. (1973). Cyclic AMP phosphodiesterase in cloned astrocytoma cells: Norepinephrine induces a specific enzyme form. *Science* **180**, 304–305.

van Calker, D. (1977). Untersuchungen zur Charakterisierung und Fraktionierung von Primärkulturen des Zentralnervensystems. Ph.D. Thesis, University of Munich.

van Calker, D., Müller, M., and Hamprecht, B. (1977). Accumulation of cyclic AMP in primary cultures of perinatal mouse brain is simultaneously not only stimulated but also inhibited by various neurohormones. *Hoppe-Seyler's Z. Physiol. Chem.* **358**, 1188.

van Calker, D., Müller, M., and Hamprecht, B. (1978a). Adrenergic α- and β-receptors expressed by the same cell type in primary culture of perinatal mouse brain. *J. Neurochem.* **30**, 713–718.

van Calker, D., Müller, M., and Hamprecht, B. (1978b). Adenosine inhibits the accumulation of cyclic AMP in cultured brain cells. *Nature (London)* **276**, 839–841.

van Calker, D., Müller, M., and Hamprecht, B. (1979a). Receptors regulating the level of cyclic AMP in primary cultures of perinatal mouse brain (1979). *In* "Neural Growth and Differentiation" (E. Meisami and M. A. B. Brazier, eds.), Int. Brain Res. Organiz. (IBRO) Monogr Ser. **5**, 11–25. Raven Press, New York.

van Calker, D., Müller, M., and Hamprecht, B. (1979b). Adenosine regulates via two different types of receptors the accumulation of cyclic AMP in cultured brain cells. *J. Neurochem.* **33**, 999–1005.

Varon, S. (1978). Macromolecular glial cell markers. *In* "Dynamic Properties of Glial Cells" (E. Schoffeniels, G. Franck, L. Hertz, and D. B. Tower, eds.), pp. 93–103. Pergamon, Oxford.

Varon, S., and Raiborn, C. W., Jr. (1969). Dissociation, fractionation, and culture of embryonic brain cells. *Brain Res.* **12**, 180–199.

Varon, S., and Somjen, G. (1979). Neuron–glia interactions. *Neurosci. Res. Program, Bull.* **17**, 1–239.

Varon, S., Raiborn, C., and Norr, S. (1974a). Association of antibody to nerve growth factor with ganglionic non-neurons (glia) and consequent interference with their neuron-supportive action. *Exp. Brain Res.* **88**, 247–256.

Varon, S., Raiborn, C. W., and Burnham, P. A. (1974b). Selective potency of homologous ganglionic non-neuronal cells for the support of dissociated ganglionic neurons in culture. *Neurobiology* **4**, 231–252.

Varon, S., Raiborn, C., and Burnham, P. A. (1974c). Implication of a nerve growth factor-like antigen in the support derived by ganglionic neurones from their homologous glia in dissociated cultures. *Neurobiology* **4**, 317–327.

Varon, S., Raiborn, C. W., and Burnham, P. A. (1974d). Comparative effects of nerve growth factor and ganglionic nonneuronal cells on purified mouse ganglionic neurons in culture. *J. Neurobiol.* **5**, 355–371.

Varon, S., and Bunge, R. P. (1978). Trophic mechanisms in the peripheral nervous system. *Annu. Rev. Neurosci.* **1**, 327–361.

Villegas, J. (1973). Effects of tubocurarine and eserine on the axon-Schwann cell relationship in the squid nerve fibre. *J. Physiol. (London)* **232**, 193–208.

Villegas, J. (1974). Effects of acetylcholine and carbamylcholine on the axon and Schwann cell electrical potentials in the squid nerve fibre. *J. Physiol. (London)* **242**, 647–659.

Villegas, J. (1975). Characterization of acetylcholine receptors in the Schwann cell membrane of the squid nerve fibre. *J. Physiol. (London)* **249**, 679–689.

Villegas, J. (1978). Cholinergic properties of satellite cells in the peripheral nervous system. *In* "Dynamic Properties of Glial Cells" (E. Schoffeniels, G. Franck, L. Hertz, and D. B. Tower, eds.), pp. 207 215. Pergamon, Oxford.

Volpe, J. J., Fujimoto, K., Marasa, J. C., and Agrawal, H. C. (1975). Relation of C-6 glial cells in culture to myelin. *Biochem. J.* **152**, 701–703.

Walum, E. (1975). Glucose uptake into cultured tumor cells from the nervous system. Thesis, University of Göteborg, Sweden (cited in Pfeiffer *et al.,* 1978).

Watson, W. E. (1974). Physiology of neuroglia. *Physiol. Rev.* **54**, 245–271.

Weiss, B., Shein, H., and Uzunov, P. (1973). Induction by norepinephrine of a specific molecular form of cyclic $3',5'$-AMP phosphodiesterase in cloned rat astrocytoma cells (C-2A). *Fed. Proc., Fed. Am. Soc. Exp. Biol.* **32**, 679.

Westermark, B., and Wasteson, Å. (1975). The response of cultured human normal glial cells to growth factors. *Adv. Metab. Disord.* **8**, 85–100.

Wicks, W. D. (1974). Regulation of protein synthesis by cyclic AMP. *Adv. Cyclic Nucleotide Res.* **4**, 335–438.

Windle, W. F. (1958). "Biology of Neuroglia." Thomas, Springfield, Illinois.

Zanetta, J. P., Benda, P., Gombos, G., and Morgan, I. G. (1972). The presence of $2',3'$-cyclic AMP $3'$-phosphohydrolase in glial cells in tissue culture. *J. Neurochem.* **19**, 881–883.

ADVANCES IN CELLULAR NEUROBIOLOGY, VOLUME 1

RETROGRADE AXONAL TRANSPORT

M. A. BISBY

Division of Medical Physiology
University of Calgary
Calgary, Alberta, Canada

I. Introduction

The study of retrograde axonal transport (RT) forms a significant part of research efforts in the neurosciences. In recent issues of a leading neuro-anatomical journal, 17% of all papers dealt with RT. The majority of these papers, however, were not concerned with the mechanism and functions of RT but with the use of RT as a means of delineating neuronal pathways in the central nervous system. In this review I hope to convince the reader that, though RT is useful to the neuroanatomist, it is still more useful to the neuron. Readers desiring a review of technical details of RT techniques in neuroanatomy should proceed no further and refer instead to, for example, LaVail (1975).

There are two fundamentally important points of which every reader should be reminded. First, the neuron is a very elongated cell. A scale model of a human motoneuron innervating a muscle in the lower leg with a cell body the size of a tennis ball would have an axon over a kilometer long. Furthermore, over 99% of the volume and surface area of the neuron would be in the axon and its branches. Second, the neuron is internally specialized according to function. The axon and its branches have very little ability to synthesize macromolecules (see Droz, 1975), so that most macromolecules required for axon or terminal function are synthesized in the cell body and have to pass along the axon to reach their sites of action. Conversely, obsolete axonal constituents recycled by the neuron have to return to the cell body, and any "environmental" agents that influence the neuron and exert their effect by a change in macromolecule synthesis also have to ascend the axon to the cell body. Neurons thus have a unique requirement for a bidirectional intracellular transport system.

Most previous reviews of axonal transport emphasized the anterograde or cell body-to-axon component of axonal transport, relegating the retrograde or axon-to-cell body component to a few lines. This was understandable, since comparatively little was known about RT. However, recent findings, particularly concerning the RT of exogenous materials of physiological and pathological significance, have shifted the focus of attention much more toward RT and justify the presentation of a review emphasizing RT. I have not ignored anterograde transport (AT) completely. To do so would be quite artificial: RT is a continuation of AT (Abe *et al.*, 1974).

Besides examining the mechanisms of RT, I will concentrate on the known, implied, and speculative roles of RT both in the internal economy of the neuron and as a mediator of long-term trophic interactions between neurons and target cells. RT is probably a manifestation of an intracellular translocation process common to all types of cells but exaggerated in the neuron by virtue of the extended geometry of this cell. Indeed, experimental evidence obtained from other types of cell has been used to construct present hypotheses concerning the mechanism of transport. Conversely, the wide separation between axon terminals and cell bodies of peripheral neurons has facilitated the study of this particular form of intracellular translocation with techniques of limited spatial and temporal resolution. It is with a discussion of the available techniques that I will begin this review.

II. Mechanism of Retrograde Axonal Transport

A. Methods for Studying Axonal Transport

1. DIRECT OBSERVATION

The movement of particles in freshly dissected axons or neurites in tissue culture has been studied for many years. Recently, time-lapse cinemicrophotography combined with frame-by-frame computer analysis of particle movements has greatly improved the usefulness of this technique for generating quantitative rather than purely descriptive data (Cooper and Smith, 1974; Breuer et al., 1975; Forman et al., 1977a,b; Leestma and Freeman, 1977; Smith and Koles, 1976). A disadvantage is that only the largest moving particles can be resolved with the light microscope, so that conclusions about the mechanism of transport derived from these experiments may not apply to organelles smaller than the resolving power of the microscope. This limitation gives rise to the remarkable observation that the majority (up to 90%, see Cooper and Smith, 1974; Forman et al., 1977a) of particles are moving in a retrograde direction.

2. TRACER EXPERIMENTS

Radioactive tracers have been used extensively to study axonal transport, especially fast AT (Ochs, 1974). The label is attached to a precursor, such as an amino acid, and fed to the neuronal cell bodies which synthesize it into the transported macromolecule. This technique, combined with nerve ligatures (see Section III,A), has also been used to study the RT of endogenous proteins subsequent to their AT into the axon. The main use of labeled tracer in the study of RT has been in the transport of exogenous molecules taken up by axon terminals and subsequently localized in the neuronal cell bodies [e.g., ^{125}I-labeled nerve growth factor (NGF) (Hendry et al., 1974; Johnson et al., 1978), ^{125}I-labeled tetanus toxin (Price et al., 1975; Stockel et al., 1975b)]. Other applications of tracer technique have used cytochemistry or immunocytochemistry to localize the retrograde-transported exogenous macromolecule. The widely used horseradish peroxidase (HRP) neuroanatomical technique (LaVail, 1975) is an example of the first method, while an example of the immunological technique is provided by the localization through indirect immunofluorescence of retrograde-transported antibodies to dopamine-β-hydroxylase (DBH) (Fillenz et al., 1976; Ziegler et al., 1976).

3. ACCUMULATION TECHNIQUES

Injury to a nerve (produced by cutting, crushing, or ligature) results in an accumulation of materials proximal and distal to the site of injury. These accumulations have been used to study axonal transport on the assumption that they result from continued delivery of transported materials to the region of injury, at which point transport is interrupted. Local uptake and synthesis by nonneural elements in the injured nerve or delivery of axonal materials by injury currents must be eliminated as contributors to the accumulation.

The accumulation of materials distal as well as proximal to an injury was one of the earliest demonstrations of RT. That this accumulation was not merely a local response to injury was shown by the fact that a second injury, made distal to the first, reduced accumulation on the distal side of the proximal injury (e.g., Bray *et al.,* 1971). This technique has been used for investigating the transport of organelles or enzymes for which no specific markers are available. It can be used to differentiate between transported and nontransported materials and between materials transported at different velocities (e.g., Fonnum *et al.,* 1973; Wooten and Coyle, 1973). Changes in RT have been inferred from experiments in which accumulations in normal and pathological, regenerating, or stimulated nerves were compared. However, care should be taken in interpreting altered accumulations in terms of alterations in axonal transport. For example, a decrease in accumulation in a nerve trunk might be the result of (1) a decrease in the velocity of transport, (2) a decrease in the amount of material transported in each axon as a result of decreased synthesis by the cell body, or (3) a reduction in the number of axons in the nerve trunk.

Measurements of velocity can be obtained from accumulation data. If the rate of accumulation per hour (A) and the quantity of mobile material per millimeter length of nerve (B) are known, then the velocity of transport is A/B millimeters per hour. Unfortunately, for most transported materials, not all the axonal contents are mobile; for example, only about 5–15% of axonal acetylcholinesterase (AChE) undergoes transport. Use of the total axonal content instead of the "mobile fraction" will lead to an underestimate of transport velocity. To illustrate the application of accumulation techniques to the measurement of velocity, AChE is a good example (Table I). On the assumption that 100% of the AChE in the nerve was mobile, values for an RT velocity of approximately 5 mm/day were obtained (Fonnum *et al.,* 1973). After the mobile fraction was determined by dividing the amount of AChE activity accumulating at the end of an isolated segment by the total AChE activity in the segment, the RT velocity

TABLE I

REPRESENTATIVE ESTIMATES OF THE VELOCITY OF RETROGRADE AXONAL TRANSPORT

Material	Tissue	Technique	Velocity at 37°C (mm/day)	Reference
Visible particles	Chicken sciatic nerve	Direct observation	186[a]	Kirkpatrick et al., 1972
	Toad sciatic nerve	Direct observation	334[a]	Cooper and Smith, 1974
	Frog sciatic nerve	Direct observation	260	Forman et al., 1977b
	Cultured mouse embryo spinal cord neurites	Direct observation	117[a]	Leestma, 1976
	Cultured chick embryo spinal cord, dorsal root ganglion, and brain neurites	Direct observation	60–95	Breuer et al., 1975
Endogenous labeled proteins	Frog sciatic nerve	Accumulation at ligatures	342[a]	Edström and Hanson, 1973
	Frog sciatic nerve	Accumulation at ligatures	>110	Abe et al., 1974
	Rabbit vagus nerve	Accumulation at ligatures	34–280	Sjöstrand and Frizell, 1975
	Rabbit hypoglossal nerve	Accumulation at ligatures	25–200	Sjöstrand and Frizell, 1975
	Rat sciatic nerve	Accumulation at ligatures	>120	Bisby and Bulger, 1977
Endogenous enzymes				
AChE	Dog peroneal nerve	Accumulation at ligatures	134	Lubinska and Niemierko, 1971
	Frog sciatic nerve	Accumulation at ligatures	75[a]	Partlow et al., 1972
	Cat sciatic nerve	Accumulation at ligatures	220	Ranish and Ochs, 1972
	Rabbit vagus nerve	Accumulation at ligatures	123	Fonnum et al., 1973
	Rabbit hypoglossal nerve	Accumulation at ligatures	73	Fonnum et al., 1973
DBH	Rabbit sciatic nerve	Cold block and release	288	Brimijoin and Helland, 1976

(Continued)

TABLE I (Continued)

Material	Tissue	Technique	Velocity at 37°C (mm/day)	Reference
Antibody to DBH	Rat postganglionic sympathetic nerves	Time taken to reach cell body after terminal application	48–72	Fillenz et al., 1976
	Rat postganglionic sympathetic nerves	Time taken to reach cell body after terminal application	48	Ziegler et al., 1976
Cytochrome oxidase GDH, hexokinase	Cat hypogastric nerve	Accumulation at ligatures	144	Banks et al., 1969
	Frog sciatic nerve	Accumulation at ligatures	43–79[a]	Partlow et al., 1972
Exogenous materials HRP	Rabbit hypoglossal nerve	Time taken to reach cell body after terminal application	120	Kristensson et al., 1971b
	Chick optic nerve	Time taken to reach cell body after terminal application	>84	LaVail and LaVail, 1974
	Mouse, rat, rabbit sciatic nerve	Time taken to reach cell body after terminal application	>72	Kristensson and Olsson, 1975
NGF	Rat postganglionic sympathetic nerves	Time taken to reach cell body after terminal application	67	Stockel et al., 1975b
	Rat postganglionic sympathetic nerves	Time taken to reach cell body after terminal application	72	Johnson et al., 1978
	Rat sensory nerves	Time taken to reach cell body after terminal application	312	Stockel et al., 1975a

Tetanus toxin	Rat postganglionic sympathetic nerves	Time taken to reach cell body after terminal application	67	Stockel et al., 1975b
	Rat sensory nerves	Time taken to reach cell body after terminal application	312	Stockel et al., 1975b
	Rat motor nerves	Time taken to reach cell body after terminal application	180	Stockel et al., 1975b
Poliomyelitis virus	Rat motor nerves (sciatic)	Cold block and release	264	Shield et al., 1977
	Monkey sciatic nerve	Maximum time interval between terminal application and nerve crush that prevented paralysis	58	Bodian and Howe, 1941
Pseudorabies virus	Cat sensory nerves (sciatic)	Time taken to reach cell body after terminal application	>41	Field and Hill, 1974
	Mouse sensory nerves (sciatic)		>250	McCracken et al., 1973

[a] Velocity at 37°C estimated by applying a Q_{10} of 2.5 to velocity reported for temperatures less than 37°C.

was recalculated as 123 mm/day. Partlow *et al.* (1972) obtained an estimate for the mobile fraction of AChE in frog sciatic nerve from the maximum reduction in AChE content of the middle region of the isolated segment, on the assumption that all mobile AChE had drained out of the middle portion and traveled to the ends of the segment. Applying the value for the mobile fraction to the rate of accumulation of AChE at the proximal end of the isolated segment, they obtained a RT velocity of 19 mm/day at 22°C, which corresponds to roughly 75 mm/day at 37°C. Lubinska and Niemierko (1971) circumvented the problem of determining the mobile fraction. First, they measured the rate of accumulation of AChE at the ends of a transected dog peroneal nerve and obtained a regression equation for accumulation as a function of time. Next, they took isolated nerve segments of various lengths and measured the maximum accumulation of AChE at the proximal end of the segments, obtaining a second regression equation for accumulation as a function of segment length. By combining the two equations they were able to calculate the duration of accumulation in an isolated segment of any length. This represented the time taken for all the mobile AChE in the segment to travel to the end of the segment. Dividing duration of accumulation by length of segment gave velocity of transport. Ranish and Ochs (1972) used a similar technique in cat sciatic nerve, except that they determined accurately the time taken for accumulation to cease in an isolated segment of fixed length.

A refinement of the accumulation technique has been the use of a local cold block to produce a "reversible crush" on a nerve trunk. While the nerve is cooled, the transported materials accumulate adjacent to the cooled region. On rewarming of the nerve, the accumulated materials resume transport as a wave which can be detected as it moves away from the point of the cold block, giving a direct measure of velocity. It is assumed that during the period of accumulation no transformations occur in the accumulated material that would cause it to be subsequently transported at an altered velocity. This method was first applied to isolated nerve *in vitro* by Brimijoin (1975), and subsequently *in vivo* devices for local cooling have been described (Shield *et al.,* 1977; Hanson, 1978). The technique has been applied to the RT of DBH (Brimijoin and Helland, 1976) and ^{131}I-labeled tetanus toxin (Shield *et al.,* 1977). The high velocity obtained for DBH (Table I) is of interest because it is similar to the velocity obtained for AT of DBH (Brimijoin, 1975), while for most other materials RT velocity is less than AT velocity (see references in Table I). One possible explanation for this discrepancy is that the movement of wave fronts on release of a cold block gives a measure of maximum velocity, whereas calculations of velocity from accumulation data yield mean velocity (Brimijoin, 1975).

B. Phases of Axonal Transport

1. SLOW TRANSPORT

Two phases of axonal transport have been described: slow transport (ST) and fast (AT and RT). Least is known about ST: it appears to convey major structural elements of the axon such as tubulin and neurofibril protein at a velocity of 1–4 mm/day (Hoffman and Lasek, 1975). The transport of some soluble enzymes also occurs in association with ST, and there may be several components within the class of slow-transported materials (Willard *et al.,* 1974; Lasek and Hoffman, 1976; Levin, 1978; Lorenz and Willard, 1978). Almost nothing is known about the mechanism of ST. Unlike AT or RT it depends upon continuity with the cell body and does not occur in isolated axons (e.g., Frizell *et al.,* 1975). The similarity between the velocity of ST and the rate of nerve regeneration (Grafstein, 1971; Frizell and Sjöstrand, 1974; Lasek and Hoffman, 1976) has led to the suggestion that ST represents a continuous proximodistal movement of the entire axoplasm, which poses the problem of what happens to the transported materials in a nongrowing axon when they reach the nerve terminal. Models for both the mechanism of ST and the disassembly of the transported proteins at the axon terminals have been proposed (Lasek and Hoffman, 1976).

2. FAST TRANSPORT

Here we must leave ST, because it does not appear to have a retrograde component, and consider the two components of the fast transport process. Far more is known about the AT component, which has been the subject of numerous reviews (e.g., Lubinska, 1975; Bisby, 1976b; Grafstein, 1977), but the two components seem to have very similar characteristics. The range of AT velocities extends up to approximately 400 mm/day for wave fronts of labeled proteins (Ochs, 1972). There is evidence for "intermediate" velocities (McLean *et al.,* 1976) with one well-defined intermediate subcomponent consisting partly of mitochondria (Lorenz and Willard, 1978) and traveling at about 30–70 mm/day. Thus the velocities reported for RT fall within the range of AT velocities (Table I). AT is dependent upon metabolic energy and is blocked by low temperatures or metabolic inhibitors. RT also is blocked at low temperatures as shown by the use of a reversible local cold block of nerves to measure RT velocity (see Section II, A,3) and the effects of low temperature on the movement of optically detectable organelles in axons. Forman *et al.* (1977b) found that the movement of organelles in both anterograde and retrograde directions in bullfrog sciatic nerve ceased at 5°C and that between 8 and 37°C average organelle velocity had a Q_{10} of 2.5–3.5, similar to values previously reported for AT. The effects of metabolic inhibition on RT of labeled endogenous protein were studied by Edström and Hanson (1973). They found that treatment of frog sciatic nerves with sodium cyanide, 2,4-dinitrophenol, or iodoacetic acid markedly reduced the amount of label returned from a nerve ligature (the process of transport reversal at a ligature will be dealt with in Section III,A). Organelle transport is also blocked by metabolic inhibitors (Kirkpatrick *et al.,* 1972). Ischemia blocks RT of exogenous HRP (Kristensson *et al.,* 1971b) and endogenous AChE (Tucek *et al.,* 1978).

Calcium ions are a prerequisite for AT, not only for loading materials onto the transport system in the cell body but also for translocation along the axon. The ionic requirements for RT have not yet been determined, except in the case of organelle movement where calcium ions were found to be essential [Kirkpatrick *et al.,* (1972); this paper reports that most of the observed particles were moving in the anterograde direction, which is apparently an error—see Hammond and Smith (1977)]. AT is sensitive to microtubule-disrupting agents such as colchicine and vinblastine, and a similar sensitivity has been demonstrated for RT. Inhibition of retrograde movement of organelles (Hammond and Smith, 1977), endogenous labeled proteins (Edström and Hanson, 1973), and exogenous proteins (Kristensson and Sjöstrand, 1972; Bunt and Lund, 1974; Hendry *et al.,* 1974; LaVail and LaVail, 1974; Fillenz *et al.,* 1976) occurs over the same range of drug concentrations that affect AT.

Finally, neither AT or RT requires continuity with the cell body, as shown by the numerous experiments utilizing isolated nerve segments. Since most experiments demonstrating RT have used the technique of accumulation at an injury, there has necessarily been a discontinuity with the cell body. It is rather hard to demonstrate that the cell body has no influence upon RT, but the retrograde movement of organelles is initially unaffected by disconnection (R. S. Smith, personal communication). Obviously, by the time Wallerian degeneration begins in a distal nerve stump, one would expect to see alterations in RT.

Very few experimental procedures have been shown to have differential effects on AT and RT. Leestma (1976) reported that AT was more susceptible to hypoxia than RT in terms of the number of particles moving in each direction in neurites of spinal cord cultures as incubation proceeded. This effect may have been exerted at the level of formation of particles in the cell body rather than on the transport process. Overall, AT and RT operate by similar mechanisms.

C. Structural Basis of Fast Transport

In many cell types microtubules act as a "guide" for intracellular organelle translocation (e.g., Rebuhn, 1972; Goldman *et al.*, 1976). The potent inhibition of AT and RT by microtubule-disrupting agents has led to the suggestion that microtubules are closely involved in the mechanism of axonal transport, though there are a few reports of transport inhibition not associated with microtubule disruption. Other indirect evidence supporting the microtubule hypothesis includes their association with transported organelles such as mitochondria (Raine *et al.*, 1971; Cooper and Smith, 1974; Friede and Ho, 1977; Chan and Bunt, 1978), lysosomes and multivesicular bodies (Breuer *et al.*, 1975), synaptic vesicles (Smith *et al.*, 1975; Bird, 1976), and a variety of organelles loaded with HRP undergoing RT (LaVail and LaVail, 1974). There is also a correlation between microtubule density and magnitude of axonal transport (Smith, 1973). Though microtubules may be involved in fast transport, they are not themselves fast-transported but move at a slow velocity of 1–2 mm/day (Hoffman and Lasek, 1975). Thus in neurons also microtubules probably function as guides along which transported materials move at much higher velocities.

Fast-transported materials are predominately membrane-bound or associated with the membranes of subcellular organelles. In particular, autoradiographic studies have revealed that fast-transported protein is associated with the smooth endoplasmic reticulum (SER) of the axon (Schonbach *et al.*, 1971; Droz *et al.*, 1973). This complex tubular system appears to run continuously from cell body to terminal. Expanded cisternae are found in association with the plasma membrane, providing a possible route for transfer of transported molecules to the plasma membrane. In addition, formation of synaptic vesicles from SER has been described (Droz *et al.*, 1975; Tsukita and Ishikawa, 1976). It is possible, therefore, that all fast AT materials are packaged either within the lumen of the SER or in its membranes.

Which axonal organelles are involved in RT is a matter for dispute. One might argue that there is no need to postulate a separation of organelles involved in AT and RT. If there are two oppositely polarized transport systems (two sets of microtubules?), then a given organelle could move either way depending upon which transport system it is associated with. In the cell body only those organelles loaded onto the AT system will appear in the axon, and in the terminals only organelles loaded onto the RT system subsequently appear in the cell body.

Many workers have described the localization of HRP en route from terminals to the cell body as being mainly or exclusively within tubules of the SER (Krishnan and Singer, 1973; Sotelo and Riche, 1974; Nauta *et al.*, 1975; Sherlock *et al.*, 1975; Price and Fisher, 1978). The localization of other exogenous RT materials appears to be similar; both NGF coupled to HRP (Schwab, 1977) and tetanus toxin coupled to colloidal gold (Schwab and Thoenen, 1977a) were localized within smooth membrane cisternae. These findings suggest that the SER is involved in both AT and RT. However, Holtzmann (1977) argues that the elongated tubules that become filled with exogenous tracers are not part of the axonal SER. He finds that most of the tubules that accumulate HRP are shorter and of greater diameter than elements of the SER. Such tubules may be distorted vacuoles, "endocytotic tubules," or elongated multivesicular bodies. His conclusion is that the axonal tubules that accumulate exogenous markers constitute a system separate from the SER involved in AT. Bunge (1977) studied the uptake of HRP by growth cones of cultured sympathetic cells and found that there were differences between the highly branched tubular elements identified as SER and the less convoluted tubular elements that accumulated HRP and were derived from the plasma membrane. Both of the aforementioned authors point out that the difference in membrane packaging of AT and RT materials could influence the direction of transport. LaVail and LaVail (1974) also expressed reservations concerning the identification of a variety of organelles containing retrograde-transported HRP as parts of the axonal endoplasmic reticulum. Birks *et al.* (1972) have clearly differentiated in cultured sympathetic neurites between blind-end tubules that accumulate exogenous markers and the highly branched, irregular, axonal endoplasmic reticulum.

Further differentiation between organelles involved in AT and RT is provided by observations on organelles moving in living axons and on the accumulation of different organelles proximal and distal to a nerve crush. Whereas most of the visible particles are moving in a retrograde direction, endogenous tracer and enzyme accumulation experiments show that the majority of these materials are moving in an anterograde direction. This implies that AT is occurring mainly in association with organelles beyond the resolving power of the light microscope, whereas RT involves larger-sized organelles. A few retrograde-transported organelles were actually fixed for electron microscopy, a remarkable feat, and identified as lysosomes and multivesicular bodies (Breuer *et al.*, 1975). On the distal side of a nerve crush dense, lamellar, and multivesicular bodies accumulate in greater numbers than on the proximal side (Kapeller and Mayor, 1969; Griffin *et al.*, 1977; Smith, 1978). Unfortunately, there is no evidence available as to the identity of organelles involved in the RT of endogenous materials. It would be of great interest to determine whether these were "packaged" differently from the exogenous proteins taken up at axon terminals. Though there do appear to be differences in the type of organelles carried by AT and RT, this information does not really help to answer the question, What determines which way an organelle moves? The first alternative is that there are two oppositely polarized transport systems, the direction of transport being determined primarily by the site of loading. Thus SER, synthesized in the cell body, moves primarily in an anterograde direction, whereas multivesicular bodies and other remnants of pinocytosis formed mainly in the terminals move primarily in a retrograde direction. The second alternative is that there is but a single transport mechanism, the direction of transport being determined by the nature of the transported organelle and its mode of interaction with the transport mechanism.

Thus far I have implicated microtubules and various membraneous organelles in axonal transport. A model has been developed by Ochs (1970) in which transported organelles are linked through a hypothetical "transport filament" to the microtubules. Motive force is provided by an energy-requiring, calcium-dependent interaction between the microtubules and the transport filament involving cross-bridges, analogous to the events in muscle contraction (Schmitt, 1970). Microtubules from cilia and flagella contain an adenosine triphosphatase called dynein which forms cross-bridges between microtubules and provides the energy for the sliding motion between them. Similar high-MW proteins are associated with neural microtubules (Amos, 1977; Sherline and Schiavone, 1977), and side arms have been observed on neural microtubules *in vitro* and *in vivo* (e.g., Smith *et al.,* 1975; Vallee and Borisy, 1977; Amos, 1977). Adenosine triphosphatase activity has also been localized to neuronal microtubules (Fitzsimons *et al.,* 1978). The transport filament of Ochs (1970) is necessary to explain the variety of velocities for axonal transport. The fastest transported organelles have the highest affinity for the transport filament, whereas more slowly transported organelles have a lower affinity and so spend more time stationary in between periods of movement, giving an average lower velocity. This hypothesis is supported by the observations made on organelles in living nerve fibers. Every report has emphasized that these organelles move in a discontinuous or saltatory fashion with occasional periods of rest or even a transient reversal of direction (Kirkpatrick *et al.,* 1972; Cooper and Smith, 1974; Leestma and Freeman, 1977; Forman *et al.,* 1977a). Thus the average velocity of the organelles is considerably less than the maximum velocity occurring during the saltations. "Instantaneous" velocities can exceed 4 μm/sec at 28°C (Forman *et al.,* 1977a) or 2 μm/sec at 20°–22°C (Cooper and Smith, 1974). Applying a Q_{10} of 2.5, these figures translate to 824 and 758 mm/day at 37°C, roughly double the maximum values reported for fast AT of proteins and enzymes. The particles appear to be following some sort of "track," for several take roughly the same path through the axon for at least some part of their observed movement, and where the axoplasm is distorted all particles take a serpentine route along invisible pathways. The saltations might represent the attachment and detachment of organelles with the transport filaments of Ochs, or they might represent discrete sites of energy transfer to the organelles. However, successive organelles apparently traveling along the same track do not begin their saltations at the same point. Identifiable threadlike axonal mitochondria also move, though infrequently, at a velocity approximately the same as that of the smaller organelles (Forman *et al.*, 1977a). The close association of the mitochondria with the microtubules and the implied role of the microtubules in transport has led to the suggestion that the mitochondria are strategically located along the microtubules to impart metabolic energy, in the form of ATP, to the transport process (Cooper and Smith, 1974). However, no increase in organelle velocity has been detected as organelles pass close to mitochondria; in fact, their motion becomes more irregular (Leestma and Freeman, 1977).

Two alternatives to a "sliding filament" type of model have been proposed. These are the "microstream" hypothesis of Gross (1975) and the "mechanochemical" hypothesis of Samson (1976). Both utilize microtubules as the central components. In the microstream hypothesis a vectorial enzyme on the surface of the microtubule imparts force to an annulus of low-viscosity axoplasm—the microstream—which surrounds the microtubule. The velocity of the microstream decreases with distance from the microtubule surface, so that a variety of transport velocities could be accommodated by having particles with different geometries located, on average, at different positions across the microstream. In the mechanochemical model of Samson, organelles are translocated by the contraction of a network of polyanionic electrolyte molecules anchored to the microtubules. The contractions would be discontinuous and account for the saltatory movements of organelles. Neither these nor the sliding filament hypothesis are very successful at explaining the bidirectional nature of axonal transport. In each case the simplest explanation is that two oppositely polarized sets of microtubules exist. Further speculation on this topic appears in Section III,A.

In summary, the mechanism for RT is probably similar to that for AT. Both processes require metabolic energy, have the same ionic requirements and drug sensitivity, and are intrinsic properties of the axon. Transport involves membranes or membrane-bound organelles, and different organelles may be involved in AT and RT, with either the "flavor" of the organelle membrane or its site of formation determining its direction of transport. Microtubules play an important role in transport, perhaps providing a track along which organelles can move.

III. Retrograde Transport of Materials Endogenous to the Neuron

A. Labeled Proteins

If labeled precursor is injected in the vicinity of cell bodies, transported protein allowed to enter the axons, and then the nerve crushed or ligated, accumulations of activity can be observed both proximal and distal to the injury. The distal accumulation has been interpreted as evidence for the turnaround and RT of proteins originally conveyed to the axon by AT. Thus Lasek (1967) found accumulations of labeled proteins distal to sections of the cat sciatic nerve when the nerve was severed 1–6 days after the injection of labeled leucine into the vicinity of spinal cord motoneurons and the animal killed 24 hr later. Bray et al. (1971) similarly found distal accumulations at ligatures placed on chicken sciatic nerve 8 hr after precursor injections and left for a further 12 or 24 hr. Frizell and Sjöstrand (1974) estimated that as much as 50% of the AT proteins reaching the distal part of the hypoglossal nerve by 16 hr after precursor injection returned to a ligature left on the nerve for a further 5 hr. Bisby (1976a) studied the time course of RT of endogenous proteins and found that RT peaked approximately 40 hr following precursor injection in motor axons, whereas in sensory axons the magnitude of RT was considerably less though the time course was similar (Bisby, 1977). In motor axons about 50% of anterograde-transported protein was subsequently returned by RT.

Reversal and RT of endogenous proteins can be elicited by placing a distal crush on the nerve (Edström and Hanson, 1973; Frizell and Sjostrand, 1974). Bisby and Bulger (1977) found that crushed axons took about 1 hr to develop the ability to reverse transported protein, and that the reversal process took less than 2 hr. A nerve injury made at the same time as the injection of precursor led to a much earlier return of fast-transported labeled protein from the axon than would have occurred in an intact axon. This premature return of transported protein in injured nerves was considered a possible "signal for chromatolysis" (see Section V,A,1).

The time course of RT of endogenous proteins following a pulse application of label to the cell bodies is of the same order as the half-lives for axon terminal proteins (Droz, 1975). Thus the proteins returning to the cell body might represent degraded synaptic proteins. However, in a long axon a considerable proportion of the AT protein may be destined for the axon rather than the terminals (Cancalon and Beidler, 1975; Ochs, 1975), so that the contributions of axonal versus terminal protein to the protein collected distal to ligatures is uncertain. Bray et al. (1971) placed a second ligature between the proximal ligature and the terminal regions of the axon and showed that RT back to the proximal ligature was abolished. This indicated that the major site of reversal was in the distal part of motor axons but did not allow a quantitative estimate of the importance of reversal in axons and terminals.

RT of endogenous proteins has been studied in regenerating axons to see if the conversion from intact axon terminals to growth cones had any effect on the magnitude or time course of RT. Frizell *et al.* (1976) demonstrated increased accumulations of activity associated with both proteins and glycoproteins distal to crushes in regenerating hypoglossal and vagus nerves of the rabbit. This increase was greater 1 week after nerve injury than after 5 weeks. Bulger and Bisby (1978) found an increase in RT in motor axons in regenerating rat sciatic nerves between 1 and 5 days after injury. At later times after injury there was a shift in the time course of RT from the early return characteristic of acutely injured axons toward the later time course characteristic of intact axons, so that by 30 days after injury RT was normal. In axons prevented from regenerating for up to 30 days following injury, RT was maintained at a level similar to that in an acutely injured nerve. In injured axons it is likely that the major site of reversal of transport is adjacent to the injury, and the simplest interpretation of the results of Bulger and Bisby (1978) is that, as the axons regenerated, the major site of reversal occurred at the growth cones that moved further from the site of the nerve crush used to collect the retrograde-transported protein. The time taken by protein that turned around at the growth cone to reach the nerve crush increased and the time course of RT lengthened.

It has not yet been possible to apply the autoradiographic techniques that have been so useful in revealing the structural basis of AT (Droz, 1975) to the RT of endogenous proteins. The major problem is the relatively low level of activity of retrograde-transported material combined with the simultaneous presence in the axon of labeled anterograde-transported material.

Attempts to characterize retrograde-transported protein in terms of MW were made by Abe *et al.* (1974). Unfortunately, fast AT protein is made up of a multitude of polypeptide species (e. g., Stone *et al.,* 1978) which were not completely resolved with the techniques used. Nevertheless, they found that, though anterograde- and retrograde-transported proteins were similar, there was a consistent decrease in one specific protein component following reversal of transport in frog sciatic nerve. Recently a more sensitive technique, permitting the resolution of 25 separate polypeptides (Bisby, 1978), has been applied to this problem. Labeled proteins accumulating over a 2-hr period proximal and distal to crushes made on the rat sciatic nerve 9 and 22 hr after precursor injection have been compared. At 22 hr the proteins undergoing AT and RT are significantly different, but the retrograde-transported proteins at 22 hr are not significantly different from the anterograde-transported proteins 9 hr after injection (M. A. Bisby, unpublished observation). Thus it appears that the label returning toward the cell bodies is definitely associated with the same proteins that were earlier transported into the axon by AT. Reversal and RT do not seem to be necessarily associated with proteolysis in the terminal axon. So far no components of AT protein have been identified as absent from RT protein, a finding that seems inconsistent with the evidence for release of protein from nerve axons and terminals (Musick and Hubbard, 1972; Bray and Harris, 1975; Hines and Garwood, 1977). This may be a problem of inadequate resolution of quantitatively minor components of transported protein.

If the same proteins are transported in anterograde and retrograde directions, how is the direction of transport during different times in the protein's life span determined? Most likely the type of packaging of these proteins, the majority of which are membrane-associated, is the determining factor. Evidence for the involvement of different types of organelles in AT and RT has been presented in Section II, C. It remains a mystery how proteins are unloaded from the AT system along the axon or in the terminals and reloaded onto the retrograde system. At the ultrastructural level one of the problems has been that microtubules, which as previously discussed are probably vital to transport processes, could not be demonstrated within axon terminals where the reversal process presumably occurs. However, with new fixation techniques microtubules have been shown to extend right up to the presynaptic membrane (Gray, 1975; Bird, 1976), and recently an arc of microtubules associated with a horseshoe-shaped mitochondrion has been described in synaptosomes and axon terminals in cerebral cortex (Chan and Bunt, 1978). It is possible that some microtubules do not end at the axon terminal but loop back toward the cell body. Since microtubules are polarized structures whose assembly in the cell body is unidirectional (Borisy et al., 1974; Dentler et al., 1974; Osborn and Weber, 1976), this microtubule loop could provide the polarity required for bidirectional axonal transport. Alternatively, if microtubule disassembly occurs at the axon terminal as proposed by Lasek and Hoffman (1976), reassembly might occur in preterminal regions generating a set of microtubules with opposite polarity extending back to the cell body. In either case there should be a slow retrograde axonal transport of microtubule protein. Such a phenomenon has never been described, and it would not be an easy point to demonstrate.

B. *Enzymes*

1. ACETYLCHOLINESTERASE

The bidirectional transport of this enzyme has been extensively studied (Lubinska and Niemierko, 1971; Ranish and Ochs, 1972; Partlow *et al.*, 1972; Fonnum *et al.*, 1973; Tucek, 1975). AChE exists in several molecular forms in whole nerve (this includes, of course, nonneural cells). The major transported form is the 16 S type (Di Giamberardino and Courand, 1978; Festoff and Fernandez, 1978). In general, twice as much axonal AChE is free to move in the anterograde direction than in the orthograde direction, and the velocity of AT is about twice that of RT. Thus four times more AChE is leaving the cell body than is returning to it, though the mobile AChE represents only 5–15% of the total AChE in the nerve.

Histochemically, AChE reaction product has been demonstrated in nerve cell bodies, dendrites, and axons in association with SER, subsurface cisternae, and plasma membrane, as well as extracellularly (Kasa and Csillik, 1968; Kreutzberg *et al.*, 1975; Somogyi *et al.*, 1975; Jessen *et al.*, 1978). When axons are crushed, the AChE that accumulates proximal to the crush is associated with SER-derived vesicles (Kasa, 1968), and "chase" experiments where AChE was irreversibly inhibited demonstrated that newly synthesized AChE appeared first in the SER, then in the subsurface cisternae, and then at the plasma membrane (Kreutzberg *et al.*, 1975). Thus, the mobile AChE is probably associated with the SER, as is the case for other fast-transported proteins (Droz, 1975). AChE is released from dendrites (Kreutzberg *et al.*, 1975), cell bodies (Jessen *et al.*, 1978; Oh *et al.*, 1977), and axons (Somogyi *et al.*, 1975). It is released from stimulated motor terminals (Skau and Brimijoin, 1978) and into cerebrospinal fluid when peripheral nerves are stimulated (Chubb *et al.*, 1976). Release of AChE from the neuron presumably represents the fate of the majority of transported AChE, for only one-quarter of it subsequently returns to the cell body by RT though some may be returned in an inactive form. The site of reversal of AChE in nerves is unknown. Since the enzyme is incorporated into the plasma membrane, the retrograde-transported AChE may be returned from the membrane, perhaps through axonal endocytosis as observed for HRP (e.g., Tischner, 1977). Alternatively, retrograde-transported AChE may represent SER-associated enzyme not incorporated into the plasma membrane. Reversal of AChE transport also occurs at nerve crushes (Partlow *et al.*, 1972), but the process of turnaround takes much longer to develop than in the case of labeled protein.

At least some of the retrograde-transported AChE may not be endogenous but derived from postsynaptic cells. Kreutzberg *et al.* (1975) have described dendritic secretion of AChE with the incorporation of secreted AChE into synaptic vesicles of the presynaptic axon. Jessen *et al.* (1978) have also described AChE within synaptic vesicles, and Politoff *et al.* (1975) have shown that exogenous AChE can be taken up by motor terminal synaptic vesicles. Subsequent to AChE uptake the synaptic vesicles might be degraded and retrograde-transported, as described for vesicles loaded with HRP (Teichberg *et al.*, 1975). Some AChE synthesized by muscle might normally be available for uptake at the neuromuscular junction, since muscle AChE can be released by protease treatment (Hall and Kelly, 1971), and nerve stimulation has been shown to produce proteolytic activity at the end plate (Poberai and Savay, 1975). End plate-specific AChE is the same 16 S form known to be retrograde-transported (Vigny *et al.*, 1976). Thus we have a speculative chain of events suggesting that some retrograde-transported AChE is derived from postsynaptic cells. Whether this AChE has any kind of "trophic" function remains to be determined.

2. Dopamine-β-Hydroxylase

The significance of the bidirectional transport of this enzyme is that it serves as a marker for the membranes of synaptic vesicles in adrenergic neurons. It seems likely that the DBH that is retrograde-transported is associated with the partially degraded membranes of adrenergic synaptic vesicles returned from the terminals. Though DBH accumulating both proximal and distal to a crush is membrane-bound (Brimijoin, 1974), the dense-core vesicles corresponding to adrenergic synaptic vesicles only accumulate proximal to ligatures (Kapeller and Mayor, 1969). Similarly, norepinephrine contained within the dense-core vesicles accumulates distal to a ligature only to a slight degree (Dahlstrom, 1965), whereas accumulation rates of DBH proximal and distal to a ligature are similar for the first few hours after ligation (Nagatsu *et al.*, 1976). Much of the DBH that accumulates distal to a ligature is not enzymatically active, whereas the DBH accumulating proximal to the ligature is present in roughly equal quantities whether measured enzymatically or immunologically (Nagatsu *et al.*, 1976).

Circulating antibodies to DBH are retrograde-transported back to sympathetic neuron cell bodies (Ziegler *et al.*, 1976). This process is specific to adrenergic neurons and can be blocked by application of 6-hydroxydopamine which destroys adrenergic axon terminals (Fillenz *et al.*, 1976). These results suggest that the antibody combines specifically with the DBH molecule as a result of exocytosis of adrenergic synaptic vesicles at the axon terminals. As a consequence, the vesicle membrane becomes exposed to extracellular fluid containing the antibody. The vesicle membrane is subsequently returned to the cell body, probably without being reused for transmitter release (De Potter and Chubb, 1971).

3. Mitochondrial Enzymes

Bidirectional movement of long, rod-shaped mitochondria in living axons has been reported (Hammond and Smith, 1977). Mitochondria move at velocities similar to those of other organelles but less frequently and so have a lower mean velocity: some mitochondria remain motionless, and others move rapidly past them. Some of the more rapidly moving, smaller, round organelles traveling predominately in a retrograde direction are also mitochondria (Cooper and Smith, 1974). Given this complicated picture of mitochondrial movement, it is not surprising that results from studies of enzyme accumulations at nerve ligatures are confusing. Khan and Ochs (1975) reported that both AT and RT of monoamine oxidase (MAO) occurred at slow velocities. In amphibian sympathetic axons the initial rate of accumulation of MAO is greater on the distal side of the crush than on the proximal side (McLean and Burnstock, 1972). In frog nerves the RT velocity of hexokinase and glutamine dehydrogenase (GDH) was similar to that for AChE, whereas the AT velocity for the mitochondrial enzymes was about one-quarter that of AChE (Partlow et al., 1972). It is possible that RT of mitochondrial enzymes, in the form of small, round mitochondria, is more rapid than AT in the form of long, rod-shaped mitochondria. A further confusing factor is that two mitochondrial enzymes studied in the same nerve do not behave in the same manner. Schmidt and McDougal (1978) found similar increases in GDH and MAO on the proximal side of a ligature, but distally MAO increased much more than GDH. These results were discussed in terms of differential localization of the two enzymes in sympathetic (MAO) and other (GDH) axons. Khan and Ochs (1975) have pointed out that some studies show that the accumulation of mitochondrial enzymes and mitochondria at the two ends of ligated nerves is approximately equal (Zelena, 1968; Zelena et al., 1968; Banks et al., 1969), suggesting that the axon is a closed system for mitochondrial transport with anterograde-transported mitochondria subsequently returned by RT. Perhaps the retrograde-transported small mitochondria are degraded rod-shaped mitochondria returning to the cell body.

4. Lysosomal Enzymes

Acid phosphatase accumulates distal to a nerve ligature at initially the same rate that it accumulates proximally (Schmidt and McDougal, 1978). The possibility that these accumulations are a result of nonneuronal cellular reaction to injury was examined by local injection of vinblastine proximal to a ligature. This completely abolished the accumulation 48 hr after nerve injury. A crush made proximal to the ligature had the same effect but did not abolish a later increase in enzyme activity at the original ligature, which was attributed to degenerative changes. RT of lysosomal enzyme is consistent with the identification of lysosomes as one of the RT organelles in living axons (Breuer *et al.*, 1975), and with the accumulation of lysosome-like profiles and the localization of acid phosphatase in tubules distal to a nerve ligature (Holtzman and Novikoff, 1965). The role of axonal lysosomes, the transport organelles containing acid phosphatase, and the relationship between these and the degradation of endocytotic structures are discussed by Holtzman (1977).

C. Functions of Retrograde Transport of Endogenous Materials

The function of the RT of endogenous proteins and other macromolecules is generally considered a recycling process. Perhaps this might be best illustrated by what is known about the life cycle of synaptic vesicles (Holtzman, 1977). Whether these vesicles are synthesized in the cell body and transported along the axon, or synthesized in the terminals from transported SER, it is apparent that they represent one of the sites of incorporation of fast-transported proteins (Droz *et al.*, 1973). Details of the life cycle of the synaptic vesicles while in the terminals are not of concern here, but it seems that the eventual fate of vesicle membranes is to be returned to the cell body. Chase experiments with HRP during a period of high neurotransmitter release followed by a period of quiescence show that exogenous HRP is incorporated first into synaptic vesicles, then into multivesicular bodies and tubules in axons, and finally into lysosomal structures in the cell body. It is assumed that the HRP enters the axon during the endocytotic retrieval of membrane occurring subsequent to the exocytotic release of transmitter from synaptic vesicles (see also Section IV, A) and thus serves as a marker for synaptic vesicle membrane (Teichberg *et al.*, 1975). The evidence on the RT of DBH is consistent with this view. Antibody to DBH is only transported from intact terminals, presumably as a consequence of vesicle exocytosis (Fillenz *et al.*, 1976). Much of the transported DBH is enzymatically inactive (Nagatsu *et al.*, 1976) and may thus be associated with a partially degraded vesicle membrane. Degradation may have been accomplished by lysosomal enzymes, since lysosomes have been observed in close association with retrograde-transported multivesicular bodies (Breuer *et al.*, 1975).

Fast-transported proteins, at least those associated with synaptic vesicle membranes, are subsequently retrograde-transported in a different package from which they were delivered to the axon and appear to be ultimately degraded in the cell body. It is not certain whether membrane components returned to the cell body are completely broken down to their constituent "building blocks" or are salvaged in a partially degraded form and reincorporated into membranes for AT. In this respect it is interesting that antibody to DBH transported back to the cell body was later localized in cell body processes where accumulations of intact synaptic vesicles have been described (Ziegler *et al.,* 1976). Brownson *et al.* (1977) reported that HRP-containing vesicles appeared in sympathetic terminals in the ciliary body 4 hr after injecting HRP into the superior cervical ganglion. In the cell body HRP was localized mainly within lysosomal structures. This result could be interpreted to indicate the transfer of HRP from lysosomes to the AT system, indicating reuse of the membrane elements in which the HRP was enveloped. These experiments should be repeated with the appropriate controls (e.g., nerve ligature) to confirm that HRP in the terminals reached them by AT and not by uptake of HRP from the circulation. The desirability of nondestruction, or at least delayed destruction, of retrograde-transported materials of biological significance is discussed in Section IV, B.

The importance of the terminal reversal of AT to normal neuronal function has been recently demonstrated (J. R. Mendell, personal communication). Rats fed zinc pyridinethione develop a dying-back neuropathy characterized by accumulation in axon terminals of tubulovesicular profiles derived from SER. With time these accumulations fill up progressively more proximal parts of the axon. At the same time there is a delay in onset and a reduction in the amount of retrograde-transported labeled endogenous protein. An abnormality in the terminal reversal process is suggested, resulting in an accumulation of AT materials and failure of terminal function.

The suggestion has often been made that the RT of endogenous materials also serves as an information feedback system. For example, massive transmitter release might result in increased vesicle membrane return to the cell body, a signal that could stimulate increased vesicle production. An alteration in return of endogenous proteins resulting from nerve injury might signal the occurrence of injury and trigger chromatolytic changes (Bisby and Bulger, 1977). At present, there is no direct evidence to support this suggestion.

IV. Retrograde Transport of Materials Exogenous to the Neuron

A. *Mechanisms of Uptake*

This topic has been recently reviewed by Kristensson (1978). He points out that uptake of blood-borne materials can occur to a significant extent only where peripheral axons are not protected by perineurial and endoneurial diffusion barriers. Normally this occurs only at axon terminals but may occur at sites of nerve injury where the diffusion barriers have been breached (Kristensson and Olsson, 1977). The substances that can be taken up and transported appear to fall into two groups: molecules for which no "specific" membrane receptors exist (e.g., HRP, albumin) and molecules for which specific receptors are present in the presynaptic membrane (e.g., NGF, tetanus toxin, lectins, antibody to DBH) (Schwab and Thoenen, 1977a). Much lower concentrations of the specific uptake molecules are required to demonstrate RT than is the case for the nonspecific group. Thus, approximately a 100-fold greater concentration of HRP alone had to be applied to the anterior eye chamber to demonstrate its RT to the superior cervical ganglion than was necessary when the HRP was coupled to NGF (Schwab, 1977).

1. Nonspecific Uptake

The nonspecific uptake of HRP can be enhanced by several procedures; basic polyamino acids such as ornithine increase HRP uptake by neurons, as in many other cells (Trachtenberg and Hadley, 1977). Bunt and coworkers have made an extensive study of factors affecting HRP uptake and transport. Initially it seemed that the isoelectric point of various HRP isoenzymes was an important factor; basic isoenzymes were transported, while acidic ones were not (Bunt et al., 1976). More recently, it has been demonstrated that charge is not the only requirement for transport; there is a basic isoenzyme that is not transported and an acidic one that is (Bunt and Haschke, 1977). Moreover, some nontransported isoenzymes, though taken up and incorporated into synaptic and coated vesicles, are not localized in multivesicular bodies in terminals, axons, or cell bodies. Only transported isoenzymes entered multivesicular bodies. These results imply that uptake of HRP into terminals may indeed be a "nonspecific" process, but that subsequent transport involves a recognition process at the level of intracellular membrane fusion. Alternatively, HRP that is transported may be specifically incorporated into a type of endocytotic vesicle different from that which nonspecifically incorporates nontransported HRP.

Uptake of HRP can also be enhanced by direct nerve stimulation (Litchy, 1973; D'Olivo et al., 1977) or by other methods of increasing transmitter release (Teichberg et al., 1975; Cooke et al., 1975). After such transmitter release HRP and other extracellular markers are incorporated into a variety of synaptic membrane-bound structures in a time-dependent fashion, leading to the proposal of various models for the recycling of synaptic vesicles in which membrane added to the presynaptic plasma membrane as a result of synaptic vesicle exocytosis is retrieved by endocytosis and eventually reformed into synaptic vesicles (Heuser and Reese, 1973;

Ceccarelli and Hurlbut, 1975; Fried and Blaustein, 1976; Zimmerman and Denston, 1977; Gennaro *et al.,* 1978). It is thus assumed that HRP and other nonspecific uptake occurs as a consequence of synaptic vesicle exocytosis. As previously mentioned, the uptake and RT of antibody to DBH is consistent with this view, for DBH bound to the membrane of the adrenergic synaptic vesicle is presumed to be exposed to the extracellular fluid during the process of vesicle recycling. Other routes not involving vesicle recycling may exist for the terminal uptake of HRP. As mentioned above, Bunt and Haschke (1977) find that incorporation of HRP isoenzymes into synaptic vesicles does not necessarily result in their RT. Botulinum toxin, which is thought to prevent vesicle exocytosis, does not prevent HRP uptake and transport (Kristensson and Olsson, 1978). Litchy (1973) estimated that the amount of HRP appearing in stimulated motor axons was "several orders of magnitude" greater than expected on the basis of exchange of synaptic vesicle contents with extracellular fluid.

Uptake of exogenous macromolecules such as HRP also occurs in damaged axons (DeVito *et al.,* 1974; Kristensson and Olsson, 1974; Furstman *et al.,* 1975; Halperin and LaVail, 1975; Kristensson and Olsson, 1977; Oldfield and McLachlan, 1977). In peripheral nerves, though the increased permeability of the perineurial barrier that permits HRP to have access to the axons persists for some months after a nerve crush (Olsson and Kristensson, 1973), HRP entry is restricted to the first hour or so after injury, and it appears to enter the axons by diffusion through the injured axon membrane (Kristensson and Olsson, 1977). In isthmooptic neurons of the chick with terminals in the retina, injury initially decreases the ability of the neurons to take up HRP, but 6–24 hr after injury uptake by injured axons is greater than uptake by control axons (Halperin and LaVail, 1975). The ability of injured axons to take up HRP raises two unpleasant possibilities: (1) that HRP is taken up only by axons injured at the site of HRP application, and (2) that HRP injected into the CNS is taken up by neurons whose axons pass through the region of injection rather than terminate there. This could lead to misinterpretation of neuroanatomical data generated through the use of HRP. The first of these possibilities has been to some extent eliminated by the demonstration that, if large amounts of HRP are injected intravenously, labeling of motoneuron cell bodies through RT will occur. The HRP can diffuse between endothelial cell borders in muscle capillaries and reach the unprotected motor nerve terminals. Thus uptake can occur without mechanical trauma to axons (Broadwell and Brightman, 1976; Kristensson, 1977).

RT of HRP has been observed in developing chick embryos during the period of axonal outgrowth prior to innervation (Oppenheim and Heaton, 1975; Heaton, 1977), leading to the suggestion that information about the environment of the growing axon could be fed back to the cell body by this route. Uptake most likely occurs by pinocytosis at the growth cone, as observed in living cultures (Hughes, 1953). Growth cone uptake of HRP in cultured neurons by pinocytosis has been described (Bunge, 1977; Tischner, 1977). Given the ability of growth cones to take up HRP, it is rather surprising that regenerating motor neurons *in vivo* were not able to take up intravenously injected HRP until the reestablishment of functional neuromuscular contact (Olsson *et al.,* 1978). At this time there was an increased uptake of another extracellular marker, Evan's blue–albumin, which later subsided to control levels (Kristensson and Sjöstrand, 1972).

2. SPECIFIC UPTAKE

Specificity of uptake and transport has been well demonstrated for NGF. This substance is taken up by postganglionic sympathetic neurons and dorsal root ganglion neurons, but not by motor neurons (Stockel and Thoenen, 1975; Stockel *et al.,* 1975a,b). Recently, transport in parasympathetic neurons has also been demonstrated (Max *et al.,* 1978). NGF is also transported by central nervous system neurons, and here also cell type specificity was demonstrated (Ebbott and Hendry, 1978). Since receptors for NGF have been localized on the surface of cells in which it is transported (Banerjee *et al.,* 1973; Herrup and Shooter, 1973; Frazier *et al.,* 1974), uptake of NGF probably occurs first by interaction with the receptor and then by endocytosis of the plasma membrane containing the receptor–NGF complex (Norr and Varon, 1975). Alterations in the structure of NGF produce a parallel loss in RT ability and biological activity (Stockel *et al.,* 1974). The major site of NGF uptake appears to be the axon terminals, for applications of drugs that destroy adrenergic terminals prevent the RT of NGF (Johnson, 1978). Specificity of transport for the antibody to DBH has also been demonstrated; it is restricted to adrenergic neurons containing DBH (Fillenz *et al.,* 1976).

Although specific receptors for NGF and antibody to DBH are restricted to specific neuron types, some materials are taken up by specific receptors common to all types of neurons. Thus cholera toxin, tetanus toxin, and wheat germ agglutinin (WGA) are transported in all types of peripheral neurons (Stockel *et al.,* 1977). WGA and other lectins (e.g., concanavalin A) bind to cell surface glycoproteins which may then be internalized by endocytosis. Koda and Partlow (1976) described a phenomenon they termed "retrograde axolemmal transport." Concanavalin-A-coated red blood cells bound to sympathetic neurons in culture moved along the neurite surface in a retrograde direction at a velocity of 1.2 mm/day, considerably less than that of RT. The significance of this membrane movement to the economy of the neuron is quite unknown. Presumably the lectin–glycoprotein complexes were unable to internalize under these circumstances. Gangliosides are the membrane receptors for cholera and tetanus toxins, and RT of these toxins, but not of NGF or WGA, was blocked by preincubation of the toxins with appropriate gangliosides (Stockel *et al.*, 1977).

It is not definitely known whether the organelles involved in RT of exogenous macromolecules are the same for the specific and nonspecific types of molecules. The similar velocity of RT and the ultrastructural localization of transported molecules strongly suggest that all molecules follow a common pathway. The exact identity of the organelles involved in RT was discussed in Section II, C.

B. *Fate of Retrograde-Transported Exogenous Materials*

Labeled retrograde-transported materials disappear within 1 or 2 days from cell bodies (e.g., Stockel *et al.*, 1974; Johnson *et al.*, 1978) and are localized mainly in lysosomes or related structures within the cell body. However, since some retrograde-transported molecules have physiological or pathological significance to the transporting neuron, it is important that they reach their sites of action prior to, or in addition to, destruction by lysosomes. One of the effects of NGF on sympathetic neurons is to induce the synthesis of catecholamine-synthesizing enzymes. This effect of NGF is blocked by actinomycin, which suggests that NGF is exerting its effect at the level of transcription (Stockel *et al.*, 1974). Nuclear receptors for NGF have been identified in chick dorsal root ganglion and sympathetic neurons in addition to cell surface receptors (Andres *et al.*, 1977), and up to 30% of retrograde-transported NGF may be bound to the purified nuclear fraction from superior cervical ganglion (Johnson *et al.*, 1978). It has been suggested, then, that some NGF can be routed from the RT system to the nucleus. There is as yet no ultrastructural evidence in support of this suggestion (Schwab and Thoenen, 1977a). Other retrograde-transported materials that must reach the nucleus of the neuron are the nucleic acids of viruses, since RT is the route by which many neurotropic viruses enter the nervous system (e.g., Kristensson *et al.*, 1971a; Field and Hill, 1974).

Holtzman (1977) points out that interaction of retrograde-transported materials with lysosomes may not be all bad; some viruses and toxins only become active after hydrolysis. In cultured fibroblasts endocytosed polypeptide hormones enter in combination with α_2-macroglobulin, a serum protein, which is a protease inhibitor and may protect the hormones from lysosomal attack (Kolata, 1978; Willingham and Pastan, 1978). It would be most interesting to determine whether α_2-macroglobulin can also be taken up and retrograde-transported by neurons. Successful evasion of lysosomal destruction is best illustrated by the remarkable example of tetanus toxin, whose intraaxonal transport has been demonstrated by a variety of methods (Erdmann *et al.,* 1975; Price *et al.,* 1975; Price and Griffin, 1977; Carroll *et al.,* 1978). Tetanus toxin exerts its action at inhibitory synapses on spinal cord motoneurons by reducing inhibitory neurotransmitter release (e.g., Osborne and Bradford, 1973), and it does this by transsynaptic migration from the motoneuron to the presynaptic terminals (Schwab and Thoenen, 1976). After intramuscular injection of ^{125}I-labeled tetanus toxin autoradiographic localization showed label over motoneuron cell bodies and presynaptic boutons, while only scant labeling was seen over adjacent glia and other neurons. The time course of labeling of presynaptic terminals was consistent with the onset of tetanic rigidity in the injected rats. Price *et al.* (1977) demonstrated a similar autoradiographic localization of ^{125}I-labeled tetanus toxin and [^{3}H]glycine, an inhibitory neurotransmitter, at presynaptic terminals on motoneurons. Transsynaptic migration of tetanus toxin also occurs in sympathetic ganglia, but this process is specific for tetanus toxin and does not occur with NGF (Schwab and Thoenen, 1977b). It is not known how tetanus toxin manages to circumnavigate the cell body. Perhaps it becomes involved in the mechanisms of dendritic transport, of which the dendritic release of AChE is an example (Kreutzberg *et al.,* 1975).

V. Functions of Retrograde Transport of Exogenous Materials

A. Trophic Functions

There are a number of "trophic" phenomena that appear to be manifestations of some influence exerted by the target cell upon the neurons that innervate it. In the following sections I will examine evidence for the involvement of RT as the pathway through which the target cell influence is exerted.

1. CHROMATOLYSIS

If a peripheral nerve axon is severed, the cell body undergoes a series of structural and biochemical changes known as chromatolysis, which are thought to represent a reprogramming of the neuron's synthetic ability and to be necessary for axon regeneration. How does the cell body detect that its axon has been damaged? what is "the signal for chromatolysis"? This topic has been the subject of several excellent reviews (Cragg, 1970; Lieberman, 1974; Grafstein, 1975; Grafstein and McQuarrie, 1978; Torvik, 1976), and the consensus of opinion is that RT is involved in carrying the signal. The major lines of evidence supporting this opinion are:

1. The onset and severity of chromatolysis depend upon the distance between the cell body and the axon injury.
2. The velocity of RT of the signal is close to that for the RT of HRP (Kristensson and Olsson, 1975).
3. Chromatolysis can be produced by colchicine application to the axon, which also blocks RT (Pilar and Landmesser, 1972; Cull, 1975; Purves, 1976).

If RT is the carrier of the signal, there are three possible types of signal:

1. Abnormal axonal uptake of extracellular proteins at the site of injury. As previously mentioned, exogenous proteins can enter the damaged axon for a period of approximately 1 hr following injury (Olsson *et al.,* 1978).
2. Cessation of transport of a trophic factor derived from the target tissue and normally taken up by the intact terminals.
3. Reversal of axonal transport at the site of injury, resulting in return to the cell body of endogenous material normally destined for the terminal axon (Bisby and Bulger, 1977; Bulger and Bisby, 1978).

As yet, there is no evidence upon which to reject any of these mechanisms; quite likely they are all involved. Types 2 and 3 would provide a continuous signal, switched off only when normal axon terminal relations are established, and type 1 would provide only a "trigger," signaling the event of injury. Watson's results (1968, 1969, 1970) show that there are probably at least two signals, one resulting from axonal damage and the other from functional disconnection from the target cell. For example, when a cut nerve was implanted in an already innervated muscle so that motoneurons did not form functional synapses, there was a decline in the initial motoneuron response to axotomy. If the nerve was sectioned again, a chromatolytic reaction occurred. Since there was no functional contact with the muscle, uptake of trophic factors from the muscle was unlikely, so the response in this instance was probably due to axonal damage. On the other hand, some chromatolytic changes are only reversed if functional contact is reestablished. The best example of this is provided by the retraction of presynaptic boutons from the cell bodies of axotomized neurons (Cull, 1974; Matthews and Nelson, 1975; Sumner, 1975). In sympathetic postganglionic neurons of the superior cervical ganglion Purves (1975) found that reestablishment of presynaptic terminals occurred at the same time as the postganglionic neurons were reforming their terminals on effector tissues. The reestablishment of presynaptic terminals did not occur if the postganglionic axons were prevented from reaching their target cells. Similar results have been reported for hypoglossal motoneurons (Sumner, 1976).

Direct support for the role of trophic substances derived from the target tissue in chromatolysis has been provided in the case of NGF action on sympathetic postganglionic cells. NGF is synthesized by the target tissues of these cells (Johnson *et al.,* 1972; Schwab *et al.,* 1976) and, as previously discussed, exogenous NGF applied at the terminals is retrograde-transported to the cell body. Many of the chromatolytic reactions of the postganglionic neurons can be prevented by direct application of NGF to the axotomized cell bodies, particularly the loss of presynaptic boutons which, as previously discussed, seems to be a response to loss of contact with the target tissue (Nja and Purves, 1978). Conversely, application of antiserum to NGF to unaxotomized neurons produces the typical postaxotomy loss of synapses. In immature animals, postganglionic axotomy produces marked ganglion cell death which can be prevented by the application of NGF to the ganglia (Hendry, 1975; Hendry and Campbell, 1976; Banks and Walter, 1977).

Neither "wound substances" taken up at the site of injury nor loss of a terminal trophic substance explains why the chromatolytic reaction is more severe with lesions closer to the cell body. However, the quantity of anterograde-transported materials returned from the site of injury is presumably greater for a proximal lesion than for a distal one, since some anterograde-transported materials are incorporated along the length of the axon. The return of materials normally destined for the terminals and distal axon to the cell body might suppress their synthesis by some type of feedback inhibition. Grafstein and McQuarrie (1978) point out that shortly after axotomy there is a transient increase in adrenergic cell bodies of transported materials related to transmitter function, followed by a rapid decline, which is consistent with this hypothesis. After axotomy there is a decrease in the AT of labeled protein (Bisby, 1978; Bulger and Bisby, 1978) and AChE (Heiwall *et al.,* 1978; Schmidt and McDougal, 1978), both of which are known to reverse the direction of transport at a nerve crush (Partlow *et al.,* 1972; Bisby and Bulger, 1977). AChE content in the cell body also decreases rapidly after axotomy (Flumerfelt and Lewis, 1975). These changes reverse as regeneration proceeds but, if regeneration is prevented, AT of both labeled protein and AChE remains depressed (Bulger and Bisby, 1978; Heiwall *et al.,* 1978).

A most surprising finding that cannot be accommodated within the accepted scheme of signals for chromatolysis was made by Dziegielewska *et al.* (1976). They were studying AT of [^{3}H]choline-labeled phospholipids in rat sciatic nerve and found that axotomy produced a considerable increase in the incorporation of labeled phospholipids into the nerve. The increased output of phospholipids occurred within *1 hr* of axotomy, far too early to be accounted for by RT of a signal from the site of axotomy.

In summary, there is good evidence that RT carries some "signal(s) for chromatolysis." The nature of the signal(s) is unknown, except for post-ganglionic sympathetic neurons, where lack of NGF appears to be one signal.

2. Cell Death during Development

Cell death during development is a widespread phenomenon identified in many well-defined nuclear groups. For example, motoneurons in the lumbar ventral horn of *Xenopus* decrease from a maximum of about 6000 to 1000–2000 at metamorphosis (Hughes, 1961). In the chicken ciliary ganglion the cells that subsequently die are as well differentiated as surviving cells (Pilar and Landmesser, 1976), and all cells, including nonsurvivors, receive preganglionic innervation (Landmesser and Pilar, 1972). Because cell death occurs at about the time that connections with the target tissue are being made (Prestige, 1974), cell death is regarded as a manifestation of competition between neurons either for a limited number of synaptic sites on the target tissue and/or a limited amount of trophic "survival factor" emanating from it. The survival factor is then presumably retrograde-transported to the cell body. Thus, increasing the size of the peripheral field for innervation by grafting a supernumerary leg bud to a chick embryo resulted in a significant increase in the number of motoneurons surviving to maturity (Hollyday and Hamburger, 1976). Cell death might be perceived as a means of eliminating neurons that have made inappropriate peripheral contacts and as a result do not receive the correct survival factor. Unfortunately, nearly all neurons, including the nonsurvivors, succeed in reaching the correct target tissue, as shown by their ability to transport retrogradely HRP injected into the target tissue (Clarke and Cowan, 1976; Oppenheim and Chu-Wang, 1977; Chu-Wang and Oppenheim, 1978) and other criteria (Pilar and Landmesser, 1976; Prestige, 1976). It appears then that cell death befalls neurons whose peripheral connections fall below a certain undefined level of "adequacy." Pilar and Landmesser (1976) suggested that failure to form or maintain adequate synaptic contacts resulted in the accumulation of transmission-related proteins in the cell body, leading to a kind of fatal constipation. Another function neuronal cell death does *not* subserve is the elimination of polyneuronal innervation of skeletal muscle.

In the chick polyneuronal innervation disappears after hatching, whereas motoneuron cell death occurs in the embryonic stages (Oppenheim and Majors-Willard, 1978).

Recently it has been shown that blockage of neuromuscular transmission with both pre- and postsynaptically acting agents during the period of cell death increases motoneuron survival (Pittman and Oppenheim, 1978). It is suggested that the increased survival results from an increased number of available synaptic sites under the conditions of blockage where muscles may accept additional innervation (Fex et al., 1966; Tonge, 1977).

Evidence that RT of trophic survival factors occurs is provided, once again, by NGF. Cell death in the superior cervical ganglion is increased by axotomy, which prevents NGF from reaching the neurons (Hill and Hendry, 1976). The increased cell death can be prevented by administration of NGF to the ganglion, and continuous administration of NGF to the normal ganglion results in a decrease in cell death (Hendry, 1977).

3. RETROGRADE TRANSNEURONAL REGENERATION

This obscure phenomenon occurs in neuronally "closed" systems where axotomy produces not only chromatolysis in the damaged cell but also degeneration among the cells synapsing upon the damaged cell (Cowan, 1970). An example of such degeneration is in ganglion cells of the retina after lesions of the occipital cortex. Extensive cell death occurs among the axotomized neurons, reducing the number of postsynaptic sites for the degenerating neurons. One might speculate that, as in cell death during development, transneuronal degeneration occurs as a result of deprivation of a survival factor elaborated by the target cell, picked up, and retrograde-transported by the neuron.

4. SPROUTING AND GROWTH OF AXONS

Intact nerve terminals will sprout in response to changes in their environment, and growing axons are subject to environmental influences. For example, motor axons will sprout in response to partial denervation of muscle, and a "sprouting factor" has been isolated from denervated muscle (Hoffman and Springell, 1951; Tweedle and Kabra, 1977). Denervated iris grafts exert a powerful growth stimulus on the adrenergic neurons of the locus coerulus transplanted into the anterior chamber of the eye (Seiger and Olson, 1977). In vitro, target organs stimulate the growth of parasympathetic (Coughlin, 1975) and sympathetic ganglion cell neurites (Chamley et al., 1973; Chamley and Dowell, 1975; Eränkö and Lahtinen, 1978). Factors other than trophic influences emanating from the target tissue are important in regulating axonal growth, especially "contact guidance" and "contact inhibition" between growing axons and their local substrate (e.g. Ebendal, 1976), an effect which may involve differential adhesiveness between growth cones and various substrates (LeTourneau, 1975). It is not known definitely whether sprouting or growth acceleration depend upon purely local interaction between the growing axon and environment, or involve ascent of retrograde signals from the axon to the cell body to produce metabolic changes required to initiate, sustain, and terminate axonal growth. Grafstein and McQuarrie (1978) have reviewed the evidence for the role of the nerve cell body in regulating axonal growth, emphasizing the "conditioning lesion effect" as evidence for cell body control of outgrowth. In axons previously injured (the conditioning lesion), the rate of axonal outgrowth after a second (testing) lesion is greater than in a previously uninjured axon (e.g., McQuarrie et al., 1977). This effect is thought to be due to an increased availability of material required for regeneration whose synthesis is stimulated by the conditioning lesion.

Yet again, NGF is the most-studied example of a trophic factor controlling growth or regeneration of neurons. NGF has a general stimulating effect on adrenergic axon regeneration and growth, both in vitro (e.g., Chun and Patterson, 1977) and in vivo (e.g., Stenevi et al., 1974). Regenerating or growing adrenergic axons grow preferentially towards sources of NGF, both in vitro (Charlwood et al., 1972; Ebendal and Jacobson, 1977, LeTourneau, 1978), and in vivo (Menesini-Chen et al., 1978).There are three possible explanations for these results:

1. Neurons located nearest the source of NGF are exposed to a higher concentration of NGF at their cell bodies and so grow more vigorously, resulting in increased axon outgrowth.

2. Axons extending initially in a random fashion towards the source of NGF are exposed to a higher concentration of NGF: this is taken up and retrograde-transported back to the cell bodies, resulting in stimulated growth. If the axon deviates away from the source of NGF, the amount of transported NGF reaching the cell body falls and growth slows.

3. There is a true chemotaxis in which high concentrations of NGF affect the growth cone directly, leading to an increased rate of elongation, without indirectly involving RT of NGF and its trophic effects on the cell body.

Some ingenious recent experiments allow us to differentiate between these possibilities. Campenot (1977) grew newborn rat superior cervical ganglia in a three-compartment chamber, which permitted independent regulation of the environment in the three compartments, so that the effects of NGF on cell bodies and growing neurites could be compared. With NGF present in the ganglion chamber, neurites grew into the second chamber containing NGF but never into the third chamber which did not contain NGF, even though it was the same distance from the ganglion as chamber 2. When NGF was withdrawn from chamber 2 after the neurites had grown into it, they stopped growing, even though NGF was still present in the ganglion chamber. However, when NGF was withdrawn from the ganglion chamber after the neurites had penetrated NGF-containing chamber 2, the ganglion cells survived quite nicely. These results show that the survival of the sympathetic neurons was dependent upon NGF that could be provided either by direct application to the ganglia or by application to the growing axons, from whence it was presumably retrograde-transported to the cell body. On the other hand, the stimulating effect of NGF on neurite elongation was a local interaction at the growth cone level, since it could not be produced by application of NGF to the ganglion.

LeTourneau (1978) grew dissociated chick embryo dorsal root ganglion cells in a gradient of NGF. He found that, irrespective of their initial direction of outgrowth from the cell body, neurites tended to turn and grow up the NGF gradient. This finding eliminates possibility 1. He found that the growth-orienting effect of NGF occurred over a wide range of concentrations higher than those required to stimulate neurite outgrowth maximally in this preparation. Thus it is unlikely that growth toward NGF sources is a result of stimulation of protein synthesis at the cell body, either through interaction with the cell body or RT from the extending growth cones. It is concluded that the evidence from these studies with NGF does not support the involvement of RT in conveying to the cell body trophic factors responsible for inducing axon sprouting or directed growth.

5. Unknown Trophic Effects—The Retrograde Transport of Nerve Growth Factor as a Stimulus to Exploration

Repeated reference has been made to NGF in this review. This is not really surprising, as it is the only known example of a trophic substance that is retrograde-transported. NGF is one of the factors (Patterson *et al.,* 1978) essential for the growth, differentiation, and survival of the developing sympathetic nervous system. It is synthesized in sympathetically innervated tissues. Applied to sympathetic nerve terminals it is retrograde-transported back to the cell body where it stimulates the synthesis of catecholamine-synthesizing enzymes. Blockage of RT in developing ganglia by colchicine application, axotomy, or drugs that destroy adrenergic terminals produces cell death which can be prevented by administration of NGF to the ganglia (Thoenen *et al.,* 1978). RT of NGF from the target tissue may not be its only route of action on neurons. The stimulatory effect of NGF on fiber outgrowth is exerted at approximately a 100-fold lower concentration than is required to induce synthesis of catecholamine-synthesizing enzymes (Hill and Hendry, 1976). The differential sensitivity of the two actions of NGF to transcription- and translation-blocking drugs and the existence of two classes of NGF receptors (Andres *et al.,* 1977) also attest to a dual mode of action for NGF. It is suggested that uptake of high concentrations of NGF from target tissue is followed by RT to the cell body, interaction with nuclear receptors, and induction of enzymes, whereas neurite outgrowth may be stimulated by interaction with cell surface receptors only (Thoenen *et al.,* 1978). NGF for this effect, requiring a lower concentration, might originate from nonneuronal cells adjacent to the neuron (e.g., Young *et al.,* 1975) or from circulating NGF. In the normal sequence of development the effect on neurite outgrowth would be expressed before the effect on enzyme induction (Hendry, 1976).

The example of NGF has stimulated many investigators to wonder if there are other retrograde trophic molecules for other classes of neurons. Indeed, the widespread effects of applied NGF and its antiserum on a variety of tissues have led to the suggestion that it may be acting as an analog to other closely related growth factors (Bradshaw, 1978). Retrograde trophic relationships between nerve and muscle have been demonstrated *in vitro.* The survival of chick spinal cord neurons was improved when muscle cells were added to the culture (Bird and James, 1973). Choline acetyltransferase activity in spinal cord cultures increased 20-fold when cocultured with muscle, or with cell-free medium in which muscles had been cultured. The agent in the conditioned medium was a molecule with a MW exceeding 50,000 (Giller *et al.,* 1977). Interestingly, cultured muscle cells have been shown to synthesize and secrete a NGF-like molecule with a MW of 140,000–160,000 (Murphy *et al.,* 1977). Neurons of the chick ciliary ganglion depend for their survival upon innervation of the optic cup; removal of the target tissue prior to innervation results in cell death. Death of dissociated chick embryo ciliary ganglion neurons in culture can be reduced by coculture with myotubes (Nishi and Berg, 1977); most of the cells formed functional synaptic contacts with the myotubes so that the "survival factor" might be either some aspect of the process of synapse formation or the RT of a trophic survival factor originating in the myotubes. Evidence for a cholinergic factor promoting growth and survival of parasympathetic neurons was provided by Helfand *et al.* (1976) who found that heart-conditioned medium was required for the survival of cultured ciliary ganglion neurons and for the growth of neurites. Hendry and colleagues (McLennan and Hendry, 1978; McLennan et al., 1978) have found three factors in cardiac muscle extracts that affect the development of ciliary ganglion neurons. Factors with MWs of 20,000 and 60,000 increased choline acetyltransferase levels, and a 30,000-MW factor promoted neurite outgrowth. In agreement with earlier work (Patterson and Chun, 1977) it was found that heart-conditioned medium induced the expression of cholinergic characteristics (specifically choline acetyltransferase activity) in cultured neurons of the superior cervical ganglion that normally become adrenergic. However, heart-conditioned medium was more effective in inducing choline acetyltransferase activity in ganglia than in spinal cord, and muscle-conditioned medium was more effective on spinal cord than on ganglia. Thus there may be at least two cholinergic factors, inducing choline acetyltransferase activity, specific to parasympathetic and somatic motor neurons and produced by the appropriate target tissue. Recently, selective RT of NGF from the eye to the ciliary ganglion has been demonstrated (Max et al., 1978).

It appears quite likely, then, that there are other target tissue-derived trophic substances that, like NGF, exert some of their effects via terminal uptake and RT.

B. Pathological Functions

The entry of viruses and bacterial toxins into the nervous system by way of RT along peripheral nerves is clearly of greater functional significance for the pathogen than for the neuron. Nevertheless, these agents are useful as "markers" of pathways which may be normally followed by nonpathogenic molecules. The example of the nucleus as the ultimate destination of retrograde-transported viral nucleic acids has already been given. The transsynaptic transfer of tetanus toxin from motoneuron to presynaptic terminals may demonstrate the existence of a route followed by informational molecules through multisynaptic neuronal pathways.

If pathogens have been "cunning" in utilizing RT pathways to gain access to the central nervous system, human ingenuity may be yet more cunning in using the same pathways to fight the pathogens. Thus Bizzini et al. (1977) isolated a polypeptide fragment from tetanus toxin which, though 100,000 times less toxic than the original toxin, had a greater affinity for gangliosides and was retrograde-transported. They suggested that this fragment could be bound to chemotherapeutic or pharmacological agents and used to carry them specifically into the central nervous system by RT.

VI. Conclusion

Neurons possess a system for translocation of materials from their periphery to the cell body. This RT system conveys not only materials endogenous to the neuron, but also exogenous materials, some of which have known physiological or pathological significance. The velocity of RT and its sensitivity to metabolic inhibitors and microtubule-disrupting agents show that it is the same type of transport mechanism involved in the more extensively studied anterograde axonal transport system. Thus many of the problems outstanding in our appreciation of axonal transport are common to both AT and RT. These problems fall into two groups: (1) What is the mechanism of axonal transport? and (2) What is the function of axonal transport?

Problems of mechanism involve both the nature of the translocation process in which metabolic energy is converted into organelle movement, and the exact nature of the translocated organelles. The bidirectional property of axonal transport is a puzzle. Is the direction of transport determined by the set of oppositely polarized microtubules onto which the organelle is loaded, or does the nature of the organelle determine its direction of transport? Evidence from the direct observation of particles shows only that these particles are dedicated to moving in one direction; reversals are infrequent and temporary. The mechanism of loading and unloading is also unknown. Since the composition of axonal and terminal axoplasm is very different with regard to the occurrence of mitochondria, synaptic vesicles, microtubules, etc., it is likely that loading and unloading are not random processes but are regulated by the local environment, perhaps by the concentration of calcium ions in the axoplasm, as suggested by Lasek and Hoffman (1976). It is not known whether endogenous and exogenous molecules are returned to the cell body packaged in the same organelle, or how some transported materials can bypass, at least temporarily, lysosomal destruction.

Problems of function include the importance of "recycling" to the neuron, the extent to which endogenous materials returned by RT are completely or incompletely broken down in the cell body, and whether RT of endogenous materials provides any feedback of information to the cell body about the status of the axon. The example of NGF has demonstrated the reality of retrograde trophic effects mediated through RT and has stimulated a search for similar trophic molecules acting on other types of neurons. The distinction between specific and nonspecific types of terminal uptake needs to be clarified. Is nonspecific uptake, as illustrated by the case of HRP, merely an incidental sampling of extracellular fluid occurring as part of a vesicle recycling process, or does this environmental sampling have an informational role?

I have tried to identify some of the more significant problems confronting workers in the field of retrograde axonal transport; the list could be continued for several pages. As in all fields of science, progress has provided some answers but more often has only permitted us to ask more specific questions.

References

Abe, T., Haga, T., and Kurokawa, M. (1974). Retrograde axoplasmic transport: Its continuation as anterograde transport. *FEBS Lett.* **47**, 272-275.

Amos, L. A. (1977). Arrangement of high molecular weight associated proteins on purified mammalian brain microtubules. *J. Cell Biol.* **72**, 642-654.

Andres, R. Y., Jeng, I., and Bradshaw, R. A. (1977). Nerve growth factor receptors: Identification of distinct classes in plasma membranes and nuclei of embryonic dorsal root neurons. *Proc. Natl. Acad. Sci. U.S.A.* **74**, 2785-2789.

Banerjee, S. P., Snyder, S. A., Cuatrecases, P., and Green, L. A. (1973). Binding of NGF receptor in sympathetic ganglia. *Proc. Natl. Acad. Sci. U.S.A.* **70**, 2519-2523.

Banks, B. E. C., and Walter, S. J. (1977). The effects of postganglionic axotomy and nerve growth factor in the superior cervical ganglia of developing mice. *J. Neurocytol.* **6**, 287-297.

Banks, P., Mangnall, D., and Mayor, D. (1969). The redistribution of cytochrome oxidase, noradrenaline and ATP in adrenergic nerves constricted at two points. *J. Physiol. (London)* **200**, 745-762.

Bird, M. M. (1976). Microtubule-synaptic vesicle associations in cultured rat spinal cord neurons. *Cell Tissue Res.* **168**, 101-115.

Bird, M. M., and James, D. W. (1973). The development of synapses *in vitro* between previously dissociated chick spinal cord neurons. *Z. Zellforsch. Mikrosk. Anat.* **140**, 203-216.

Birks, R. J., MacKay, M. C., and Welson, P. B. (1972). Organelle formation from pinocytotic elements in neurites of cultured sympathetic ganglia. *J. Neurocytol.* **1**, 311-340.

Bisby, M. A. (1976a). Orthograde and retrograde axonal transport of labeled protein in motoneurons. *Exp. Neurol.* **50**, 628-640.

Bisby, M. A. (1976b). Axonal transport. (Mini review.) *Gen. Pharmacol.* **7**, 387-393.

Bisby, M. A. (1977). Differences in retrograde axonal transport of labeled protein in motor and sensory axons. *J. Neurochem.* **28**, 249-251.

Bisby, M. A. (1978). Fast axonal transport of labeled protein in sensory axons during regeneration. *Exp. Neurol.* **61**, 281-300.

Bisby, M. A., and Bulger, V. T. (1977). Reversal of axonal transport at a nerve crush. *J. Neurochem.* **29**, 313-320.

Bizzini, B., Stoeckel, K., and Schwab, M. (1977). An antigenic polypeptide fragment isolated from tetanus toxin: Chemical characterization, binding to gangliosides and retrograde axonal transport in various neuron systems. *J. Neurochem.* **28**, 529-542.

Bodian, D., and Howe, H. A. (1941). The rate of progression of poliomyelitis virus in nerves. *Bull. Johns Hopkins Hosp.* **69**, 79-85.

Borisy, G. G., Olmsted, J. B., Marcum, J. M., and Allen, C. (1974). Microtubule assembly *in vitro*. *Fed. Proc., Fed. Am. Soc. Exp. Biol.* **32**, 167–174.

Bradshaw, R. A. (1978). Nerve growth factor. *Annu. Rev. Biochem.* **47**, 191–216.

Bray, J. J., and Harris, A. J. (1975). Dissociation between nerve-muscle transmission and nerve trophic effects on rat diaphragm using type D botulinum toxin. *J. Physiol. (London)* **253**, 53–77.

Bray, J. J., Kon, C. M., and Breckenridge, B. M. (1971). Reversed polarity of rapid axonal transport in chicken motoneurons. *Brain Res.* **33**, 560–564.

Breuer, A. C., Christian, C. N., Henkart, M., and Nelson, P. G. (1975). Computer analysis of organelle translocation of primary neuronal cultures and continuous cell lines. *J. Cell Biol.* **65**, 562–576.

Brimijoin, S. (1974). Local changes in subcellular distribution of dopamine-β-hydroxylase after blockade of axonal transport. *J. Neurochem.* **22**, 347–353.

Brimijoin, S. (1975). Stop flow: A new technique for measuring axonal transport, and its application to the transport of dopamine-β-hydroxylase. *J. Neurobiol.* **6**, 379–394.

Brimijoin, S., and Helland, L. (1976). Rapid retrograde transport of dopamine-β-hydroxylase as examined by the stop-flow technique. *Brain Res.* **102**, 217–228.

Broadwell, R. D., and Brightman, M. W. (1976). Entry of peroxidase into neurons of the central and peripheral nervous system from extracerebral and cerebral blood. *J. Comp. Neurol.* **166**, 257–284.

Brownson, R. H., Uusitalo, R., and Palkama, A. (1977). Intra-axonal transport of horseradish peroxidase in the sympathetic nervous system. *Brain Res.* **84**, 279–291.

Bulger, V. T., and Bisby, M. A. (1978). Reversal of axonal transport in regenerating nerves. *J. Neurochem.* **31**, 1411–1418.

Bunge, M. B. (1977). Initial endocytosis of peroxidase or ferritin by growth cones of cultured nerve cells. *J. Neurocytol.* **6**, 407–439.

Bunt, A. H., and Haschke, R. H. (1977). Morphologic analysis of uptake and retrograde transport of peroxidases by axons. *Soc. Neurosci. Abstr.* **3**, 29.

Bunt, A. H., and Lund, R. J. (1974). Vinblastine-induced blockage of orthograde and retrograde axonal transport of protein in retinal ganglion cells. *Exp. Neurol.* **45**, 288–297.

Bunt, A. H., Haschke, R. H., Lund, R. D., and Collins, D. F. (1976). Factors affecting retrograde axonal transport of horseradish peroxidase in the visual system. *Brain Res.* **102**, 152–155.

Campenot, R. B. (1977). Local control of neurite development by nerve growth factor. *Proc. Natl. Acad. Sci. U.S.A.* **74**, 4516–4519.

Cancalon, P., and Beidler, L. M. (1975). Distribution along the axon and into various subcellular fractions of molecules labeled with ^{3}H-leucine and rapidly transported in the garfish olfactory nerve. *Brain Res.* **89**, 225–244.

Carroll, P. T., Price, D. L., Griffin, J. W., and Morris, J. (1978). Tetanus toxin: Immunocytochemical evidence for retrograde transport. *Neurosci. Lett.* **8**, 335–339.

Ceccarelli, B., and Hurlburt, W. P. (1975). The effects of prolonged repetitive stimulation in hemicholinium on the frog neuromuscular junction. *J. Physiol. (London)* **247**, 163–188.

Chamley, J. H., and Dowell, J. J. (1975). Specificity of nerve fibres attraction to autonomic effector organs in tissue culture. *Exp. Cell Res.* **90**, 1–7.

Chamley, J. H., Goller, I., and Burnstock, G. (1973). Selective growth of sympathetic nerve fibres to explants of normally densely innervated effector organs in tissue culture. *Dev. Biol.* **31**, 362–379.

Chan, K. Y., and Bunt, A. H. (1978). Association between mitochondria and microtubules in synaptosomes and axon terminals of cerebral cortex. *J. Neurocytol.* **7**, 137–144.

Charlwood, K. A., Lamont, D. M., and Banks, B. E. C. (1972). Apparent orienting effect produced by nerve growth factor. *In* "Nerve Growth Factor and Its Antiserum" (E. Zaimis and J. Knight, eds.), pp. 102–107. Oxford Univ. Press (Athlone), London and New York.

Chubb, I. W., Goodman, S., and Smith, A. D. (1976). Is acetylcholinesterase secreted from central neurones into cerebrospinal fluid? *Neuroscience* **1**, 57–62.

Chun, L. L. Y., and Patterson, P. H. (1977). Role of nerve growth factor in the development of rat sympathetic neurons *in vitro*. I. Survival, growth and differentiation of catecholamine production. *J. Cell Biol.* **75**, 694–704.

Chu-Wang, I. W., and Oppenheim, R. W. (1978). Cell death of motoneurons in the chick embryo spinal cord. II. A quantitative and qualitative analysis of degeneration in the ventral roots, including evidence for axon outgrowth and limb innervation prior to cell death. *J. Comp. Neurol.* **177**, 59–86.

Clarke, P. G. H., and Cowan, W. M. (1976). The development of the isthmo-optic tract in the chick with special reference to the occurrence and correction of developmental errors in the location and connections of isthmo-optic neurons. *J. Comp. Neurol.* **167**, 143–164.

Cooke, C. T., Cameron, P. U., and Jones, D. J. (1975). Stimulation-induced uptake of horseradish peroxidase by rat cortical synapses. *Neurosci. Lett.* **1**, 15–18.

Cooper, P. D., and Smith, R. S. (1974). The movement of optically detectable organelles in myelinated axons of *Xenopus laevis*. *J. Physiol. (London)* **242**, 77–99.

Coughlin, M. D. (1975). Target organ stimulation of parasympathetic nerves in the mouse submandibular gland. *Dev. Biol.* **43**, 140–158.

Cowan, W. M. (1970). Anterograde and retrograde transneuronal degeneration in the central and peripheral nervous system. *In* "Contemporary Research Methods in Neuroanatomy" (W. J. H. Nauta and S. O. E. Ebbesson, eds.), pp. 217–251. Springer-Verlag, Berlin and New York.

Cragg, B. G. (1970). What is the signal for chromatolysis? *Brain Res.* **23**, 1–21.

Cull, R. E. (1974). Role of nerve-muscle contact in maintaining synaptic connections. *Exp. Brain Res.* **20**, 307–310.

Cull, R. E. (1975). Role of axonal transport in maintaining central synaptic connections. *Exp. Brain Res.* **24**, 97–106.

Dahlstrom, A. (1965). Observations on the accumulation of noradrenaline in the proximal and distal parts of peripheral adrenergic nerves after compression. *J. Anat.* **99**, 677–689.

Dentler, W. L., Granett, S., Wltman, G. B., and Rosenbaum, I. I., (1974). Directionality of brain microtubule assembly *in vitro*. *Proc. Natl. Acad. Sci. U.S.A.* **71**, 1710–1714.

De Potter, W. P., and Chubb, I. W. (1971). The turnover rate of noradrenergic vesicles. *Biochem. J.* **125**, 375–376.

DeVito, J. L., Clausing, K. W., and Smith, O. A. (1974). Uptake and transport of horseradish peroxidase by a cut end of the vagus nerve. *Brain Res.* **82**, 269–271.

Di Giamberardino, L., and Courand, J. Y. (1978). Rapid accumulation of high molecular weight acetylcholinesterase in transected sciatic nerve. *Nature (London)* **271**, 170–171.

D'Olivo, M., Meurant, C., and Verdan, C. (1977). Retrograde flow depends on neuronal activity. *Experientia* **33**, 779.

Droz, B. (1975). Synthetic machinery and axoplasmic transport: Maintenance of neuronal connectivity. *In* "The Nervous System" (D. B. Tower, ed.), Vol. 1, pp. 111–127. Raven Press, New York.

Droz, B., Koeing, H., and Di Giamberardino, L. (1973). Axonal migration of protein and glycoproteins to nerve endings. I. Renewal of protein in nerve endings of chicken ciliary ganglion after intracerebral injection of ^{3}H-lysine. *Brain Res.* **60**, 93–127.

Droz, B., Rambourg, A., and Koenig, H. L. (1975). The smooth endoplasmic reticulum:

Structure and role in the renewal of axonal membrane and synaptic vesicles by fast axonal transport. *Brain Res.* **93,** 1–13.

Dziegielewska, K. M., Evans, C. A. N., and Saunders, N. R. (1976). Effect of axotomy on rapid axonal transport of phospholipids in peripheral nerve of the rat. *J. Physiol. (London)* **263,** 200P.

Ebbott, S., and Hendry, I. (1978) Retrograde transport of nerve growth factor in the rat central nervous system. *Brain Res.* **139,** 160–163.

Ebendal, T. (1976). The relative roles of contact inhibition and contact guidance in orientation of axons extending on aligned collagen fibrils *in vitro. Exp. Cell Res.* **98,** 159–169.

Ebendal, T., and Jacobson, C. (1977). Tissue explants affecting extension and orientation of axons in cultured chick embryo ganglia. *Exp. Cell Res.* **105,** 379–387.

Edström, A., and Hanson, M. (1973). Retrograde axonal transport of proteins *in vitro* in frog sciatic nerves. *Brain Res.* **61,** 311–320.

Eränkö, O., and Lahtinen, T. (1978). Attraction of nerve fibres outgrowth from sympathetic ganglia to heart auricles in tissue culture. *Acta Physiol. Scand.* **103,** 394–403.

Erdmann, G., Wiegand, H., and Wellhoner, H. H. (1975). Intra-axonal and extra-axonal transport of ^{125}I tetanus toxin in early local tetanus. *Naunyn-Schmiedebergs Arch. Pharmacol.* **290,** 357–373.

Festoff, B. W., and Fernandez, H. L. (1978). Bidirectional axonal transport of macromolecular (16S) acetylcholinesterase. *Soc. Neurosci. Abstr.* **4,** 32.

Fex, S., Sonesson, B., Thesleff, S., and Zelena, J. (1966). Nerve implants in botulinum-poisoned mammalian muscles. *J. Physiol. (London)* **184,** 872–882.

Field, H. J., and Hill, T. J. (1974). The pathogenesis of pseudorabies in mice following peripheral inoculation. *J. Gen. Virol.* **23,** 145–157.

Fillenz, M., Gagnon, C., Stockel, K., and Thoenen, H. (1976). Selective uptake and retrograde axonal transport of dopamine-β-hydroxylase antibodies in peripheral adrenergic neurons. *Brain Res.* **114,** 293–303.

Fitzsimons, J. T. R., Kerkut, G. A., and Sharp, G. A. (1978). Localization of microtubular ATPase in rat, snail and crab nerves. *J. Physiol. (London)* **278,** 6P.

Flumerfelt, B. A., and Lewis, P. R. (1975). Cholinesterase activity in the hypoglossal nucleus of the rat and the changes produced by axotomy: A light and electron microscopic study. *J. Anat.* **119,** 309–331.

Fonnum, F., Frizell, M., and Sjöstrand, J. (1973). Transport, turnover and distribution of choline acetyltransferase and acetylcholinesterase in the vagus and hypoglossal nerves of the rabbit. *J. Neurochem.* **21,** 1109–1120.

Forman, D. S., Padjen, A. L., and Siggins, G. R. (1977a). Axonal transport of organelles visualized by light microscopy: Cinemicrographic and computer analysis. *Brain Res.* **136,** 197–213.

Forman, D. S., Padjen, A. L., and Siggins, G. R. (1977b). Effect of temperature on the rapid retrograde transport of microscopically visible intra-axonal organelles. *Brain Res.* **136,** 215–226.

Frazier, W. A., Boyd, L. F., and Bradshaw, R. A. (1974). Interaction of nerve growth factor with surface membranes: Biological competence of insolubilized nerve growth factor. *Proc. Natl. Acad. Sci. U.S.A.* **70,** 2931–2935.

Fried, C., and Blaustein, M. P. (1976). Synaptic vesicle recycling in synaptosomes *in vitro. Nature (London)* **261,** 255–256.

Friede, R. L., and Ho, K. C. (1977). The relation of axonal transport of mitochondria with microtubules and other axoplasmic organelles. *J. Physiol. (London)* **265,** 507–519.

Frizell, M., and Sjöstrand, J. (1974). The axonal transport of slowly migrating [³H]leucine-labeled proteins and the regeneration rate in regenerating hypoglossal and vagus nerves of the rabbit. *Brain Res.* **81,** 267–283.

Frizell, M., McLean, G. W., and Sjöstrand, J. (1975). Slow axonal transport of proteins: Blockade by interruption of contact between cell body and axon. *Brain Res.* **86**, 67–73.

Frizell, M., McLean, W. G., and Sjöstrand, J. (1976). Retrograde axonal transport of rapidly migrating labeled proteins and glycoproteins in regenerating peripheral nerves. *J. Neurochem.* **27**, 191–196.

Furstman, L., Saporta, S., and Kruger, L. (1975). Retrograde axonal transport of horseradish peroxidase in sensory nerves and ganglia of the rat. *Brain Res.* **84**, 320–324.

Gennaro, J. F., Nastuk, W. L., and Rutherford, D. T. (1978). Reversible depletion of synaptic vesicles induced by application of high external potassium to the frog neuromuscular junction. *J. Physiol. (London)* **280**, 237–247.

Giller, E. L., Neale, J. H., Bullock, P. N., Schrier, B. K., and Nelson, P. G. (1977). Choline acetyltransferase activity of spinal cord cell cultures increased by co-culture with muscle and by muscle-conditioned medium. *J. Cell Biol.* **74**, 16–29.

Goldman, R., Pollard, T., and Rosenbaum, J., eds. (1976). "Cell Motility: Microtubules and Related Proteins." Cold Spring Harbor Lab., Cold Spring Harbor, New York.

Grafstein, B. (1971). Role of slow axonal transport in nerve regeneration. *Acta Neuropathol., Suppl.* **5**, 144–152.

Grafstein, B. (1975). The nerve cell body response to axotomy. *Exp. Neurol.* **48**, 32–51.

Grafstein, B. (1977). Axonal transport: The intracellular traffic of the neurone. *In* "Handbook of Physiology" (E. R. Kandel, ed.), Sect. 1, Vol. 1, pp. 691–717. Am. Physiol. Soc., Bethesda, Maryland.

Grafstein, B., and McQuarrie, I. G. (1978). Role of the nerve cell body in axonal regeneration. *In* "Neuronal Plasticity" (C. W. Cotman, ed.), pp. 155–195. Raven Press, New York.

Gray, E. G. (1975). Presynaptic microtubules and their association with synaptic vesicles. *Proc. R. Soc. London, Ser. B* **190**, 369–372.

Griffin, J. W., Price, D. L., Engel, W. K., and Drachman, D. B. (1977). Pathogenesis of reactive axonal swellings—Role of axonal transport. *J. Neuropathol. Exp. Neurol.* **36**, 214–227.

Gross, G. W. (1975). The microstream concept of axoplasmic and dendritic transport. *In* "Physiology and Pathology of Dendrites" (G. W. Krentzberg, ed.), pp. 283–296. Raven Press, New York.

Hall, Z. W., and Kelly, R. B. (1971). Enzymatic detachment of end plate acetylcholinesterase from muscle. *Nature (London), New Biol.* **232**, 62–63.

Halperin, J. J., and LaVail, J. H. (1975). A study of the dynamics of retrograde transport and accumulation of horseradish peroxidase in injured neurons. *Brain Res.* **100**, 253–269.

Hammond, G. R., and Smith, R. S. (1977). Inhibition of the rapid movement of optically detectable axonal particles by colchicine and vinblastine. *Brain Res.* **128**, 227–242.

Hanson, M. (1978). A new method to study fast axonal transport *in vivo*. *Brain Res.* **153**, 121–126.

Heaton, M. B. (1977). Retrograde axonal transport in lateral motor neurons of the chick embryo prior to limb bud innervation. *Dev. Biol.* **58**, 421–427.

Heiwall, P. O., Dahlstrom, A., Larsson, P. A., and Booj, S. (1978). The intra-axonal transport of acetylcholine and cholinergic enzymes in rat sciatic nerve during regeneration after various types of axonal trauma. *J. Neurobiol.* **10**, 119–136.

Helfand, S. L., Smith, G. A., and Wessells, N. K. (1976). Survival and development in culture of dissociated parasympathetic neurons from ciliary ganglia. *Dev. Biol.* **50**, 541–547.

Hendry, I. A. (1975). The response of adrenergic neurones to axotomy and nerve growth factor. *Brain Res.* **94**, 87–97.

Hendry, I. A. (1976). Control in the development of the vertebrate sympathetic nervous system. *Rev. Neurosci.* **2**, 149–194.

Hendry, I. A. (1977). Cell division in the developing sympathetic nervous system. *J. Neurocytol.* **6**, 299–310.

Hendry, I. A., and Campbell, J. (1976). Morphometric analysis of the rat superior cervical ganglion after axotomy and nerve growth factor treatment. *J. Neurocytol.* **5**, 351–360.

Hendry, I. A., Stockel, K., Thoenen, H., and Iversen, L. L. (1974). The retrograde transport of nerve growth factor. *Brain Res.* **68**, 103–121.

Herrup, K., and Shooter, E. M. (1973). Properties of the nerve growth factor receptor of avian dorsal root ganglia. *Proc. Natl. Acad. Sci. U.S.A.* **70**, 3884–3888.

Heuser, J. E., and Reese, T. S. (1973). Evidence for recycling of synaptic vesicle membrane during transmitter release at the frog neuromuscular junction. *J. Cell Biol.* **57**, 315–344.

Hill, C. E., and Hendry, I. A. (1976). Differences in sensitivity to nerve growth factor of axon formation and tyrosine hydroxylase induction in cultured sympathetic neurons. *Neuroscience* **1**, 489–496.

Hines, J. F., and Garwood, M. M. (1977). Release of protein from axons during rapid axonal transport: An *in vitro* preparation. *Brain Res.* **125**, 141–148.

Hoffman, H., and Springell, P. H. (1951). An attempt at the chemical identification of "neurocletin" (the substance evoking axon sprouting). *Aust. J. Exp. Biol. Med. Sci.* **29**, 417–424.

Hoffman, P. N., and Lasek, R. J. (1975). The slow component of axonal transport. Identification of major structural polypeptides of the axon and their generality among mammalian neurons. *J. Cell Biol.* **66**, 351–366.

Hollyday, M., and Hamburger, V. (1976). Reduction of the naturally occurring motor neuron loss by enlargement of the periphery. *J. Comp. Neurol.* **170**, 311–319.

Holtzman, E. (1977). The origin and fate of secretory packages, especially synaptic vesicles. *Neuroscience* **2**, 327–355.

Holtzman, E., and Novikoff, A. B. (1965). Lysosomes in the rat sciatic nerve following crush. *J. Cell Biol.* **27**, 657–669.

Hughes, A. F. (1961). Cell degeneration in the larval ventral horn of *Xenopus laevis* (Daudin). *J. Embryol. Exp. Morphol.* **9**, 269–284.

Hughes, A. F. (1953). The growth of embryonic neurites. A study on cultures of chick neural tissue. *J. Anat.* **87**, 150–162.

Jessen, K. R., Chubb, I. W., and Smith, A. D. (1978). Intracellular localization of acetylcholinesterase in nerve terminals and capillaries of rat superior cervical ganglion. *J. Neurocytol.* **7**, 145–154.

Johnson, D. G., Silberstein, S. R., Hanbauer, I., and Kopin, I. J. (1972). The role of nerve growth factor in the ramification of sympathetic nerve fibres into rat iris in organ culture. *J. Neurochem.* **19**, 2025–2029.

Johnson, E. M. (1978). Destruction of the sympathetic nervous system in neonatal rats and hamsters by vinblastine: Prevention by concomitant administration of nerve growth factor. *Brain Res.* **141**, 105–118.

Johnson, E. M., Andres, R. Y., and Bradshaw, R. A. (1978). Characterization of the retrograde transport of nerve growth factor (NGF) using high specific activity [^{125}I]NGF. *Brain Res.* **150**, 319–331.

Kapeller, K., and Mayor, D. (1969). An electron microscopic study of the early changes distal to a constriction in sympathetic nerves. *Proc. R. Soc. London, Ser. B* **172**, 53–63.

Kasa, P. (1968). Acetylcholinesterase transport in the central and peripheral nervous tissue: The role of tubules in enzyme transport. *Nature (London)* **218**, 1265–1267.

Kasa, P., and Csillik, B. (1968). Acetylcholinesterase synthesis in cholinergic neurones: Electron histochemistry of enzyme translocation. *Histochemie* **12**, 175–183.

Khan, M. A., and Ochs, S. (1975). Slow axoplasmic transport of mitochondria (MAO) and lactic dehydrogenase in mammalian nerve fibres. *Brain Res.* **96**, 267–277.

Kirkpatrick, J. B., Bray, J. J., and Palmer, S. M. (1972). Visualization of axoplasmic flow *in vitro* by Nomarski microscopy. Comparison to rapid flow of radioactive proteins. *Brain Res.* **43**, 1–10.

Koda, L. Y., and Partlow, L. M. (1976). Membrane marker movement on sympathetic axons in tissue culture. *J. Neurobiol.* **7**, 157–172.

Kolata, G. B. (1978). Polypeptide hormones: What are they doing in cells? *Science* **201**, 895–897.

Kreutzberg, G. W., Toth, L., and Kaiya, H. (1975). Acetylcholinesterase as a marker for dendritic transport and dendritic secretion. *In* "Physiology and Pathology of Dendrites" (G. W. Kreutzberg, ed.), pp. 269–281. Raven Press, New York.

Krishnan, N., and Singer, M. (1973). Penetration of peroxidase into peripheral nerve fibres. *Am. J. Anat.* **136**, 1–14.

Kristensson, K., and Olsson, Y. (1974). Retrograde transport of horseradish peroxidase in transected axons. I. Time relationships between transport and induction of chromatolysis. *Brain Res.* **79**, 101–111.

Kristensson, K. (1977). Retrograde axonal transport of horseradish peroxidase. Uptake at mouse neuromuscular junction following systemic injection. *Acta Neuropathol.* **38**, 143–146.

Kristensson, K. (1978). Retrograde transport of macromolecules in axons. *Annu. Rev. Pharmacol. Toxicol.* **18**, 97–110.

Kristensson, K., and Olsson, Y. (1975). Retrograde transport of horseradish peroxidase in transected axons. II. Relations between rate of transfer from the site of injury to the perikaryon and the onset of chromatolysis. *J. Neurocytol.* **4**, 653–661.

Kristensson, K., and Olsson, Y. (1977). Retrograde transport of horseradish peroxidase in transected axons. 3. Entry into injured axons and subsequent localization in perikaryon. *Brain Res.* **126**, 154–159.

Kristensson, K., and Olsson, Y. (1978). Uptake and retrograde axonal transport of horseradish peroxidase in botulinum-intoxicated mice. *Brain Res.* **155**, 118–123.

Kristensson, K., and Sjöstrand, J. (1972). Retrograde transport of protein tracer in the rabbit hypoglossal nerve during regeneration. *Brain Res.* **45**, 175–182.

Kristensson, K., Lycke, E., and Sjöstrand, J. (1971a). Spread of *Herpes simplex* virus in peripheral nerves. *Acta Neuropathol.* **17**, 44–53.

Kristensson, K., Olsson, Y., and Sjöstrand, J. (1971b). Axonal uptake and retrograde transport of exogenous proteins in the hypoglossal nerve. *Brain Res.* **32**, 399–406.

Landmesser, L., and Pilar, G. (1972). The onset and development of transmission in the chick ciliary ganglion. *J. Physiol. (London)* **222**, 691–713.

Lasek, R. (1967). Bidirectional transport of radioactively labelled axoplasmic components. *Nature (London)* **216**, 1212–1214.

Lasek, R. J., and Hoffman, P. N. (1976). The neuronal cytoskeleton, axonal transport and axonal growth. *In* "Cell Motility; Microtubules and Related Proteins" (R. Goldman, T. Pollard, and J. Rosenbaum, eds.), pp. 1021–1051. Cold Spring Harbor Lab., Cold Spring Harbor, New York.

LaVail, J. H. (1975). The retrograde transport method. *Fed. Proc., Fed. Am. Soc. Exp. Biol.* **34**, 1618–1624.

LaVail, M. M., and LaVail, J. (1974). The retrograde intra-axonal transport of horseradish peroxidase in the chick visual system: A light and electron microscope study. *J. Comp. Neurol.* **157**, 303–357.

Leestma, J. E. (1976). Velocity measurements of particulate neuroplasmic flow in organized mammalian CNS tissue cultures. *J. Neurobiol.* **7**, 173–183.

Leestma, J. E., and Freeman, S. S. (1977). Computer-assisted analysis of particulate axoplasmic flow in organized CNS tissue cultures. *J. Neurobiol.* **8**, 453–467.

LeTourneau, P. C. (1975). Cell to substratum adhesion and guidance of axonal elongation. *Dev. Biol.* **44**, 92-101.

LeTourneau, P. C. (1978). Chemotactic response of nerve fibre elongation to nerve growth factor. *Dev. Biol.* **66**, 183-196.

Levin, B. E. (1978). Axonal transport of [³H] proteins in a noradrenergic system of the rat brain. *Brain Res.* **150**, 55-68.

Lieberman, A. R. (1974). Some factors affecting retrograde neuronal responses to axonal lesions. *In* "Essays on the Nervous System" (R. Bellairs and E. G. Gray, eds.), pp. 71-105. Oxford Univ. Press (Clarendon), London and New York.

Litchy, W. J. (1973). Uptake and retrograde transport of horseradish peroxidase in frog sartorius nerve *in vitro. Brain Res.* **56**, 377-381.

Lorenz, T., and Willard, M. (1978). Subcellular fractionation of intra-axonally transported polypeptides in the rabbit visual system. *Proc. Natl. Acad. Sci. U.S.A.* **75**, 505-509.

Lubinska, L. (1975). On axoplasmic flow. *Int. Rev. Neurobiol.* **17**, 241-296.

Lubinska, L., and Niemierko, S. (1971). Velocity and intensity of bidirectional migration of acetylcholinesterase in transected nerves. *Brain Res.* **27**, 329-342.

McCracken, R. M., McFerran, J. B., and Dow, C. (1973). The neural spread of pseudorabies virus in calves. *J. Gen. Virol.* **20**, 17-28.

McLean, J. R., and Burnstock, G. (1972). Axoplasmic flow of adrenaline and monoamine oxidase in amphibian sympathetic nerves. *Z. Zellforsch. Mikrosk. Anat.* **124**, 44-56.

McLean, W. G., Frizell, M., and Sjöstrand, J. (1976). Labeled proteins in the rabbit vagus nerve between the fast and slow phases of axonal transport. *J. Neurochem.* **26**, 77-82.

McLennan, I. S., and Hendry, I. A. (1978). Multiple factors affecting development of cholinergic neurones? *Proc. Aust. Physiol. Pharmacol. Soc.* **9**, 31P.

McLennan, I. S., Bonyhady, R. E., and Hendry, I. A. (1978). Separation of factors affecting cholinergic development. *Proc. Aust. Physiol. Pharmacol. Soc.* **9**, 210P.

McQuarrie, I. G., Grafstein, B., and Gershon, M. D. (1977). Axonal regeneration in the rat sciatic nerve: Effect of a conditioning lesion and of dbcAMP. *Brain Res.* **132**, 443-453.

Matthews, M. R., and Nelson, V. (1975). Detachment of structurally intact nerve endings from chromatolytic neurones of the rat superior cervical ganglion during depression of synaptic transmission induced by postganglionic axotomy. *J. Physiol. (London)* **245**, 91-135.

Max, S. R., Schwab, M., Dumas, M., and Thoenen, H. (1978). Retrograde axonal transport of nerve growth factor (NGF) in the ciliary ganglion of the chick. *Soc. Neurosci. Abstr.* **4**, 35.

Menesini-Chen, M. G., Chen, J. S., and Levi-Montalcini, R. (1978). Sympathetic fiber ingrowth in the central nervous system of neonatal rodent upon intracerebral NGF injection. *Arch. Ital. Biol.* **116**. 53-84.

Murphy, R. A., Singer, R. H., Saide, J. D., Pantazis, N. J., Blanchard, M. H., Byron, K. S., Arnason, B. G. W., and Young, M. (1977). Synthesis and secretion of high molecular weight form of nerve growth factor by skeletal muscle cells in culture. *Proc. Natl. Acad. Sci. U.S.A.* **74**, 4496-4500.

Musick, J., and Hubbard, J. I. (1972). Release of protein from mouse motor nerve terminals. *Nature (London)* **237**, 279-281.

Nagatsu, I., Kondo, Y., and Nagatsu, T. (1976). Retrograde axoplasmic transport of inactive dopamine-β-hydroxylase in sciatic nerves. *Brain Res.* **116**, 277-286.

Nauta, H. J. W., Raiserman-Abramof, I. R., and Lasek, R. J. (1975). Electron microscopic observation of horseradish peroxidase transported from the caudatoputamen to the subtantia nigra in the rat: Possible involvement of the agranuler reticulum. *Brain Res.* **85**, 373-384.

Nishi, R., and Berg, D. K. (1977). Dissociated ciliary ganglion neurons *in vitro:* Survival and synapse formation. *Proc. Natl. Acad. Sci. U.S.A.* **74,** 5171-5175.

Nja, A., and Purvés, D. (1978). The effects of nerve growth factor and its antiserum on synapses in the superior cervical ganglion of the guinea pig. *J. Physiol. (London)* **277,** 53-75.

Norr, S. C., and Varon, S. (1975). Dynamic, temperature-sensitive association of ^{125}I-NGF *in vitro* with ganglionic and non-ganglionic cells from embryonic chick. *Neurobiology* **5,** 101-118.

Ochs, S. (1970). Characteristics and a model for fast axoplasmic transport in nerve. *J. Neurobiol.* **2,** 331-345.

Ochs, S. (1972). Rate of fast axoplasmic transport in mammalian nerve fibres. *J. Physiol. (London)* **227,** 627-645.

Ochs, S. (1974). Systems of material transport in nerve fibres (axoplasmic transport) related to nerve function and trophic control. *Ann. N.Y. Acad. Sci.* **228,** 202-223.

Ochs, S. (1975). Retention and redistribution of proteins in mammalian nerve fibres by axoplasmic transport. *J. Physiol. (London)* **253,** 459-476.

Oh, T. H., Chiu, J. Y., and Max, S. R. (1977). Release of acetylcholinesterase by cultured spinal cord cells. *J. Neurobiol.* **8,** 469-476.

Oldfield, B. J., and McLachlan, E. M. (1977). Uptake and retrograde transport of HRP by axons of intact and damaged peripheral nerve trunks. *Neurosci. Lett.* **6,** 135-141.

Olsson, T. P., Forsberg, I., and Kristensson, K. (1978). Uptake and retrograde axonal transport of horseradish peroxidase in regenerating facial motor neurons of mouse. *J. Neurocytol.* **7,** 323-336.

Olsson, Y., and Kristensson, K. (1973). The perineurium as a diffusion barrier to protein tracers following trauma to nerves. *Acta Neuropathol.* **23,** 105-111.

Oppenheim, R. W., and Chu-Wang, I. W. (1977). Spontaneous cell death of spinal motoneurons following peripheral innervation in the chick embryo. *Brain Res.* **125,** 154-160.

Oppenheim, R. W., and Heaton, M. B. (1975). The retrograde transport of horseradish peroxidase from the developing limb of the chick embryo. *Brain Res.* **98,** 291-302.

Oppenheim, R. W., and Majors-Willard, C. (1978). Neuronal cell death in the brachial spinal cord of the chick is unrelated to the loss of polyneuronal innervation in wing muscle. *Brain Res.* **154,** 148-152.

Osborn, M., and Weber, K. (1976). Cytoplasmic microtubules in tissue culture appear to grow from an organizing structure towards plasma membrane. *Proc. Natl. Acad. Sci. U.S.A.* **73,** 867-871.

Osborne, R. H., and Bradford, H. F. (1973). Tetanus toxin inhibits amino acid release from nerve endings *in vitro. Nature (London), New Biol.* **244,** 157-158.

Partlow, L. M., Ross, C. D., Motwani, R., and McDougall, D. B. (1972). Transport of axonal enzymes in surviving segments of frog sciatic nerve. *J. Gen. Physiol.* **60,** 388-405.

Patterson, P. H., and Chun, L. L. Y. (1977). The induction of acetylcholine synthesis in primary cultures of dissociated rat sympathetic neurons. I. Effects of conditioned medium. *Dev. Biol.* **56,** 263-280.

Patterson, P. H., Potter, D., and Furshpan, E. J. (1978). The chemical differentiation of nerve cells. *Sci. Am.* **239,** 50-59.

Pilar, G., and Landmesser, L. (1972). Axotomy mimicked by localized colchicine application. *Science* **177,** 1116-1118.

Pilar, G., and Landmesser, L. (1976). Ultrastructural differences during embryonic cell death in normal and peripherally deprived ciliary ganglia. *J. Cell Biol.* **68,** 339-356.

Pittman, R. H., and Oppenheim, R. W. (1978). Neuromuscular blockade increases motoneurone survival during normal cell death in the chick embryo. *Nature (London)* **271,** 364-366.

Poberai, M., and Savay, G. (1975). Time course of proteolytic enzyme alterations in the motor end-plates after stimulation. *Acta Histochem.* **57**, 44–48.

Politoff, A., Blitz, A. L., and Rose, S. (1975). Incorporation of acetylcholinesterase is associated with blockade of synaptic transmission. *Nature (London)* **256**, 324–325.

Prestige, M. C. (1974). Axon and cell numbers in the developing nervous system. *Br. Med. Bull.* **30**, 107–111.

Prestige, M. C. (1976). Evidence that at least some of the motor nerve cells that die during development have first made peripheral connections. *J. Comp. Neurol.* **170**, 123–133.

Price, D. L., and Griffin, J. W. (1977). Tetanus toxin: Retrograde axonal transport of systemically administered toxin. *Neurosci. Lett.* **4**, 61–66.

Price, D. L., Griffin, J. W., and Peck, K. (1977). Tetanus toxin: Evidence for binding at presynaptic nerve endings. *Brain Res.* **121**, 379–384.

Price, D. L., Griffin, J., Young, A., Peck, K., and Stocks, A. (1975). Tetanus toxin: Direct evidence for retrograde intra-axonal transport. *Science* **188**, 945–946.

Price, P., and Fisher, A. W. F. (1978). Electron microscopical study of retrograde axonal transport of horseradish peroxidase in the supra-optico-hypophyseal tract in the rat. *J. Anat.* **125**, 137–147.

Purves, D. (1975). Functional and structural changes of mammalian sympathetic neurones following interruption of their axons. *J. Physiol. (London)* **252**, 429–463.

Purves, D. (1976). Functional and structural changes in mammalian sympathetic nerves following colchicine application to postganglionic nerves. *J. Physiol. (London)* **259**, 159–176.

Raine, C. S., Ghetti, B., and Shelanski, M. L. (1971). On the association between microtubules and mitochondria within axons. *Brain Res.* **34**, 389–393.

Ranish, N., and Ochs, S. (1972). Fast axoplasmic transport of acetylcholinesterase in mammalian nerve fibres. *J. Neurochem.* **19**, 2641–2649.

Rebuhn, L. I. (1972). Polarized intracellular particle transport: Saltatory movements and cytoplasmic streaming. *Int. Rev. Cytol.* **32**, 93–137.

Samson, F. E. (1976). Pharmacology of drugs that affect intracellular movement. *Annu. Rev. Pharm. Toxicol.* **16**, 143–159.

Schmidt, R. E., and McDougal, D. B. (1978). Axonal transport of selected particle-specific enzymes in rat sciatic nerve *in vivo* and its response to injury. *J. Neurochem.* **30**, 527–535.

Schmitt, F. O. (1970). Fibrous proteins and neuronal dynamics. *Symp. Int. Soc. Cell Biol.* **8**, 95–111.

Schonbach, J., Schonbach, C. H., and Cuénod, M. (1971). Rapid phase of axoplasmic flow and synaptic proteins: An electron microscopical autoradiographic study. *J. Comp. Neurol.* **141**, 485–498.

Schwab, M. E. (1977). Ultrastructural localization of a nerve growth factor—Horseradish peroxidase (NGF-HRP) coupling product after retrograde axonal transport in adrenergic neurons. *Brain Res.* **130**, 190–196.

Schwab, M. E., and Thoenen, H. (1976). Electron microscopic evidence for a transsynaptic migration of tetanus toxin in spinal cord motoneurons, an autoradiographic and morphometric study. *Brain Res.* **105**, 213–228.

Schwab, M. E., and Thoenen, H. (1977a). Retrograde axonal transport and transynaptic transport of macromolecules: Physiological and pathophysiological importance. *Agents Actions* **7**, 361–368.

Schwab, M. E., and Thoenen, H. (1977b). Selective trans-synaptic migration of tetanus toxin after retrograde axonal transport in peripheral sympathetic nerves: A comparison with nerve growth factor. *Brain Res.* **122**, 459–474.

Schwab, M. E., Stockel, K., and Thoenen, H. (1976). Immunocytochemical localization of nerve growth factor (NGF) in the submandibular gland of adult mice by light and electron microscopy. *Cell Tissue Res.* **169**, 289–300.

Seiger, A., and Olson, L. (1977). Reinitiation of directed nerve fibre growth in central monoamine neurons after intraocular maturation. *Exp. Brain Res.* **29**, 15–44.

Sherline, P., and Schiavione, K. (1977). Immunofluorescence localization of proteins of high molecular weight along intracellular microtubules. *Science* **198**, 1038–1040.

Sherlock, D. A., Field, P. M., and Raisman, G. (1975). Retrograde transport of horseradish peroxidase in the magnocellular neurosecretory system of the rat. *Brain Res.* **88**, 403–414.

Shield, L. K., Griffin, J. W., Drachman, D. B., and Price, D. L. (1977). Retrograde axonal transport: A direct method for measurement of rate. *Neurology* **27**, 393.

Sjöstrand, J., and Frizell, M. (1975). Retrograde axonal transport of rapidly migrating proteins in peripheral nerves. *Brain Res.* **85**, 325–330.

Skau, K. A., and Brimijoin, S. (1978). Release of acetylcholinesterase from rat hemidiaphragm preparations stimulated through the phrenic nerve. *Nature (London)* **275**, 224–226.

Smith, D. S., Jarlfors, V., and Cameron, B. F. (1975). Morphological evidence for the participation of microtubules in axonal transport. *Ann. N.Y. Acad. Sci.* **253**, 472–506.

Smith, R. S. (1973). Microtubule and neurofilament densities in amphibian spinal root nerve fibres: Relationship to axoplasmic transport. *Can. J. Physiol. Pharmacol.* **51**, 798–806.

Smith, R. S. (1978). Morphological characterization of organelles undergoing rapid retrograde transport in isolated axons. *Abstr. Int. Congr. Neuromuscular Disord., 4th, 1978*, p. 354.

Smith, R. S., and Koles, Z. J. (1976). Mean velocity of optically detected intra-axonal particles measured by a cross-correlation method. *Can. J. Physiol. Pharmacol.* **54**, 859–869.

Somogyi, P., Chubb, I. W., and Smith, A. D. (1975). A possible structural basis for the extracellular release of acetylcholinesterase. *Proc. R. Soc. London, Ser. B* **191**, 271–283.

Sotelo, C., and Riche, D. (1974). The smooth endoplasmic reticulum and the retrograde and fast orthograde transport of horseradish peroxidase in the straito-nigral loop. *Anat. Embryol.* **146**, 209–218.

Stenevi, U., Bjerre, B., Bjorklund, A., and Mobley, W. (1974). Effects of localized intracerebral injections of nerve growth factor on the regenerative growth of lesioned central noradrenergic neurones. *Brain Res.* **69**, 217–234.

Stockel, K., and Thoenen, H. (1975). Retrograde axonal transport of nerve growth factor: Specificity and biological importance. *Brain Res.* **85**, 337–341.

Stockel, K., Paravicini, V., and Thoenen, H. (1974). Specificity of the transport of nerve growth factor. *Brain Res.* **76**, 413–421.

Stockel, K., Schwab, M., and Thoenen, H. (1975a). Specificity of the retrograde transport of nerve growth factor (NGF) in sensory neurons: A biochemical and morphological study. *Brain Res.* **89**, 1–14.

Stockel, K., Schwab, M., and Thoenen, H. (1975b). Comparison between the retrograde axonal transport of nerve growth factor and tetanus toxin in motor, sensory and adrenergic neurons. *Brain Res.* **99**, 1–16.

Stockel, K., Schwab, M., and Thoenen, H. (1977). Role of gangliosides in the uptake and retrograde axonal transport of cholera and tetanus toxin as compared to nerve growth factor and wheat germ agglutinin. *Brain Res.* **132**, 273–286.

Stone, G. C., Wilson, D. C., and Hall, M. E. (1978). Two-dimensional gel electrophoresis of proteins in rapid axoplasmic transport. *Brain Res.* **144**, 287–302.

Sumner, B. E. H. (1975). A quantitative analysis of the response of presynaptic boutons to postsynaptic motor neuron axotomy. *Exp. Neurol.* **46**, 605–615.

Sumner, B. E.H. (1976). Quantitative ultrastructural observations on the inhibited recovery of the hypoglossal nucleus from the axotomy response when regeneration of the hypoglossal nerve is prevented. *Exp. Brain Res.* **26**, 141-150.

Teichberg, S., Holtzmann, E., Crain, S. M., and Peterson, E. R. (1975) Circulation and turnover of synaptic vesicle membrane in cultured spinal cord neurons. *J. Cell Biol.* **57**, 88-108.

Thoenen, H., Schwab, M., and Otten, U. (1978). Nerve growth factor as a mediator of information between effector organs and innervating neurons. *In* "Molecular Control of Proliferation and Differentiation" (J. Papaconstantinou and W. J. Rutter, eds.), pp. 101-118. Academic Press, New York.

Tischner, K. (1977). Uptake of horseradish peroxidase by sensory nerve fibres *in vitro. J. Anat.* **124**, 83-97.

Tonge, D. A. (1977). Effect of implantation of an extra nerve on the recovery of neuromuscular transmission from botulinum toxin. *J. Physiol. (London)* **265**, 809-820.

Torvik, A. (1976). Central chromatolysis and the axon: A reappraisal. *Neuropathol. Appl. Neurobiol.* **2**, 423-432.

Trachtenberg, M. C., and Hadley, R. T. (1977). Poly-L-ornithine improves light microscopic visualization of horseradish peroxidase. *Soc. Neurosci. Abstr.* **3**, 32.

Tsukita, S., and Ishikawa, H. (1976). Three-dimensional distribution of smooth endoplasmic reticulum in myelinated axons. *J. Electron Micros.* **25**, 141-149.

Tucek, S. (1975). Transport of choline acetyltransferase and acetylcholinesterase in the central stump and isolated segments of a peripheral nerve. *Brain Res.* **86**, 259-270.

Tucek, S., Hanzlikova, V., and Stranikova, D. (1978). Effect of ischemia on axonal transport of choline acetyltransferase and acetylcholinesterase, and on ultrastructural changes of isolated segments of rabbit nerves *in situ. J. Neurol. Sci.* **36**, 237-245.

Tweedle, C. D., and Kabara, J. J. (1977). Lipophilic nerve sprouting factor(s) isolated from denervated muscle. *Neurosci. Lett.* **6**, 41-46.

Vallee, R. B., and Borisy, G. G. (1977). Removal of the projections from cytoplasmic microtubules *in vitro* by digestion with trypsin. *J. Biol. Chem.* **252**, 377-382.

Vigny, M., Koenig, J., and Rieger, F. (1976). The motor end plate specific form of acetylcholinesterase: Appearance during embryogenesis and reinnervation of rat muscle. *J. Neurochem.* **27**, 1347-1354.

Watson, W. E. (1968). Observations on the nucleolar and total cell body nucleic acid of injured nerve cells. *J. Physiol. (London)* **196**, 655-676.

Watson, W. E. (1969). The response of motor neurons to intramuscular injection of botulinum toxin. *J. Physiol. (London)* **202**, 611-630.

Watson, W. E. (1970). Some metabolic responses of axotomized neurons to contact between their axons and denervated muscle. *J. Physiol. (London)* **210**, 321-344.

Willard, M., Cowan, W. M., and Vagelos, P. R. (1974). The polypeptide composition of intra-axonally transported polypeptides: Evidence for four transport velocities. *Proc. Natl. Acad. Sci. U.S.A.* **71**, 2183-2187.

Willingham, M. C., and Pastan, I. (1978). The visualization of fluorescent proteins in living cells by video intensification microscopy (V.I.M.). *Cell* **13**, 501-507.

Wooten, G. F., and Coyle, J. T. (1973). Axonal transport of catecholamine-synthesizing and -metabolizing enzymes. *J. Neurochem.* **20**, 1361-1371.

Young, M., Oger, J., Blanchard, M., Asdourian, H., Amos, H., and Arnason, B. G. W. (1975). Secretion of a nerve growth factor by primary chick fibroblast cultures. *Science* **187**, 361-362.

Zelena, J. (1968). Bidirectional movements of mitochondria along axons of an isolated nerve. *Z. Zellforsch. Mikrosk. Anat.* **92**, 186-196.

Zelena, J., Lubinska, L., and Gutmann, E. (1968). Accumulation of organelles at the ends of interrupted axons. *Z. Zellforsch. Mikrosk. Anat.* **91**, 200–219.

Ziegler, M. G., Thomas, J. A., and Jacobowitz, D. M. (1976). Retrograde axonal transport of antibody to dopamine-β-hydroxylase. *Brain Res.* **104**, 390–395.

Zimmerman, H., and Denston, C. R. (1977). Recycling of synaptic vesicles in the cholinergic synapses of the *Torpedo* electric organ during induced transmitter release. *Neuroscience* **2**, 695–714.

BIOCHEMICAL CHARACTERISTICS OF INDIVIDUAL NEURONS

TAKAHIKO KATO

Department of Biochemistry, Institute of Brain Research
University of Tokyo Faculty of Medicine, Tokyo, Japan

I. Introduction

In spite of recent developments in the bulk isolation technique, nerve cell body samples prepared by this method are usually contaminated to various

119

extents with other cellular components of the nervous tissue (glia and capillary endothelial cells). Therefore microchemical analysis of a single neuron is indispensable to determine its exact properties. Besides the general characteristics of a cell, the mammalian neuron has the following unique properties: Neurons of the same cell type group themselves as a nucleus, and each cell type has a characteristic morphology and function, depending on the location of the nucleus in the central nervous system. A network built of individual neurons belonging to different nuclei is thought to form the basis for the function of the nervous system by integration of bioelectrical activities of individual cell units. In addition, some functional correlations are evidently presumed between the neurons and the glial cells surrounding them. The analysis of neurons isolated in bulk can provide information about their average properties; however, microchemical analysis can give knowledge on the individual features of neurons, which gives a basis for integrating their characteristics at the biochemical level. Methods other than microchemical analysis are not able to solve the problem of the biochemical interrelations between neurons and between neurons and glial cells.

In the last two decades, microchemical techniques ranging from sample preparation to chemical determination have been developed, and a fairly large body of findings has been accumulated by using these techniques. In this review, the focus is on recent results concerning mammalian differentiated neurons, with occasional references to invertebrate neurons. In the case of some substances and enzymes, biochemical analysis has been performed only on the giant neurons from invertebrates; these results are briefly discussed as supplementary knowledge without detailed explanation of analytical methods. Earlier reports are also cited as introductory explanation or with critical comments, in spite of some overlap with the previous reviews (e.g., Brand and Lehrer, 1971; Giacobini, 1969, 1975; Hydén, 1973, 1976). Some data obtained by other techniques (large-scale biochemistry, histochemistry, autoradiography, and cell culture) are reported to challenge the results from the single-cell analysis.

First, the characteristics of single-cell samples are compared in freeze-dried and fresh neurons; then, the biochemical characteristics are described in the following on substances and enzymes analyzed by the use of microchemical techniques at the cellular level. For a more detailed description of the analytical methods, the reader should refer to monographs on the methodological aspects (Lowry and Passonneau, 1972; Neuhoff, 1973; Osborne, 1974) and recent reviews (Giacobini, 1975; Glick, 1977; Kato, 1978; Wu, 1978).

II. Single-Cell Samples

A. Freeze-Dried Samples

Nervous tissue is plunged into liquid nitrogen (or Freon chilled with liquid nitrogen) and frozen instantly. Similarly the whole body or the decapitated head of a small animal (mouse or rat) is easily frozen, and the frozen brain and spinal cord are removed as tissue blocks (Lowry, 1953). From the frozen tissue, tissue slices (10–30 μm thick) are cut in a cryostat at $-23°C$ and freeze-dried at $-30°C$ by placing them under vacuum. Neuronal cell bodies and neuronal tracts are clearly observable under the microscope in unstained tissue sections (Fig. 1a and b). Single cell bodies can be dissected from the dried sections with a microknife under a dissection microscope at 20- to 100-fold magnification. The dissected cell bodies (Fig. 1c) are used as assay samples after the dry weights are measured on a very sensitive quartz fiber balance (sensitivity up to ± 1 pg, Lowry, 1953; Kato and Lowry, 1973a).

The dissected cell sample included in a tissue section is only part of the perikaryon. The plasma membrane, having nerve endings and a small amount of surrounding neuropil, remains fragmentarily around the contour. Part of the cell nucleus is contained in the sample with various volume ratios of nuclear to cytoplasmic components. To obtain a pure cell body sample, the dissected cell body can be further trimmed closely inside the contour to eliminate the peripheral portion. The trimmed peripheral portion can be used as a sample containing the plasma membrane and synaptic boutons. The instant freezing of the tissue minimizes postmortem changes in substrate concentrations and enzymes in the samples. Therefore dried samples are most suitable for determining low-MW substances that turn over rapidly *in vivo* (e.g., ATP, glucose, acetylcholine, amino acids) and thus for estimating the concentrations of these compounds *in situ* in the living animal. A slow freezing process, such as in a deep freezer, gradually produces large ice crystals in the tissue, and expansion of the crystals deforms the surrounding structure of the tissue. Rapid freezing prevents deformation of the cellular structure by forming very fine ice crystals. Under vacuum, these crystals evaporate and leave very small cavities in cells, resulting in destruction of the fine structure of the neuron. The fine cavities left after evaporation of ice crystals give a spongy appearance to the sample, but the intracellular structures (nucleus and nucleolus) are clearly visible (Fig. 1b and c). The cell body sample can be dissected further for subcellular analysis to prepare cytoplasmic and

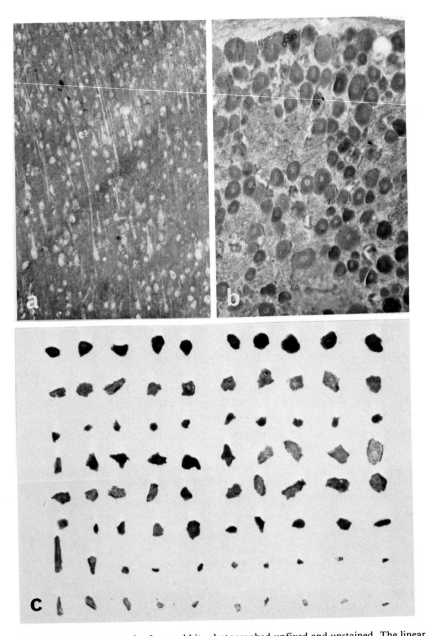

FIG. 1. Freeze-dried samples from rabbit, photographed unfixed and unstained. The linear magnification is × 175. (a) Cerebral cortex. White pyramidal cells are observed with apical dendrites. (b) Dorsal root ganglion of the spinal cord. White cell nuclei are present in the center of dark neuronal perikarya. (c) Dissected nerve cell bodies. From the top: dorsal root ganglion cells of spinal cord, anterior horn cells of spinal cord, Purkinje cells, giant cells in reticular formation, Deiters' nucleus cells, facial nucleus cells, pyramidal cells in cerebral cortex, and

nuclear samples. The porous nature of the sample promotes absorption of extraction medium or reaction mixture, and enzymes bound to the cellular structure are easily solubilized without the aid of detergents in the assay.

One advantage in using freeze-dried samples is that a sample is weighed and the same weighed sample can be directly analyzed to determine the amounts of substrates and enzyme activities in it. On the basis of the dry weight, the concentrations of substances and specific activities of enzymes can be expressed and compared quantitatively in different cell samples. Glial cells are so small that freeze-dried individual perikarya have not been dissected out. Often, a portion of perineuronal neuropil is dissected out from the tissue slice and used as a sample to estimate the characteristics of glial cells.

B. Isolated Fresh Neurons

Under the dissection microscope (60- to 100-fold magnification), a neuronal perikaryon is dissected out from a section of the tissue block with the tip of a knife-shaped stainless steel wire (18 μm in diameter). From the cell body with a 100- to 200-μm-long axon and dendrites, the surrounding neuropil (glial cells, axons and dendrites extending from other neurons, and capillaries) is cleaned off by using the stainless steel wire in sucrose solution or buffer solution (Hydén, 1959; Cummins and Hydén, 1962). The conglomerate of neuropil tissue takes a spherical form (Fig. 2c); it is composed mainly of cell bodies of oligodendroglial and astroglial cells (Hydén, 1959; Egyházi and Hydén, 1961), but the extent of contamination with neuronal components (neuronal axons and dendrites) is not quantitatively measurable. Therefore, a certain reservation is needed to interpret as glial characteristics the biochemical data obtained from analysis of these fresh glial samples or of the freeze-dried neuropil samples.

The neuronal cell body is enveloped in the plasma membrane (Fig. 2a and b). Electron microscopy has shown that well-preserved synaptic boutons and junctional complexes are retained on the outside of the plasma membrane (Fig. 2d). The plasma membrane is cut open by using the stainless steel knife, and the cell contents can be removed to prepare a membrane sample (Hansson and Hydén, 1974). The oxygen uptake or respiration *in vitro* of an isolated neuron was doubled by increasing the K^+ concentration from 5 to 54 mM in the medium bathing the perikaryon

pyramidal cells in Ammon's horn. For illustrative purposes some of the surrounding neuropil has been intentionally left attached to some of the cell bodies. Note spongy appearance of cytoplasmic portion in some cells. (From Kato and Lowry, 1973a.)

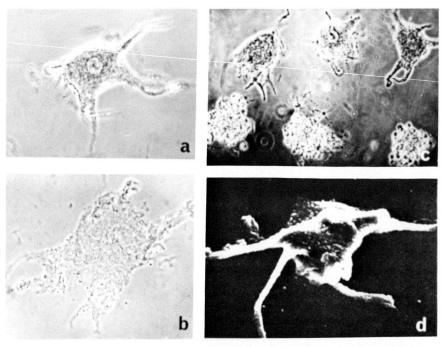

FIG. 2. Fresh neurons and glial cell samples prepared from rabbit. (a) Deiters' cell isolated from the brain stem and stained in diluted methylene blue solution. Synaptic boutons from other neurons are observable on the surface of the perikaryon and dendrites. (b) Sample of plasma membrane prepared from the isolated nerve cell in (a) by dissecting its plasma membrane and eliminating the cytoplasmic content and nucleus. (a and b, from Cummins and Hydén, 1962). (c) Deiters' neurons (top row) and glial cell samples (bottom row) prepared from neuropil immediately around the above neurons (from Hydén and Pigon, 1960). (a-c) Phase-contrast microscopy. (d) Scanning electron micrograph of isolated hypoglossal neuron with perikaryal surface covered by spherical particles (mainly synaptic boutons). The linear magnification is × 950. (From Hamberger *et al.*, 1970.)

(Hultborn and Hydén, 1974). About a 40-mV membrane potential was found across the plasma membrane, and the potential was influenced by Na⁺ and K⁺ ions. An action potential could, however, not be demonstrated (Hillman and Hydén, 1965a). These studies indicate that the plasma membrane may be almost intact even in the portion not covered by synaptic boutons. However, an electron microscopic observation indicated that the plasma membrane was defective (Roots and Johnston, 1964, 1965) and that the mitochondria were also damaged (Bondareff and Hydén, 1969). Based on the latter findings, it is safer not to consider the isolated fresh neuron completely intact. The cell content can be separated into a cell nucleus and a cytoplasmic portion. These are used as subcellular samples from single neurons in the same way as in the case of freeze-dried material. As it takes a few minutes or more to prepare fresh samples, even if prepared in a cold medium, these samples are not suitable for measuring low-MW substances with rapid turnover rates but for analyzing more stable high-MW substances (RNA, protein, enzymes, etc.).

The fresh neuron is dried instantly when exposed to the air; then direct weighing is impossible. The cell body is too small to determine its protein content by using the methods available at present. Thus, the data are not expressed on the basis of weight or protein content. The biochemical data are usually given on the basis of a whole cell body and compared in different samples. The volume of the cell body is estimated from microscopic photography of the sample by measuring its diameters, and the concentrations of substrates and the specific activities of enzymes are calculated to be those per unit volume (or expressed as those per unit wet weight on the assumption that the specific gravity of the fresh neuron is 1.00). The dry mass is known to be approximately 20% of the total wet weight in rabbit and puffer fish neurons (see below; Hydén, 1959; Lehrer *et al.,* 1970); then, a rough estimate of the concentrations and specific activities on the basis of the dry weight is obtained by multiplying the value per unit volume or wet weight by 5.

As shown in the following discussion fresh perikaryon and membrane are most suitable for analyzing the morphological characteristics of the cell structure, part of which is well-preserved and easily observable in the absence of the surrounding components *in situ.* Fresh neurons are furthermore useful in studying metabolic or functional activities *in vitro* (oxygen consumption, ion transport across the plasma membrane, energy metabolism, etc.; see the review by Hertz and Schousboe, 1975), although the whole cell structure is not completely intact. These approaches are not feasible with freeze-dried samples.

A large cell body of an invertebrate neuron is easily isolated as a fresh sample, as in the case of colossal neurons from *Aplysia californica* (McCaman and Dewhurst, 1970). Because of their large size, biochemical measurements have also been easily performed with invertebrate neurons. The samples of plasma membrane and cytoplasmic content can be prepared in the same way as described for mammalian neurons. However, the cell structures of invertebrate neurons are markedly different from those of vertebrate neurons. Generally, synaptic junctions are mainly formed on the surface of dendrites, which are structurally separated from the cell bodies (cockroach, Cohen, 1970; *Aplysia,* Frazier *et al.,* 1967; snail, Bocharova *et al.,* 1975). Much thinner membranes are generally formed at the junctional structures of the synapses (Coggeshall, 1967), and the surface of perikaryal membranes is not smooth but covered with glial scraps and fibrous processes that come from other neurons (Bocharova *et al.,* 1975).

The amount of DNA in cell nuclei of a colossal *Aplysia* neuron is more than 200,000-fold higher than the haploid DNA content in sperm (Lasek and Dower, 1971). The biochemical characteristics might be essentially the same at the molecular level in both vertebrate (especially mammalian) and invertebrate neurons, but the findings mentioned above signify that the invertebrate neurons have biological differences in cellular organization. One should keep these differences in mind when invertebrate neurons are used as a model of general neurons and be cautious in extending their properties directly to estimate the characteristics of vertebrate neurons (see Section V).

III. Cell Structure

A. Cell Size and Mass Density

As the shapes of vertebrate neurons are variable, their sizes are extremely variable depending on their location in the central nervous system. For instance, measurement of the volumes of fresh neurons gave the following values for cell bodies of rabbit neurons: dorsal root ganglion cells of spinal cord, 180 pl (Edström and Pigon, 1958); Deiters' cells, 93.2 pl (Hydén, 1959); anterior horn cells, 22 pl (Edström, 1956); hypoglossal nucleus cells, 13.3 pl (Brattgård et al., 1957). Dry masses of freeze-dried samples of eight types of rabbit neurons have also been determined directly on a quartz fiber balance, and the total volumes of these neuronal perikarya were calculated by using the mass density (about 0.2) on the assumption that the samples were meridian sections of spherical somata. The following results were obtained: dorsal root ganglion cells, 125–375 pl; anterior horn cells, 150–200 pl; Deiters' cells, 200–375 pl; Purkinje cells, 25–50 pl; and pyramidal cells of cerebral cortex and Ammon's horn, about 5 pl (Kato and Lowry, 1973a). Finally, fresh neuronal perikarya were dissected out from the third cortical layer of human cerebrum (Brodmann's area 9), and the change in the volumes of the pyramidal neurons was analyzed with eight brains from 8 months to 94 yr of age. The mean volume was 1.4 pl at 8 months, and it was maintained with surprising constancy (2.76 pl) from 9 to 68 yr of age; thereafter, it gradually dropped to 1.9 pl at the age of 92 yr (Uemura and Hartmann, 1978).

In supramedullary neurons of puffer fish, the mass density of cytoplasm (0.22 gm/ml) was higher than that of the cell nucleus (0.14); and the neuropil had a density (0.15) similar to that of the nucleus (Lehrer et al., 1970). The density of the cytoplasm (0.23) was similar to that of the nucleus, but the nucleolus had a higher density (0.38). The axon of a hypoglossal cell had a very low density (0.06, Brattgård *et al.*, 1957). In the case of rabbit Deiters' neuron, a decreasing gradient of mass density was observed from the perinuclear portion to the peripheral portion of the cytoplasm, and the decrease amounted to about 25% of the perinuclear value (Hydén and Pigon, 1960).

B. Nucleus

The ratios between the volumes of the nucleus and those of the cytoplasm were calculated to be 0.096 for the cell body of a cat cervical ganglion cell (Pevzner, 1965) and 0.41 for a rabbit Deiters' cell (Hydén, 1959). The nucleus of the vertebrate neuron is so small that the DNA content in a single nucleus is not measurable. In the giant neurons (R2, L6, and P1) of *A. californica,* the amount of DNA was determined and the cells could be divided into two groups having average values of 67 and 131 ng per nucleus (Lasek and Dower, 1971). When the amount of DNA in *Aplysia* sperm (1 pg) is taken as the haploid value, the above neuronal DNA contents correspond to 67,000- and 131,000-fold the haploid amount. Because 67 ng of DNA in a neuron is approximately 2^{16}-fold larger than the diploid value (2 pg) and the number of nucleoli was found to be in a fixed ratio to the DNA content in the neuron, synchronous replication of DNA was thought to occur during maturation of the neuron, which arrested cell division (Lasek *et al.,* 1972).

With respect to mammalian neurons, microspectrophotometric measurement by means of Feulgen staining of DNA was reported to show that some large neurons in the cat contained a tetraploid amount of DNA (Herman and Lapham, 1968, 1969). An increase from diploid to tetraploid level was found with postnatal maturation of Purkinje cells (Lentz and Lapham, 1969, 1970; Mareš *et al.,* 1973). In contrast to these results, an improved cytophotometric method demonstrated that large human neurons contained the diploid amount of DNA: Purkinje cells (Mann and Yates, 1973; Fujita, 1974); and neuroblasts and large neurons in the spinal cord (Fujita, 1974). Quantitative measurement of DNA content in bulk-isolated neuronal fractions indicated that samples enriched in Purkinje perikarya contained 7.2 pg of DNA per cell as compared with 7.6 pg of DNA per cell in mixed cerebellar cell samples (Cohen *et al.,* 1973). The cell nuclei isolated from large neurons collected by the bulk isolation technique from ventral spinal cord contained 5.3 pg of DNA per cell (McIlwain and Capps-Covey, 1976). These values are similar to the diploid DNA content in general mammalian cells (see Sober, 1973). Thus the above results support the concept that polyploidy is not present in large mammalian neurons like Purkinje cells and anterior horn cells. A final conclusion will be drawn if the determination of DNA becomes possible in single isolated neuronal nuclei. This is an example of the observation that microscopic cytochemistry at times leads to rather misleading conclusions and that quantitative biochemical analysis is necessary to obtain unambiguous results.

C. Structural Components in Cytoplasm

Cell contents of fresh Deiters' neurons were spread over grids and observed by electron microscopy (Ekholm and Hydén, 1965). This provided images of neuronal polyribosomes not contaminated with glial polysomes. The filamentous network was found electron microscopically to cover the inner surface of the plasma membrane samples of fresh Deiters' neurons (Hansson and Hydén, 1974; Metuzals and Mushynski, 1974). The membrane sample was spread over a carbon-coated grid, exposing the inner surface upward. The samples were stained by uranyl acetate and observed in an electron microscope. The peripheral portion of cytoplasm (ectoplasm) adhered to the membrane surface, and the ectoplasm included filaments interwoven and associated in a semilattice pattern. Hansson and Hydén (1974) showed that the filaments were about 90 Å wide and composed of two 20-Å unit filaments which were helically intercoiled with different pitches. They claimed that these filaments could be considered "microfilaments" containing actin, since they were dressed with heavy meromyosin (HMM) and formed an arrowhead pattern (arrowhead structure; see Ishikawa et al., 1969). However, Metuzals and Mushynski (1974) stated that the arrowhead formation was never observed in a fresh ectoplasm sample, although they demonstrated a clear photograph of HMM-dressed microfilaments in a preterminal region and in postsynaptic areas of Dieters' cell bodies in a thin section. Metuzals and Mushynski showed the filamentous network in thin sections of fixed single cells and of fixed Deiters' nucleus as well as in ectoplasm spreads obtained from fresh membranes. These "neurofilaments" were digested by trypsin and pronase but were resistant to other enzymes (DNAase, RNAase, hyaluronidase, neuraminidase, and phospholipase). Calcium chloride (2 mM) uncoiled the filaments, but K$^+$ did not affect them. Colchicine had no visible effect, and cytochalasin B, which generally disrupts microfilaments (Wessells et al., 1971), produced loosening of the filaments but no disruption occurred (Hansson and Hydén, 1974; Metuzals and Mushynski, 1974). Cytochalasin B led to the formation of round microfilamentous aggregates which appeared to be intimately associated with the neurofilaments, thus producing berry-like patterns. However, this change did not occur constantly (Metuzals and Mushynski, 1974). A similar neurofilamentous network was found on the inner surface of an axolemma prepared from a squid giant axon (Metuzals and Tasaki, 1978). The axoplasm of a single axon (15 mm long) was removed by perfusion with seawater, and the axolemma was cut longitudinally and opened with the aid of a broken razor blade in a manner similar to that used for the preparation of membrane samples of fresh neurons. Scanning electron microscopy showed the filamentous network extending three-dimensionally into the axoplasm. In the network, 10- to 20-nm filaments were found connected by ~ 7-nm-wide filaments; the latter could be decorated with HMM. The presence of actin-like proteins in axoplasm was confirmed by polyacrylamide gel electrophoresis, which showed that the axoplasm contained a protein with a MW of 46,000, corresponding to muscle actin. The actin-like protein was found biochemically in the synaptosomal fraction (Blitz and Fine, 1974), and electron microscopically it could be observed from the perikaryon to the axonal endings (growth cone) (LeBeux and Willemot, 1975a,b). At the light microscopic level, the ubiquitous distribution of microfilaments composed of actin-like protein was shown immunohistochemically in neuroblastoma cell bodies (Sanger, 1975). These results support the above observations on the filamentous structures in single-cell samples, which have been discussed.

D. Plasma Membrane and Growth Cone

Single fresh neurons isolated from rabbit vestibular nucleus and rat dorsal root ganglion were cultured for 14–28 days in the presence of nerve growth factor. The outgrowth of new processes with growth cones was observed (Hillman, 1966; Hillman and Sheikh, 1968). The growth cone of neurons from 1-day-old rat cervical ganglia seemed to be a veillike ruffling membrane sheet extending slender microspikes (filopodia, Bray, 1970). Microtubules, neurofilaments, and microfilaments were observed in the processes and growth cones. The peripheral flange and filopodia contained a filamentous network composed of 5-nm microfilaments (Yamada et al., 1971; Bunge, 1973), and the microfilaments were densely blocked when treated with cytochalasin B, which caused rounding up of the growth cones (Wessells et al., 1971; Bunge, 1973).

Lectins (concanavalin A, soybean aggulutinin, wheat germ aggulutinin, etc.) are known to bind to glycans (heteropolysaccharides) which are constituents of glycoprotein or glycolipid forming plasma membrane. Isolated fresh neurons from the vestibular nucleus of young adult and senescent rats were incubated with fluorescamine-conjugated concanavalin A to investigate its binding to mannoside residues in glycoprotein (Bennett and Bondareff, 1977). Fluorescence microscopy showed that senescent neurons were significantly more fluorescent than the neurons from young adult rats. In contrast to the observations on uniformly labeled young adult neurons, a formation of patches and caps of fluorescamine–concanavalin A was found on the surface of senescent neurons. The lectin-binding capacity was absent in dorsal root ganglion cells isolated from 5- to 6-day-old chick embryos (Denis-Donini *et al.*, 1978). As the outgrowth of processes increased in number with the age of the embryo, the plasma membrane of the perikarya began to bind concanavalin A on day 7, but the plasma membrane of processes did not show such binding. At this stage soybean aggulutinin bound to both the surface of perikarya and processes. Neurons from 8- to 12-day-old chick embryos formed processes with a high degree of branching, and concanavalin A as well as soybean and wheat germ aggulutinin bound intensely to the surface of perikarya and processes. These lectin-binding characteristics are only detectable with isolated neurons and clearly demonstrate the changing properties of the plasma membrane during maturation *in vivo.* The change may result from differentiation of the membrane for its function, since concanavalin A and other lectins were reported to bind preferentially to postsynaptic membranes (Kelly *et al.*, 1976). Concanavalin A binding induced a new depolarizing response to L-glutamate on all neurons from both *Aplysia* and *Helix* tested (Kehoe, 1978). In the case of medial cells of the pleural ganglion of *A. californica,* this response was pharmacologically different from the three cell-specific glutamate responses and was not thought to be due to the lectin's capacity to produce redistribution of receptor sites originally present in the plasma membrane.

IV. Biochemical Components

A. Ribonucleic Acid

1. SPECIES AND COMPOSITION OF RNA

The RNA content of isolated neuronal cell bodies is higher in larger neurons. Thus, the average amount of RNA per single perikaryon of vertebrate neurons ranges from 27 pg in a small neuron (in the superior frontal gyrus of human cerebral cortex, Uemura and Hartmann, 1978) to 1800 pg in a large Mauthner neuron (goldfish, Edström *et al.*, 1962).

Values for RNA content in various neurons have been presented in a review by Shashoua (1974). In groups of anterior horn cells (average RNA content = 545 pg/cell, Edström, 1956) and dorsal root ganglion cells (average RNA content = 1070 pg/cell, Edström and Pigon, 1958) from rabbit spinal cord, a proportionality was found between the surface area and RNA content of the neuronal cell bodies. During aging up to 60 yr of age, the RNA content in cell bodies from the human central nervous system is known to increase gradually (anterior horn cells, Hydén, 1973; cortical neurons in Brodmann area 9, Uemura and Hartmann, 1978). In individual neurons in cultured dorsal root ganglia of fetal rats, RNA content increased more slowly than in the neurons *in vivo* (Sobkowicz *et al.*, 1973). A few days after explantation of ganglia, the cells increased rapidly in size, with formation of a "fibrillogenous zone" from which the neurofibrillar network and the axon seemed to derive. At this stage the Nissl granules containing ribosomes were found at the periphery of the cell bodies, but in later stages the Nissl bodies invaded the whole cell body, producing a simultaneous increase in RNA content.

The average base composition of neuronal perikarya is known to be of the GC type, which indicates that the ratio of guanine plus cytosine $(G + C)$ to adenine plus uridine $(A + U)$ ranges as follows: hypoglossal nucleus cells from rabbit, 1.31 (Daneholt and Brattgård, 1966); Deiters' cells from rat, 1.57 (Hydén and Egyházi, 1962); and Deiters' cells from rabbit, 1.67 (Egyházi and Hydén, 1961). The base composition of nuclear RNA of rabbit Deiters' cells was of the GC type $[(G + C)/(A + U) = 1.39$, Hydén and Egyházi, 1962], although the average composition of nuclear RNA prepared from fractionated rat brain cell nuclei was reported to be of the AU type $[(G + C)/(A + U) = 0.9$, Bondy and Roberts, 1968]. The glial samples collected from around neurons did not have a constant pattern of base composition [AU type in rabbit hypoglossal nucleus, $(G + C)/(A + U) = 0.83$, Daneholt and Brattgård, 1966; GC type in Deiters' nucleus from rabbit, $(G + C)/(A + U) = 1.54$, Egyházi and Hydén, 1961; and GC type in Deiters' nucleus from rat, $(G + C)/(A + U) = 1.22$, Hydén, 1973]. The $(G + C)/(A + U)$ ratios for 4, 17, and 28 S RNAs from rat cerebral cortex were reported to be about 1.5, 1.7, and 1.8, respectively (Mahler *et al.*, 1966). The tRNA is considered 4 S RNA, and the rRNA is 17 and 28 S RNA. The ratios were also 1.74 in ribosomes from rat cerebral cortex (Mahler *et al.*, 1966) and 1.85 in polysomes from rat brain (Campagnoni *et al.*, 1971). These ratios of neuronal RNA may reflect rRNA and tRNA in the cytoplasm, which occupies a large portion of the perikaryon (see Section III,B). In glial samples, the volume ratio of cytoplasm to nucleus might not be constant because of different extents of contamination with neuronal components.

Recently, RNA was extracted from the CA3 region in rat hippocampus containing packed pyramidal cells (Cupello and Hydén, 1975) and from neuronal cell bodies isolated in bulk from pig brain stem and rat cerebral cortex (Okazaki *et al.,* 1978). Micro gel electrophoresis showed that the CA3 sample contained 4, 18, and 28 S peaks of RNA and that the RNA species in the isolated cell bodies were mainly of the 18 and 28 S classes. The $(G + C)/(A + U)$ ratios of these 18 and 28 S RNAs were similar in the above two types of neurons (1.34–1.35 for 18 S RNA and 1.81–1.82 for 28 S RNA). Compared with these classes, the relative amount of 4 S RNA was larger in pig brain stem neurons than in rat cerebral cortex neurons. RNA classes larger than 28 S and those between 18 and 28 S were also found in much larger amounts in neurons from pig brain stem. When the giant single neurons (R2 and R14) of *A. californica* were incubated in a medium containing [³H]uridine and [³H]cytidine, Peterson (1970) demonstrated the presence of a high-MW, rapidly labeled RNA in the cell nucleus. In cytoplasm, 4 S RNA was labeled earlier. The sequence of labeling was 38 S, then 31 and 21 S in the nucleus, and finally 27 and 18 S in the cytoplasm, which was thought to provide evidence for processing of RNA from precursors to rRNA in the neuron. A similar study on RNA processing was performed on 0.4-gm samples of rabbit lateral vestibular nucleus after the injection of [³H]orotic acid into the fourth ventricle (Egyházi and Hydén, 1966). Nuclear RNA classes below 16 S were labeled 15 min after the injection, and labeled 10–12 S RNAs were found in cytoplasm 15 min thereafter.

In the axoplasm of a squid axon (*Loligo pealii,* Lasek, 1970), as well as in a myelin-free axon preparation from an accessory nerve root of the cat (Koenig, 1965) RNA was found at low concentrations. More than 80% of RNA in the squid and polychaete (*Myxicola infundibulum*) axoplasm was found in soluble form in the supernatant obtained by centrifugation at 27,000 g for 20 min, and the RNA was mainly of the 4 S class (Lasek *et al.,* 1973). When RNA was labeled in cell bodies of motor neurons by injecting [¹⁴C]uridine into the spinal cord, the axon of chicken sciatic nerve contained 4 S RNA in the distal portion, but 29 and 48 S classes were observed in the proximal portion (Por *et al.,* 1978). A single axon of a goldfish giant neuron (Mauthner cell) contained 6000 pg of RNA, and the surrounding myelin sheath contained 8000 pg of RNA (Edström, 1964). This amount of axoplasmic RNA was approximately three times larger than that in the cell body. The concentration curve of RNA in the axon was similar to a concave parabola with a minimum value (0.03%) at a point corresponding to three-fifths of the length of the whole axon from the cell body (where it was 0.05%) to the caudal tip (where it was 0.07%). The $(G + C)/(A + U)$ ratio was about 1.6 in the axon and myelin sheath. In a preliminary report,

Koenig (1978) found by means of microelectrophoresis that a goldfish Mauthner axon contained major RNA classes of 18 and 28 S indistinguishable from rRNA in fish brain. In addition, nonribosomal 15 and 4 S RNAs were also found in the axon. Although ribosomes are generally believed to be absent in the axon of mammalian neurons as well as in Mauthner axons (Edström, 1964), the above results indicate that several classes of RNA are possibly present in the axon, but the nature of these RNAs and their biological roles remain to be elucidated.

2. CAPACITY FOR RNA SYNTHESIS

Four hours after the injection of [³H]adenine and [³H]cytidine into the fourth ventricle of a rabbit, fresh nerve cells and glial cell samples were isolated from a hypoglossal nucleus and the base composition of extracted RNA corresponding to the amount in 25–50 neurons was analyzed by microelectrophoresis (Daneholt and Brattgård, 1966). [³H]adenine was incorporated into two purines and [³H]cytidine into two pyrimidines in neuronal and glial RNA. The incorporation into adenine, guanine, and cytosine, expressed as picomoles of labeled base per mole of extracted base, was about twofold greater in glial cell samples than in neuronal cells. In experiments in which bulk-isolated neuronal and glial cell fractions were incubated with [³H]uridine or [³H]guanine, the glial cell fraction incorporated the label more rapidly than the neuronal fraction because of a more active uptake of precursors into glial cells (Yanagihara, 1979). However, in view of the higher activity of RNA polymerase found in isolated neuronal nuclei than in glial nuclei, neurons were thought to have, on an average, a higher capacity for RNA synthesis (Kato and Kurokawa, 1970; Austoker et al., 1972). When an isolated single axon segment surrounded by a myelin sheath was incubated for 21.5 hr in a medium containing [³H]uridine and [³H]cytidine, the axonal RNA was labeled and sucrose density gradient analysis showed a markedly high single peak in the 4 S region (Edström et al., 1969). This RNA synthesis was inhibited 80% by actinomycin D (10 μg/ml). A similar RNA synthesis in vitro and its inhibition by actinomycin D were observed with myelin-free axons of rabbit accessory nerves (Koenig, 1967).

3. STIMULATION OF RNA SYNTHESIS BY SEVERAL FACTORS

Various kinds of stimulation of the animal body are known to influence RNA synthesis in vertebrate neurons. Vestibular stimulation, produced by rotating immobilized animals horizontally and vertically promoted RNA synthesis in Deiters' cell bodies of rat (5.7% increase, Hydén and Egyházi,

1962) and of rabbit (4.3% increase, Hydén and Pigon, 1960). The same stimulation led to a 11–12% increase in the Purkinje cells in the vermian and hemispherical parts of cerebellar central lobules, and a 22% decrease in the Purkinje cells in nodules of rat cerebellum (Jarlstedt, 1966b). In Deiters' neurons of rats exposed to gravitational changes (1.65–3.65 g) in a centrifuge, the amount of RNA decreased 30% transiently during the stimulation period (less than 1 hr) and then returned to the nonstimulated level (648 pg/cell) until 12 hr after the beginning of stimulation. Stimulation for longer periods (1–30 days) resulted in a 50% decrease (Grenell *et al.*, 1968). The rotation stimulation was given to unrestrained rats, producing proprio- and exteroceptive stimulation and contraction of muscles and tendons, since the rats had to cling to the rotating cage to prevent themselves from falling. The stimulation induced about a 10% increase in the RNA content of the Purkinje cells in the vermian part of the central lobule and in the paraflocculus, and about a 25% increase in the pyramis and copula pyramidis (Jarlstedt, 1966a). Increased RNA synthesis induced by similar proprio- and exteroceptive stimulation was also demonstrated in the Purkinje cells of cat cerebellum by cytophotometric analysis (Vraa-Jensen, 1971), as well as by autoradiography (Vraa-Jensen, 1972). Irrigation with warm water (48°C) for 30 min in the left outer ear of rabbits induced a 46% increase of RNA in neurons from the ipsilateral dorsal part of the vestibular nucleus, compared with contralateral neurons (Jarlstedt, 1966b). The same stimulation led to a 30–60% increase in the ipsilateral Purkinje cells, compared with contralateral Purkinje cells, in the vermian and lateral parts of the central lobule and nodules of rabbit cerebellum. Stimulation by irrigation for 30 min with cold water (20°C) similarly induced 23 and 27% increases in RNA in Purkinje cells in the vermian and lateral parts of the central lobule, respectively. Neurotomy of the right vestibular nerves of rabbit also resulted in about a 20% increase in the RNA content of the Purkinje cells in the contralateral vermian and lateral parts of the central lobule (Jarlstedt, 1966b). Chemical stimulation (intravenous injection of 1,1,3-tricyano-2-amino-1-propane, a dimer of malononitrile) induced more than a 25% increase in the RNA content of Deiters' cells from rabbit, and a corresponding amount of RNA was lost from the neuroglial samples immediately surrounding Deiters' neurons (Egyházi and Hydén, 1961). The ratio of purine to pyrimidine [(A + G)/(C + U)] was simultaneously found to be raised 13–16% in ipsilateral Purkinje cells of the central lobule by warm stimulation, and found to be decreased 12% in contralateral Purkinje cells of the central lobule by right neurotomy of the vestibular nerve (Jarlstedt, 1966b). Vestibular stimulation led to an 8% decrease in the same ratio in Purkinje cells in the vermian part of the central lobule (Jarlstedt, 1966a). The chemical stimulation mentioned above brought about no significant change in the purine/pyrimidine ratio (Egyházi and Hydén, 1961). Cyto-photometry indicated that hypoxia gave rise to an increase in the content of RNA both in the Purkinje cells and in the surrounding neuroglia (Pevzner, 1972); and swimming for 3- and 4-hr periods led to an increase in the RNA content of the anterior horn cells of rat spinal cord but no change in the surrounding neuroglia (Pevzner, 1971).

The above-mentioned stimulations were not given directly to the neurons analyzed but influenced them through actions on the sensory organs of the animals. In contrast, the invertebrate neuron is good material for direct stimulation. The slowly adapting nerve cells of the crustacean abdominal stretch receptor organ were made to generate 10,000 spike potentials during several hours by stretching the abdominal muscle. The total amount of RNA and the incorporation of ^{32}P into RNA were not changed during stimulation (Grampp and Edström, 1963; Edström and Grampp, 1965). The ratios of adenine to uridine and purine to pyrimidine were, however, significantly increased (Grampp and Edström, 1963). Actinomycin D (2 $\mu g/ml$) completely inhibited RNA synthesis but had no effect on continuous generation of action potentials for 12–24 hr. A giant neuron (R2) in the abdominal ganglion of *A. californica* was stimulated transsynaptically in a medium containing [^{3}H]uridine by applying the stimulation electrode to the left connective and other nerves (see Fig. 3), including the axons forming synapses on the R2 neuron. The R2 cell body with its adjacent glia was isolated after stimulation and RNA incorporation, and the labeled RNA was extracted for analysis (Berry, 1969; Berry and Cohen, 1972; Kernell and Peterson, 1970). The normalized incorporation ratio, i.e., the ratio of radioisotope counts in the RNA fraction to those in the acid-soluble fraction, increased linearly with increasing number of spikes from 20 (nonstimulated level) to 500 spikes (stimulated level) per hour (Berry, 1969). Autoradiography showed that 30% of the total label in the isolated cell sample was incorporated into the peripheral glia, 20% into the cytoplasm, and 50% into the nucleus of the neuron. The precursors (uridine, UMP, and UDPG plus UTP) in the acid-soluble fraction were present in similar fraction ratios in stimulated and nonstimulated neurons (Berry and Cohen, 1972). The label was incorporated into the RNA regions (18 and 28 S) in the nucleus and cytoplasm and was also found in the nonribosomal classes. The nonribosomal incorporation, especially in classes over 28 S, was higher in the cytoplasm than in the nucleus after an incubation period of 86 min (Peterson and Kernell, 1970). Spikes, directly elicited by current pulse injected into the cell body through an intracellular microelectrode had no significant effect on RNA labeling, but weak stimulation of the neuron producing postsynaptic potentials but few spikes

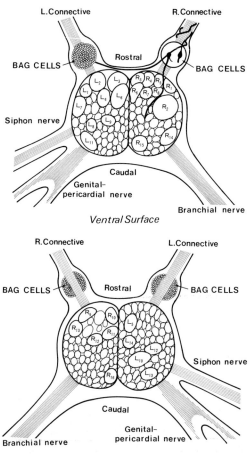

FIG. 3. Schematic presentation of localization of large neurons in the abdominal ganglion of *A. californica*. The numbered letters R and L are referred to in the text. Two cells are shown in the right cluster of bag cells in the dorsal view in order to indicate the relationship of the bag cell processes to the connective and the main part of the ganglion. (Based on Frazier *et al.*, 1967, and Kupfermann and Kandel, 1970.)

brought about a small but significant increase in RNA labeling (Kernell and Peterson, 1970). The stimulation-linked increase in the label incorporation was thought to be caused by increased radiospecific activity of uridine in the precursor pool rather than by activation of RNA synthesis (Wilson and Berry, 1972). However, Peterson and Erulkar (1973) demonstrated that the newly synthesized RNA in transsynaptically stimulated R2 neurons contained less methylated fraction, hybridizing more easily to reiterated DNA sequences; this was thought to imply that the transsynaptic stimulation induced the transcription of molluscan genome. As in the case of the invertebrate neurons discussed above, cytophotometric studies showed that electrical stimulation *in situ* of the superior cervical sympathetic ganglion of the cat resulted in a marked increase in RNA concentration in ganglion cells (Pevzner, 1965).

There seems to be no comprehensive principle underlying all the above results. With regard to invertebrate neurons, electrical activity appears to have no direct correlation with RNA synthesis, but the stimulation could provoke some alterations in RNA metabolism only when transsynaptic routes in terms of neurotransmitters are involved. The stimulation *in vivo* of vertebrate neurons might be more complicated than the stimulation *in vitro* of invertebrate neurons in view of the fact that a vertebrate neuron is being continuously influenced *in situ* by transsynaptic stimulation from other neurons, neuroregulators, hormones, and other metabolic intermediates supplied from blood. All these influencing factors are modulated by the above-mentioned physical and chemical stimulation of the animal body.

4. RNA SYNTHESIS IN BEHAVIORAL TASKS

RNA synthesis during acquisition of new behavioral patterns ("learning" and "short-term memory") has been one of the most interesting subjects pursued at the cellular level by Hydén and his co-workers. The results were summarized and well documented in the reviews by Hydén (1973) and, from a slightly different viewpoint, by Shashoua (1974). The detailed description is hence omitted here, but some other work which is partly very recent is included to supplement and extend the range of the previous reviews.

According to the reviews by Hydén and Shashoua, animals trained to establish a new behavior were thought to synthesize a specific species of RNA (presumably mRNA) in their neurons. These RNA species were regarded as having template activity for synthesizing neuronal proteins which must form and consolidate "long-term memory" for an acquired behavior.

In the experiments leading to this hypothesis, there are three points which may deserve consideration and criticism. Firstly, the neuron regarded as functionally involved in the behavior has been selected as material for the analysis. The criteria for the selection have a basis in other scientific fields such as physiology, neurology, behavioral psychology, and psychiatry. For example, Deiters' neurons, involved in postural equilibrium, were selected for analysis in rats learning to balance on a thin steel wire (Hydén and Egyházi, 1962); the cerebral neurons in the fifth and sixth layers of the parietal cortex on the right side, which were thought to control the left paws, were selected in the case of rats compelled to learn to use nonpreferred left paws ("transfer of handedness," Hydén and Egyházi, 1964). In a simpler experimental system (shock avoidance training of cockroaches, Kerkut *et al.*, 1970), quantitative and autoradiographic analyses showed that [³H]uridine was incorporated much more, as compared with the yoked control, into neurons in the region of the ganglion, which was involved in the avoidance movement. In the same avoidance experiment with mice, autoradiography indicated that [³H]uridine was incorporated into the neuronal nuclei in the hippocampus and other parts of the limbic system of the trained mice but not into the same neurons as in the yoked control mice (Kahan *et al.*, 1970). When rats were trained for brightness discrimination in a Y chamber, [³H]uridine and [³H]orotic acid were incorporated significantly more into neurons from the hippocampal regions (CA1-CA4) and some cortical areas (lamina polymorpha, visual cortex, and cingulate cortex) of trained animals than into corresponding neurons from sham-trained active control animals (Pohle and Matthies, 1971, 1974). No marked incorporation was observed into glial cells of trained rats. These data support the above-mentioned criteria for selection of neurons as experimental material and suggest strongly that hippocampal neurons participate in acquiring various behavioral patterns. As pointed out previously by Kuffler and Nicholls (1966), the neurons selected for analysis were not known to have been in an electrophysiologically excitatory or inhibitory state during behavioral tasks. The results of the experiments by Hydén and Egyházi (1962, 1963, 1964) should therefore be considered as indicating an increased amount of RNA (2–10%) and a change in its base composition in neurons sampled at random from an area of brain presumed to participate in such a training task. (For the meaning of random sampling, see Section V.)

The second point is how to establish a control for the purpose of biochemical analysis. The yoked control and sham-trained animals (Hydén *et al.*, 1974; and some experiments mentioned above) are used as controls at the behavioral level, which does not necessarily mean controls at the physiological level (see below). In the case of rats balancing on a wire (Hydén and Egyházi, 1962, 1963), Deiters' neurons stimulated by rotating movements of the animal's body were used as controls for "trained" Deiters' neurons. In the experiment on the transfer of handedness, control neurons were obtained in the trained animals from the cerebral area believed to control the preferred paws, the use of which was abolished (Hydén and Egyházi, 1964). These control neurons were not trained but believed to be stimulated only "physiologically" to a similar extent as the trained neurons. In the case of goldfish acquiring a new swimming skill (Shashoua, 1970), the control brain was obtained similarly from the animals stimulated physiologically but not trained. In this experiment, the control is plausible because a whole brain, inevitably involved in any behavioral task, was analyzed in the animals stimulated differently from the trained animals. In the experiments on single neurons, there is no guarantee that the control neurons were stimulated in a similar manner and to a similar extent as the trained neurons except for participating in the learning process. Thus, the criteria for the controls are rather arbitrary and depend on the experimental design.

The third point is the nature of RNA newly synthesized as a result of training. Hydén and Lange (1965a, 1966) thought that only the increased portion of RNA (ΔRNA) after training was newly formed, and the base composition of this ΔRNA was calculated from the difference in the average base composition of RNAs in the control and the trained neurons. As a result, the ΔRNA was found to have an "asymmetric" base composition characterized by a high value of the adenine/uridine (A/U) ratio (1.3–5.9). Based on these values, they concluded that mRNA containing poly(A) [poly(A)$^+$ mRNA] was synthesized in the neuronal nucleus by the learning process. At present, it is known that all the heterogeneous nuclear RNA of mouse and rat brains which contains poly(A) [poly(A)$^+$ hnRNA] includes a messenger sequence of polysomal poly(A)$^+$ mRNA (Hahn et al., 1978), and that the base compositions of polysomal mRNA and nuclear RNA are similar (Bondy and Roberts, 1968; Zomzely et al., 1970). Therefore, as suggested by Hydén and his co-workers, the increase in nuclear RNA in neurons may be attributable to mRNA synthesis. On the other hand, the portion of poly(A) was only 3.1% of the total poly(A)$^+$ hnRNA, and this poly(A)$^+$ hnRNA was maximally 22% of the total nuclear RNA in mouse brain (Bantle and Hahn, 1976); therefore the portion of poly(A) is only 0.66% of the total nuclear RNA. In addition, it is known from nonneuronal cells that the A/U ratio of poly(A)$^+$ mRNA is not so much different from that of poly(A)$^-$ mRNA (0.77–1.2, sea urchin embryo, Nemer et al., 1974), and that the pool of poly(A)$^-$ mRNA precursor is about twice that of poly(A)$^+$ mRNA precursor in nuclei (mouse L cells, Latorre and Perry, 1973). These data suggest that the amount of adenine component of the poly(A)$^+$ nuclear RNA, which was increased maximally by 10% in the above experiments (Hydén and Egyházi, 1962, 1963, 1964), is too small to explain the increase in the A/U ratio from 1.3 to 5.6 in the ΔRNA (see the preceding discussion). Thus, the postulate by Hydén and Lange (1965a, 1966) that the newly synthesized RNA was added to the immutable RNA fractions similar to those in the control neurons appears unreasonable. The change in RNA species occurring in the neurons of trained animals must be more complicated than merely addition of newly synthesized RNA as the ΔRNA to the RNA fraction in the control neurons.

Rhesus monkeys were subjected to visual discrimination and delayed alternation training, and the amount of RNA and its base composition were determined in the pyramidal cells from three different regions of their cerebral cortices (CA3 region of the hippocampus, and the fifth layers of the inferior temporal gyrus and gyrus principalis, Hydén et al., 1974). A significant increase (11%) in RNA content was found only in the hippocampal neurons in the case of short-term visual discrimination tasks. The base composition was changed in the neurons from the hippocampus and inferior temporal gyrus by visual discrimination behavior, which was known to involve both these regions. In the delayed alternation experiment, in which the gyrus principalis was thought to play an important role, the base composition changed to some extent in the neurons from all three regions. However, these changes did not include any increase in the A/U ratio and therefore do not support the concept of promoted mRNA synthesis derived from the earlier experiments (see the preceding discussion). In order to analyze unambiguously the RNA change induced by behavioral tasks, experiments using radiolabeled precursors are more effective because the label is incorporated into the newly formed RNA fraction. [14C]orotic acid and [3H]uridine were injected into the ventricles of rats trained for 4 days to use unpreferred paws for obtaining food (Cupello and Hydén, 1976a,b, 1978). On the fourth day, successful performance reached a plateau level (70–90% of the total trials). The active control rats were allowed to obtain food with the preferred paws. Five micrograms of tissue mass containing pyramidal cells was dissected from each side of the hippocampal CA3 region, and total RNA and poly(A)+ RNA were extracted. Microelectrophoresis showed that incorporation of the label from [14C]orotic acid, expressed as a percentage of total label in all fractions, was higher in trained rats than in controls in the fractions of 8–9 S and 16–17 S but lower in the regions around 4 S (Cupello and Hydén, 1976a). In the poly(A)+ RNA fractions, [3H]uridine was incorporated to a greater extent in the RNA classes between 18 and 28 S extracted from the trained rats (Cupello and Hydén, 1976b, 1978). These workers speculated again that the 8–9 S and 16–17 S RNAs, which increased in total RNA analysis but not in poly(A)+ RNA analysis, represented poly(A)- mRNA. The clearcut demonstration of the nature of mRNA seems difficult in single-cell analysis because the amount of obtainable RNA is so small that its stimulation or template activity is almost impossible to demonstrate by the techniques available at present.

B. Proteins

1. Species and Composition of Neuronal Proteins

The protein content of a fresh hypoglossal neuron from rabbit (2770 pg of dry weight) was found to be 2200 pg (about 80% of the total dry mass) with the aid of X-ray microradiography (Brattgård et al., 1957). A micro immunoprecipitation reaction indicated that the brain-specific protein S-100 (about 0.6% of the total soluble protein of beef brain, Moore, 1965) was contained in glial samples from rabbit Deiters' nucleus and that Deiters' neurons did not contain detectable amounts of S-100 (Hydén and McEwen, 1966). In cryostat sections of Deiters' nucleus, however, the S-100 protein was detected by an immunofluorescence technique in cell nuclei of Deiters' neurons and in cytoplasm of glial cells. This finding that S-100 protein was preferentially localized to glial cells and occurred only to a small extent in the neuron has been confirmed by other analytical methods (Cicero et al., 1970; Packman et al., 1971; Haglid et al., 1976). Other workers obtained indications by immunofluorescence microscopy that this protein was present in neuronal nuclei (Sviridov et al., 1972; Michetti et al., 1974). The cell bodies of several types of neurons were exposed by making a superficial incision on a tissue block and widening the fissure using a microknife. To prevent damage to the cell surface, care was taken not to touch it (Hydén and Rönnbäck, 1975a). The cell bodies were incubated in the presence of anti-S-100 conjugated with fluorescein isothianate (direct staining). The large rabbit neurons (Deiters' cells, trigeminal nerve cells, anterior horn cells, hypoglossal cells, and Purkinje cells) became unevenly fluorescent on the surface, and the small neurons (cortical pyramidal cells) were labeled by interrupted rings of fluorescence in the equator plane of the cell bodies. In the case of large neurons, the dendrites and axons were also unevenly fluorescent. Membrane preparations from Deiters' neurons formed patches of fluorescein staining, and neurons with a damaged plasma membrane showed specific fluorescence in the nuclei. Several factors (carbonyl cyanide, m-chlorophenylhydrazone, vinblastine, and cytochalasin B) known to disturb cap formation of lymphocytes did not influence the polarity of the fluorescence distribution. Incubation with univalent F_{ab} fragments of fluorescein-conjugated anti-S-100 antiserum (indirect staining) induced no alteration of fluorescence distribution, indicating the static distribution of S-100 rather than a cluster formation produced by the antibody. During postnatal development, S-100 protein appeared in glial cell samples from 4- to 5-day-old rats, 2- to 3-day-old rabbits, and newborn guinea pigs (Hydén and Rönnbäck, 1975b). The membrane-bound S-100 appeared later in rat neurons: at 10–14 days for

Deiters' cells, Purkinje cells, dorsal root ganglion cells, and trigeminal cells, but at 15–16 days for small neurons from the sensory-motor cortex. The uneven distribution of S–100 developed during the maturation process of Purkinje cells. Finally, the distribution patterns of these neurons approached the adult pattern but not until 1 month after birth. A species difference was found in this postnatal development: Deiters' neurons from newborn guinea pig already showed the heterogeneous distribution, and those from 10-day-old rabbit showed the first S–100 protein on their surface. These findings were thought to indicate that the heterogeneous and polar distribution of S–100 was a sign of unique differentiation of membrane proteins in each type of neuron.

The presence of S–100 on the surface of neurons was further confirmed by showing that fresh Deiters' neurons from rabbit bound to the surface of Sepharose 4B or methylacrylate particles to which purified S–100 antibody was conjugated (Hydén and Rönnbäck, 1978). S–100 antiserum was labeled with horseradish peroxidase and incubated with isolated fresh neurons from Deiters' nucleus (Haglid et al., 1974, 1976; Hansson et al., 1975). An uneven peroxidase reaction was observed by electron microscopy on the external surface of the neuronal plasma membrane; the postsynaptic membrane especially showed strong peroxidase activity. The nuclear membrane and nucleoplasm of the neurons exhibited peroxidase activity. The mitochondrial membrane also showed staining with peroxidase reaction product. Matus and Mughal (1975) demonstrated the immunoperoxidase reaction only in glial cell bodies in rat brain tissue sections, but a biochemical study proved that S–100 protein was tightly bound to the membranes of a synaptosomal fraction as well as to those of a mitochondrial fraction obtained from rat brain (Rusca et al., 1972). Therefore it seems likely that S–100 protein is present on the neuronal plasma membrane and synaptic membranous structure. In the neurons of the subesophageal ganglion of Helix pomatia, S–100 protein was found to be present on neuronal plasma membranes and in neuronal nuclei (Piven′ and Shtark, 1976).

S–100 seems to affect membrane permeability, as seen from the following experiments: Two microchambers were separated by a thin glass plate with a hole (25 μm in diameter) which was covered with a plasma membrane prepared from Deiters' neurons of rabbit (Hydén and Lange, 1977). [^{3}H]-γ-aminobutyric acid (GABA) was added in solution in the chamber on one side, and the transport of [^{3}H]GABA was studied by measuring the label in both chambers after incubation for 3 min at 29°C. S–100 protein, added with Ca$^+$, increased the transport rate, and this stimulated transport was further enhanced by the addition of K$^+$. A previous finding was that S–100 and Ca$^+$ increased the release of ^{86}Rb$^+$ from liposomes through an artificial lipid membrane (25% phosphatidylcholine, 75% phosphatidyl-serine) forming the liposomes (Calissano and Bangham, 1971).

During the last decade, it has become a general approach to analyze protein components of single invertebrate neurons, since their sizes are suitable for micro gel electrophoretic analysis. Protein components have been extracted with aqueous media from neurons in an abdominal ganglion of *A. californica* (Fig. 3) and separated on several kinds of gels depending on the sizes of the proteins. The ganglion was incubated in a medium containing [^{3}H]- and [^{14}C]amino acids for a long period (3–12 hr), and the radiolabeled proteins were detected by radioautography after separation on micro gel electrophoresis. Sodium dodecyl sulfate (SDS)- and acid urea–polyacrylamide gel electrophoresis revealed that many components of different molecular sizes were present in phenotypically different neurons. The R2 neuron (a large cholinergic and electrically silent neuron presumed not to be neurosecretory) contained a high relative amount of protein with a MW of 25,000 (25K) (Gainer, 1971; Wilson, 1971; Gainer and Wollberg, 1974; Loh and Peterson, 1974; Loh and Gainer, 1975a; Rüchel, 1976; Aswad, 1978). Similar silent neurons, R1 and left pleural ganglion cells, contained high-MW proteins and no peak of low-MW classes (Wilson, 1974). In some nonsecretory cells having endogenous electrical activities, a peak of 12K-MW protein was found (L3, L4, L6, L7, L10, L11; Loh and Peterson, 1974; Loh and Gainer, 1975a; Wilson, 1971, 1974; Berry, 1975). However, other endogenously firing cells with no secretory granules (L8 and L9) did not contain a 12K peptide; therefore it seemed unlikely that there was any correlation between pacemaker activity and the synthesis of 12K peptide (Berry, 1975). In silent cells (R3–R13, R14, and bag cells) and a firing cell (R15), which have putative neurosecretory granules, high relative amounts of 12K and 6K–9K classes of proteins were detectable (Gainer, 1971; Wilson, 1971, 1974; Gainer and Wollberg, 1974; Loh and Peterson, 1974; Berry, 1975; Loh and Gainer, 1975a; Rüchel, 1976; Aswad, 1978). The L5 neuron also contained presumptive neurosecretory granules and peptides lower than 12K in MW (Berry, 1975). With respect to bag cells, 25K protein was synthesized and degraded to 12K and 6K proteins in cell bodies (see the following discussion). The 6K protein had a pI of about 9.3 and was transported to the axon terminals of bag cells in the surrounding connective tissues. This protein was released together with other 12K–15K proteins by highly concentrated (about 110 mM) K$^+$ in the presence of Ca$^+$ (Arch, 1972; Loh *et al.*, 1975; Arch *et al.*, 1976a). The 6K protein was bioassayed and estimated to induce egg laying when injected into the animal's body (Toevs and Brackenbury, 1969; Arch *et al.*, 1976a). The R15 neuron also contained proteins between the 15K and 25K classes. Injection of these proteins into the hemocele of *Aplysia* was found to induce water intake into

the animal's body, suggesting that these classes of proteins were active hormones, or hormone and carrier molecules, involved in ionic regulation or
water balance (Kupfermann and Weiss, 1976). The low-MW proteins
($\leq$ 12K) were generally present in the cytoplasmic samples prepared from
the secretory neurons (Loh and Peterson, 1974; Berry, 1976a), and the proteins of R3–R13 neurons were included in neurosecretory granules (Gainer
and Wollberg, 1974). The 12K protein of L11 was isolated at a point corresponding to a MW on disc gel electrophoresis slightly higher than that of
R15 neurons (Wilson, 1974; Berry, 1976a), but slub gel electrophoresis gave
a clearer separation and the two components (13.1K and 12.5K) were found
in R15 cells as compared with a single component (12.8K) in L11 cells
(Aswad, 1978). Acid urea gel electrophoresis showed that differently
charged proteins were present in peaks of similar size (about 12K) in R15,
R3–13 cells, and bag cells (Loh and Gainer, 1975a; Rüchel *et al.*, 1977). An
isoelectric focusing study with a gradient of pH 3–10 indicated markedly
different patterns with R2, R3–R13, R14, R15, L11, and bag cells, (Gainer
and Wollberg, 1974). The 6K–9K proteins in bag cells had two peaks at pI
4.6 and 7. In R3–R13 cells, proteins of similar sizes had a large peak at pI
8.9. Proteins of higher MWs in all the neurons mentioned above were
thought to be actin-like protein (43K), tubulin (55K–58K), and presumptive
neurofilament protein (77K) (Loh and Peterson, 1974).

Axonal proteins from myelin-free axons of rabbit hypoglossal and dorsal
root nerves were solubilized in a 1% SDS solution (Frankel and Koenig,
1977). SDS micro gel electrophoresis showed that five main peaks (I, 620K;
II, 200K; III, 100K; IV, 62K; and V, 52K) were included in the solubilized
proteins. When a segment of dorsal root nerve was incubated in a medium
containing Ca^{2+}, the 100K component was converted to the smaller proteins
in the 50K–60K region. This conversion suggested that a calcium-activated
protease degraded the 100K component of neurofilament to 50K–60K components.

2. SYNTHESIS AND PROCESSING OF PROTEINS

An increase in protein content was previously observed in physiologically
stimulated single Deiters' neurons from rabbit (Hydén and Pigon, 1960),
but protein metabolism has not been investigated in single vertebrate
neurons because the amount of protein is too small to analyze at the single-
cell level. Bulk-isolated mammalian neurons, however, were found to have
a more active capacity than glial cells for protein synthesis *in vitro* (Lisý
and Lodin, 1973) and *in vivo* (Yanagihara, 1979), whereas glial cells were
reported to have a higher protein synthesis activity than neurons at certain
stages in development (Johnson and Sellinger, 1971). Autoradiography had

earlier demonstrated that proteins were synthesized more actively in rat and mouse neurons than in their glial cells, and that some of the synthesized proteins were catabolized in the perikarya, whereas others were transported out of the perikarya via axons (Droz and Leblond, 1963). In single invertebrate neurons (mainly *Aplysia* neurons) the modes of protein synthesis and processing were clearly demonstrated. Isolated abdominal ganglia were incubated for 1-2 hr in medium containing [^{3}H]- or [^{14}C]amino acids (leucine, arginine, etc.). This incubation to label newly synthesized proteins was followed by a chase incubation, where excess nonlabeled amino acids were added to determine how the labeled proteins were transformed and degraded (processing). The patterns of proteins synthesized *in vitro* in the R2 neuron were similar to those obtained from the R2 neuron after labeling of the whole animal (Wilson, 1971). The protein synthesis in *Aplysia* neurons was effectively inhibited by anisomycin, sparsomycin, and pactamycin, whereas other antibiotics effective on mammalian cells were ineffective (streptomycin, chloramphenicol, cycloheximide, and puromycin, Schwartz *et al.*, 1971). RNA synthesis inhibitors (actinomycin D and camptomycin) did not affect the pattern of protein synthesis during a short-term incubation (< 4 hr, Wilson, 1976). Dopamine (10 mM) caused a selective decrease in the synthesis of the 12K class of proteins, whereas an increased concentration of K$^+$ (60 mM) caused a selective increase in this class (Gainer and Barker, 1975).

In R2 neurons, relatively intense isotope labeling of proteins was observed in the region larger than 12K MW (Gainer and Wollberg, 1974; Loh and Peterson, 1974; Gainer and Barker, 1975). An elevated temperature (from 13°-15°C to 22°-30°C) changed the pattern of protein synthesis (Wilson, 1976), but no processing of proteins was observed in R2 neurons (Loh *et al.*, 1977). In some endogenously firing cells (L3, L6, and L7), there was a high rate of incorporation of labeled leucine into 12K protein (Loh and Peterson, 1974; Gainer and Wollberg, 1974; Berry, 1976a). In subcellular fractions, the incorporation was specifically high in the membrane fraction of L11 neurons (Loh and Peterson, 1974). In this cell type, 12K protein was found to be processed as follows: 12.8K → 12.3K → 9.2K-11K (Aswad, 1978).

With respect to the neurosecretory cells (R3-R13, R14, R15, and bag cells), labeled amino acids were incorporated dominantly into small proteins (< 12K classes, Loh and Peterson, 1974; Gainer and Wollberg, 1974; Wilson, 1974). In these neurons, similar processing patterns were observed in 12K and 9K-6K classes as follows: R15 cells: Loh and Gainer (1975b), Berry (1976b), Strumwasser and Wilson (1976), and Aswad (1977, 1978); R14 cells: Loh and Gainer (1975b), Loh *et al.* (1977), and Aswad (1978); L2-L6 cells: Loh and Gainer (1975b); bag cells: Arch *et al.* (1976b) and

Loh *et al.* (1977). The following schemes can be taken as examples for the processing of small proteins: R15 cells: 13.1K → 12.6K → 8.3K — 9.0K → 3K–6K (Aswad, 1977); R14 cells: 12.1K → 6.2K and 4.6K (Aswad, 1978); and bag cells: 29K → 11.3K → two 6K classes (Arch *et al.*, 1976b). The small classes of proteins were found in both soluble and membrane-bound forms in bag cells, and the soluble form increased with incubation time (Arch *et al.*, 1976b). The processing proved to occur in the neurosecretory granule fraction in R15 and R3–R14 neurons (Loh *et al.*, 1977). Addition of a protein synthesis inhibitor, anisomycin, to the incubation medium did not influence the processing (Strumwasser and Wilson, 1976; Berry, 1976b). The proteins accumulated in cell bodies when colchicine and vinblastine were added to the incubation medium (Loh and Gainer, 1975b; Loh *et al.*, 1977) and, when the cell body of a R15 neuron was separated from its axon and then chased in isolation, no decrease in 6K proteins was detectable (Loh and Gainer, 1975b). These findings suggested that the small proteins were transported through the axon as neurosecretory hormones in general neurosecretory cells as well as in bag cells (Arch *et al.*, 1976b). Acetate blocked an early step of the processing (Wilson, 1976), and Ca^{2+} had little effect on the processing or synthesis of proteins in R15 neurons (Aswad, 1978). Similar processing and transport of low-MW proteins were observed in a no. 11 neurosecretory cell of the land snail *Octa lactea* (Gainer, 1971; Loh *et al.*, 1976).

In a segment of single unmyelinated giant axon from squid, quantitative and autoradiographic studies indicated that proteins were not synthesized in the axon but were synthesized in sheath cells (Schwann cells) and transported to the axoplasm (Lasek *et al.*, 1974). This was thought to offer an example of macromolecule transfer between glial cells and neurons (Gainer, 1978). Microsamples of pure axons were obtained by extruding axoplasm from mammalian nerve stumps (Koenig, 1967). When myelinated roots of hypoglossal nerves from rabbits were incubated in [³H]leucine-containing medium, the label was incorporated into the proteins of the pure axons, and this incorporation was inhibited by cycloheximide but not by chloramphenicol (Koenig, 1967; Tobias and Koenig, 1975a), which indicated that an extramitochondrial mechanism for protein synthesis was present in the axon and operated independently of protein synthesis in the cell body. This synthesis activity was enhanced more than 20-fold in the neurotomized nerve stumps. Separation of the cell body at the time of neurotomy did not influence this synthesis; and colchicine had no effect when added to the incubation medium (Tobias and Koenig, 1975b). In non-neurotomized hypoglossal nerves the axonal proteins synthesized *in vivo* had main peaks on SDS micro gel electrophoresis at MWs of 100K, 52K, and 18K. In contrast, the 100K peak disappeared and a new peak at 41K was detected in the neurotomized nerves (Frankel and Koenig, 1977, 1978). When [³H]leucine was given *in vivo* to the neurotomized stump by microinjection, the newly synthesized proteins were not transported back to the cell body by retrograde axonal transport (Frankel and Koenig, 1978). This suggested that these locally synthesized proteins were components of neurofilaments and microfilaments. In a single axon of *Aplysia* neuron R2, a glycoprotein, labeled by microinjection of [³H]fucose, was transported along the axon in several discrete waves (Ambron *et al.,* 1974). This mode of transport contrasted with the axonal transport pattern in a smooth wave front, which was observed in mammalian nerve composed of thousands of individual axons.

3. Synthesis in Stimulated Neurons

In addition to the quantitative study of protein synthesis in stimulated single Deiters' neurons of rabbit mentioned above (Hydén and Pigon, 1960), the protein content in the neurons of the sympathetic ganglion of the cat was found by microscopic photometry to increase after electrical stimulation for 3 hr and to decrease 2 weeks after epinephrine injection (Pevzner, 1965). Acute Cardiazol convulsion induced a change in protein content level in anterior horn cells of rat spinal cord and, finally, the content was decreased by 20% during the rest period after convulsion. Reciprocal changes were observed in perineuronal glia cells (Pevzner, 1971).

In invertebrate neurons, the relations between protein synthesis and neurophysiological activity of neurons were investigated. An abdominal ganglion of *A. californica* was incubated in medium containing [³H]leucine and a protein synthesis inhibitor (anisomycin). Whereas protein synthesis was arrested, several electrophysiological properties of R2, R15, and L7 cells (resting potentials, spike generation, endogenously firing patterns, and synaptic transmission) were not changed (Schwartz *et al.,* 1971). Two presumptive short-term plastic changes in synaptic function (postsynaptic potentiation in R15 neurons and habituation and dishabituation observed in excitatory postsynaptic potential amplitude of L7 neurons) also were not changed under the same conditions. The protein synthesis rate of an R2 neuron was studied as above in the absence of the inhibitor, and the neuron was stimulated transsynaptically by applying electric current to the left connective (Wilson and Berry, 1972). The pattern of incorporation of label into the protein fraction on SDS gel electrophoresis was not different in the stimulated and unstimulated neurons. In contrast, when inhibitory postsynaptic potentials were evoked transsynaptically in R15 neurons by stimulating the branchial nerve, the incorporation rates of [³H]leucine into

12K peptides were found to be decreased significantly. In a medium containing 10^{-3} M dopamine, which hyperpolarized the membrane potential, the synthesis rate of 12K peptide was decreased (Gainer and Barker, 1974). Under the same conditions, the 6K proteins showed a tendency to decrease in synthesis rate. The decay rate (2 hr half-time) of 12K protein was unchanged or decreased; therefore the synthesis rates of these proteins were thought to decrease in the presence of dopamine. This suggested that these proteins with a high turnover rate may be modified transsynaptically.

4. METABOLISM IN BEHAVIORAL TASKS

A mass of 300 neuronal cell bodies (0.5 μg of protein) in the CA3 region of rat hippocampus was analyzed in rats trained to use their unpreferred paws to obtain food pills (see Section IV,A,4). [^{3}H]leucine, injected into the lateral ventricles, was incorporated much more into the two protein fractions on micro gel electrophoresis in trained rats as compared with active control rats using their preferred paws (Hydén and Lange, 1968, 1970a). These fractions migrated close to S–100 protein on 25% acrylamide micro gel electrophoresis and were actively synthesized in an early period of training. This synthesis was more dominant in pyramidal cells in the CA3 region than in the CA1 and CA4 regions in the same hippocampus (Yanagihara and Hydén, 1971), and it was again enhanced when the animals were forced to use their original preferred paws after being trained to use their unpreferred paws (Hydén and Lange, 1972). These fractions were thought to contain a brain-specific protein, 14–3–2. Synthesis of S–100 protein was also stimulated in CA3 region pyramidal cells of rats trained for a long period (Hydén and Lange, 1970a). As a result of training, a new S–100 peak appeared ahead of the frontal band of untrained rats containing S–100 and other proteins (Hydén and Lange, 1970b). S–100 protein was demonstrated in the nuclei and cytoplasm of pyramidal cells in the CA3 region by using antibody against S–100 (Coon's double-layer method) injected into the lateral ventricles. Antiserum against S–100 protein, when injected into the lateral ventricles, specifically inhibited the acquisition of new behavior in the rats mentioned above (Hydén and Lange, 1970b) and in rats tested on a maze-learning task (Karpiak *et al.*, 1976). These findings at the cellular level are in good accordance with the increased incorporation of intraperitoneally injected [^{3}H]leucine into the subcellular fractions (microsomes, synapotosomes, and synaptosomal and mitochondrial membranes, Levitan *et al.*, 1972) and into 30K and 80K classes of membrane-bound proteins of synapses (Hydén *et al.*, 1977) when these fractions were prepared from the hippocampus of trained rats. Activation of protein synthesis in special classes was also demonstrated in the whole brain of goldfish (Shashoua, 1976) and in the whole ganglion of *Helix aspersa* (Emson *et al.,* 1971) when these animals were trained in new behavioral tasks; and this supports the concept of synthesis of special proteins in neurons during training experiments.

C. Lipids

One to eight fresh neurons, isolated from the hypoglossal nucleus of beef brain stem, were assayed for total phospholipid content in terms of fluorescence developed in a reaction with rhodamine 6GO. The contents were distributed in a wide range from 7.2 to 32.2 ng per cell, since each cell sample retained dendrites to a different extent (Schiefer and Neuhoff, 1971). Deiters' neurons were isolated from ox brain stem, and 50–70 cell bodies and glial samples were collected. Simultaneously the neuropil samples, being rich in terminal axons and synapses, were teased from immediately around the neuronal perikarya (Derry and Wolfe, 1967). The collected samples were dried and weighed. Ganglioside was extracted from the weighed samples with organic solvents. The highly fluorescent quinaldines formed in the reaction between ganglioside and 3,5-diaminobenzoic acid were measured to determine the ganglioside content of these samples: neuronal perikarya, 10.7 mg/gm of dry weight (corresponding to 358 ng/cell if the average dry weight of the cell is 36 ng); neuropil sample, 13.0 mg/gm; glial cell sample, 1.5–2.9 mg/gm. The high concentrations in neuronal perikarya and neuropil samples suggested that ganglioside was concentrated in their main component, synaptic membrane. Since the size of the hypoglossal neuron is about one-fifth that of Deiters' cell in ox brain (see Section III,A), the ratio of ganglioside to phospholipid content is 0.01–0.002. The same ratio calculated from results obtained from bulk-isolated neuron-rich samples is between 0.01 (Hamberger and Svennerholm, 1971) and 0.004 (Norton and Poduslo, 1971). These values coincide well with each other.

D. Enzymes

1. Enzymes Involved in Cellular Metabolism

A large number of enzymes have been determined in single neurons as indices of functions and properties of neurons (for a tabulation of data, see Brand and Lehrer, 1971). Sometimes the enzymes were selected for analysis only because microdetermination methods were available at the cellular level.

Glycolytic and other enzymes related to energy metabolism were determined with freeze-dried neuronal perikarya from rabbits (Lowry *et al.*, 1956; Lowry, 1957; Kato and Lowry, 1973a) and monkeys (Robins, 1960). The specific activities of nine enzymes on a lipid-free dry weight basis (hexokinase; phosphoglucoisomerase; lactate, glucose 6-phosphate, 6-phosphogluconate, isocitrate, malate, and glutamate dehydrogenases; and glutamate oxalacetate transaminase) were remarkably uniform in rabbit and monkey spinal cord anterior horn cells (Robins, 1960). Eight species of rabbit neurons had characteristic activity patterns of enzymes, but these patterns appeared to be divided into four types (Fig. 4). One type was represented by motor neurons (anterior horn cells and facial nucleus cells) and dorsal root ganglion cells, the latter showing an exceptional high acti-

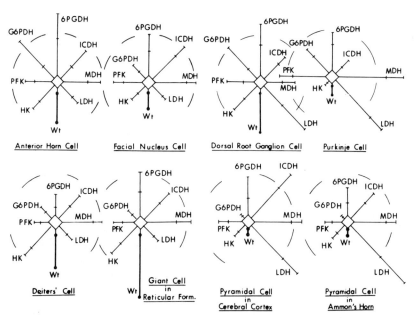

Fig. 4. Relative enzyme activities in single nerve cell bodies from rabbit (from Kato and Lowry, 1973a). The enzyme activities of the molecular layer of rabbit cerebellum were used as the basis of comparison and are represented by the dashed circles. The lengths of the straight lines from the central square to the middle points between the cross marks represent relative activities of the type of cell bodies shown. The cross marks indicate plus and minus standard deviations. The weight of the samples is given on the vertical line as the distance between two dots indicating the range of dry weights. The length of the diagonal line of the central square corresponds to 2 ng. HK, Hexokinase; PFK, phosphofructokinase; G6PDH, glucose-6-phosphate dehydrogenase; 6PGDH, 6-phospho-gluconate dehydrogenase; ICDH, isocitrate dehydrogenase; MDH, malate dehydrogenase; and LDH, lactate dehydrogenase. Eight types of neurons seem to belong to the four different categories of enzyme patterns (see text).

vity of lactate dehydrogenase. The second type was exhibited by pyramidal cells from cerebral cortex and Ammon's horn, which were similar in morphology and developed from telencephalon phylogenetically and ontogenetically. As the third type, a similarity in the enzyme pattern was seen between giant cells in reticular formation and Deiters' cells. The last unique pattern was observed in Purkinje cells. This individuality of neurons found in enzymes catalyzing fundamental processes in energy metabolism may detract from the value of data on bulk-isolated neuronal fractions. Two kinds of dorsal root ganglion cells were observed as light and dark in freeze-dried tissue sections, but no essential difference was detectable in the specific activities of the above-mentioned nine enzymes (Lowry, 1957). The specific activities of isocitrate and glucose-6-phosphate dehydrogenases in individual perikarya of anterior horn cells had bimodal or trimodal distributions, whereas a single peak of these activities was observed in the unimodal distribution in facial nucleus cells (Fig. 5; Kato and Lowry, 1973a). This distribution of activities in anterior horn cells coincides with the finding by Fernandez-Campa and Engel (1970) that there were two groups of alpha motor neurons, as judged by staining for succinate dehydrogenase and phosphorylase. When the above-mentioned enzyme activities were compared in neuronal perikarya and the surrounding neuropils from rabbit (dorsal root ganglion cells and the ganglion cell capsule or fibers, Lowry, 1957; Purkinje cells and the cerebellar molecular layer, Kato and Lowry, 1973a), there was no constant relation between the magnitudes of their activity values. Four components of lactate dehydrogenase isozymes were found on disc electrophoresis in fresh nerve cell bodies from rat (pyramidal cells in Ammon's horn; cells in the trigeminal motor nucleus and Deiters' cells) but, in contrast, the granular layer of the cerebellum (including granule cells, neuropil, and glial cells) contained the fifth component of the isozyme (Hazama and Uchimura, 1970).

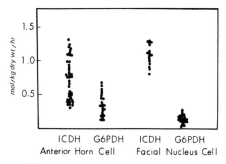

FIG. 5. Distribution of the activities of isocitrate dehydrogenase (ICDH) and glucose-6-phosphate dehydrogenase (G6PDH) in anterior horn cells of the spinal cord and facial nucleus cells from rabbit. (From Kato and Lowry, 1973a.)

The activities of lysosomal enzymes (β-galactosidase, β-glucuronidase, and β-glucosidase) were measured in freeze-dried perikarya of anterior horn cells and dorsal root ganglion cells from monkeys. Dorsal root ganglion cells had higher activities of these enzymes than anterior horn cells (Robins and Hirsch, 1968). The activities of acid phosphatase (Hirsch, 1968) and arylsulfatase (Hirsch, 1969) were 10-fold higher in the anterior horn cells of monkey and human lumbar spinal cords than in their surrounding neuropil. These data support the view based on microscopic histochemistry that lysosomes are assembled in the neuronal perikaryon (Koenig, 1969).

Carbonic anhydrase was specifically contained in fresh glial samples prepared from around Deiters' cells of hooded rats (Giacobini, 1961). The activity per unit volume was 120-fold higher in glial cell samples than in neuronal perikarya. Therefore this enzyme has been used as a marker enzyme for glial cell contamination in the bulk-isolated neuronal fraction.

Vestibular stimulation, by rotating rabbits 25 min every day for 7 days, enhanced succinoxidase and cytochrome oxidase activities in fresh Deiters' neurons and reduced both activities in glial cells (Hydén and Pigon, 1960). Simultaneously, anaerobic glycolysis was decreased by 23% in neuronal perikarya and increased by 72% in glial samples (Hamberger and Hydén, 1963). Rabbits were placed in a mixture of 8% oxygen and 92% nitrogen gasses for 15 min to be subjected to hypoxic conditions. Under these conditions, aerobic glycolysis in both neuronal and glial samples was decreased. Concomitantly cytochrome oxidase activity was increased in Deiters' neurons but was not changed in the glial samples. This enzyme activity showed a circadian oscillation (high activity during sleep and low activity during wakefulness) in neurons isolated from the rostral and caudal parts of the reticular formation in rabbit brain stem (Hydén and Lange, 1965b). In the glial samples from caudal reticular formation, the reverse change was observed. Succinoxidase activity of rabbit Deiters' cells was decreased after the animal was injected with GABA but was increased by the injection of hydroxylamine and thioisosemicarbazide (Aleksidze and Blomstrand, 1968). This activity in the glial cell samples from around Deiters' neurons was decreased by GABA injection. When a rabbit was given tricyanoaminopropane, which increased RNA content, cytochrome oxidase activity of Deiters' neuron was increased to 300% of the activity of the control neuron (Hamberger, 1963). These results indicate that the activities of the oxidases reflect the functional states of neurons.

Whole cell bodies of Deiters' neurons, carefully isolated from rabbit brain stem, were incubated with ATP. ATP was not hydrolyzed in a detectable amount. When the cell bodies were homogenized, ATPase activity was detectable in these neuronal samples (Cummins and Hydén, 1962). A plasma membrane sample, weighing about 10% of the whole cell body, contained 25% of the total activity in the whole cell body. The glial samples prepared from around Deiters' neurons or around the capillaries contained higher ATPase activity per unit volume than neuronal perikarya (Hillman and Hydén, 1965b). The ATPase activity of neuronal plasma membrane was activated by Mg^{2+}, and further activation was observed in the presence of Na^+ and K^+ (Cummins and Hydén, 1962). The activity was completely inhibited by $2 \times 10^{-5} M$ ouabain. These findings suggested that Na^+, K^+-ATPase participating in ion transport was localized to neuronal plasma membrane.

2. ENZYMES RELATED TO METABOLISM OF NEUROTRANSMITTER CANDIDATES

The enzymes catalyzing the synthesis reactions of neurotransmitter candidates have been analyzed as indices of which transmitter is utilized by a specific neuron for chemical transmission. Sometimes the enzymes degrading transmitters were determined for the same purpose, although the degradation enzymes proved to be nonspecific as such indices.

Among the fresh neurons from cat L7 sympathetic ganglia, 2 of 20 cells contained choline acetyltransferase [CAT, acetylcholine (ACh)-synthesizing enzyme] and were considered cholinergic neurons (Buckley et al., 1967). Almost all the cells contained acetylcholinesterase (AChE), but the specific activities were distributed in a wide range from 0 to 400 pmol of ACh hydrolyzed per hour per cell (Koslow and Giacobini, 1969). This indicates that AChE is not a specific marker for cholinergic transmission and that the sympathetic ganglion cells are heterogeneous. These cells from rat and frog had a wide variety of AChE activities and were also composed of a heterogeneous population (Giacobini, 1957). Markedly high CAT activity was found in microsamples of the inner nuclear layer and the inner plexiform layer of retina from eight different animals (cat, monkey, mouse, rabbit, frog, turtle, garfish, and goldfish, Ross and McDougal, 1976). Therefore the cholinergic neurons in the retina were thought to be found predominantly among amacrine cell types.

ACh is synthesized by CAT with acetyl-CoA and choline as substrates. Therefore, five enzymes synthesizing acetyl-CoA were determined in freeze-dried samples of the anterior horn cells of rabbit spinal cord chosen as typical cholinergic neurons. For comparison, the same enzyme activities were measured in dorsal root ganglion cells as representatives of noncholinergic neurons (Hayashi and Kato, 1978). ATP citrate lyase activity was higher in anterior horn cells than in dorsal root ganglion cells (Fig. 6). This enzyme activity was high in the stratium of mouse and the cerebellar

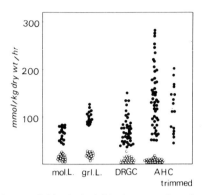

FIG. 6. ATP citrate lyase activities in individual nerve cell bodies and cerebellar layers of rabbit (from Hayashi and Kato, 1978). The activities of the molecular layer (mol.L.) and the granular layer (grl.L.) are plotted for comparison. These plots show the distribution of activities in samples of layers that are structurally homogeneous. The activities of dorsal root ganglion cells (DRGC) and anterior horn cells (AHC) of the spinal cord disperse in a wider range than those of the layer samples, which suggests that there are individual differences in biological properties in the case of single neurons. The peripheral portions of anterior horn cell bodies were trimmed to eliminate the plasma membrane and the synapses on the membrane. The average activity of untrimmed cell samples (134 ± 9 mmol/kg dry weight per hour) was statistically not different from that of trimmed cell samples (116 ± 12 mmol/kg dry weight per hour). ○, Blank activity value; ●, enzyme activity plus blank activity. Therefore the true enzyme activity equals the difference between the two values. (For the meaning of the blank value, see the original paper.)

granular layers of mouse and rabbit. These structures are known to be abundant in cholinergic interneurons, synapses, and fibers (Goldberg and McCaman, 1967; McGeer et al., 1971; Butcher and Butcher, 1974; Kan et al., 1977). The high concentration of ATP citrate lyase in cholinergic neurons suggested that this enzyme may play an important role in supplying acetyl-CoA for ACh synthesis in these neurons. The widely scattered specific activities of ATP citrate lyase in a group of anterior horn cells were thought to reflect the various levels of cell activity in these motor neurons (Fig. 6).

The activity of glutamate decarboxylase (GAD, an enzyme synthesizing GABA) was concentrated in freeze-dried samples of the peripheral portion of the cell bodies of dorsal Deiters' neurons of rabbit (Y. Okada, unpublished; see the data in Table 1 cited in Wu, 1978). This result indicated the termination on Deiters' neurons of GABA-ergic axons from Purkinje cells.

The marker enzymes for synthesis of neurotransmitter candidates have been measured in single invertebrate neurons, since the sensitivity of the assay methods is high enough for such analyses. CAT activity was found to be high in R2, L10, L11, and left pleural ganglion cells of *A. californica* (McCaman and Dewhurst, 1970; Giller and Schwartz, 1968, 1971a) and in the sensory nerve fibers of the shore crab (Emson *et al.*, 1976) and lobster (Barker *et al.*, 1972). This enzyme was localized to the sensory neurons of metathoracic ganglion of the locust (Emson *et al.*, 1974). The large neurons found near Retzius cells in the leech ganglion contained relatively high CAT activity (Coggeshall *et al.*, 1972). When [^{3}H]choline in 1–3 nl of solution was injected into neuronal cell bodies of *Aplysia*, [^{3}H]ACh was synthesized in these ACh-containing cells (R2, L10, and, L11 cells) and in the LD cluster composed of motor neurons to heart and gills, but [^{3}H]ACh was not synthesized in L1, L3, L14, and R15 neurons lacking ACh (Eisenstadt *et al.*, 1973). [^{3}H]choline injected into the axon of an R2 neuron was converted to [^{3}H]ACh (Treistman and Schwartz, 1974, 1977). Aromatic amino acid decarboxylase [AAD, an enzyme responsible for the synthesis of dopamine and serotonin (5–HT) from 3,4-dihydroxyphenylalanine (dopa) and 5-hydroxytryptophan (5–HTP), respectively] was present at highly active levels in 5–HT-containing neurons [Retzius cells of leech ganglion, Coggeshall *et al.*, 1972; giant serotonin-containing (GSC) cells of *H. pomatia*, Cottrell and Powell, 1971; cerebral neuron 1 (C–1) and pedal neurons (Pd-1) of *Tritonia diomedia*, Weinreich *et al.*, 1973]. Among *Aplysia* neurons, AAD activity was ubiquitously found (Weinreich *et al.*, 1972), but cerebral neuron 1, containing 5–HT, had a 10-fold higher activity than other neurons lacking 5–HT (Weinreich *et al.*, 1973). [^{3}H]Tryptophan, injected into 5–HT-containing neurons (the RB cluster composed of 20–25 cells present in the right caudal quadrant of the abdominal ganglion of *Aplysia*) was converted to [^{3}H]5–HT effectively, though [^{3}H]tryptophan was incorporated concurrently into proteins (Eisenstadt *et al.*, 1973). Among the GSC cells of *H. pomatia*, some neurons contained both CAT and AAD, indicating the possibility that two transmitters were utilized in a single neuron (Hanley *et al.*, 1974). In contrast, Osborne (1977) reported that the GSC cells contained only 2% of the CAT activity of cell 21 in the visceral ganglion of *Helix* and concluded that such trace activity of CAT in GSC cells might be attributable to the contamination of cell samples by neuropil or glia cells. GAD activity was specifically high in single axons of inhibitory nerve from lobster walking legs (Hall *et al.*, 1970) and in promotor and remotor nerves of the shore crab (Emson *et al.*, 1976). This enzyme was also high in inhibiting motor neurons of locust ganglia (Emson *et al.*, 1974).

Two enzymes degrading two putative transmitters (AChE and catechol-*O*-methyl transferase) were not specifically distributed in neurons of *A. californica* (McCaman and Dewhurst, 1971; Giller and Schwartz, 1971b). AChE was also found ubiquitously among motor neurons of the locust

(Emson *et al.*, 1974) and in sensory and motor neurons of the shore crab (Emson *et al.*, 1976). These findings support the fact that the degrading enzyme is nonspecific as a marker for neurotransmitter utilized in neurons.

3. SUBCELLULAR ANALYSIS

Direct analysis of subcellular components of single neurons can avoid the redistribution of enzymes and substrates brought about by the cell fractionation technique as a result of homogenization. The nucleoplasm of anterior horn cells from rat spinal cord contained on a volume basis about 1/10 the AChE activity found in the cytoplasm (Giacobini, 1959a,b). Little activity was found in the nucleolus. The axon and dendrite contained lower activities than the cell body. The enzymes involved in energy-converting systems were present in freeze-dried samples of a single nucleus as well as in samples of cytoplasm of rabbit dorsal root ganglion cells (Lowry, 1963; Kato and Lowry, 1973b). The activities of hexokinase, 6-phosphogluconate dehydrogenase and isocitrate dehydrogenase were higher in the nucleus than in the cytoplasm, but the activities of phosphofructokinase and glucose-6-phosphate dehydrogenase were found to be lower in the nucleus. Malate and lactate dehydrogenase activities showed no differences in nucleus and cytoplasm (Kato and Lowry, 1973b). Phosphofructokinase activity also did not differ (Lowry, 1963). Glutamate dehydrogenase proved to be present in the cell nucleus, although this enzyme had been generally thought to be a mitochondrial enzyme. ATP:NMN adenylyltransferase catalyzing the formation of NAD^+ was found to be more active in the peripheral portion of cytoplasm than in the portion close to the nucleus, and a very low activity was found in the nucleus (Fig. 7). Some of the above-mentioned enzymes were found in a single nucleus of freeze-dried supramedullary neurons from the puffer fish (Lehrer *et al.*, 1968). β-Glucuronidase activity was not present in the nucleus.

4. PATHOLOGICAL CHANGES

The lumbosacral motor root of monkey spinal cord was cut, and the freeze-dried anterior horn cells were isolated 7 or 35 days after root section (Robins *et al.*, 1961). Compared with the cells from the unoperated side, used as controls, malate, lactate, and isocitrate dehydrogenase activities were not changed. In contrast, the level of glucose-6-phosphate dehydrogenase was increased about threefold at day 35 when functional recovery of the axon was thought to begin, whereas it was not yet changed at day 7 when new axons were thought to grow out through the degenerating tissue and the scar (Brattgård *et al.*, 1957).

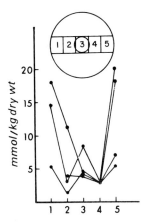

FIG. 7. Intracellular distribution of ATP:NMN adenylyltransferase activity in single dorsal root ganglion cell bodies from rabbit spinal cord. Each cell body was trimmed to eliminate the plasma membrane, and a strip was dissected out along the axis through the center as shown in the diagram. The numbers on the abscissa indicate the position of pieces along the strip (number 3 is the nucleus). The sum of the NAD$^+$ formed in a 1-hr reaction and the native NAD$^+$ is shown on the ordinate. (From Kato and Lowry, 1973b.)

In the spinal motor neurons of patients with amyotrophic lateral sclerosis, no marked or reproducible changes were found in the levels of eight enzymes (malate, lactate, glutamate, and glucose-6-phosphate dehydrogenases; hexokinase; β-galactosidase; β-glucuronidase; and acid phosphatase; Hirsch and Chen, 1969).

E. Low-Molecular-Weight Substances

1. METABOLITES WITH A RAPID TURNOVER RATE AND AMINO ACIDS

The concentration levels of ATP, phosphocreatine, glucose, and glycogen were determined in freeze-dried samples of anterior horn cells, dorsal root ganglion cells, and neuropil around both types of neurons. In order to stop the rapid turnover of these substances, the whole body of a mouse was immersed in Freon-12 chilled with liquid nitrogen at 0 and 40 sec after decapitation, by which means the blood supply was cut off to induce ischemia in the spinal cord. The animals were anesthetized with pentobarbital to slow down the metabolic rates, and the samples from anesthetized animals were analyzed simultaneously (Passonneau and Lowry, 1971). Generally, the levels of the above four substrates were higher in neuronal perikarya of anesthetized animals, where they decreased scarcely or to a lesser extent during ischemia. In anterior horn cells, phosphocreatine (32 mmol/kg of lipid-free dry weight) and glucose (68 mmol/kg) were almost consumed during 40 sec of ischemia. In contrast, the levels of ATP and glycogen (17 and 34 mmol/kg, respectively) decreased to about half the original level during ischemia. In the neuropil samples around anterior horn cells, the levels of the four substances were similar to those in neuronal cell bodies. The changes in high-energy phosphate during ischemia were calculated from the changes in phosphocreatine concentration plus 1.4 times the ATP concentration.(This gives a close approximation for early changes in the energy level.) Anterior horn cells lost high-energy phosphate more than twice as fast as dorsal root ganglion cells. In neuropil samples, the change in energy level took place to an extent similar to that in dorsal root ganglion cells.

The concentrations of glycine, glutamate, GABA, and aspartate were measured in freeze-dried samples of anterior horn cells, dorsal root ganglion cells, and neuropil from rabbit spinal cords (Berger *et al.*, 1977). Both neurons contained similar levels of glycine (61–65 mmol/kg of lipid-free dry weight) and glutamate (51–52 mmol/kg). GABA and aspartate were more concentrated in the anterior horn cells (42 and 8.7 mmol/kg, respectively) than in the dorsal root ganglion cells (20 and 1.7 mmol/kg, respectively). The neuropil samples around these neurons contained concentrations of the four amino acids similar to those in the neuronal cell bodies. GABA was present at a higher level in fresh cerebellar nucleus cells (8.0 mM) and Purkinje cells (5.8 mM) of cats (Otsuka *et al.*, 1971). GABA concentration in dorsal Deiters' cells (6.3 mM) was higher than in ventral Deiters' cells (2.7 mM). The removal of cerebellar vermis, whose Purkinje cells form inhibiting synapses on only dorsal Deiters' cells, resulted in a 73% reduction in the GABA concentration in dorsal Deiters' cells but had no effect on ventral Deiters' cells. This indicated that GABA was an inhibiting neurotransmitter in the synaptic terminals of Purkinje cells on dorsal Deiters' cells. The freeze-dried samples of the peripheral portion of rabbit dorsal Deiters' neurons (including synapses from Purkinje cells) proved to contain a higher concentration of GABA (32.2 mmol/kg of dry weight) than the central portion of the cell bodies (16.4 mmol/kg, Okada and Shimada, 1976). This supports the results reported by Otsuka *et al.* (1971) on fresh neurons. When Deiters' neurons were incubated with several amino acids and oxygen consumption by the neurons determined, glutamate stimulated oxygen consumption more than twice as much as pyruvate and malate (Hamberger, 1961). Inversely, pyruvate and malate stimulated the oxygen consumption of glial cell samples surrounding the neurons more than glutamate. The extent of stimulation by α-ketoglutarate was approximately equal in both the neuronal and glial samples.

Metabolic changes in electrically stimulated neurons have been analyzed in invertebrate neurons. During a period of prolonged mechanical stimulation, i.e., 8–16 × 10^4 impulses, of the slowly adapting cells of the crayfish stretch receptor organ, the concentration levels of glycogen and ATP gradually decreased by 20%, with a rapid increment in inorganic phosphate and lactate (Giacobini and Grasso, 1966). The ADP level increased until 12 × 10^4 pulses were delivered but thereafter decreased. A marked increase (200%) in glucose-6-phosphate and a rather constant level of glucose suggested that the glycolytic control might work at the phosphorylase step. When the stretch receptor cells were incubated with several tricarboxylic acid cycle intermediates and glutamate, four intermediates (citrate, isocitrate, α-ketoglutarate, and succinate) increased the frequency of spikes, but glutamate reduced the frequency. Fumarate, malate, oxalacetate, and pyruvate did not influence the frequency (Giacobini and Marchisio, 1966). The above-mentioned four intermediates that increased the frequency, as well as malate and glucose, stimulated the oxygen consumption of this neuron. The other intermediates had no effect on the oxygen consumption. The GSC cell of *H. pomatia* was stimulated by applying electric current to the external lip nerve, and simultaneously the neuron was perfused *in situ* with snail saline containing [^{14}C]glucose and [^{14}C]glutamate (Osborne, 1972). The electrical stimulation resulted in two new spots on two-dimensional microchromatography and increased incorporation of the label from [^{14}C]glucose into an unknown substance; incorporation into alanine and into another unknown spot was also increased. When [^{14}C]glutamate was used as the substrate, the intensity of radiolabel in the spots of alanine, glutamine, and an unknown substance were increased by electrical stimulation.

Among *Aplysia* neurons, glutamate and glutamine were distributed ubiquitously (Borys *et al.*, 1973; Iliffe *et al.*, 1977), whereas high amounts of aspartate were found specifically in R2 and R14 cells. The concentration of aspartate was higher in L1, B3, and R4–R9 cells than in other neurons (Zeman and Carpenter, 1975). The most remarkable finding was that R3–R13 and R14 cells contained an extremely high amount of glycine (580 and 1400 pmol/cell, respectively, Iliffe *et al.*, 1977). These R3–R14 cells were shown autoradiographically to incorporate glycine from the bathing medium actively (Price *et al.*, 1978). Uptake did not occur in the absence of Na^+ and was inhibited by Hg^{2+} (McAdoo *et al.*, 1978). GABA was taken up into *Aplysia* ganglion, but light microscope autoradiography showed that accumulation sites of silver grains were predominantly extraneuronal (Zeman *et al.*, 1975). GABA was found specifically in motor nerves and scarcely in sensory nerves of the shore crab (Emson *et al.*, 1976). Seven other amino acids (glutamate, glutamine, aspartate, alanine, glycine, proline, and taurine) were present ubiquitously in sensory and motor nerves.

2. ACETYLCHOLINE AND BIOGENIC AMINES

These substances have been determined to be neurotransmitter candidates in invertebrate neurons, because assay methods are not sensitive
enough to analyze vertebrate neurons quantitatively. Especially, the
necessity has not been felt to analyze the amines in vertebrate neurons
because the histochemical fluorescence method (Falck, 1962) was successfully applied to the mapping of monoamine neurons and pathways in
the mammalian central nervous system (Ungerstedt, 1971).

ACh was found in *Aplysia* neurons containing CAT (R2, L10, L11, and
left pleural ganglion cells), and its concentration was nearly the same (0.35
mM) in all these neurons (R. E. McCaman *et al.,* 1973). Other neurons
(R14, R15, L2–L6, and C–1 cells) did not contain ACh at an assayable
level, but choline concentrations were similar in the above neurons.
[^{3}H]ACh was synthesized from [^{3}H]choline and [^{3}H]acetyl-CoA in the R2
neuronal cell body, and a small portion of synthesized ACh was included in
particles corresponding in size and density to presumed cholinergic vesicles
(Eisenstadt and Schwartz, 1975). The [^{3}H]ACh in a sedimentable fraction
was transported intraaxonally in apparently two components at rates of 17
and 55 mm/day (Koike *et al.,* 1972). [^{3}H]ACh, formed in the axon by
microinjection of [^{3}H]choline, moved in the axon with the kinetics expected
for diffusion (Treistman and Schwartz, 1977). Choline was known to be incorporated for ACh synthesis by a high-affinity process into primarily
neuropil and nerves and not into the R2 cell body when the ganglion was
incubated with [^{3}H]choline (Eisenstadt *et al.,* 1975). In the GSC cell of *H.
pomatia,* ACh was found at 1/10 the level of the concentration of 5–HT
(Hanley and Cottrell, 1974). About 74% of the total ACh in the isolated
cell was detected in the contents (cytoplasm plus nucleus) of the cell bodies.
The remainder was found in the plasma membrane sample, probably contaminated with surrounding small neurons and glial cells (Cottrell, 1977).
This is an example of the possible presence of two kinds of neurotransmitters in a single neuron (see Section V).

Amines have been well studied at the cellular level in molluscan neurons.
5–HT was found to be present at a high concentration in the giant neurons
in the cerebral ganglion (GSC cell, metacerebral cell, or C–1 cell) of *A.
californica* (Weinreich *et al.,* 1973; Brownstein *et al.,* 1974), *A. dactylomela* (Cottrell, 1974) and *T. diomedia* (Weinreich *et al.,* 1973). The
pedal neuron of *Tritonia* (Pd–1) contained a high amount of 5–HT
(Weinreich *et al.,* 1973). In the C–1 cell of *A. californica,* 5–HT was synthesized from 5–HTP and stored in particles which could be obtained by
cell fractionation (Goldman and Schwartz, 1974). This particulate form
was the dense-core vesicle which could be shown to be transported along

axons (Goldman et al., 1976). The transport velocity was 70–120 mm/day and dependent on the movement of vesicles, which could be approached theoretically by modification of Michaelis–Menten kinetics (Goldberg et al., 1976). 5–HT was synthesized from 5–HTP in the ACh-containing neuron R2, but it remained soluble in the cytoplasm and diffused through axons (Goldman and Schwartz, 1974). 5–HTP was also transported by diffusion along axons (Goldman et al., 1976). Direct electrical stimulation of the usually silent C–1 neurons resulted in the remarkable release of 5–HT, and such a release was also induced in high-K^+ media (Gerschenfeld et al., 1978). These results imply that 5–HT is the synaptic transmitter belonging to the C–1 neuron. A high amount of 5–HT was found in identified neurons of other species, i.e., the colossal neuron of Retzius in the leech (*Hirudo medicinalis*) ganglion (Rude et al., 1969; M. W. McCaman et al., 1973; McAdoo and Coggeshall, 1976) and giant cells from the metacerebral ganglion of *H. pomatia* (Osborne, 1971, 1977) and *Limax maximus* (Cottrell and Osborne, 1970; Osborne and Cottrell, 1971). Dopamine was detected in *Aplysia* ganglia (M. W. McCaman et al., 1973) but was not measurable in R2, R14, and L11 cells (Brownstein et al., 1974). In the subesophageal ganglion of *H. pomatia,* the presence of a high-affinity uptake system of dopamine as well as of 5–HT suggested that dopamine was present in neurons as a neurotransmitter. Cell 21 in the visceral ganglion with a high activity of CAT contained dopamine at about a four-fold higher level than that of dopamine in GSC cells (Osborne, 1977). Indications were also found that dopamine was present in the giant cells in the left pedal ganglion of the water snail *Planorbis corneus* (Powell and Cottrell, 1974; Osborne et al., 1975). 5–HT and dopamine added to the bathing medium stimulated the formation *in vitro* of adenosine-3,5'-cyclic monophosphate (cyclic AMP) in abdominal and pleuropedal ganglion of *A. californica,* and 5–HT was found to stimulate cyclic AMP formation in single neurons (R2, R15, and left pleural ganglion cells) as well as in nerves and neuropil (Cedar and Schwartz, 1972). Norepinephrine was not found in single neurons of *Aplysia* (R2, R14, and L11 cells; Brownstein et al., 1974), *H. pomatia* (GSC cell and cell 21; Osborne, 1977), and *H. medicinalis* (Retzius cell; McAdoo and Coggeshall, 1976).

Octopamine was distributed at different levels in single neurons of *A. californica* and was found at relatively high concentration levels in R14 cells of the abdominal ganglion and in B5, B7, and B10 cells of the buccal ganglion (390–89 μM) compared with other neurons (R2, L2–L6, L11, L13, B2, B3, and B4 cells, Saavedra et al., 1974). Since the presence of an octopamine receptor was demonstrated electrophysiologically in identified neurons from the cerebral ganglion, it was thought to act as a transmitter (Carpenter and Gaubatz, 1974). Phenylethanolamine was present in the ganglion neuropil, the region of neuronal synaptic contact, and the receptor was found electrophysiologically in identified neurons from buccal and cerebral ganglia of *A. californica* (Saavedra and Ribas, 1977). In contrast, McCaman and McCaman (1978) reported that they could not measure octopamine in single neurons (R2, R14, R15, L2–L6, L11, B7, and C–1 cells) or phenylethanolamine in the ganglia from *A. californica.*

V. Conclusion

Single-cell analysis is underway to accumulate fragmentary knowledge on fundamental properties of neurons and on the differences among several types of neurons. As an example of a unique result from such analysis, it can be mentioned that a debate has been provoked about the well-known "Dale's principle" that each neuron synthesizes and releases only one transmitter (Dale, 1935; for detailed discussion, see Burnstock, 1976). As described in Section IV, more than one putative transmitter was detected in single neurons of *Aplysia* and *Helix* (Brownstein *et al.,* 1974; Osborne, 1977), and some *Helix* neurons contained two enzymes synthesizing different transmitter candidates (Hanley *et al.,* 1974). The possibility was suggested that GSC neurons of *H. pomatia* and *H. aspersa* release ACh in addition to serotonin to induce a depolarizing response in a following neuron M (Cottrell, 1977). One type of large molluscan neuron (Lasek and Dower, 1971) contained on an average 67,000 to 131,000-fold as much DNA as the haploid amount contained in sperm. This suggests that multiple copies of the genomes with different properties are present in a molluscan neuron and possibly utilized to express multiple phenotypes in a single neuron. Among vertebrate neurons having a diploid amount of DNA (see Section III,B), 5–HT and substance P have been shown histochemically to coexist in the same neurons in the raphe nuclei and in the interfascicularis hypoglossus nucleus (Chan-Palay *et al.,* 1978; Hökfelt *et al.,* 1978). Quantitative analysis of transmitter candidates in mammalian individual neurons requires more sensitive determination methods than are available at present.

Similar types of mammalian neurons from the same nucleus in the central nervous system are thought to support as a whole a certain physiological function in animals. As pointed out by Kuffler and Nicholls (1966), a nucleus is composed of neuronal populations at different bioelectrical levels—some are active and others are suppressed. Sampling of these neurons is usually performed at random, and the analyzed neurons generally include all the neurons at different and random activity levels. Therefore the obtained parameters or values (e.g., substrate concentrations and enzyme activities) are reasonably thought to be present according to a normal or a quasi-normal distribution. Thus the arithmetical average of the parameters generally reflects the characteristics of these neuronal groups; and the standard deviation of the parameters is thought also to represent the characteristic functional state (e.g., parameter dispersion) of the neuronal group (see Figs. 5 and 6). In conclusion, the results from individual cell analyses should be assessed in terms of the standard deviations, even though the average value is the same or constant in different species or in neurons that differ in the state of their activities.

Finally, as mentioned in Section I, one of the purposes of single-cell analysis is to understand the biochemical and functional correlation between different single neurons (integration of each neuronal function) and that between neurons and glia cells. However, the results indicate only the parallel characteristics and changes in these cells at some time point. A sequential analysis of the same cell is generally impossible in microchemical investigation, since a cell is destroyed once it is used for a certain analysis. For example, the reciprocal change in RNA content in neurons and perineuronal glia found by Hydén and his co-workers (Egyházi and Hydén, 1961; Hydén and Lange, 1966) represents evidently parallel alterations in both neighboring cell units. The transfer of a macromolecule (e.g., RNA) between a neuron and a perineuronal glia cell is, although very interesting and attractive, only speculation. In this respect, other techniques that correlate the data of single-cell analysis (e.g., autoradiography, histochemistry, electrophysiological analysis; see Lasek *et al.,* 1974) are indispensable for interpretation of the results.

Further development and refinement of microchemical methods can be expected to extend the field of knowledge of the biochemical properties and functions of individual neurons.

Acknowledgment

The author thanks Dr. N. Inoue for his help in preparing the manuscript for this article.

References

Aleksidze, N., and Blomstrand, C. (1968). The influence of hydroxylamine, thiosemicarbazide and gamma-aminobutyric acid upon succinate oxidation by Deiters' nerve cells and neuroglia. *Brain Res.* **11,** 717–719.

Ambron, R. T., Goldman, J. E., and Schwartz, J. H. (1974). Axonal transport of newly synthesized glycoproteins in a single identified neuron of *Aplysia californica*. *J. Cell Biol.* **61**, 665–675.

Arch, S. (1972). Biosynthesis of the egg-laying hormone (ELH) in the bag cell neurons of *Aplysia californica*. *J. Gen. Physiol.* **60**, 102–119.

Arch, S., Earley, P., and Smock, T. (1976a). Biochemical isolation and physiological identification of the egg-laying hormone in *Aplysia californica*. *J. Gen. Physiol.* **68**, 197–210.

Arch, S., Smock, T., and Earley, P. (1976b). Precursor and product processing in the bag cell neurons of *Aplysia californica*. *J. Gen. Physiol.* **68**, 211–225.

Aswad, D. W. (1977). Heterogeneity and processing of low molecular weight protein in cell R15 of *Aplysia californica*. *J. Neurochem.* **28**, 1137–1140.

Aswad, D. W. (1978). Biosynthesis and processing of presumed neurosecretory proteins in single identified neurons of *Aplysia californica*. *J. Neurobiol.* **9**, 267–284.

Austoker, J., Cox, D., and Mathias, A. P. (1972). Fraction of nuclei from brain by zonal centrifugation and a study of the ribonucleic acid polymerase activity in the various classes of nuclei. *Biochem. J.* **129**, 1139–1155.

Bantle, J. A., and Hahn, W. E. (1976). Complexity and characterization of polyadenylated RNA in the mouse brain. *Cell* **8**, 139–150.

Barker, D. L., Herbert, E., Hildebrand, J. G., and Kravitz, E. A. (1972). Acetylcholine and lobster sensory neurons. *J. Physiol. (London)* **226**, 205–229.

Bennett, K. D., and Bondareff, W. (1977). Age-related differences in binding of concanavalin A to plasma membranes of isolated neurons. *Am. J. Anat.* **150**, 175–184.

Berger, S. J., Carter, J. G., and Lowry, O. H. (1977). The distribution of glycine, GABA, glutamate and aspartate in rabbit spinal cord, cerebellum and hippocampus. *J. Neurochem.* **28**, 149–158.

Berry, R. W. (1969). Ribonucleic acid metabolism of a single neuron: Correlation with electrical activity. *Science* **166**, 1021–1023.

Berry, R. W. (1975). Functional correlates of low molecular weight peptide synthesis in *Aplysia* neurons. *Brain Res.* **86**, 323–333.

Berry, R. W. (1976a). A comparison of the 12,000 dalton proteins synthesized by *Aplysia* neurons L11 and R15. *Brain Res.* **115**, 457–466.

Berry, R. W. (1976b). Processing of low molecular weight proteins by identified neurons of *Aplysia*. *J. Neurochem.* **26**, 229–231.

Berry, R. W., and Cohen, M. J. (1972). Synaptic stimulation of RNA metabolism in the giant neuron of *Aplysia californica*. *J. Neurobiol.* **3**, 209–222.

Blitz, A. L., and Fine, R. E. (1974). Muscle-like contractile proteins and tubulin in synaptosomes. *Proc. Natl. Acad. Sci. U.S.A.* **71**, 4472–4476.

Bocharova, L. S., Kostenko, M. A., Veprintsev, B. N., and Allachverdov, B. L. (1975). Completely isolated molluscan neurons. An ultrastructural study. *Brain Res.* **101**, 185–198.

Bondareff, W., and Hydén, H. (1969). Submicroscopic structure of single neurons isolated from rabbit lateral vestibular nucleus. *J. Ultrastruct. Res.* **26**, 399–411.

Bondy, S. C., and Roberts, S. (1968). Hybridizable ribonucleic acid of rat brain. *Biochem. J.* **109**, 533–541.

Borys, H. K., Weinreich, D., and McCaman, R. E. (1973). Determination of glutamate and glutamine in individual neurons of *Aplysia californica*. *J. Neurochem.* **21**, 1349–1351.

Brand, M. M., and Lehrer, G. M. (1971). The biochemistry of single nerve cell bodies. *In* "Handbook of Neurochemistry" (A. Lajtha, ed.), Vol. 5A, pp. 337–371. Plenum, New York.

Brattgård, S.-O., Edström, J.-E., and Hydén, H. (1957). The chemical changes in regenerating neurons. *J. Neurochem.* **1**, 316–325.

Bray, D. (1970). Surface movements during the growth of single explanted neurons. *Proc. Natl. Acad. Sci. U.S.A.* **65**, 905-910.

Brownstein, M. J., Saavedra, J. M., Axelrod, J., Zeman, G. H., and Carpenter, D. O. (1974). Coexistence of several putative neurotransmitters in single identified neurons of *Aplysia*. *Proc. Natl. Acad. Sci. U.S.A.* **71**, 4662-4665.

Buckley, G., Consolo, S., Giacobini, E., and McCaman, R. (1967). A micromethod for the determination of choline acetylase in individual cells. *Acta Physiol. Scand.* **71**, 341-347.

Bunge, M. B. (1973). Fine structure of nerve fibers and growth cones of isolated sympathetic neurons in culture. *J. Cell Biol.* **56**, 713-735.

Burnstock, G. (1976). Do some nerve cells release more than one transmitter? *Neurosci.* **1**, 239-248.

Butcher, S. G., and Butcher, L. L. (1974). Origin and modulation of acetylcholine activity in the neostriatum. *Brain Res.* **71**, 167-171.

Calissano, P., and Bangham, A. D. (1971). Effect of two brain specific proteins (S100 and 14.3.2) on cation diffusion across artificial lipid membranes. *Biochem. Biophys. Res. Commun.* **43**, 504-509.

Campagnoni, A. T., Dutton, G. R., Mahler, H. R., and Moore, W. J. (1971). Fractionation of the RNA components of rat brain polysomes. *J. Neurochem.* **18**, 601-611.

Carpenter, D. O., and Gaubatz, G. L. (1974). Octopamine receptors on *Aplysia* neurones mediate hyperpolarisation by increasing membrane conductance. *Nature (London)* **252**, 483-485.

Cedar, H., and Schwartz, J. H. (1972). Cyclic adenosine monophosphate in the nervous system of *Aplysia californica*. II. Effect of serotonin and dopamine. *J. Gen. Physiol.* **60**, 570-587.

Chan-Palay, V., Jonsson, G., and Palay, S. L. (1978). Serotonin and substance P coexist in neurons of the rat's central nervous system. *Proc. Natl. Acad. Sci. U.S.A.* **75**, 1582-1586.

Cicero, T. J., Cowan, W. M., Moore, B. W., and Suntzeff, V. (1970). The cellular localization of the two brain specific proteins, S-100 and 14-3-2. *Brain Res.* **18**, 25-34.

Coggeshall, R. E. (1967). A light and electron microscope study of the abdominal ganglion of *Aplysia californica*. *J. Neurophysiol.* **30**, 1263-1278.

Coggeshall, R. E., Dewhurst, S. A., Weinreich, D., and McCaman, R. E. (1972). Aromatic acid decarboxylase and choline acetylase activities in a single identified 5-HT containing cell of the leech. *J. Neurobiol.* **3**, 259-265.

Cohen, J., Mareš, V., and Lodin, Z. (1973). DNA content of purified preparations of mouse Purkinje neurons isolated by a velocity sedimentation technique. *J. Neurochem.* **20**, 651-657.

Cohen, M. J. (1970). A comparison of invertebrate and vertebrate central neurons. *In* "The Neurosciences" (F. O. Schmitt, ed.), pp. 798-812. Rockefeller Univ. Press, New York.

Cottrell, G. A. (1974). Serotonin and free amino acid analysis of ganglia and isolated neurones of *Aplysia dactylomela*. *J. Neurochem.* **22**, 557-559.

Cottrell, G. A. (1977). Identified amine-containing neurones and their synaptic connexions. *Neurosci.* **2**, 1-18.

Cottrell, G. A., and Osborne, N. N. (1970). Subcellular localization of serotonin in an identified serotonin-containing neurone. *Nature (London)* **255**, 470-472.

Cottrell, G. A., and Powell, B. (1971). Formation of serotonin by isolated serotonin-containing neurons and by isolated non-amine-containing neurons. *J. Neurochem.* **18**, 1695-1697.

Cummins, J., and Hydén, H. (1962). Adenosine triphosphate levels and adenosine triphosphatases in neurons, glia and neuronal membranes of the vestibular nucleus. *Biochim. Biophys. Acta* **60**, 271-283.

Cupello, A., and Hydén, H. (1975). Separation of brain RNA by micro-electrophoresis on agarose-acrylamide gels. *Neurobiology* 5, 129–136.

Cupello, A., and Hydén, H. (1976a). Alteration of the pattern of hippocampal nerve cell RNA labelling during training in rats. *Brain Res.* 114, 453–460.

Cupello, A., and Hydén, H. (1976b). Pattern of labelling of poly(A)-associated RNA in the CA₃ region of rat hippocampus during training. *J. Neurosci. Res.* 2, 255–260.

Cupello, A., and Hydén, H. (1978). Studies on RNA metabolism in the nerve cells of hippocampus during training in rats. *Exp. Brain Res.* 31, 143–152.

Dale, H. (1935). Pharmacology and nerve endings. *Proc. R. Soc. Med.* 28, 319–332.

Daneholt, B., and Brattgård, S.-O. (1966). A comparison between RNA metabolism of nerve cells and glia in the hypoglossal nucleus of the rabbit. *J. Neurochem.* 13, 913–921.

Denis-Donini, S., Estenoz, M., and Augusti-Tocco, G. (1978). Cell surface modifications in neuronal maturation. *Cell Differ.* 7, 193–201.

Derry, D. M. and Wolfe, L. S. (1967). Gangliosides in isolated neurons and glial cells. *Science* 158, 1450–1452.

Droz, B., and Leblond, C. P. (1963). Axonal migration of proteins in the central nervous system and peripheral nerves as shown by radioautography. *J. Comp. Neurol.* 121, 325–346.

Edström, A. (1964). The ribonucleic acid in the Mauthner neuron of the goldfish. *J. Neurochem.* 11, 309–314.

Edström, A., Edström, J.-E. and Hökfelt, T. (1969). Sedimentation analysis of ribonucleic acid extracted from isolated Mauthner nerve fibre components. *J. Neurochem.* 16, 53–66.

Edström, J.-E. (1956). The content and the concentration of ribonucleic acid in motor anterior horn cells from the rabbit. *J. Neurochem.* 1, 159–165.

Edström, J.-E., and Grampp, W. (1965). Nervous activity and metabolism of ribonucleic acids in the crustacean stretch receptor neuron. *J. Neurochem.* 12, 735–741.

Edström, J.-E., and Pigon, A. (1958). Relation between surface, ribonucleic acid content and nuclear volume in encapsulated spinal ganglion cells. *J. Neurochem.* 3, 95–99.

Edström, J.-E., Eichner, D., and Edström, A. (1962). The ribonucleic acid of axons and myelin sheaths from Mauthner neurons. *Biochim. Biophys. Acta* 61, 178–184.

Egyházi, E., and Hydén, H. (1961). Experimentally induced changes in the base composition of the ribonucleic acids of isolated nerve cells and their oligodendroglial cells. *J. Biophys. Biochem. Cytol.* 10, 403–410.

Egyházi, E., and Hydén, H. (1966). Biosynthesis of rapidly labeled RNA in brain cells. *Life Sci.* 5, 1215–1223.

Eisenstadt, M., and Schwartz, J. H. (1975). Metabolism of acetylcholine in the nervous system of *Aplysia californica*. III. Studies of an identified cholinergic neuron. *J. Gen. Physiol.* 65, 293–313.

Eisenstadt, M., Goldman, J. E., Kandel, E. R., Koike, H., Koester, J., and Schwartz, J. H. (1973). Intrasomatic injection of radioactive precursors for studying transmitter synthesis in identified neurons of *Aplysia californica*. *Proc. Natl. Acad. Sci. U.S.A.* 70, 3371–3375.

Eisenstadt, M. L., Treistman, S. N., and Schwartz, J. H. (1975). Metabolism of acetylcholine in the nervous system of *Aplysia californica*. II. Regional localization and characterization of choline uptake. *J. Gen. Physiol.* 65, 275–291.

Ekholm, R., and Hydén, H. (1965). Polysomes from microdissected fresh neurons. *J. Ultrastruct. Res.* 13, 269–280.

Emson, P., Walker, R. J., and Kerkut, G. A. (1971). Chemical changes in a molluscan ganglion associated with learning. *Comp. Biochem. Physiol. B* 40, 223–239.

Emson, P. C., Burrows, M., and Fonnum, F. (1974). Levels of glutamate decarboxylase, choline acetyltransferase and acetylcholinesterase in identified motor neurons of the locust. *J. Neurobiol.* 5, 33–42.

Emson, P. C., Bush, B. M. H., and Joseph, M. H. (1976). Transmitter-metabolizing enzymes and free amino acid levels in sensory and motor nerves and ganglia of the shore crab (*Carcinus maenas*). *J. Neurochem.* **26**, 779–783.

Falck, B. (1962). Observations on the possibilities of the cellular localization of monoamines by a fluorescence method. *Acta Physiol. Scand. Suppl.* **197**, 3–25.

Fernandez-Campa, J., and Engel, W. K. (1970). Histochemical differentiation of motor neurones and interneurones in the anterior horn of the cat spinal cord. *Nature (London)* **225**, 748–749.

Frankel, R. D., and Koenig, E. (1977). Identification of major indigenous protein components in mammalian axons and locally synthesized axonal protein in hypoglossal nerve. *Exp. Neurol.* **57**, 282–295.

Frankel, R. D., and Koenig, E. (1978). Identification of locally synthesized proteins in proximal stump axons of the neurotomized hypoglossal nerve. *Brain Res.* **141**, 67–76.

Frazier, W. T., Kandel, E. R., Kupfermann, I., Waziri, R., and Coggeshall, R. E. (1967). Morphological and functional properties of identified neurons in the abdominal ganglion of *Aplysia californica*. *J. Neurophysiol.* **30**, 1288–1351.

Fujita, S. (1974). DNA constancy in neurons of the human cerebellum and spinal cord as revealed by Feulgen cytophotometry and cytofluorometry *J. Comp. Neurol.* **155**, 195–202.

Gainer, H. (1971). Micro disc electrophoresis in sodium dodecyl sulfate: An application to the study of protein synthesis in individual, identified neurons. *Anal. Biochem.* **44**, 589–605.

Gainer, H. (1978). Intercellular transfer of proteins from glial cells to axons. *TINS* **1**, 93–96.

Gainer, H., and Barker, J. L. (1974). Synaptic regulation of specific protein synthesis in an identified neuron. *Brain Res.* **78**, 314–319.

Gainer, H., and Barker, J. L. (1975). Selective modulation and turnover of proteins in identified neurons of *Aplysia*. *Comp. Biochem. Physiol. B* **51**, 221–227.

Gainer, H., and Wollberg, Z. (1974). Specific protein metabolism in identifiable neurons of *Aplysia californica*. *J. Neurobiol.* **5**, 243–261.

Gerschenfeld, H. M., Hamon, M., and Paupardin-Tritsch, D. (1978). Release of endogenous serotonin from two identified serotonin-containing neurones and the physiological role of serotonin re-uptake. *J. Physiol. (London)* **274**, 265–278.

Giacobini, E. (1957). Quantitative determination of cholinesterase in individual sympathetic cells. *J. Neurochem.* **1**, 234–244.

Giacobini, E. (1959a). Determination of cholinesterase in the cellular components of neurones. *Acta Physiol. Scand.* **45**, 311–327.

Giacobini, E. (1959b). The distribution and localization of cholinesterases in nerve cells. *Acta Physiol. Scand., Suppl.* **156**, 1–45.

Giacobini, E. (1961). Localization of carbonic anhydrase in the nervous system. *Science* **134**, 1524–1525.

Giacobini, E. (1969). Chemistry of isolated invertebrate neurons. *In* "Handbook of Neurochemistry" (A. Lajtha, ed.), Vol. 2, pp. 195–239. Plenum, New York.

Giacobini, E. (1975). The use of microchemical techniques for the identification of new transmitter molecules in neurons. *J. Neurosci. Res.* **1**, 1–18.

Giacobini, E., and Grasso, A. (1966). Variations of glycolytic intermediates, phosphate compounds and pyridine nucleotides after prolonged stimulation of an isolated crustacean neurone. *Acta Physiol. Scand.* **66**, 49–57.

Giacobini, E., and Marchisio, P. C. (1966). The action of tricarboxylic acid cycle intermediates and glutamate on the impulse activity and respiration of the crayfish stretch receptor neurone. *Acta Physiol. Scand.* **66**, 58–66.

Giller, E., Jr., and Schwartz, J. H. (1968). Choline acetyltransferase: Regional distribution in the abdominal ganglion of *Aplysia*. *Science* **161**, 908–911.

Giller, E., Jr., and Schwartz, J. H. (1971a). Choline acetyltransferase in identified neurons of abdominal ganglion of *Aplysia californica. J. Neurophysiol.* **34**, 93-107.

Giller, E., Jr., and Schwartz, J. H. (1971b). Acetylcholinesterase in identified neurons of abdominal ganglion of *Aplysia californica. J. Neurophysiol.* **34**, 108-115.

Glick, D. (1977). The contribution of microchemical methods of histochemistry to the biological sciences. *J. Histochem. Cytochem.* **25**, 1087-1101.

Goldberg, A. M., and McCaman, R. E. (1967). A quantitative microchemical study of choline acetyltransferase and acetylcholinesterase in the cerebellum of several species. *Life Sci.* **6**, 1493-1500.

Goldberg, D. J., Goldman, J. E., and Schwartz, J. H. (1976). Alterations in amounts and rates of serotonin transported in an axon of the giant cerebral neurone of *Aplysia californica. J. Physiol. (London)* **259**, 473-490.

Goldman, J. E., and Schwartz, J. H. (1974). Cellular specificity of serotonin storage and axonal transport in identified neurones of *Aplysia californica. J. Physiol. (London)* **242**, 61-76.

Goldman, J. E., Kim, K. S., and Schwartz, J. H. (1976). Axonal transport of [³H]serotonin in an identified neuron of *Aplysia californica. J. Cell Biol.* **70**, 304-318.

Grampp, W., and Edström, J. E. (1963). The effect of nervous activity on ribonucleic acid of the crustacean receptor neuron. *J. Neurochem.* **10**, 725-731.

Grenell, R. G., Hazama, H., Nakazawa, M., and Einberg, E. (1968). Effects of gravitational changes on RNA of cerebral neurons and glia. I. RNA changes of Deiters' cells and glia. *Brain Res.* **9**, 115-125.

Haglid, K. G., Hamberger, A., Hansson, H.-A., Hydén, H., Persson, L., and Rönnbäck, L. (1974). S-100 protein in synapses of the central nervous system. *Nature (London)* **251**, 532-534.

Haglid, K. G., Hamberger, A., Hansson, H.-A., Hydén, H., Persson, L., and Rönnbäck, L. (1976). Cellular and subcellular distribution of the S-100 protein in rabbit and rat central nervous system. *J. Neurosci. Res.* **2**, 175-191.

Hahn, W. E., Van Ness, J., and Maxwell, I. H. (1978). Presence of mRNA sequences in large polyadenylated hnRNA molecules in the mouse brain. *Fed. Proc., Fed. Am. Soc. Exp. Biol.* **37**, 1504.

Hall, Z. W., Bownds, M. D., and Kravitz, E. A. (1970). The metabolism of gamma aminobutyric acid in the lobster nervous system. *J. Cell Biol.* **46**, 290-299.

Hamberger, A. (1961). Oxidation of tricarboxylic acid cycle intermediates by nerve cell bodies and glial cells. *J. Neurochem.* **8**, 31-35.

Hamberger, A. (1963). Difference between isolated neuronal and vascular glia with respect to respiratory activity. *Acta Physiol. Scand., Suppl.* **203**, 1-58.

Hamberger, A., and Hydén, H. (1963). Inverse enzymatic changes in neurons and glia during increased function and hypoxia. *J. Cell Biol.* **16**, 521-525.

Hamberger, A., and Svennerholm, L. (1971). Composition of gangliosides and phospholipids of neuronal and glial cell-enriched fractions. *J. Neurochem.* **18**, 1821-1829.

Hamberger, A., Hansson, H.-A., and Sjöstrand, J. (1970). Surface structure of isolated neurons. Detachment of nerve terminals during axon regeneration. *J. Cell Biol.* **47**, 319-331.

Hanley, M. R., and Cottrell, G. A. (1974). Acetylcholine activity in an identified 5-hydroxytryptamine-containing neuron. *J. Pharm. Pharmacol.* **26**, 980.

Hanley, M. R., Cottrell, G. A., Emson, P. C., and Fonnum, F. (1974). Enzymatic synthesis of acetylcholine by a serotonin-containing neurone from *Helix. Nature (London)* **251**, 631-633.

Hansson, H.-A., and Hydén, H. (1974). A membrane-associated network of protein filaments in nerve cells. *Neurobiology* **4**, 364-375.

Hansson, H.-A., Hydén, H., and Rönnbäck, L. (1975). Localization of S-100 protein in isolated nerve cells by immunoelectron microscopy. *Brain Res.* **93**, 349-352.

Hayashi, H., and Kato, T. (1978). Acetyl-CoA-synthesizing enzyme activities in single nerve cell bodies of rabbit. *J. Neurochem.* **31**, 861-869.

Hazama, H., and Uchimura, H. (1970). Separation of lactate dehydrogenase isozymes of nerve cells in the central nervous system by micro-disc electrophoresis on polyacrylamide gels. *Biochim. Biophys. Acta* **200**, 414-417.

Herman, C. J., and Lapham, L. W. (1968). DNA content of neurons in the cat hippocampus. *Science* **160**, 537.

Herman, C. J., and Lapham, L. W. (1969). Neuronal polyploidy and nuclear volumes in the cat central nervous system. *Brain Res.* **15**, 35-48.

Hertz, L., and Schousboe, A. (1975). Ion and energy metabolism of the brain at the cellular level. *Int. Rev. Neurobiol.* **18**, 141-211.

Hillman, H. (1966). Growth of processes from single isolated dorsal root ganglion cells of young rats. *Nature (London)* **209**, 102-103.

Hillman, H., and Hydén, H. (1965a). Membrane potentials in isolated neurones *in vitro* from Deiters' nucleus of rabbit. *J. Physiol. (London)* **177**, 398-410.

Hillman, H., and Hydén, H. (1965b). Characteristics of the ATP-ase activity of isolated neurons of rabbit. *Histochemie* **4**, 446-450.

Hillman, H., and Sheikh, K. (1968). The growth *in vitro* of new processes from vestibular neurons isolated from adult and young rabbits. *Exp. Cell Res.* **50**, 315-322.

Hirsch, H. E. (1968). Acid phosphatase localization in individual neurons by a quantitative histochemical method. *J. Neurochem.* **15**, 123-130.

Hirsch, H. E. (1969). Localization of arylsulphatase in neurons. *J. Neurochem.* **16**, 1147-1155.

Hirsch, H. E., and Chen, K.-M. (1969). Enzyme activities in individual motor neurons in amyotrophic lateral sclerosis. A quantitative histochemical study. *J. Neuropathol. Exp. Neurol.* **28**, 267-277.

Hökfelt, T., Ljungdahl, Å., Steinbusch, H., Verhofstad, A., Nilsson, G., Brodin, E., Pernow, B., and Goldstein, M. (1978). Immunohistochemical evidence of substance P-like immunoreactivity in some 5-hydroxytryptamine containing neurons in the rat central nervous system. *Neurosci.* **3**, 517-538.

Hultborn, R., and Hydén, H. (1974). Microspectrophotometric determination of nerve cell respiration at high potassium concentration. *Exp. Cell Res.* **87**, 346-350.

Hydén, H. (1959). Quantitative assay of compounds in isolated, fresh nerve cells and glial cells from control and stimulated animals. *Nature (London)* **184**, 433-435.

Hydén, H. (1973). RNA changes in brain cells during changes in behavior and function. *In* "Macromolecules and Behavior" (G. B. Ansell and P. B. Brady, eds.), pp. 51-75. Macmillan, New York.

Hydén, H. (1976). Plastic changes of neurons during acquisition of new behavior as a problem of protein differentiation. *Prog. Brain Res.* **45**, 83-100.

Hydén, H., and Egyházi, E. (1962). Nuclear RNA changes of nerve cells during a learning experiment in rats. *Proc. Natl. Acad. Sci. U.S.A.* **48**, 1366-1373.

Hydén, H., and Egyházi, E. (1963). Glial RNA changes during a learning experiment in rats. *Proc. Natl. Acad. Sci. U.S.A.* **49**, 618-624.

Hydén, H., and Egyházi, E. (1964). Changes in RNA content and base composition in cortical neurons of rats in a learning experiment involving transfer of handedness. *Proc. Natl. Acad. Sci. U.S.A.* **52**, 1030-1035.

Hydén, H., and Lange, P. W. (1965a). A differentiation in RNA response in neurons early and rate during learning. *Proc. Natl. Acad. Sci. U.S.A.* **53**, 946-952.

Hydén, H., and Lange, P. W. (1965b). Rhythmic enzyme changes in neurons and glia during sleep. *Science* **149**, 654-656.

Hydén, H., and Lange, P. W. (1966). A genetic stimulation with production of adenic-uracil rich RNA in neurons and glia in learning. *Naturwissenschaften* 53, 64–70.

Hydén, H., and Lange, P. W. (1968). Protein synthesis in the hippocampal pyramidal cells of rats during a behavioral test. *Science* 159, 1370–1373.

Hydén, H., and Lange, P. W. (1970a). Brain-cell protein synthesis specifically related to learning. *Proc. Natl. Acad. Sci. U.S.A.* 65, 898–904.

Hydén, H., and Lange, P. W. (1970b). Correlation of the S100 brain protein with behavior. *Exp. Cell Res.* 62, 125–132.

Hydén, H., and Lange, P. W. (1972). Protein synthesis in hippocampal nerve cells during re-reversal of handedness in rats. *Brain Res.* 45, 314–317.

Hydén, H., and Lange, P. W. (1977). The effect of S100 protein on transport phenomena across freshly prepared nerve cell and glial membranes. *Proc. Int. Sci. Neurochem.* 6, 616.

Hydén, H., and McEwen, B. (1966). A glial protein specific for the nervous system. *Proc. Natl. Acad. Sci. U.S.A.* 55, 354–358.

Hydén, H., and Pigon, A. (1960). A cytophysiological study of the functional relationship between oligodendroglial cells and nerve cells of Deiters' nucleus. *J. Neurochem.* 6, 57–72.

Hydén, H., and Rönnbäck, L. (1975a). Membrane-bound S–100 protein on nerve cells and its distribution. *Brain Res.* 100, 615–628.

Hydén, H., and Rönnbäck, L. (1975b). S100 on isolated neurons and glial cells from rat, rabbit and guinea pig during early postnatal development. *Neurobiology*, 5, 219–302.

Hydén, H., and Rönnbäck, L. (1978). The brain-specific S–100 protein on neuronal cell membranes. *J. Neurobiol.* 9, 489–492.

Hydén, H., Lange, P. W., Mihailović, L. J., and Petrović-Minić, B. (1974). Changes of RNA base composition in nerve cells of monkeys subjected to visual discrimination and delayed alternation performance. *Brain Res.* 65, 215–230.

Hydén, H., Lange, P. W., and Perrin, C. L. (1977). Protein pattern alterations in hippocampal and cortical cells as a function of training in rats. *Brain Res.* 119, 427–437.

Iliffe, T. M., McAdoo, D. J., Beyer, C. B., and Haber, B. (1977). Amino acid concentrations in the *Aplysia* nervous system: Neurons with high glycine concentrations. *J. Neurochem.* 28, 1037–1042.

Ishikawa, H., Bischoff, R., and Holtzer, H. (1969). Formation of arrowhead complexes with heavy meromyosin in a variety of cell types. *J. Cell Biol.* 43, 312–328.

Jarlstedt, J. (1966a). Functional localization in the cerebellar cortex studied by quantitative determinations of Purkinje cell RNA. I. RNA changes in rat cerebellar Purkinje cells after proprio- and exteroceptive and vestibular stimulation. *Acta Physiol. Scand.* 67, 243–252.

Jarlstedt, J. (1966b). Functional localization in the cerebellar cortex studied by quantitative determinations of Purkinje cell RNA. *Acta Physiol. Scand., Suppl.* 271, 1–24.

Johnson, D. E., and Sellinger, O. Z. (1971). Protein synthesis in neurons and glial cells of the developing rat brain: An *in vivo* study. *J. Neurochem.* 18, 1445–1460.

Kahan, B. E., Krigman, M. R., Wilson, J. E., and Glassman, E. (1970). Brain function and macromolecules. VI. Autoradiographic analysis of the effect of a brief training experience on the incorporation of uridine into mouse brain. *Proc. Natl. Acad. Sci. U.S.A.* 65, 300–303.

Kan, K.-S., Chao, L.-P., Wolfgram, F., and Eng., L. F. (1977). Immunohistochemical localization of choline acetyltransferase in rabbit CNS. *Proc. Int. Soc. Neurochem.* 6, 142.

Karpiak, S. E., Serokosz, M., and Rapport, M. M. (1976). Effects of antisera to S–100 protein and to synaptic membrane fraction on maze performance and EEG. *Brain Res.* **102**, 313–321.

Kato, T. (1978). Ultramicro analysis of enzymes and substrates by enzymatic amplification reactions. "Enzymatic cycling." *Pure Appl. Chem.* **50**, 1069–1073.

Kato, T., and Kurokawa, M. (1970). Studies on ribonucleic acid and homopolyribonucleotide formation in neuronal, glial and liver nuclei. *Biochem. J.* **116**, 599–609.

Kato, T., and Lowry, O. H. (1973a). Enzymes of energy-converting systems in individual mammalian nerve cell bodies. *J. Neurochem.* **20**, 151–163.

Kato, T., and Lowry, O. H. (1973b). Distribution of enzymes between nucleus and cytoplasm of single nerve cell bodies. *J. Biol. Chem.* **248**, 2044–2048.

Kehoe, J. S. (1978). Transformation by concanavalin A of the response of molluscan neurones to L-glutamate. *Nature (London)* **274**, 866–869.

Kelly, P., Cotman, C. W., Gentry, C., and Nicolson, G. L. (1976). Distribution and mobility of lectin receptors on synaptic membranes of identified neurons in the central nervous system. *J. Cell Biol.* **71**, 487–496.

Kerkut, G. A., Oliver, G., Rick, J. T., and Walker, R. J. (1970). Biochemical changes during learning in an insect ganglion. *Nature (London)* **227**, 722–723.

Kernell, D., and Peterson, R. P. (1970). The effect of spike activity versus synaptic activation on the metabolism of ribonucleic acid in a molluscan giant neurone. *J. Neurochem.* **17**, 1087–1094.

Koenig, E. (1965). Synthetic mechanisms in the axon. II. RNA in myelin-free axons of the cat. *J. Neurochem.* **12**, 357–361.

Koenig, E. (1967). Synthetic mechanisms in the axon. IV. *In vitro* incorporation of [³H] precursors into axonal protein and RNA. *J. Neurochem.* **14**, 437–446.

Koenig, E. (1978). Analysis of major RNA classes in the Mauthner axon. *Neurosci. Abstr. Soc. Neurosci.* **4**, 318.

Koenig, H. (1969). Lysosomes. *In* "Handbook of Neurochemistry" (A. Lajtha, ed.), Vol. 2, pp. 255–301. Plenum, New York.

Koike, H., Eisenstadt, M., and Schwartz, J. H. (1972). Axonal transport of newly synthesized acetylcholine in an identified neuron of *Aplysia*. *Brain Res.* **37**, 152–159.

Koslow, S. H., and Giacobini, E. (1969). An isotopic micromethod for the measurement of cholinesterase activity in individual cells. *J. Neurochem.* **16**, 1523–1528.

Kuffler, S. W., and Nicholls, J. G. (1966). The physiology of neuroglial cells. *Ergeb. Physiol., Biol. Chem. Exp. Pharmakol.* **57**, 1–90.

Kupfermann, I., and Kandel, E. R. (1970). Electrophysiological properties and functional interconnections of two symmetrical neurosecretory clusters (bag cells) in abdominal ganglion of *Aplysia*. *J. Neurophysiol.* **33**, 865–876.

Kupfermann, I., and Weiss, K. R. (1976). Water regulation by a presumptive hormone contained in identified neurosecretory cell R15 of *Aplysia*. *J. Gen. Physiol.* **67**, 113–123.

Lasek, R. J. (1970). The distribution of nucleic acids in the giant axon of the squid (*Loligo pealii*). *J. Neurochem.* **17**, 103–109.

Lasek, R. J., and Dower, W. J. (1971). *Aplysia californica:* Analysis of nuclear DNA in individual nuclei of giant neurons. *Science* **172**, 278–280.

Lasek, R. J., Lee, C. K., and Przybylski, R. J. (1972). Granular extensions of the nucleoli in giant neurons of *Aplysia californica*. *J. Cell Biol.* **55**, 237–242.

Lasek, R. J., Dabrowski, C., and Nordlander, R. (1973). Analysis of axoplasmic RNA from invertebrate giant axons. *Nature (London), New Biol.* **244**, 162–165.

Lasek, R. J., Gainer, H., and Przybylski, R. J. (1974). Transfer of newly synthesized proteins from Schwann cells to the squid giant axon. *Proc. Natl. Acad. Sci. U.S.A.* **71**, 1188–1192.

Latorre, J., and Perry, R. P. (1973). The relationship between polyadenylated heterogenous nuclear RNA and messenger RNA: Studies with actinomycin D and cordycepin. *Biochim. Biophys. Acta* **335**, 92–101.

LeBeux, Y. J., and Willemot, J. (1975a). An ultrastructural study of the microfilaments in rat brain by means of heavy meromyosin labeling. I. Perikaryon, the dendrites and the axon. *Cell Tissue Res.* **160**, 1–36.

LeBeux, Y. J., and Willemot, J. (1975b). An ultrastructural study of the microfilaments in rat brain by means of E-PTA staining and heavy meromyosin labeling. II. The synapses. *Cell Tissue Res.* **160**, 37–68.

Lehrer, G. M., Weiss, C., Silides, D. J., Lichtman, C., Furman, M., and Mathewson, R. F. (1968). The quantitative histochemistry of supramedullary neurons of puffer fishes. *J. Cell Biol.* **37**, 575–579.

Lehrer, G. M., Katzman, R., and Wilson, C. E. (1970). Volume and density measurements in subcellular portions of single nerve cell bodies. *J. Histochem. Cytochem.* **18**, 44–48.

Lentz, R. D., and Lapham, L. W. (1969). A quantitative cytochemical study of the DNA content of neurons of rat cerebellar cortex. *J. Neurochem.* **16**, 379–384.

Lentz, R. D., and Lapham, L. W. (1970). Postnatal development of tetraploid DNA content in rat Purkinje cells: A quantitative cytochemical study. *J. Neuropathol. Exp. Neurol.* **29**, 43–56.

Levitan, I. B., Ramirez, G., and Mushynski, W. E. (1972). Amino acid incorporation in the brains of rats trained to use the non-preferred paw in retrieving food. *Brain Res.* **47**, 147–156.

Lisý, V., and Lodin, Z. (1973). Incorporation of radioactive leucine into neuronal and glial proteins during postnatal development. *Neurobiology* **3**, 320–326.

Loh, Y. P., and Gainer, H. (1975a). Low molecular weight specific proteins in identified molluscan neurons. I. Synthesis and storage. *Brain Res.* **92**, 181–192.

Loh, Y. P., and Gainer, H. (1975b). Low molecular weight specific proteins in identified molluscan neurons. II. Processing, turnover and transport. *Brain Res.* **92**, 193–205.

Loh, Y. P., and Peterson, R. P. (1974). Protein synthesis in phenotypically different, single neurons of *Aplysia*. *Brain Res.* **78**, 83–98.

Loh, Y. P., Sarne, Y., and Gainer, H. (1975). Heterogeneity of proteins synthesized, stored, and released by the bag cells of *Aplysia californica*. *J. Comp. Physiol.* **100**, 283–295.

Loh, Y. P., Barker, J. L., and Gainer, H. (1976). Neurosecretory cell protein metabolism in the land snail, *Otala lactea*. *J. Neurochem.* **26**, 25–30.

Loh, Y. P., Sarne, Y., Daniels, M. P., and Gainer, H. (1977). Subcellular fractionation studies related to the processing of neurosecretory proteins in *Aplysia* neurons. *J. Neurochem.* **29**, 135–139.

Lowry, O. H. (1953). The quantitative histochemistry of the brain: Histological sampling. *J. Histochem. Cytochem.* **1**, 420–428.

Lowry, O. H. (1957). Enzyme concentrations in individual nerve cell bodies. *In* "Metabolism of the Nervous System" (D. Richter, ed.), pp. 323–328. Macmillan (Pergamon), New York.

Lowry, O. H. (1963). The chemical study of single neurons. *Harvey Lect.* **58**, 1–19.

Lowry, O. H., and Passonneau, J. V. (1972). "A Flexible System of Enzymatic Analysis." Academic Press, New York.

Lowry, O. H., Roberts, N. R., and Chang, M.-L. W. (1956). The analysis of single cells. *J. Biol. Chem.* **222**, 97–107.

McAdoo, D. J., and Coggeshall, R. E. (1976). Gas chromatographic-mass spectrometric analysis of biogenic amines in identified neurons and tissues of *Hirudo medicinalis*. *J. Neurochem.* **26**, 163–167.

McAdoo, D. J., Iliffe, T. M., Price, C. H., and Novak, R. A. (1978). Specific glycine uptake by identified neurons of *Aplysia californica*. II. Biochemistry. *Brain Res.* **154**, 41–51.

McCaman, M. W., and McCaman, R. E. (1978). Octopamine and phenylethanolamine in *Aplysia* ganglia and in individual neurons. *Brain Res.* **141**, 347-352.

McCaman, R. E., and Dewhurst, S. A. (1970). Choline acetyltransferase in individual neurons of *Aplysia californica*. *J. Neurochem.* **17**, 1421-1426.

McCaman, R. E., and Dewhurst, S. A. (1971). Metabolism of putative transmitters in individual neurons of *Aplysia californica*. Acetylcholinesterase and catechol-*O*-methyltransferase. *J. Neurochem.* **18**, 1329-1335.

McCaman, M. W., Weinreich, D., and McCaman, R. E. (1973). The determination of picomole levels of 5-hydroxytryptamine and dopamine in *Aplysia, Tritonia* and leech nervous tissues. *Brain Res.* **53**, 129-137.

McCaman, R. E., Weinreich, D., and Borys, H. (1973). Endogenous levels of acetylcholine and choline in individual neurons of *Aplysia*. *J. Neurochem.* **21**, 473-476.

McGeer, P. L., McGeer, E. G., Fibiger, H. C., and Wickson, V. (1971). Neostriatal choline acetylase and cholinesterase following selective brain lesions. *Brain Res.* **35**, 308-314.

McIlwain, D. L., and Capps-Covey, P. (1976). The nuclear DNA content of large ventral spinal neurons. *J. Neurochem.* **27**, 109-112.

Mahler, H. R., Moore, W. J., and Thompson, R. J. (1966). Isolation and characterization of ribonucleic acid from cerebral cortex of rat. *J. Biol. Chem.* **241**, 1283-1289.

Mann, D. M. A., and Yates, P. O. (1973). Polyploidy in the human nervous system. Part 1. The DNA content of neurones and glia of the cerebellum. *J. Neurol. Sci.* **18**, 183-196.

Mareš, V. L., Lodin, Z., and Sácha, J. (1973). A cytochemical and autoradiographic study of nuclear DNA in mouse Purkinje cells. *Brain Res.* **53**, 273-289.

Matus, A., and Mughal, S. (1975). Immunohistochemical localization of S-100 protein in brain. *Nature (London)* **258**, 746-748.

Metuzals, J., and Mushynski, W. E. (1974). Electron microscope and experimental investigations of the neurofilamentous network in Deiters' neurons. *J. Cell Biol.* **61**, 701-722.

Metuzals, J., and Tasaki, I. (1978). Subaxolemmal filamentous network in the giant nerve fiber of the squid. (*Loligo pealei* L.) and its possible role in excitability. *J. Cell Biol.* **78**, 597-621.

Michetti, F., Miani, N., De Renzis, G., Caniglia, A., and Correr, S. (1974). Nuclear localization of S-100 protein. *J. Neurochem.* **22**, 239-244.

Moore, B. W. (1965). A soluble protein characteristic of the nervous system. *Biochem. Biophys. Res. Commun.* **19**, 739-744.

Nemer, M., Graham, M., and Dubroff, L. M. (1974). Co-existence of non-histone messenger RNA species lacking and containing polyadenylic acid in sea urchin embryo. *J. Mol. Biol.* **89**, 435-454.

Neuhoff, V. (1973). Micromethods in molecular biology. *Mol. Biol. Biochem. Biophys.* Vol. 14. Springer, Berlin.

Norton, W. T., and Poduslo, S. E. (1971). Neuronal perikarya and astroglia of rat brain: Chemical composition during myelination. *J. Lipid Res.* **12**, 84-90.

Okada, Y., and Shimada, C. (1976). Gamma-aminobutyric acid (GABA) concentration in a single neuron—Localization of GABA in Deiters' neuron. *Brain Res.* **107**, 658-662.

Okazaki, H., Abe, S., and Satake, M. (1978). Biochemical characterization of the neuron: Ribonucleic acid composition of the neuronal cell bodies. *J. Neurochem.* **31**, 1149-1155.

Osborne, N. N. (1971). A micro-chromatographic method for the detection of biologically active monoamines isolated neurons. *Experientia 27*, 1502-1503.

Osborne, N. N. (1972). Effect of electrical stimulation on the *in vivo* metabolism of glucose and glutamic acid in an identified neuron. *Brain Res.* **41**, 237-241.

Osborne, N. N. (1974). Microchemical analysis of nervous tissue. "Methods in Life Sciences," Vol. 1. Pergamon, Oxford.

Osborne, N. N. (1977). Do snail neurones contain more than one neurotransmitter? *Nature (London)* **270**, 622-624.

Osborne, N. N., and Cottrell, G. A. (1971). Distribution of biogenic amines in the slug, *Limax maximus. Z. Zellforsch. Mikrosk. Anat.* **112**, 15-30.

Osborne, N. N., Priggemeier, E., and Neuhoff, V. (1975). Dopamine metabolism in characterized neurones of *Planorbis corneus. Brain Res.* **90**, 261-271.

Otsuka, M., Obata, K., Miyata, Y., and Tanaka, Y. (1971). Measurement of γ-aminobutyric acid in isolated nerve cells of cat central nervous system. *J. Neurochem.* **18**, 287-295.

Packman, P. M., Blomstrand, C., and Hamberger, A. (1971). Disc electrophoretic separation of proteins in neuronal, glial and subcellular fractions from cerebral cortex. *J. Neurochem.* **18**, 479-487.

Passonneau, J. V., and Lowry, O. H. (1971). Metabolite flux in single neurons during ischemia and anesthesia. *In* "Recent Advances in Quantitative Histo- and Cytochemistry. Methods and Applications" (U. C. Dubach and U. Schmidt, eds.), pp. 198-212. Huber, Bern.

Peterson, R. P. (1970). RNA in single identified neurons of *Aplysia. J. Neurochem.* **17**, 325-338.

Peterson, R. P., and Erulkar, S. D. (1973). Parameters of stimulation of RNA synthesis and characterization by hybridization in a molluscan neuron. *Brain Res.* **60**, 177-190.

Peterson, R. P., and Kernell, D. (1970). Effects of nerve stimulation on the metabolism of ribonucleic acid in a molluscan giant neurone. *J. Neurochem.* **17**, 1075-1085.

Pevzner, L. Z. (1965). Topochemical aspects of nucleic acid and protein metabolism within the neuron-neuroglia unit of the superior cervical ganglion. *J. Neurochem.* **12**, 993-1002.

Pevzner, L. Z. (1971). Topochemical aspects of nucleic acid and protein metabolism within the neuron-neuroglia unit of the spinal cord anterior horn. *J. Neurochem.* **18**, 895-907.

Pevzner, L. Z. (1972). Topochemical aspects of nucleic acid metabolism within the neuronal-neuroglial unit of cerebellum Purkinje cells. *Brain Res.* **46**, 329-339.

Piven', N. V., and Shtark, M. B. (1976). Immunohistochemical investigation of protein S-100 in neurons and glia of *Helix pomatia. Bull. Exp. Biol. Med. (Engl. Transl.)* **82**, 1876-1878.

Pohle, W., and Matthies, H. (1971). The incorporation of [³H]uridine monophosphate into the rat brain during the training period. A microautoradiographic study. *Brain Res.* **29**, 123-127.

Pohle, W., and Matthies, H. (1974). Incorporation of RNA precursors into neuronal and glial cells of rat brain during a learning experiment. *Brain Res.* **65**, 231-237.

Por, S. B., Komiya, Y., McGregor, A., Jeffrey, P. L., Gunning, P. W., and Austin, L. (1978). The axoplasmic transport of 4S RNA within the sciatic nerve of the chicken. *Neurosci. Lett.* **8**, 165-169.

Powell, B., and Cottrell, G. A. (1974). Dopamine in an identified neuron of *Planorbus corneus. J. Neurochem.* **22**, 605-606.

Price, C. H., Coggeshall, R. E., and McAdoo, D. J. (1978). Specific glycine uptake by identified neurons of *Aplysia californica.* I. Autoradiography. *Brain Res.* **154**, 25-40.

Robins, E. (1960). The chemical composition of central tracts and of nerve cell bodies. *J. Histochem. Cytochem.* **8**, 431-436.

Robins, E., and Hirsch, H. E. (1968). Glycosidases in the nervous system. II. Localization of β-galactosidase, β-glucuronidase, and β-glucosidase in individual nerve cell bodies. *J. Biol. Chem.* **243**, 4253-4257.

Robins, E., Kissane, J. M., and Lowe, I. P. (1961). Quantitative biochemical studies of chromatolysis. *In* "Chemical Pathology of the Nervous System" (J. Holchi-Pi ed.), pp. 244-248. Pergamon, Oxford.

Roots, B. I., and Johnston, P. V. (1964). Neurons of ox brain nuclei: Their isolation and appearance by light and electron microscopy. *J. Ultrastruct. Res.* **10**, 350-361.

Roots, B. I., and Johnston, P. V. (1965). Isolated rabbit neurones: Electron microscopical observations. *Nature (London)* **207**, 315-316.

Ross, C. D., and McDougal, D. B., Jr. (1976). The distribution of choline acetyltransferase activity in vertebrate retina. *J. Neurochem.* **26**, 521-526.

Rüchel, R. (1976). Sequential protein analysis from single identified neurons of *Aplysia californica*. A microelectrophoretic technique involving polyacrylamide gradient gels and isoelectric focusing. *J. Histochem. Cytochem.* **24**, 773-791.

Rüchel, R., Loh, Y. P., and Gainer, H. (1977). A technique for the selective extraction of water-soluble polypeptides from identified neurons of *Aplysia californica*. *Hoppe-Seyler's Z. Physiol. Chem.* **358**, 659-665.

Rude, S., Coggeshall, R. E., and Van Orden, L. S., 3rd (1969). Chemical and ultrastructural identification of 5-hydroxytryptamine in an identified neuron. *J. Cell Biol.* **41**, 832-854.

Rusca, G., Calissano, P., and Alema, S. (1972). Identification of a membrane-bound fraction of the S-100 protein. *Brain Res.* **49**, 223-227.

Saavedra, J. M., and Ribas, J. (1977). Phenylethanolamine: A new putative neurotransmitter in *Aplysia*. *Science* **195**, 1004-1006.

Saavedra, J. M., Brownstein, M. J., Carpenter, D. O., and Axelrod, J. (1974). Octopamine: Presence in single neurons of *Aplysia* suggests neurotransmitter function. *Science* **185**, 364-365.

Sanger, J. W. (1975). Intracellular localization of actin with fluorescently labelled heavy meromyosin. *Cell Tissue Res.* **161**, 431-444.

Schiefer, H.-G., and Neuhoff, V. (1971). Fluorometric microdetermination of phospholipids on the cellular level. *Hoppe-Seyler's Z. Physiol. Chem.* **352**, 913-926.

Schwartz, J. H., Castellucci, V. F., and Kandel, E. R. (1971). Functioning of identified neurons and synapses in abdominal ganglion of *Aplysia* in absence of protein synthesis. *J. Neurophysiol.* **34**, 939-953.

Shashoua, V. E. (1970). RNA metabolism in goldfish brain during acquisition of new behavioral patterns. *Proc. Natl. Acad. Sci. U.S.A.* **65**, 160-167.

Shashoua, V. E. (1974). RNA metabolism in the brain. *Int. Rev. Neurobiol.* **16**, 183-231.

Shashoua, V. E. (1976). Brain metabolism and the acquisition of new behaviors. I. Evidence for specific changes in the pattern of protein synthesis. *Brain Res.* **111**, 347-364.

Sober, H. A. (1973). "Handbook of Biochemistry, Selected Data for Molecular Biology," 2nd ed., H112-H113. CRC Press, Cleveland, Ohio.

Sobkowicz, H. M., Hartmann, H. A., Monzain, R., and Desnoyers, P. (1973). Growth, differentiation and ribonucleic acid content of the fetal rat spinal ganglion cells in culture. *J. Comp. Neurol.* **148**, 249-284.

Strumwasser, F., and Wilson, D. L. (1976). Patterns of proteins synthesized in the R15 neuron of *Aplysia*. Temporal studies and evidence for processing. *J. Gen. Physiol.* **67**, 691-702.

Sviridov, S. M., Korochkin, L. I., Ivanov, V. N., Maletskaya, E. I., and Bakhtina, T. K. (1972). Immunohistochemical studies of S-100 protein during postnatal ontogenesis of the brain of two strains of rats. *J. Neurochem.* **19**, 713-718.

Tobias, G. S., and Koenig, E. (1975a). Axonal protein-synthesizing activity during the early outgrowth period following neurotomy. *Exp. Neurol.* **49**, 221-234.

Tobias, G. S., and Koenig, E. (1975b). Influence of nerve cell body and neurolemma cell on local axonal protein synthesis following neurotomy. *Exp. Neurol.* **49**, 235-245.

Toevs, L. A., and Brackenbury, R. W. (1969). Bag cell-specific proteins and the humoral control of egg laying in *Aplysia californica*. *Comp. Biochem. Physiol.* **29**, 207-216.

Treistman, S. N., and Schwartz, J. H. (1974). Injection of radioactive materials into an identified axon of *Aplysia. Brain Res.* **68**, 358–364.

Treistman, S. N., and Schwartz, J. H. (1977). Metabolism of acetylcholine in the nervous system of *Aplysia californica.* IV. Studies of an identified cholinergic axon. *J. Gen. Physiol.* **69**, 725–741.

Uemura, E., and Hartmann, H. A. (1978). RNA content and volume of nerve cell bodies in human brain. I. Prefrontal cortex in aging normal and demented patients. *J. Neuropathol. Exp. Neurol.* **37**, 487–496.

Ungerstedt, U. (1971). I. Stereotaxic mapping of the monoamine pathways in the rat brain. *Acta Physiol. Scand., Suppl.* **367**, 1–48.

Vraa-Jensen, J. (1971). RNA in the cerebellum of the cat during postnatal development and the effect of specific stimulation. I. Cytoplasmic basophilia, volume and enzymes in the Purkinje cells. *Neurobiology* **1**, 191–202.

Vraa-Jensen, J. (1972). RNA in the Purkinje cells of the cat during specific physiological stimulation: An autoradiographic study. *Brain Res.* **42**, 525–528.

Weinreich, D., Dewhurst, S. A., and McCaman, R. E. (1972). Metabolism of putative transmitters in individual neurons of *Aplysia californica:* Aromatic amino acid decarboxylase. *J. Neurochem.* **19**, 1125–1130.

Weinreich, D., McCaman, M. W., McCaman, R. E., and Vaughn, J. E. (1973). Chemical, enzymatic and ultrastructural characterization of 5-hydroxytryptamine-containing neurons from the ganglia of *Aplysia californica* and *Tritonia diomedia. J. Neurochem.* **20**, 969–976.

Wessells, N. K., Spooner, B. S., Ash, J. F., Bradley, M. O., Luduena, M. A., Taylor, E. L., Wrenn, J. T., and Yamada, K. M. (1971). Microfilaments in cellular and developmental processes. *Science* **171**, 135–143.

Wilson, D. L. (1971). Molecular weight distribution of proteins synthesized in single, identified neurons of *Aplysia. J. Gen. Physiol.* **57**, 26–40.

Wilson, D. L. (1974). Protein synthesis and nerve cell specificity. *J. Neurochem.* **22**, 465–467.

Wilson, D. L. (1976). Alteration of protein metabolism in individual, identified neurons from *Aplysia. J. Neurobiol.* **7**, 407–416.

Wilson, D. L., and Berry, R. W. (1972). The effect of synaptic stimulation on RNA and protein metabolism in the R2 soma of *Aplysia. J. Neurobiol.* **3**, 369–379.

Wu, J.-Y. (1978). Microanalytical methods for neuronal analysis. *Physiol. Rev.* **58**, 863–904.

Yamada, K. M., Spooner, B. S., and Wessells, N. K. (1971). Ultrastructure and function of growth cones and axons of cultured nerve cells. *J. Cell Biol.* **49**, 614–635.

Yanagihara, T. (1979). Protein and RNA synthesis and precursor uptake with isolated nerve and glial cells. *J. Neurochem.* **32**, 169–177.

Yanagihara, T., and Hydén, H. (1971). Protein synthesis in various regions of rat hippocampus during learning. *Exp. Neurol.* **31**, 151–164.

Zeman, G. H., and Carpenter, D. O. (1975). Asymmetric distribution of aspartate in ganglia and single neurons of *Aplysia. Comp. Biochem. Physiol.* C **52**, 23–26.

Zeman, G. H., Myer, P. R., and Dalton, T. K. (1975). Gamma-aminobutyric acid uptake and metabolism in *Aplysia dactylomela. Comp. Biochem. Physiol.* C **51**, 291–299.

Zomzely, C. E., Roberts, S., and Peache, S. (1970). Isolation of RNA with properties of messenger RNA from cerebral polyribosomes. *Proc. Natl. Acad. Sci. U.S.A.* **67**, 644–651.

Section 2

AGING AND PATHOLOGY

ADVANCES IN CELLULAR NEUROBIOLOGY, VOLUME 1

CEREBELLAR GRANULE CELLS IN NORMAL AND NEUROLOGICAL MUTANTS OF MICE

ANNE MESSER

Division of Laboratories and Research, New York State Department of Health
Albany, New York

I. Introduction

The cerebellum is probably the most comprehensively studied area of the brain, beginning with the cytological work of Ramón y Cajal and continuing through extensive and elegant work such as the physiological studies of Eccles *et al.* (1967) and the morphological studies of Palay and Chan-Palay (1974). The cerebellar cortex is characterized by a trilaminar (granule, Purkinje, and molecular) structure and contains only five different neuronal cell classes (granule, Purkinje, basket, stellate, and Golgi), which interconnect in a rigid geometry of repeating groups presumed to be responsible for specific information processing (Eccles *et al.*, 1967).

Whereas the large, elaborate Purkinje cells are the most prominent cell type in the cerebellum, the tiny (5–8 μm) granule cells are by far the most numerous, with large increases in the granule/Purkinje cell ratio in the brains of higher animals (e.g., 140:1 in the mouse, 250:1 in the rat, 950:1 in the monkey, and 1100:1 in the human; Blinkov and Glezer, 1968). These numbers portend an enormous increase in the complexity of interconnections, since, physiologically, granule cells are in contact with all the other neuronal cell types in the cerebellum. Granule cell dendrites receive afferent input from mossy and climbing fibers, as well as Golgi and Purkinje cell feedback fibers, and granule cell axons synapse on all the inhibitory interneurons (basket, stellate, and Golgi) in the cerebellum, as well as the Purkinje cells (Fig. 1, see p. 184).

The development of this highly organized area of the brain has received a great deal of attention, partly because of its well-defined physiology, cell types, and architecture, and partly because of the availability of mouse mutations (see Table I) which perturb the normal sequence of development, allowing analysis of individual events (reviewed in Caviness and Rakic, 1978).

Purkinje cells are generated relatively early in development (embryonic days 11–13 in the mouse), and the precursor cells migrate from the roof of the fourth ventricle in the cerebellar anlage. Postnatally, granule cell precursors proliferate on the external surface of the developing cerebellum. Postmitotic granule cells then migrate inward along a Bergmann glial fiber to the granule layer, while an axon remains in the molecular layer and con-

TABLE I

SUMMARY OF MUTANT CHARACTERISTICS[a,b]

Mutant	Granule Cell				Purkinje Cell		Bergmann glial fibers (early ages)
	Number present at 4 weeks	Proliferation	Migration	Number present at 4 weeks	Shape		
Staggerer (sg/sg)	Almost none	Reduced	N once started	~75% reduction overall	Very abnormal—reduced arbor and no tertiary spines		N
Weaver (wv/wv)	Almost none	Somewhat reduced (?)	Very abnormal	N(?)	Slightly abnormal		Abnormal
(+/wv)	~20% reduction	N	Slow	N(?)	N		Slightly abnormal
Reeler (rl/rl)	>80% reduction	Reduced(?)	Along oblique fibers(?)	N(?)	Fairly abnormal		Oblique

[a] From Hatten and Messer, 1978.
[b] Data given in this table are discussed and referenced in the text. N = appears normal.

tacts the growing dendrites of the earlier formed Purkinje cells (Miale and Sidman, 1961; Rakic, 1971; Sidman, 1972).

A number of questions can be asked about the development of the cerebellum, most of which have wide implications for the development and functioning of the remainder of the brain as well. For example, what controls the proliferation of the external granule cell, especially in proportion to the other cell types? (This is of interest both from an ontogenetic and a phylogenetic point of view.) What influences the final position of the cells? What influences the formation of specific synapses by the granule cells? What is the transmitter at these synapses?

These are rather broad questions. Answers to more specific aspects of these questions have been sought by studying a few neurological mutants (Table I), utilizing a variety of technical approaches. With the exception of the question regarding the transmitter, cellular and biochemical studies of the phenomena described morphologically are just beginning. Some data do exist, however, and the technology becoming available from other fields should allow increasingly sophisticated examination of developmental processes in the cerebellum.

The first three sections of this review will concentrate on two neurological mutants, staggerer *(sg/sg)* and weaver *(wv/wv)*, in which granule cells degenerate almost completely. It will focus on approaches that lead to information on what is transpiring at the *cellular* level in the mutant cerebella and, by inference, on what may be happening in the normal case. In addition to the microscopic structure of brains perfused *in vivo,* studies that use newer techniques for dissection of cellular processes will be highlighted. Ideally, in order to examine function at the cellular level, single cells should be followed in their normal milieu and then in situations where proper interactions with their neighbors are disrupted in defined and reproducible ways. Although this degree of precision is not feasible given the limits of present technology, *in vitro* studies of cell types isolated from mutant brains, or mixing mutant cells with wild-type cells *in vivo* in tetraparental chimeras, offer some valuable insights and data. Wolf (1978) has written a recent review on some of the tissue culture experiments, and Mullen (1977) has reviewed his work on chimeras.

Data on the reeler mutant, in which the organization of the cortex is disrupted, have not yet been generated at a cellular level. The results with experimental chimeras do, however, allow some conclusions. Caviness and Rakic (1978) have recently published an extensive review of mouse cortical development, highlighting the literature on reeler and weaver mutants.

As to the identity of the granule cell transmitter, investigations using cerebellar mutants as well as virally depleted cerebella, isolation of cerebellar cells, and kainic acid injections will be cited.

II. Possible Effect of Purkinje Cells on Proliferation of Granule Cells

A. Staggerer Mutant Mice as a Model System

1. MORPHOLOGICAL STUDIES

The neurological mutant staggerer *(sg/sg)* was described by Sidman *et al.* (1962) as an autosomal recessive, ataxic mouse with cerebellar pathology characterized by extensive degeneration of the granule cells shortly after their migration into the granule layer. Extensive abnormalities of the Purkinje cells have also been revealed by morphological studies showing that (1) the Purkinje cells are ectopic with skimpy dendritic arbors (Sidman, 1968); (2) their tertiary branchlet spines, which normally contain the postsynaptic sites for synapses with granule cell parallel fibers, are absent (Hirano and Dembitzer, 1975; Landis and Sidman, 1978; Sotelo and Changeux, 1974a); and (3) there is a regional variation in severity along the mediolateral axis, and cell counts reveal that about 75% of the medium-to-large neurons—which may include Golgi as well as abnormal Purkinje cells—are actually missing (Herrup and Mullen, 1979b). The intrinsic nature of the postnatal *sg/sg* Purkinje cell defect has been substantiated using two investigative systems that allow mixing of staggerer with nonstaggerer influences. Yoon (1976, 1977b) examined reeler–staggerer *(rl/rl–sg/sg)* double mutants in which the effects of the *sg* gene are superimposed on the abnormally organized cortex of the reeler *(rl/rl)* mutant. Despite location of the Purkinje cells in positions that seem to be determined as much by the *rl* as by the *sg* gene, the resulting double-mutant Purkinje cells are readily identifiable by fine-structure analysis as characteristic of staggerer Purkinje cells. In another genetic approach, experimental (tetraparental) chimeras are created, allowing the investigation of interactions of genotypically wild-type and genotypically mutant cells within a single brain. Studies of *sg/sg* ←→ wild-type (+/+) chimeras reveal the presence of phenotypically staggerer Purkinje cells, identified as being *sg/sg* in genotype both by means of their morphology and by means of an independent marker gene, in relatively normal regions of the cerebellum (Herrup and Mullen, 1976, 1979a; Mullen, 1977).

Given the nature of the Purkinje cell defects, particularly the absence of the granule cell postsynaptic sites, it has been suggested that the primary site of gene action is in the Purkinje cells and that the granule cells degenerate secondary to this, a transsynaptic degeneration *en cascade* (Sotelo and Changeux, 1974a; Sidman, 1968). However, the granule cells,

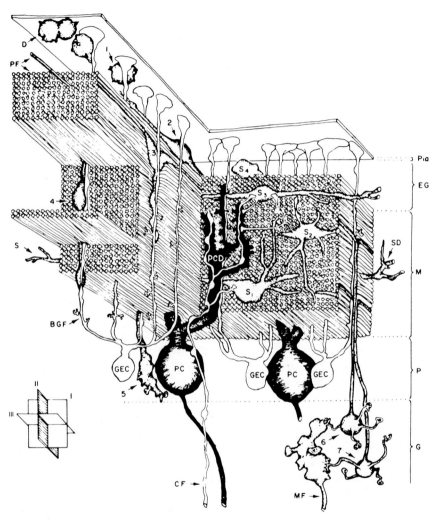

FIG. 1. A "four-dimensional" (time and space) diagrammatic reconstruction of the histogenesis of the cerebellar cortex. The orientation of the drawing is indicated by the figure on the lower left. I, Transverse to the folium; II, longitudinal to the folium; III, parallel to the pial surface. BGF, Bergmann glial fiber; D, dividing external granule cell; EG, external granule layer; GEC, Golgi epithelial cell (Bergmann glia); G, granular layer; M, molecular layer; MF, mossy fiber; P, Purkinje layer; PC, Purkinje cell; PCD, Purkinje cell dendrite; PF, parallel fiber; S^{1-4}, stellate cells; SD, stellate cell dendrite. (From Rakic, 1974.)

although they appear to migrate normally, actually exhibit a reduced rate of proliferation in, and premature emigration from, the external granule layer (Yoon, 1976). Furthermore, the external granule layer is more severely affected in the *sg/sg–rl/rl* double mutant than in either of the mutants separately (Yoon, 1977a).

2. MONOLAYER CELL CULTURE STUDIES

In order to examine more precisely the intrinsic cellular effects of the cerebellar mutants, Messer (1977a) developed a monolayer cell culture system based on the work of Barkley *et al.* (1973), Lasher and Zagon (1972), Nelson and Peacock (1973), and Lasher (1974) in which a mixture of cerebellar cells, including identifiable granule cells, could survive *in vitro* for periods of up to several weeks. Cells from postnatal day 7 mice, which have been dissociated with trypsin, are plated in a special high-potassium medium (Lasher and Zagon, 1972) with added fetal calf serum, and their morphological differentiation is monitored by means of phase-contrast microscopy. It appears that granule cells, inhibitory interneurons, and a variety of glial cells survive under these conditions, although there is no evidence for the persistence of Purkinje cells. (This may be due to the fact that the Purkinje cells are relatively well differentiated by the time of dissociation, whereas most of the surviving cells are either still dividing or recently postmitotic.) The most crucial point is that the cerebellar granule cells survive in large numbers and can be positively identified by utilizing a combination of criteria which include (1) size, shape, and relative proportion of the total cell population, as determined by phase-contrast and scanning electron microscopy (Fig. 2); (2) nuclear morphology, demonstrated by transmission electron microscopy (Fig. 3b); and (3) failure to take up [^{3}H]γ-aminobutyric acid (GABA) in the presence of several other cell types that do, shown by autoradiography.

Under these culture conditions, cerebellar granule cells from *sg/sg* mutants both clump less (Figs. 2 and 3a) and survive considerably longer (Fig. 4) than their wild-type counterparts (Messer and Smith, 1977). This indicates that, while there is no irreversible programmed cell death of the mutant cells, there is a measurable difference between mutant and wild-type behavior under some conditions. Further experiments have revealed that, when the substrate on which the cells are plated is altered, these cell behavior differences also vary. Among conditions tested, polylysine coating (Yavin and Yavin, 1974; Letourneau, 1975), which in this system promotes cell attachment to the substrate, diminishing intracellular adhesion, shows the smallest differences in survival between mutant and control. However, in order to eliminate the difference entirely it is necessary to

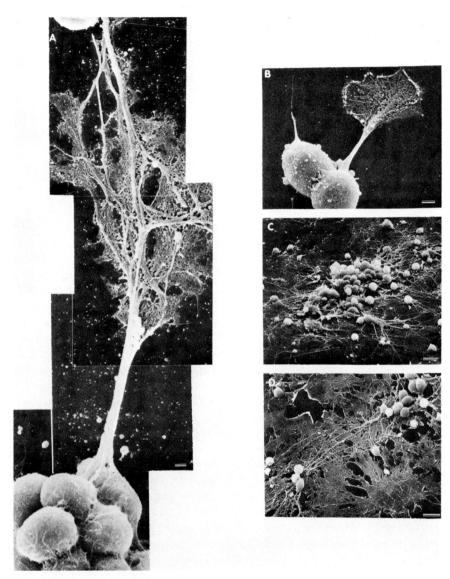

FIG. 2.   Scanning electron micrographs of fixed, coated cells. (A) One day *in vitro* (DIV), bar = 1 μm; (B) 0.5 DIV, bar = 1 μm; (C and D) 8 DIV, bar = 10 μm. (From Messer, 1977a.)

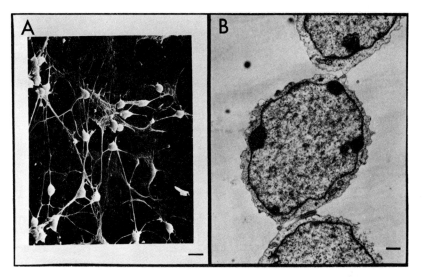

FIG. 3.   Electron micrographs of *sg/sg*, 3 DIV. (A) Scanning, bar = 10 μm; (B) transmission, bar = 1 μm (by James Kaye). (From Messer and Smith, 1977.)

change the media supplement from fetal calf serum to horse serum, in addition to using the polylysine coating (Fig. 4). It was concluded that there were intrinsic differences between mutants and controls in their postnatal granule cell–cell interactions and that these differences could be demonstrated qualitatively by manipulation of culture conditions (Messer, 1977b, 1978). (The question of granule cell survival is discussed in Section III.)

3. HYPOTHESES OF STAGGERER GENE ACTION

Yoon (1976, 1977a) favors the hypothesis that the effects of the *sg* gene are pleiotropic, affecting the granule and the Purkinje cells separately, while Sotelo and Changeux (1974a), Mallet *et al.* (1976), Landis and Sidman (1978), and Herrup and Mullen (1979b) favor the interpretation that abnormal Purkinje cells may modulate the development of the granule cells. The latter two articles, in particular, discuss the possible role of abnormal cell interactions during crucial early stages of development.

Landis and Sidman (1978) present a model whereby the Purkinje cell abnormality can lead to a block in Purkinje cell–granule cell interactions. This, in turn, may be responsible for granule cell hypoplasia and perhaps even some of the other Purkinje cell defects that appear later. They suggest that the initial (cerebellar) effect of the gene is probably expressed prenatally, which would be consistent with the observations of Yoon (1972) that the external granule cell layer is hypoplasic at birth.

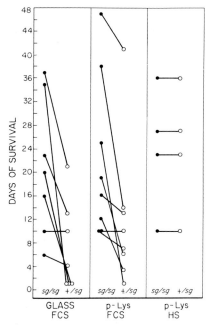

FIG. 4.   Comparative cell survival. Survival of cells observed by phase-contrast microscopy of living cultures prepared from cerebella of staggerer *(sg/sg)* mutants and their littermate controls *(+/sg)*. Cultures from littermates are joined by solid lines. Cells are grown as described in the text, on glass or polylysine (p-Lys) substrata, in media supplemented with fetal calf serum (FCS) or horse serum (HS). (From Messer, 1978.)

Herrup and Mullen (1979b) also make a time-dependent argument that early abnormalities in the Purkinje cells affect the later-generated granule cells rather than vice versa. Given the number of morphological studies on staggerer brains, they suggest that the enormous loss of large neurons they report should have been observed as degenerating cells unless (1) the cells are never generated, or (2) cell death is taking place very early in development. In fact, Herrup and Mullen (1979b) cite evidence from their own work on reduced numbers of medium-to-large neurons in the cortex, as well as that of Roffler-Tarlov and Sidman (1978), showing reduced weights of *sg/sg* deep nuclei, which suggests that all the derivatives of the ventricular zone of the roof of the fourth ventricle may be affected in *sg/sg*.

4. Cell Surface Characteristics of Granule Cells in Staggerer Mutants

If the normal granule cell precursors are dependent on some signal from the developing normal Purkinje cells to direct their further correct development, then there may be a demonstrable lack of such a signal on mutant granule cells. Qualitatively, *sg/sg* granule cells can be seen, as noted above, to exhibit intracellular adhesion patterns different from those of their normal counterparts. Two studies that attempt to quantitate cell surface characteristics conclude that postnatal *sg/sg* cells seem to retain embryonic cell surface properties. One set of experiments has revealed that wheat germ agglutinin (WGA) will agglutinate normal embryonic but not normal postnatal cells (Hatten and Sidman, 1978). Staggerer postnatal cells, however, *are* agglutinated by WGA. This agglutination is specific both with respect to hapten, since it is inhibited by *N*-acetyl-D-glucosamine, and with respect to lectin, since neither concanavalin A nor *Ricinus communis* gives similar results (Table II) (Messer and Hatten, 1977; Hatten and Messer, 1978). Trenkner *et al.* (1977) reported similar results with a totally different assay system. They studied the effects of an antisialic acid antibody on (1) outgrowth of fibers in microwell cultures, and (2) fluorescence of cryostat sections of cerebellum, after the antibody had been conjugated with fluorescein isothiocyanate. Normal embryonic and postnatal day-7 *sg/sg* cerebellar cells bind the antibody, whereas normal postnatal cells do not.

While neither of these studies offers firm proof that *sg/sg* granule cells have failed to undergo a change from their embryonic states, they are certainly both consistent with such a hypothesis. Further studies are needed to

TABLE II

AGGLUTINATION OF STAGGERER AND WILD-TYPE MOUSE CEREBELLAR CELLS WITH
CONCANAVALIN A AND WGA[a]

Genotype	Age	WGA	Concanavalin A
Wild-type C57BL/6J	E13	50	200
	P0	450	800
	P3	500	1000
	P7	500	1000
	P10	500	1000
Wild-type +*dse*/+*dse*	P7	500	1000
	P10	500	1000
Staggerer *sg*+ +/*sg*+ +	P7	50	1000
	P10	50	1000

[a] Values are given as the lectin concentration (micrograms per milliliter) required for half-maximal agglutination (75%). Agglutination assays were performed in quadruplicate, and results given are averaged values from four sets of assays, each set having been performed with animals from a different litter. (From Hatten and Messer, 1978.)

specify exactly which cell types are responsible for the differences seen, followed by a more complete time course to establish that multiple changes are not taking place within the times that have not been examined. Agglutination studies of cerebellar cells from *wv/wv* and *rl/rl* mice are planned to test the specificity of this assay. The existence of positive results obtained using two different kinds of measurements here, plus the qualitative monolayer cell culture data on differences in mutant cell interactions that can be more definitively assigned to granule cells, is certainly very suggestive of a change in the granule cells.

B. Other Model Systems

There are a few other animal systems in which there seems to be a link between Purkinje cell abnormalities and granule cell loss. The most prominent of these is the Gunn rat, in which there is marked cerebellar hypoplasia with hereditary hyperbilirubinemia. The major pathological change in the cerebellar cortex is seen in the Purkinje cell, but a decrease in the postnatal DNA synthesis in the external granule layer has also been observed, as well as a decrease in the activity of the enzyme thymidine kinase. Since the levels of bilirubin found in the postnatal mutant rat cerebellum are not sufficient to inhibit the activity of the thymidine kinase enzyme directly, and no improvement of enzyme activity is observed in photoirradiated mutant cerebellar cytosol fractions, a decrease in the induction of the enzyme is implicated (Yamada *et al.*, 1977). Purkinje cell abnormalities have been noted as early as 2–3 days after birth (Schutta and Johnson, 1967); therefore, this mutant provides another example of a mutation in which an early abnormality in Purkinje cells is followed by decreased proliferation in the external granule layer. It is particularly interesting because there is a defined metabolic defect for this mutant as well as a focal cerebellar defect in the level (probably induction) of an enzyme crucial to DNA synthesis.

Thyroid dysfunction in rats may offer an experimentally induced model of a similar effect, although the precise development and pathology of the defect varies from that of the mutants. Hypothyroidism results in early reduction, followed by prolonged cell proliferation, in the external granule layer, in abnormalities of Purkinje cell arborizations, and in increased cell death of granule cells (Nicholson and Altman, 1972; Rabié *et al.*, 1977). This is similar to the description of *sg/sg*, although the reduced rate of proliferation in the external granule layer is temporary and the prolonged survival of this layer, allowing the cerebellum to "catch up," resembles x-irradiated rather than the mutant animals. However, the initial defective signal may still have to do with the Purkinje cell abnormality in this case.

Experiments that directly test the cell surface properties of cerebellar cells from these rat models might prove extremely interesting.

C. Conclusion

The hypothesis that developing Purkinje cells effect a specific cell surface signal controlling the proliferation of external granule cells is both attractive and testable. Data already generated are very promising in support of such a hypothesis, and the techniques just described extended to additional mutants of the mouse and the rat, as well as possibly animals with experimental cerebellar hypoplasia, should allow further analysis of the system.

III. Effect of Glial Cells on Granule Cell Migration

A. Weaver Mutant Mice as a Model System

1. MORPHOLOGICAL STUDIES

Homozygous weaver mutants *(wv/wv)* are characterized clinically by severe ataxia and tremor. Histologically, there is an almost total depletion of the granule cell population of the cerebellum, due to cell death in the external granule cell layer during the first two postnatal weeks. The remainder of the cerebellum is somewhat disorganized and reduced in volume (Sidman *et al.,* 1965; Sidman, 1968). Although the original mutation was considered to be inherited as an autosomal recessive, Rezai and Yoon (1972) established that cerebellar development was also abnormal in the heterozygotes, giving a semidominant effect at the morphological level. These authors reported that the major defect in heterozygotes was a reduced rate of postmitotic granule cell migration following normal proliferation. They suggested that the extensive cell death in the homozygote could be due to failure of migration. This work was confirmed and expanded by Rakic and Sidman (1973a,b), who cited not only retarded granule cell migration but also abnormalities of the Bergmann glial fibers, along which the granule cells appear to migrate during normal development. These abnormalities, and particularly the gene dosage effect—fewer abnormalities among glia corresponding to greater survival of granule cells in the heterozygote—led to the conclusion that granule cells were dying secondary to a failure to migrate along the abnormal glial fibers. The authors also noted that, even in the homozygote, there was a decreased severity in the lateral parts of the cerebellum (Rakic and Sidman, 1973b).

Bergmann glial fibers studied by immunofluorescence with glial fibrillary acidic (GFA) protein antiserum, however, show only modest changes (Bignami and Dahl, 1974). Also, Sotelo and Changeux (1974b) confirmed structural abnormalities but not decreased numbers of Bergmann fibers. In addition, they observed that those few granule cells that did successfully migrate also died. They therefore postulated a direct effect of the mutation on the granule cell neurons. If the granule cells are directly and severely affected by the mutation, a modest glial abnormality may have nothing to do with their failure to migrate.

Since Rakic and Sidman used the inbred C57BL/6J weaver strain and Sotelo and Changeux used the hybrid B6CBA weaver strain, it is possible that the differences observed, and the conclusions drawn, from the two groups are due to genetic variation. However, given the small number of surviving granule cells being considered, and the lateral variation reported by Rakic and Sidman, the precise location of the sections may also be a crucial variable. While some granule cells that migrate successfully die, others, in the more lateral parts of the cerebellum, both migrate and survive. Thus, although granule cells seem more severely affected than glial cells, the morphological data do not exclude a role for glial cells in the migratory defect.

2. CELL CULTURE STUDIES

Messer and Smith (1977) examined the behavior of weaver granule cells in a monolayer cell culture system (described in Section II,A,2). They found that the mutant cells could survive under these conditions and appeared similar to control cells in culture. However, processes emerging from small clumps of cells in the cultures do not seem to form the large fiber bundles that those from normal clumps do, suggesting that different culture conditions might reveal further differences in cell behavior.

A microwell culture system was developed in order to analyze more precisely cell interactions and migrations *in vitro* (Trenkner and Sidman, 1977). Cells dissociated from the cerebella of postnatal day-7 mice are plated at high density and allowed to reaggregate into small clusters. After about 1 day, the 10–15 aggregates per well develop interconnections of either sheets of migrating cells and cell processes or cables of fiber bundles with cells migrating along their surfaces. Scanning and transmission electron microscopy are used to identify granule cells, basket and/or stellate cells, Purkinje cells, and two types of glial cells, as well as synapses. Differences in the patterns of aggregation and interconnection are noted under different culture conditions. In particular, horse serum generates clumping and two kinds of interconnections, whereas fetal calf serum promotes adhesion to the surface with formation of a monolayer.

When cells from homozygous weaver mice are tested in this system, few cables of fiber bundles are seen under conditions in which many such cables form in control cultures. There is also more degeneration of granule cells in mutant than in wild-type cultures, as well as a reduction in the number of migrating granule cells seen. When ethanol–ether-extracted horse serum is used in place of ordinary horse serum, control cultures double their production of cables, whereas mutant cultures increase their production 10-fold. Death and migration defects are also reduced in treated mutant cultures. Addition of the ether-soluble components of the serum can reverse this effect, although purified cholesterol, stearate, palmitate, oleate, lineoleate, testosterone, hydrocortisone, or thyroxine, added back individually, cannot (Trenkner *et al.,* 1978).

These results indicate that, by using appropriate manipulations of culture conditions, large differences can be observed in the behavior of mutant and control cells. While the granule cells appear to be directly and severely affected in this system, it is not possible to identify unequivocally the mutant cell type specifically responsible for the differences in behavior. The aggregates probably contain glial as well as granule cells, and the fiber bundles and sheets of processes probably consist of a mixture of fiber types. This system does, however, offer great promise and perhaps a combination of growth of separated cell types (Campbell and Williams, 1978), and better individual cell markers will allow a distinction between those mutant properties intrinsic to granule cells and those intrinsic to the Bergmann glia. The definitive experiment will require mixing of separated and identified wild-type granule cells with *wv/wv* glial cells, and vice versa.

B. Conclusion

Granule cells appear to migrate along radial glial fibers during the cerebellar development of normal animals (Rakic, 1971), and along skewed glial fibers during development of the misaligned cortex of *rl/rl* mice (Rakic, 1976). They also appear to migrate at a reduced rate along the glial fibers of the *+/wv* heterozygote and appear not to associate at all with the abnormal glial fibers of *wv/wv*. Cell culture data point to a biochemical basis for the association phenomenon, but the mixed cell culture system does not actually allow identification of the cell type(s) involved. A hypothesis that *wv/wv* cells are abnormally sensitive to some component of serum that inhibits the normal interaction between granule and glial cells is most consistent with the results. The weight of the evidence favors a granule cell defect, but this has not been demonstrated rigorously.

IV. Survival of Granule Cells in Mutant Mice (Staggerer and Weaver)

A. *Advantages and Limitations of Tissue Culture Methods*

It is of great interest that there are cell culture conditions under which the *wv/wv* and *sg/sg* granule cells, which degenerate *in vivo,* survive *in vitro.* The fact that these mutant granule cells can survive *in vitro* suggests that there is no preprogrammed irreversible cell death imposed by the mutant genes. It does not mean, however, that the cells are identical to wild type. They may, in fact, harbor intrinsic differences that predispose them to degeneration in their normal milieu. A careful evaluation of these phenomena should yield information on the causes of the cell deaths *in vivo.*

When survival is used as a criterion, the two kinds of mutant cells appear to behave in two different ways when compared to wild-type under specific culture conditions (i.e., there are conditions under which *sg/sg* cells survive *longer* than controls, and there are other conditions under which *wv/wv* cells degenerate much *sooner* than controls). However, the absence of one behavior or the other might simply be a function of failure to find the right culture conditions; therefore, negative results are not too meaningful. Assays of survival are also qualitative, so that small differences could easily be missed.

Both of the mutant cell types are responsive to differences in the serum used to supplement the growth media. Since the serum in tissue culture medium supplies a large collection of proteins, hormones, and trace elements, this is hardly surprising and does not allow a great deal of additional speculation. Both responses also seem to exhibit differences in the number and types of cell interactions. The analysis of the data is made more difficult by the fact that the cell cultures are heterogeneous not just for the major categories of cerebellar cell types but for the degrees of differentiation at the time of dissociation as well. The mutants have the additional heterogeneity of the extent to which the cells are affected, since both *sg/sg* and *wv/wv* show variations in the severity of the disorder in different parts of the cerebellum. Thus the population of "mutant" granule cells probably includes those that are premitotic, postmitotic, migrating, and mature, with "normal" and "mutant" phenotypes superimposed. Nevertheless, the data that have been generated show that there are still distinctions between mutant and wild-type behavior, and that it is even possible to quantitate some effects (such as agglutination).

B. Survival of Granule Cells from Staggerer Mutant Mice in Vitro

The possible reasons for the increase in the survival of sg/sg cells when compared to $+/+$ fall into two categories. One reason may be that the cells in culture are not receiving signals to complete the maturation process and that such signals force the mutant cells to degenerate *in vivo*. This explanation of the culture behavior is somewhat unlikely, since the cells do develop a mature appearance *in vitro* at the electron microscopic level—they have rounded shapes, small amounts of cytoplasm, vesiculated processes and, occasionally, synapses that seem to involve granule cell processes (Messer, 1977a; Messer and Smith, 1977; J. Kaye and A. Messer, unpublished observations; Trenkner and Sidman, 1977; Nelson and Peacock, 1973). However, because of their very small size, it is difficult to examine the synaptic activity physiologically, and so the latter must be regarded as inconclusive. It is also difficult to measure maturation of transmitter function, since glutamic acid, the putative transmitter (Section V), is ubiquitous.

The second theory of *in vitro* granule cell survival is that there are changes in the patterns of glyco conjugates on the cell surface, which engender a fortuitous change in cell–serum–cell and/or cell–serum–substrate interactions. This in turn promotes survival of mutant cultures. Such a change appears to be intrinsic to granule cells but could also be due to some other cell type that is affecting the mixed cell culture behavior. It is not likely that this other cell type is the Purkinje cell, since such cells do not appear to be present in these cultures at all (mutant or wild type). Glial cells may play a role. In fact, preliminary data indicate that the factor excreted by the C6 glioma cell line, which causes neuroblastoma cells to send out processes, also can help to "rescue" wild-type cell cultures degenerating under culture conditions using fetal calf serum (Messer, 1976).

In any case, the culture experiments show that normal granule cells outside their normal milieu do not degenerate because of a failure to form synapses with Purkinje cells.

C. Survival of Granule Cells from Weaver Mutant Mice in Vitro

Granule cells from weaver mutant mice *(wv/wv)* survive in culture when a serum fraction is removed from the medium. This implies the presence of a specific toxic agent in that fraction. (Normal cells also do better without the fraction, but only to a small extent, and they will survive in its presence.) As Trenkner *et al.* (1978) note, it is hard to determine whether it is the granule cell degeneration or the lack of cable formation that comes first *in vitro,* or whether the two are, in fact, concurrent and/or independent. The data from the monolayer studies (Messer and Smith, 1977) weakly favor the possibility of a differential interaction with fiber formation without degeneration. This differential sensitivity to a normal component of serum in turn imposes some sort of differential reception mechanism (broadly defined). Such a mechanism could be a property of either neurons or glia or both. The results of experiments using hyperthyroid rats may be of interest, since the hyperthyroid neonatal rat seems to terminate the proliferative phase of cerebellar development early, leading to a decreased overall number of granule cells (Rabié *et al.,* 1977). Removal of a hormonal pressure on particularly sensitive mutant cell populations could be occurring either for the neurons or the glia (which are known to have their own set of hormonal differentiation signals).

It could also be that a normal cell interaction is being inhibited in *wv/wv* by some normal component of serum, such as a lipid. This could be due to an aberrant receptor on either neurons or glial cells. Removal of this component would then restore normal behavior. Again, it is probably significant that, whatever the pressure causing the degeneration *in vivo,* it does not seem to be present *in vitro* in monolayers, despite some evidence of abnormal interactions.

D. Conclusion

Survival of granule cells from both mutant and wild-type mice can be altered by manipulation of culture conditions. Since monolayers do not contain cells that can be identified as normal Purkinje cells, Purkinje cell synaptic contact must not be an absolute requirement for survival, although it is possible that, as a result of chronic immaturity, the *in vitro* granule cells differ from their *in vivo* counterparts. In the case of *wv/wv,* the *in vitro* data suggest that mutant cells *in vivo* may be abnormally sensitive to some component of normal serum.

V. Parallel Fiber–Purkinje Cell Synapses

A. *Reeler Mutant Mice as a Model System*

1. MORPHOLOGICAL STUDIES

The motor dysfunction and genetics of the reeler mutant *(rl/rl)* were first described by Falconer (1951). The cerebellum was later recognized to be greatly reduced in size and lacking the cellular orientation and alignment that characterize the normal cerebellum (Hamburgh, 1963; Sidman, 1968). Although the ratio of granule to Purkinje cells is greatly reduced, the former are not missing entirely, as is the case in *sg/sg* and *wv/wv*. Many surviving granule cells of *rl/rl* lie external to the Purkinje cells, which are scattered throughout the white matter of the cerebellum rather than lined up neatly between the granule and molecular layers. The dendritic arbors of these ectopic Purkinje cells are relatively normal in form but appear to have a random orientation. Unlike the other two mutant genes described, the *rl* gene is also expressed in other areas of the brain, which exhibit a systematic malposition of cells analogous to that seen in the cerebellum (review in Caviness, 1977).

2. CHIMERA STUDIES

In the case of the *sg/sg* ⟷ +/+ chimera, it is possible to demonstrate that there is an intrinsic defect in the mutant Purkinje cells. When a similar experiment is done using *rl/rl,* the opposite result is obtained. The chimeric cerebellum exhibits a mixture of normal and mutant organization; and within each of these areas both wild-type and *rl/rl* Purkinje cells are found (the Purkinje cells that carry the mutant gene can be identified by means of their staining for β-glucouronidase). The finding of normal Purkinje cells in abnormal positions, as well as *rl/rl* Purkinje cells that are properly located, implies that the instructions for the final positioning of the Purkinje cell lie extrinsic to the cells themselves (Mullen, 1977). Unfortunately, the genotype of the granule cells is difficult to determine using this technique (Mullen, 1977).

3. SYNAPTIC SPECIFICITY

Rakic and Sidman (1972) reported that in the areas of the *rl/rl* cerebellum that contain a relatively normal molecular layer there is a moderate density of normal parallel fiber–Purkinje cell spine synapses. Rakic (1976) and Mariani *et al.* (1977) carefully examined the *rl/rl* cortex for evidence of abnormal synaptic connections. They found that most of the Purkinje cells in the very disorganized areas contained empty postsynaptic tertiary branchlet spines (possibly because these are also the areas that contain few parallel fibers), although there were occasional mossy fiber synapses. Thus, it appears that the spines are usually specific for the parallel fiber synapses in that no other class of synapse regularly takes over these tertiary sites. (Most of the other cell types in the cerebellum, as well as the climbing fiber afferents, synapse on other parts of the Purkinje cell.) In cases where abnormally positioned Purkinje cells invade the granule layer, Mariani *et al.* (1977) report synapses between granule cell *soma* and Purkinje cell dendritic spines. This suggests that the entire granule cell surface can recognize specific synaptic targets.

In the cerebellar cortex of *rl/rl,* many Bergmann glial fibers run in an abnormally oriented, oblique path. Nevertheless, the migrating granule cells are apposed to them in electron micrographs (Caviness and Rakic, 1978). This implies that the granule cells are not guided by their postsynaptic targets and that they can only synapse on cells that they find in their immediate area. There does not seem to be a mechanism attracting the parallel fibers to the Purkinje cells in severely ectopic positions. However, since the total number of granule cells in *rl/rl* is very low (probably as a result of agenesis), no absolute statement can be made about the ability of the axons to seek out their proper target in abnormal conditions on the basis of this mutant.

B. Conclusion

The Purkinje cell postsynaptic site for the granule cell synapse seems to exclude most anomolous innervation, although synapses with nonaxonal parts of granule cells, and an occasional mossy fiber synapse, are found. The failure of granule cell parallel fibers to locate their targets on ectopic Purkinje cell dendrites may be due to a lack of enough positional information to get them within target range.

VI. Granule Cell Transmitter

A. Transmitter Criteria

There are two major requirements for proving that a given compound is acting as a neurotransmitter at a specific class of synapse. One is to demonstrate specific release of the compound after firing, and the other is to show that it can elicit a physiological response identical to the natural firing (Hammerschlag and Roberts, 1976). Both of these are difficult to do in an intact cerebellum. Additional criteria of distribution, storage, and transport (inactivation) have therefore been substituted in this area. These tend to support the hypothesis that glutamic acid is the excitatory neurotransmitter at granule cell parallel fiber–Purkinje cell dendritic spine synapses (glutamic acid hypothesis).

B. Mutant Mice as Model Systems

A correlation of decreased levels of glutamic acid with a decrease in the number of cerebellar granule cells provides one line of evidence consistent with the glutamic acid hypothesis. The granuloprival mutants *(sg/sg, wv/wv,* and *rl/rl)* offer a large and consistent depletion of granule cells, and there have been several studies of glutamic acid levels in these mutant animals. McBride *et al.* (1976a) report decreases of about 30–45% in the glutamic acid levels of *sg/sg* and *wv/wv,* whereas Glu levels in the nervous *(nr/nr)* mutant, which has a near-normal complement of granule cells, are unchanged. Hudson *et al.* (1976) found similar reductions for *wv/wv* as well as for *rl/rl.* (One might have expected a lesser reduction for *rl/rl,* since the granule cell deficiency in *rl/rl* is less severe.) Assay of whole cerebella (with the vermis intact) from weaver heterozygotes *(+ /wv)* indicates that in these cases the reduction in glutamate levels is roughly proportional to the loss of granule cells (S. Roffler-Tarlov, personal communication).

Roffler-Tarlov and Sidman (1978) reexamined the levels of glutamic acid in *wv/wv, sg/sg,* and *nr/nr* mutants, assaying both the cerebellar cortex, which contains the granule cells and their processes, and the cerebellar deep nuclei, which contain no granule cells. They found that, for *sg/sg* and *wv/wv* mutants, the glutamate concentration was reduced to less than two-thirds of control values in *both* the cortex and the deep nuclei. They therefore concluded that glutamate reduction in the cerebella of these mutants could not be taken as sufficient evidence in favor of the glutamate transmitter hypothesis, since the reduction was not specific to those areas containing the granule cells. (There may be some involvement of the excitatory mossy fibers.)

C. Experimental Depletion of Granule Cells in Hamsters and Rats

When neonatal hamsters are infected with the rat virus strain PRE 308, the resulting adult cerebella contain only 5% of the normal complement of granule cells and about half of the normal endogenous levels of glutamate (Young *et al.,* 1974). Repeated X-irradiation of neonatal rats also depletes granule cells by killing off the precursor cells dividing at that time. Reduced levels of glutamate have been reported to correlate with a reduction in granule cells in such animals (Valcana *et al.*, 1972; McBride *et al.*, 1976b). Differences in the exact amount of reduction depended upon the exact schedule and dose of X-irradiation.

Recently, Rea and McBride (1978) compared glutamic acid levels in cerebellar cortex, deep nuclei, and white matter of irradiated rats. Unlike the Roffler-Tarlov and Sidman (1978) mutant cases, here glutamate reduction was found only in the cortex and not in the white matter or in the deep nuclei. These authors suggest that the mutant results were due to a more widespread deficit in the mutant animals, rendering them a less specific test of the hypothesis than the irradiated animals. While this may be true, there seem to be some quantitative inconsistencies among the experiments with irradiated animals as well, and they may be subject to the same problems of substantial metabolic and/or development effects on other cell types. In the final analysis, the mutant mouse and irradiated rat studies may not be comparable at all. Extrapolations from the published tables of data show that the deep nuclei/cortex glutamic acid ratios for the controls are about 0.75 for the mice and only 0.50 for the rats. The GABA ratios are even more disparate, averaging just over 1.1 in the mouse and 3.0 in the rat. These differences may not be meaningful, since statistically significant comparisons would require using ratios from individual animals, but they suggest that there is either a species or a methodological difference in the cell populations examined.

D. Glutamic Acid Distribution in Different Layers of the Cerebellum

A major problem with the interpretation of correlations between the loss of granule cells and a reduction in the levels of glutamic acid is that the results do not strictly imply that glutamate is a transmitter in these cells, only that the substance is concentrated in them. It is possible that there are metabolic differences among cell types in the cerebellum, which could also account for differences in the levels of a compound with as important a metabolic role as glutamate. The results of Berger *et al.* (1977) showing that, in the rabbit, the highest concentrations of glutamic acid per kilogram of lipid-free dry weight are found in the molecular layer (which contains the granule cell axons), whereas the lowest are found in the granular cell layer (which contains the cell bodies), may be important in this regard, since they focus on (presumably) synaptic levels of glutamic acid. The actual ratio of the averages is about 1.5. However, the concentration of glutamic acid was also high in the Purkinje cell layer. Using a less elaborate dissection technique, Nadi *et al.* (1977) found a much lower molecular/ granule layer glutamic acid ratio (1.1, expressed per milligram dry weight). Given that the Purkinje cell layer, which contains almost as much glutamic acid as the molecular layer in the assay of Berger *et al.*, was present in *both* molecular and granule layers of the Nadi *et al.* dissection, the studies are hard to compare on any level. The potential differences in expressing values as a function of lipid-free versus total dry weight must also play a role, since the molecular/granule layer lipid average ratio is about 1.3. Thus, while the data of Berger *et al.* are consistent with the glutamic acid hypothesis, more experiments are required to establish the phenomena.

E. Uptake of Glutamic Acid

The existence (in brain slices and synaptosomes) of a specific, sodium-dependent, high-affinity uptake system for glutamic acid has been cited as support for its role as a neurotransmitter (Balcar and Johnston, 1972; Kuhar and Snyder, 1970; Logan and Snyder, 1972). Localizing such a system to granule cells would provide additional evidence for the glutamic acid hypothesis. Young *et al.* (1974) report a 65–70% reduction in the high-affinity uptake of glutamic acid into synaptosomes from hamsters with a virally depleted granule cell population. Homogenates of isolated bovine molecular and granule layers take up labeled amino acids, and the velocity of glutamic acid incorporation is nine times greater in the molecular than in the granule layer (Mikoshiba and Changeux, 1978). Recently, Campbell and Shank (1978) reported a high-affinity uptake of glutamate in an isolated, enriched population containing 95% granule cells from 10-day-old mice. In fact, two systems were reported, one with an apparent K_m of about 15×10^{-6} and one with a K_m of about 3×10^{-6}. Since a synaptosomal preparation gives only the former, it was hypothesized that the higher-affinity system was specific to the cell soma and might be used to clear glutamate that had leaked out of the synaptic region.

F. Studies with Kainic Acid

One final set of experiments indicates that, when the neurotoxic rigid analog of glutamic acid, kainic acid, is injected into the cerebellum, it causes rapid degeneration of Purkinje, basket, stellate, and Golgi, but not granule, cells (Herndon and Coyle, 1977). Experiments using decorticated striatum and striatal brain slices suggest that kainic acid may require intact presynaptic glutamatergic terminals in order to exert its toxic effect (McGeer *et al.*, 1978; Biziere and Coyle, 1978a,b). If this can be proved to be the mechanism of action of kainic acid, then the Herndon and Coyle experiment provides further evidence for the glutamic acid hypothesis.

G. Conclusion

In conclusion, there is a great deal of evidence consistent with the glutamic acid hypothesis, although no single set of experiments is sufficient to prove this and there are a few discrepancies that remain to be resolved. As yet, however, there have been no direct demonstrations of the major criteria for absolute proof of transmitter identity. Certainly no other identified neurotransmitter is a reasonable candidate, although a peptide containing or mimicking glutamic acid is a theoretical possibility. *In vitro* growth and differentiation of a purified population of granule cells may offer a system in which an even more rigorous proof is possible.

VII. Comments on the Use of Neurological Mutants as Model Systems

A. Mutant Genes as Perturbants

In the studies described in this review, mutant genes were used in a rather specific and limited sense as consistent and reproducible perturbants of a particular developmental process rather than as probes of a specific molecular product. Although in many bacterial systems the enzymatic or regulatory defect being studied is itself the gene product, in mammalian systems even the earliest recognizable effects of a genetic alteration may be many steps removed from the "primary gene defect." Thus, while one could ask whether the Purkinje or the granule cell is the "primary" defect in the staggerer mutant, the actual defective product of the gene itself probably lies many steps earlier in a cause-and-effect sequence. In fact, there is some evidence of a more generalized effect of the *sg* gene that affects cells other than those in the brain. Trenkner (1979) reports alterations of cell surface carbohydrates and cell function of *sg/sg* T lymphocytes.

B. Recessive or Semidominant Genes

The level at which a gene action is being observed (i.e., behavioral, morphological, or biochemical) becomes particularly important when considering the difference between dominant and recessive mutations. Although dominant mutations are generally thought to involve structural proteins and recessive mutations to involve enzymes, the lowered levels of a normal enzyme resulting from the heterozygous effect of a recessive gene may be sufficient to affect a small number of cells during a delicate, metastatic stage in development. It is quite likely that the differences among dominant, semidominant, and recessive are, in most cases of these neurological mutants, a function of the sensitivity of the probes used to measure minor changes, as in the case of the weaver mutant. It may be that really important cellular clues to development will lie in the more normal, slightly affected heterozygotes rather than in the severely disorganized homozygotes.

This point may also be crucial when considering the use of "recessive" mutants as animal models of "dominant" human disorders—both systems may, in reality, be semidominant, with the lack of recognition of heterozygotes of recessive mutants being due to inadequate technology, and the lack of recognition of homozygotes of dominant mutants being due to early (perhaps fetal) death, as well as infrequent occurrence. Phenylketonuria, in which the effect of a single dose of the gene is clearly recognized at the biochemical level, is an example of the former (Menkes and Koch, 1977), and Huntington's disease, in which a double dose of a dominant gene seems to lead to a very early onset and an unusually rapid and severe course, may exemplify the latter (I. Tellez-Nagel, personal communication).

Acknowledgments

Helpful discussions (and occasional arguments) with Drs. Verne Caviness, Mary Beth Hatten, Karl Herrup, Richard Mullen, and Suzanne Roffler-Tarlov are gratefully acknowledged. I wish to thank Drs. Herrup, Mullen, and Roffler-Tarlov, and William J. McBride, for making manuscripts and data available prior to publication. The author's work reported here was done during the tenure of a postdoctoral fellowship from the Helen Hay Whitney Foundation and a special fellowship from the Medical Foundation of Boston. Research expenses were covered by USPHS Grant No. 11327 to Dr. Richard L. Sidman.

References

Balcar, V. J., and Johnston, G. A. R. (1972). Glutamate uptake by brain slices and its relation to depolarization of neurones by acidic amino acids. *J. Neurobiol.* **3**, 295–301.
Barkley, D. S., Rakic, L. L., Chaffee, J. K., and Wong, D. L. (1973). Cell separation by velocity sedimentation of postnatal mouse cerebellum. *J. Cell. Physiol.* **81**, 271–280.

Berger, S. J., Carter, J. G., and Lowry, O. H. (1977). The distribution of glycine, GABA, glutamate and aspartate in rabbit spinal cord, cerebellum and hippocampus. *J. Neurochem.* **28**, 149–158.

Bignami, A., and Dahl, D. (1974). The development of Bergmann glia in mutant mice with cerebellar malformations: Reeler, staggerer and weaver. Immunofluorescence study with antibodies to the glial fibrillary acidic protein. *J. Comp. Neurol.* **155**, 219–230.

Biziere, K., and Coyle, J. T. (1978a). Influence of cortico-striatal afferents on striatal kainic acid neurotoxicity. *Neurosci. Lett.* **8**, 303–310.

Biziere, K., and Coyle, J. T. (1978b). Effects of kainic acid on ion distribution and ATP levels of striatal slices incubated *in vitro*. *J. Neurochem.* **31**, 513–520.

Blinkov, S. M., and Glezer, I. I. (1968). In "The Human Brain in Figures and Tables" (B. Haigh, ed.), p. 159. Plenum, New York.

Campbell, G. LeM., and Shank, R. P. (1978). Glutamate and GABA uptake by cerebellar granule and glial cell-enriched populations. *Brain Res.* **153**, 618–622.

Campbell, G. LeM., and Williams, M. (1978). *In vitro* growth of glial cell-enriched and depleted populations from mouse cerebellum. *Brain Res.* **156**, 227–239.

Caviness, V. S. (1977). Reeler mutant mouse: A genetic experiment in developing mammalian cortex. *Soc. Neurosci. Symp.* **2**, 27–46.

Caviness, V. S., and Rakic, P. (1978). Mechanisms of cortical development: A view from mutations in mice. *Annu. Rev. Neurosci.* **1**, 297–326.

Eccles, J. C., Ito, M., and Szentagothai, J. (1967). "The Cerebellum as Neuronal Machine." Springer-Verlag, Berlin and New York.

Falconer, D. S. (1951). Two new mutants "trembler" and "reeler" with neurological actions in the house mouse *(Mus musculus). J. Genet.* **50**, 192–201.

Hamburgh, M. (1963). Analysis of the postnatal development of "reeler," a neurological mutation in mice. A study in developmental genetics. *Dev. Biol.* **8**, 165–185.

Hammersclag, R., and Roberts, E. (1976). Overview of chemical transmission. *In* "Basic Neurochemistry" (G. J. Siegel, R. W. Albers, R. Katzman, and B. W. Agranoff, Eds.), pp. 167–179. Little, Brown, Boston, Massachusetts.

Hatten, M. E., and Messer, A. (1978). Postnatal cerebellar cells from staggerer mutant mice express embryonic cell surface characteristics. *Nature (London)* **276**, 504–506.

Hatten, M. E., and Sidman, R. L. (1978). Cell reassociation behavior and lectin-induced agglutination of embryonic mouse cells from different brain regions. *Exp. Cell Res.* **113**, 111–125.

Herndon, R. M., and Coyle, J. T. (1977). Selective destruction of neurons by a transmitter agonist. *Science* **198**, 71–72.

Herrup, K., and Mullen, R. J. (1976). Intrinsic Purkinje cell abnormalities in staggerer mutant mice revealed by analysis of a staggerer normal chimera. *Soc. Neurosci. Annu. Meet.* Abstract. No. 141.

Herrup, K., and Mullen, R. J. (1979a). Staggerer chimeras: Intrinsic nature of Purkinje cell defects and implications for normal cerebellar development. *Brain Res.* **178**, 443–457.

Herrup, K., and Mullen, R. J. (1979b). Regional variation and absence of large neurons in the cerebellum of the staggerer mouse. *Brain Res.* **172**, 1–12.

Hirano, A., and Dembitzer, H. M. (1975). The fine structure of staggerer cerebellum. *J. Neuropathol. Exp. Neurol.* **34**, 1–11.

Hudson, D. B., Valcana, T., Bean, G., and Timiras, P. S. (1976). Glutamic acid: A strong candidate as the neurotransmitter of the cerebellar granule cell. *Neurochem. Res.* **1**, 73–81.

Kuhar, M. J., and Snyder, S. H. (1970). The subcellular distribution of free ^{3}H-glutamic acid in rat cerebral cortical slices. *J. Pharmacol. Exp. Ther.* **171**, 141–152.

Landis, D. M. D., and Sidman, R. L. (1978). Electron microscopic analysis of postnatal histogenesis in the cerebellar cortex of staggerer mutant mice. *J. Comp. Neurol.* **179**, 831–864.

Lasher, R. S. (1974). The uptake of ³H-GABA and differentiation of stellate neurons in cultures of dissociated postnatal rat cerebellum. *Brain Res.* **69**, 235–254.

Lasher, R. S., and Zagon, I. S. (1972). The effect of potassium on neuronal differentiation in cultures of dissociated newborn rat cerebellum. *Brain Res.* **41**, 482–488.

Letourneau, P. C. (1975). Possible roles for cell-to-substratum adhesions in neuronal morphogenesis. *Dev. Biol.* **44**, 77–91.

Logan, W. J., and Snyder, S. H. (1972). High affinity uptake systems for glycine, glutamic acid and aspartic acids in synaptosomes of rat central nervous tissues. *Brain Res.* **42**, 413–431.

McBride, W. J., Aprison, M. H., and Kusano, K. (1976a). Contents of several amino acids in the cerebellum, brain stem and cerebrum of the "staggerer," "weaver" and "nervous" neurologically mutant mice. *J. Neurochem.* **26**, 867–870.

McBride, W. J., Nadi, N. S., Altman, J., and Aprison, M. H. (1976b). Effects of selective doses of X-irradiation on the levels of several amino acids in the cerebellum of the rat. *Neurochem. Res.* **1**, 141–152.

McGeer, E. G., McGeer, P. L., and Singh, K. (1978). Kainate-induced degeneration of neostriatal neurons: Dependency upon corticostriatal tract. *Brain Res.* **139**, 381–383.

Mallet, J., Huchet, M., Pougeois, R., and Changeux, J.-P. (1976). Anatomical, physiological and biochemical studies on the cerebellum from mutant mice. III. Protein differences associated with the weaver, staggerer and nervous mutations. *Brain Res.* **103**, 291–312.

Mariani, J., Creppel, F., Mikoshiba, K., Changeux, J.-P., and Sotelo, C. (1977). Anatomical, physiological and biochemical studies of the cerebellum from reeler mice. *Philos. Trans. Soc. London* **281**, 1–28.

Menkes, J. H., and Koch, R. (1977). Phenylketonuria. *In* "Handbook of Clinical Neurology" (P. J. Vinken and G. W. Bruyn, eds.), pp. 29–51. North-Holland Publ., Amsterdam.

Messer, A. (1976). An additional factor influencing monolayer cell culture morphology and survival of cerebellar granule cells. *Proc. 2nd Annu. Maine Biomed. Symp.* Vol. 2, pp. 484–487.

Messer, A. (1977a). The maintenance and identification of mouse cerebellar granule cells in monolayer culture. *Brain Res.* **130**, 1–12.

Messer, A. (1977b). Factors influencing monolayer cell culture morphology of granule cells from wild-type and mutant mice. *In* "Cellular Neurobiology" (Z. Hall, R. Kelly, and C. F. Fox, eds.), pp. 251–257. Alan R. Liss, Inc., New York.

Messer, A. (1978). Abnormal staggerer cerebellar cell interactions and survival *in vitro*. *Neurosci. Lett.* **9**, 185–188.

Messer, A., and Hatten, M. (1977). *In vitro* studies of interactions and survival of cerebellar cells from staggerer versus wild-type mice. *Soc. Neurosci. 7 Meet.* Abstract No. 175.

Messer, A., and Smith, D. (1977). *In vitro* behavior of granule cells from agranular neurological mutants of mice. *Brain Res.* **130**, 13–23.

Miale, I. L., and Sidman, R. L. (1961). An autoradiographic analysis of histogenesis in the mouse cerebellum. *Exp. Neurol.* **4**, 227–296.

Mikoshiba, K., and Changeux, J.-P. (1978). Morphological and biochemical studies on isolated molecular and granular layers from bovine cerebellum. *Brain Res.* **142**, 487–504.

Mullen, R. J. (1977). Genetic dissection of the CNS with mutant-normal mouse and rat chimeras. *Soc. Neurosci. Symp.* **2**, 47–65.

Nadi, N. S., McBride, W. J., and Aprison, M. H. (1977). Distribution of several amino acids in the regions of the cerebellum of the rat. *J. Neurochem.* **28**, 453-455.

Nelson, P. G., and Peacock, J. H. (1973). Electrical activity in dissociated cell cultures from fetal mouse cerebellum. *Brain Res.* **61**, 163-174.

Nicholson, J. L., and Altman, J. (1972). The effects of early hypo- and hyperthyroidism on the development of rat cerebellar cortex. I. Cell proliferation and differentiation. *Brain Res.* **44**, 13-23.

Palay, S. L., and Chan-Palay, V. (1974). "Cerebellar Cortex. Cytology and Organization." Springer-Verlag, Berlin and New York.

Rabié, A., Favre, C., Clavel, M., and LeGrand, J. (1977). Effects of thyroid dysfunction on the development of the rat cerebellum, with special reference to cell death within the internal granular layer. *Brain Res.* **120**, 521-531.

Rakic, P. (1971). Neuron-glia relationship during granule cell migration in developing cerebellar cortex. A Golgi and electron microscopic study in *Macacus rhesus. J. Comp. Neurol.* **141**, 283-312.

Rakic, P. (1974). Intrinsic and extrinsic factors influencing the shape of neurons and their assembly into neuronal circuits. *In* "Frontiers in Neurology and Neuroscience Research" (P. Seeman and G. M. Brown, eds.), pp. 112-132. Neuroscience Institute, Toronto, Canada.

Rakic, P. (1976). Synaptic specificity in the cerebellar cortex: Study of anomalous circuits induced by single gene mutations in mice. *Cold Spring Harbor Symp. Quant. Biol.* **40**, 333-346.

Rakic, P., and Sidman, R. L. (1972). Synaptic organization of displaced and disoriented cerebellar cortical neurons in reeler mice. *J. Neuropathol. Exp. Neurol.* **31**, 192 (abstr.).

Rakic, P., and Sidman, R. L. (1973a). Sequence of developmental abnormalities leading to granule cell deficit in cerebellar cortex of weaver mutant mice. *J. Comp. Neurol.* **152**, 103-132.

Rakic, P., and Sidman, R. L. (1973b). Organization of cerebellar cortex secondary to deficit of granule cells in weaver mutant mice. *J. Comp. Neurol.* **152**, 133-162.

Rea, M. A., and McBride, W. J. (1978). Effects of X-irradiation on the levels of glutamate, aspartate and GABA in different regions of the cerebellum of the rat. *Life Sci.* **23**, 2355-2367.

Rezai, Z., and Yoon, C. H. (1972). Abnormal rate of granule cell migration in the cerebellum of "weaver" mutant mice. *Dev. Biol.* **29**, 17-26.

Roffler-Tarlov, S., and Sidman, R. L. (1978). Concentrations of glutamic acid in cerebellar cortex and deep nuclei of normal mice and weaver, staggerer and nervous mutants. *Brain Res.* **142**, 269-283.

Schutta, H. S., and Johnson, L. (1967). Bilirubin encephalopathy in the Gunn rat. A fine structure study of the cerebellar cortex. *J. Neuropathol. Exp. Neurol.* **26**, 377-396.

Sidman, R. L. (1968). Development of the interneuronal connections in brains of mutant mice. *In* "Physiological and Biochemical Correlates of Nervous Integration" (F. D. Carlson, ed.), pp. 163-193. Prentice-Hall, Englewood Cliffs, New Jersey.

Sidman, R. L. (1972). Cell interactions in developing mammalian nervous system. *In* "Cell Interactions" (L. G. Silverstri, ed.), pp. 1-13. North-Holland Publ., Amsterdam.

Sidman, R. L., Lane, P. W., and Dickie, M. M. (1962). Staggerer, a new mutation in the mouse affecting the cerebellum. *Science* **137**, 610-612.

Sidman, R. L., Green, M. S., and Appel, S. H. (1965). "Catalog of the Neurological Mutants of the Mouse." Harvard Univ. Press, Cambridge, Massachusetts.

Sotelo, C., and Changeux, J.-P. (1974a). Transsynaptic degeneration "en cascade" in the cerebellar cortex of staggerer mutant mice. *Brain Res.* **67**, 519-526.

Sotelo, C., and Changeux, J.-P. (1974b). Bergmann fibers and granular cell migration in the cerebellum of homozygous weaver mutant mouse. *Brain Res.* **77**, 484–491.

Trenkner, E. (1979). Postnatal cerebellar cells of staggerer mutant mice express immature components on their surface. *Nature (London)* **277**, 566–567.

Trenkner, E., and Sidman, R. L. (1977). Histogenesis of mouse cerebellum in microwell cultures: Cell reaggregation and migration, fiber and synapse formation. *J. Cell Biol.* **75**, 915–940.

Trenkner, E., Herrup, K., Hatten, M. E., and Sarkar, S. (1977). Staggerer mutant mice and normal littermates express different carbohydrates on postnatal cerebellar cells. *Soc. Neurosci. 7th Annu. Meet.* Abstract No. 186.

Trenkner, E., Hatten, M. E., and Sidman, R. L. (1978). Effect of ether-soluble serum components *in vitro* on the behavior of immature cerebellar cells in weaver mutant mice. *Neuroscience* **3**, 1093–1100.

Valcana, T., Hudson, D. B., and Timiras, P. S. (1972). Effects of X-irradiation on the content of amino acids in the developing rat cerebellum. *J. Neurochem.* **19**, 2229–2232.

Wolf, M. K. (1978). Cell and organotypic culture studies of neurological mutations affecting structural development. *In* "Cell, Tissue, and Organ Cultures in Neurobiology" (S. Fedoroff and L. Hertz, eds.), pp. 555–572. Academic Press, New York.

Yamada, N., Sawasaki, Y., and Nakajima, H. (1977). Impairment of DNA synthesis in the Gunn rat cerebellum. *Brain Res.* **126**, 295–307.

Yavin, E., and Yavin, Z. (1974). Attachment and culture of dissociated cells from rat embryo cerebral hemispheres on polylysine-coated surface. *J. Cell Biol.* **62**, 540–546.

Yoon, C. H. (1972). Developmental mechanism for changes in cerebellum of "staggerer" mouse, a neurological mutant of genetic origin. *Neurology* **22**, 743–754.

Yoon, C. H. (1976). Pleiotropic effect of the staggerer gene. *Brain Res.* **109**, 206–215.

Yoon, C. H. (1977a). Fine structure of the cerebellum of "staggerer-reeler," a double mutant of mice affected by staggerer and reeler conditions. I. The premature disappearance of the external granular layer and ensuing cerebellar disorganization. *J. Neuropathol. Exp. Neurol.* **36**, 413–426.

Yoon, C. H. (1977b). Fine structure of the cerebellum of "staggerer-reeler," a double mutant of mice affected by staggerer and reeler conditions. II. Purkinje cell anomalies. *J. Neuropathol. Exp. Neurol.* **36**, 427–439.

Young, A. B., Oster-Granite, M. L., Herndon, R. M., and Snyder, S. H. (1974). Glutamic acid: Selective depletion by viral-induced granule cell loss in hamster cerebellum. *Brain Res.* **73**, 1–13.

CELL GENERATION AND AGING OF NONTRANSFORMED GLIAL CELLS FROM ADULT HUMANS

JAN PONTÉN AND BENGT WESTERMARK

Department of Tumor Biology
Wallenberg Laboratory
Uppsala, Sweden

I. Cell Generation and Aging

The very concept that cells in culture may age received its strongest impetus from the now classical observations of Hayflick and Moorhead (1961) on serially cultivated human embryonic lung fibroblasts. The essential fact was that such cells eventually entered a nondividing state in spite of enjoying excellent tissue culture conditions permitting them to go through some 50 population doublings as rapidly as fibroblasts ever can. Although implicit in most experiments, it is still not clear whether incapacity for further proliferation is linked to inevitable degeneration and cell death. Since this distinction has often not been made, we will follow common practice and use the term "cellular senescence" or "aging" for the acquisition of inability to multiply further regardless of whether it was actually shown that this was accompanied by degeneration.

The paradox is that these easily cultivatable cells, in sharp contrast to malignant cells, certain higher plants, prokaryotes, and some other unicellular organisms (Kirkwood, 1977), have some kind of biological clockwork that prohibits an infinite number of entrances into the cell cycle. Hayflick and others have repeatedly suggested that this behavior reflects senescence on a cellular level and that it may be an important reason for the demise that all animals eventually undergo (Hayflick, 1966).

The central nervous system has been the focal point of many studies on senescence *in vivo*. Although much remains to be done, it seems likely that degenerative changes may occur independently of a malfunctioning circulatory system and that therefore alterations in brain cells can be considered a bona fide part of the "aging syndrome" (see Brody and Vijayashankar, 1977).

Key problems are (1) whether fibroblast senescence is peculiar to this cell type or also applicable to, for instance, brain cells; (2) whether the apparently unavoidable loss of mass cultures of normal cells *in vitro* in a sense is a statistical artifact as suggested by Kirkwood and Holliday (1975); (3) whether cellular degeneration is related to chronological time per se or to some other "clock" such as the number of cell cycles completed; (4) the nature and eventual irreversibility of the final nondividing state; (5) whether cellular senescence has any relation to malignant transformation and/or other diseases associated with old age (Pontén, 1977); (6) whether loss of division potential is related to terminal differentiation (Bell *et al.*, 1978); and (7) the nature of the precise relation between cell generation and aging.

II. Origin of Adult Human Glia-like Cell Lines

Systematic studies of brain cell generation and aging *in vitro* have been performed only with adult human tissue. From explants of such tissue serially cultivatable cell lines may be obtained with great regularity (Pontén and Macintyre, 1968). They are composed of flat, star-shaped, or epithelioid cells apparently unique to nervous tissue. Mainly by exclusion of other candidates, the cells were dubbed "glia-like" and considered "simplified astrocytes" (Westermark, 1973). However, their true derivation has not been positively proven.

Table I compares the properties of possible precursors of the glia-like cells we regularly obtain from biopsies of gray and white matter of the adult human brain dissected free of meninges and visible vessels. Fibrocytes and endothelial cells are virtually excluded on morphological criteria alone. In contrast to endothelial cells, glia-like cells do not manufacture blood-clotting factor VIII. Therefore, our conclusion is that endothelial cells and fibrocytes are very unlikely precursors of glia-like cells. Since, however, the comparisons are predominantly based on extracerebral vessels, it must be noted that small brain vessels may give rise to lines of endothelial cells or fibroblasts *in vitro*, which could be quite different from analogous lines from other organs and thus be the origin of glia-like cells.

Microglia have not been studied sufficiently *in vitro* to permit meaningful comparisons, but one would expect a macrophage-like cell with properties other than those of glia-like lines. Oligodendroglia are possible candidates but seem unlikely because our glia-like cells are morphologically different from the apparently extremely fragile oligodendrocytes from normal or neoplastic brain previously described (see Lumsden, 1959). Oligodendroglia are generally thought of as cells with little if any capacity for regeneration—an apparent sine qua non for successful prolonged proliferation *in vitro*.

Mature astrocytes differ from glia-like cells mainly in their high content of fibrils, which are sometimes positive for glial fibrillar antigen (GFA) (Bignami *et al.*, 1972), and their highly branched configuration. In addition, astrocytes are completely negative for the glycoprotein fibronectin (Schachner *et al.*, 1978), whereas glia-like cells in culture are strongly positive for this antigen (Vaheri *et al.*, 1976).

The three remaining alternatives are (1) pericytes, (2) astrocyte precursors, and (3) smooth muscle cells.

The nature of the *pericyte* is obscure, but it is most often regarded as an analog of endothelium located outside the basal lamina having unknown functions. It is presumed to be of mesodermal origin. Lumsden observed cells in explants from tumors and normal human brain, which he labeled mesoblasts and which showed a certain resemblance to our glia-like cells. Lumsden's (1959) terminology suggests that he considered pericytes possible precursors of the cells we have labeled glia-like. It should be recalled that reactive tissue in and around malignant gliomas readily gives rise to glia-like lines (Pontén, 1975)—a fact explicable by the abundance of vessels in a glioblastoma multiforme with a high content of pericytes. The objections to pericytes rest on the same grounds as the objections to endothelial cells, i.e., no synthesis of factor VIII and a growth pattern other than that of classical endothelium.

Astrocyte precursors have been observed mainly in cultures of embryonic or newborn rodent brain. A recent thorough electron microscope study by

TABLE I

Comparison of Possible Precursors of Glia-like Cells in Vitro

	Endothelial cells		Smooth muscle cells		Fibrocytes	
	In vivo	In vitro	In vivo	In vitro	In vivo	In vitro
Localization, growth pattern morphology, etc.	Contiguous sheets of flat cells on basal lamina. Weibel-Palade bodies and girdles of zonae occludentes	Monolayers of flat, polygonal cells with strong intercellular adhesions and smooth upper surface: Weibel-Palade bodies	Only in arterioli and arteries; elongated cells with myofilaments and peripheral dense bodies	Bundles of elongated cells with myofilaments	Mainly in arteries, veins and meninges; elongated, branched cells	Whorls and bundles of elongated cells with oval nuclei
Basal lamina components	Yes	Yes	Yes	—	Yes?	—
GFA	No	No	—	—	No	No
S-100	—[b]	No	—	—	—	No
GABA[a]	—	—	—	—	—	No
Factor VIII	Yes	Yes	—	—	—	No

	Pericytes In vivo	Microglia In vivo	Oligodendroglia In vivo	Astrocytes In vivo	Astrocytes In vitro	Astrocyte precursors In vitro	Glia-like cells In vitro
Localization, growth pattern, morphology, etc.	Perivascular, branched cells enveloped by basal lamina; many lysosomes and dense bodies	Perivascular or perineuronal small phagocytes with tortuous, granular endoplasmic reticulum	Neurosatellites with few fibrils and numerous microtubules; myelinogenesis	Star-shaped cells with extensive processes and perivascular end feet; many glial fibrils and dense bodies	Star-shaped cells with thin extensions and glial fibrils	Monolayered sheets of asteroid-epithelioid cells, inducible to mature astrocytes	Sheets of moderately branched asteroid cells with overlapping cytoplasm and weak intercellular adhesions; smooth upper surface
Basal lamina components	Yes	—	—	Yes?	—	—	Yes
GFA	—	—	—	Yes	—	No	No
S-100	—	—	—	Yes	Yes	No	Yes
GABA [a]	—	—	—	Yes	Yes	—	Yes
Factor VIII	—	—	—	—	—	—	No

[a] Specific high-affinity uptake.
[b] Not known.

213

Haugen and Laerum (1978) demonstrated a striking morphological similarity between flat epithelioid fetal rat brain cells and our glia-like human cells. When the rat cells were treated with adult brain extract, they matured into a differentiated astrocytoid phenotype which, inter alia, could contain 10-nm GFA-positive glial filaments. Indirectly, this supports the suggestion that our glia-like cells could be of astrocytic origin in spite of their negative immunofluorescence for glial fibrillar antigen (A.-C. Nilsson and B. Westermark, unpublished). Whether astrocyte precursors produce fibronectin has not been determined. Two strong arguments for a relation between glia-like cells and astrocytes are the presence of S-100 protein and high-affinity γ-aminobutyric acid (GABA) membrane transport sites in the cell membrane (E. Walum, B. Westermark, and J. Pontén, unpublished).

Smooth muscle cells can be obtained from vessels and can also be cultivated for several passages in vitro (Ross, 1971), as may be expected from their regenerative capacity in vivo. The general morphology of the glia-like cells is not very similar to that of smooth muscle cells in vitro (Gimbrone and Cotran, 1975). Two main arguments against a smooth muscle origin are the high-affinity uptake of GABA and the substantial capacity for regeneration in vitro, which seems to exceed that of human smooth muscle. In our experience dissected brain vessels do not give rise to lines resembling our glia-like cells.

Arguments about the histogenetic origin of glia-like cells can also be derived from comparisons with human malignant astrocytoma lines (Westermark, 1973; Pontén, 1975). Glia-like lines have certain features in common with at least some glioma lines. Glioma cells may produce fibronectin in vitro (Vaheri et al., 1976). The presence of this substance in glia-like cells (Haugen, 1978) is thus not a decisive argument against an astrocytic origin. Most glioma lines are negative for GFA (A.-C. Nilsson and B. Westermark, unpublished); the absence of this antigen cannot therefore be taken as proof of a nonastrocytic origin of glia-like cells. It may be of relevance that ethylnitrosourea-induced rat brain tumor cells displayed a more differentiated phenotype (expression of S-100 and GFA) than control fetal neuroectoderm (Haugen, 1978), reminiscent of the differences between control glia-like cells and glioblastoma multiforme cells (Pontén, 1975).

We favor astrocyte precursors (simplified astrocytes, primitive neuroectoderm) as the most likely derivation of our glia-like lines. We do not know whether they arise from a small not previously identified population of embryonic "rest" cells that persist in the adult human brain or whether they arise by "dedifferentiation" of mature astrocytes. Figure 1 depicts a dense layer of glia-like cells with well-developed, intertwined cytoplasmic processes.

The similarities between the principal events in mass cultures of human cells, the ease with which these phenomena have been reproduced in dif-

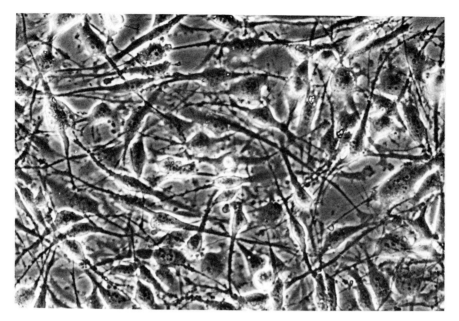

FIG. 1.   Phase-contrast micrograph of living adult human glia-like cells. Note the extensive network of fairly straight branches reminiscent of astrocytic morphology *in vivo*.

ferent laboratories, and the regular observation that embryonic cells are always capable of more divisions than their adult counterparts indicate that an important biological control mechanism is reflected that limits the multiplication of somatic cells. Since fibroblasts during the last few cell cycles show apparently irreversible signs of declining function such as decreased rate of excision repair (Mattern and Cerutti, 1975), decreasing fidelity of different enzymes (Linn *et al.*, 1976), and diminished plasma membrane motility (Blomquist *et al.*, 1978), it does not seem unreasonable to employ the term "cellular senescence" or "aging" for this terminal phenomenon. We use these conventional and convenient terms interchangeably but do not wish to imply by this that this cellular event has been proven to have any definite relation to organismic aging, senility, or senescence.

III. Theories about Cellular Aging

It is by now certain that cellular aging as exemplified by fibroblasts and glia-like cells cannot be explained by trivial medium deficiency. It occurs without exception in media of different compositions, is not prevented by the addition of young cells (Hayflick, 1965), and can be postponed but not prevented by the addition of such a compound as cortisone (Macieira-Coelho, 1966). The conclusion that a finite potential for divisions is a characteristic property of somatic cells is reinforced by the apparent inability of normal somatic tissues or cells to survive transplantation for more than a limited number of times (Daniel, 1977).

With the possible exception of Epstein–Barr virus-infected normal human B cells (Nilsson and Pontén, 1975), a population of somatic cells seems to be faced with only two alternatives: (1) to remain "normal" and nonneoplastic but to lose reproductive capacity after a rather fixed number of cell cycles, or (2) to "transform" into cell lines with an infinite potential for multiplication usually combined with heteroploidy and tumorigenicity (see Pontén, 1971). The likelihood of alternative 2 is strongly influenced by poorly understood species factors. Whereas mouse and other rodent cell populations inevitably develop spontaneous transformants, human and chicken cells, for instance, have zero probability for spontaneous transformation (Pontén, 1971; Hayflick, 1977).

Cellular aging may be either a physiological process or the result of intrinsic or extrinsic pathological factors.

According to the hypothesis of genetic preprogramming, somatic cells starting from the fertilized egg behave and differentiate according to a program unique for a particular species. Part of this program manifests itself as a tendency toward "terminal differentiation." This implies that, as soon as a cell has left the stem cell stage, it will by continuous division move toward a mature state in which it can no longer enter the cell cycle. A clear example is granulocytopoiesis, where polymorphonuclear leukocytes incapable of proliferation are produced via intermediate dividing stages. By extrapolation one may then assume that all cell populations studied *in vitro* are either already past the stem cell stage or cannot maintain their stem cell character in culture. They will therefore be driven step by step toward a stage where division is no longer possible. This physiological theory is supported by observations linking cellular senescence to comparable *in vivo* situations.

Embryonic fibroblasts have a longer remaining life span *in vitro* than adult fibroblasts (Hayflick, 1977). A rough, direct correlation exists between the life span of an animal and the regenerative capacity of its fibroblasts *in vitro* (Hayflick, 1977). It should be emphasized that regeneration of fibroblasts, for example, far exceeds the needs that could arise during the life span of a particular individual. Even if all cells could ultimately become terminally differentiated, this extreme end would rarely be reached in the living animal. Serial tissue culture may be looked upon as a way in which a biological sequence that seldom has time to become realized *in vivo* is artificially brought to completion by forcing the cells to divide almost continuously.

The "pathologic" theory has never been very precisely stated. It seems to assume that handling of cells, inadvertent exposure to irradiation and chemicals, etc., will induce irreversible damage which eventually leads to an inability to divide (Hayflick, 1977). A certain degree of support for this view may be derived from irradiation experiments where the life span of fibroblasts is shortened (Macieira-Coelho, 1973). The theory fails to differentiate between damage invoked *in vivo,* i.e., before explanation, and *in vitro.* A prediction of this theory is that animals shielded from environmental agents such as ionizing irradiation will live longer and give rise to cell lines with a longer life span. It also implies that the life span of serially cultivated cells will be prolonged by avoiding noxious elements. This has never been positively proven. The reproducibility in life span studies of such standard fibroblast stains as WI-38 and MRC-5 in different laboratories obviously under varied conditions speaks against this "pathologic" hypothesis (Hayflick, 1977).

In this review we will follow the most widely held view, namely, that cellular senescence *in vitro* probably reflects an important general biological principle. This view makes a sharp distinction between immortal sex cell and mortal somatic cell lineages in higher animals. Among somatic cells only stem cells may undergo an infinite number of divisions *in vivo.* However, positive identification of such reproductively immortal stem cells is still lacking.

The mechanism for the actual realization of cellular aging has been the subject of much speculation and experimentation. Two theories can be distinguished. One is the model proposed by Kirkwood and Holliday (1975). It says that a population of, say, fibroblasts is composed of immortal cells (stem cells) and reproductively mortal cells. At each division of the former there is a probability of less than 0.5 that the daughter cells will become committed to death at a later time. The committed cell will retain its original vigor and speed of regeneration for a defined number of cell cycles until it finally loses all capacity for further division and dies. The number of generations between commitment and cell death is referred to as the incubation time (*M*). The theory could explain a puzzling feature of all fibroblast populations, namely, the great heterogeneity in the remaining clonal life span (Holliday *et al.,* 1977), because any cell population would contain many classes of committed cells characterized by the number of divisions still remaining in the total incubation period. The theory also takes care of one obvious objection: Why does a core of immortal stem cells not remain and make the population capable of endless transfer? Instead of simply proposing a selective tissue culture disadvantage for the uncommitted cells, Kirkwood and Holliday deduced figures for the commitment probability ($p \approx 0.275$) and for the incubation time ($M \approx 60$ generations) that would reduce the proportion of stem cells to a level at which they would by chance all be discarded during routine laboratory cell culture. Evidence that uncommitted cells exist was demonstrated in a clever "bottleneck" experiment (Holliday *et al.*, 1977). Direct proof for the existence of these postulated stem cells is still not available.

Recently, a basically different alternative has been proposed by Shall and Stein (1979). They assume that reproductive death is realized instantly; i.e., the incubation time is zero generations. At each division there is a probability that a daughter cell will become unable to divide, and this probability increases as a simple function of the number of divisions. In a growing population the likelihood that each cell will give rise to "sterile" progeny therefore becomes greater until it exceeds 0.5 and division soon stops. This theory would explain why nondividing cells are found already in early passages and seem to increase steadily with each passage (Cristofalo and Scharf, 1973). It holds that malignant cells are qualitatively different from normal cells because they do not show an increasing probability of becoming nondividers as the number of completed cell cycles increases. The Kirkwood–Holliday theory, on the other hand, would explain the difference between malignant and normal cells on a quantitative basis. If, for instance, M becomes small, uncommitted cells will constitute a large proportion of any population and will thus not be easily diluted out. The malignant population will consequently display immortality.

Both theories need to be made biochemically plausible. What are the actual molecular mechanisms for commitment and inability to divide? The most elaborate theory was formulated by Orgel (1970). He has proposed, in simplified terms, that protein synthesis always results in some malformed molecules. Normally a balance is struck between the production and elimination of erroneous proteins so that cell viability remains unimpaired. However, a certain probability exists that any cell can "jump" into an unstable situation where error frequency increases because of error feedback into the proteins of the synthesizing machinery itself. Such an error catastrophe will lead to extinction of the cell. In support of this theory, experiments have shown that aged fibroblasts do indeed contain a large proportion of errors in functional enzymes (Linn *et al.*, 1976; Holliday and Tarrant, 1972). These findings were not confirmed by Pendergrass *et al.* (1976).

Only with some difficulty can error catastrophies be accommodated in Holliday's commitment theory. The theory requires that the length of the incubation period be defined by the unstable state during which error frequency increases (Kirkwood and Holliday, 1975). This implies that considerable cumulative insertion of incorrect amino acids into proteins has to take place for some 60 fibroblast cell cycles before the functional effect is noted as an inability to divide further. It is hard to imagine that uninhibited rapid cell cycles would continue up until the very last few rounds under these conditions. Furthermore, Holliday's theory assumes a very regular progression toward extinction during the incubation period, which seems difficult to realize through the unpredictable accumulation of protein synthesis errors.

Since high fidelity in protein synthesis always has to be paid for by energy expenditure, Kirkwood (1977) suggested that this luxury has been dropped in somatic cells, to the evolutionary advantage of the species in question.

IV. Basic Characteristics of Adult Human Glia-like Cells *in Vitro*

A large number of mass cultures of adult glia-like cells have been serially cultivated in our laboratory (Pontén, 1975) essentially according to the original scheme of Hayflick for embryonic lung fibroblasts. Invariably, the cells have entered an irreversible degenerative phase III in Hayflick's (1977) terminology. Figure 2 depicts schematically the behavior of a variety of human cells cultivated under comparable conditions. The capacity for regeneration *in vitro* differs tremendously. Whereas 10^6 embryonic lung cells under the assumptions of Fig. 2 can give rise to a total of 10^{18} cells (corresponding to 1000 tons), 10^6 kidney epithelium cells will only produce a maximum of about 19 million cells (i.e., 0.01 gm). Adult glia-like cells take an intermediate position with a maximum of a little less than 10^{12} cells, which is about the weight of the whole adult human brain.

In spite of this enormous span in total capacity for regeneration, the different systems have several principles in common. All are density-dependent, i.e., proliferation in any given culture dish ceases at a certain so-called terminal density which in the case of glia-like cells is 70,000 cells/cm^2 solid support. The majority of the cells are then held in the G_1 part of the cell cycle (Westermark, 1973). They need at least 12 hr of continuous stimulation by fresh serum or pure polypeptide growth factor to be forced irreversibly into one division cycle (Lindgren and Westermark, 1977). Phase III makes itself felt by a decreased terminal density (Macieira-Coelho, 1970), a prolonged intermitotic time (Macieira-Coelho *et al.*, 1966), an accumulation of lysosomes (Brunk *et al.*, (1973), sluggish membrane movements (Blomquist *et al.*, 1978), and chromosome alterations (Miller *et al.*, 1977).

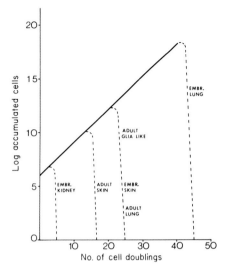

Fɪɢ. 2. Idealized total population growth curves for different varieties of human cells capable of serial *in vitro* propagation. The curves were constructed under the assumption of no cell death, a constant intermitotic time of 24 hr, and a starting population of 10^6 cells.

V. Relation between Cell Generation and Aging in Human Glia-like Cells

The precise control of density-dependent inhibition of growth and movement (Pontén *et al.*, 1969; Westermark, 1971) in glia-like cells has permitted an analysis of whether the number of divisions or chronological time is of decisive importance for entrance into phase III. Colonies were started from one or a few early-passage cells (Brunk *et al.*, 1971; Pontén, 1973). After a few generations of uninhibited division the centrally placed cells will become crowded and therefore density-arrested in G_1, but growth will continue around the periphery where cells can move outward, thus keeping the density below the critical threshold at which inhibition sets in. In this way a gradient is created where centrally placed cells have only divided a few times, whereas cells at the perimeter will have gone through many divisions. After about 1 month when the "mature" colony had reached a size of about 3 cm^2, cells in the periphery will begin to slow down their multiplication considerably. This will roughly correspond to some 25 successive cell cycles.

At this stage a colony may be wounded by removal of a strip of cells along one diameter. This procedure will stimulate division along the wound edge, provided the cells have not become senescent and thus incapable of further multiplication. In such experiments a graded response with respect to proportion of cells entering their division cycle was established, which was inversely related to the radial distance from the center of the original colony (Brunk *et al.,* 1971; Pontén, 1973). Since all cells were chronologically equally old but had undergone a different number of divisions, the result showed that loss of division potential seemed to be predominantly determined by the number of completed cell cycles rather than the mere passage of time. We still have no idea how the number of cycles can be "counted," but would like to point out that such a relation is not easily compatible with the error catastrophe theory, where metabolic (chronological) time per se would be important for senescence and loss of regenerative power rather than number of completed cell cycles.

VI. Miniclone Analysis of Glia-like Cells

A regular plastic petri dish can be rendered completely nonadhesive by coating with a thin film of agarose. Adhesiveness can then be restored to small areas by precipitation of nontoxic metal onto the agarose through a perforated template. By this technique small islands can be created carrying single glia-like cells whose multiplication as miniclones can then be analyzed (Westermark, 1978). The technique permits one to follow the fate of hundreds of defined individual cells under nonselective, undisturbed conditions. We have used it to study proliferation of glia-like cells at different passage levels *in vitro.*

The fraction of nondividers enters as an important element in any understanding of cellular senescence. In the Kirkwood–Holliday model (1975) there is at first a stage during which none of the committed cells will have divided enough times to reach the end of the incubation period. During this period no nondividers should be found. Later, when cells have passed their "incubation period," a constant fraction will be nondividers. The population will be in an equilibrium with cells regularly distributed over a number of classes characterized by their respective number of expected residual divisions left until the nondividing stage is reached. One fraction will still have the stem cell character of being "noncommitted." Once the "dilution catastrophe" occurs, no more noncommitted cells are left. The fraction of nondividers will then begin to increase. It will reach 100% after the last division of the class of cells that just became committed at the point of the dilution catastrophe. The period between the dilution catastrophe and no residual capacity for division will thus correspond to the incubation period.

According to Shall and Stein (1979), there should already be a low fraction of nondividers at the onset of serial cultivation, which steadily increases as a function of the number of divisions until it reaches 100%.

Table II demonstrates how the number of nondividers increased as the passage number increased in two experiments with a line of normal glia-like cells. In both instances the increase was approximately linear, but the starting level was higher in experiment B than in experiment A. The results tend to support the Shall–Stein model. They are, however, not incompatible with the Kirkwood–Holliday theory if one assumes that a dilution catastrophe occurred before passage 6. This in turn implies that at explantation there will already be very few (or perhaps no) uncommitted stem cells among adult human glia-like cells. The situation may possibly be clarified by miniclone analysis experiments where no cells are discarded in early passages.

Our tentative conclusion is that glia-like cells may well behave in a manner not expected from a Holliday-type model. This, in turn, could reflect the state of astrocytes in the *adult* brain where a stage may already have been reached where all cells have become committed and thus will produce sterile progeny at each successive division with increasing probability. It is not excluded that *embryonic* cells which form the basis for the stem cell concept differ from adult cells by still retaining a proportion of uncommitted cells.

TABLE II

Occurrence of Nondividers at Different Passage
Levels in Human Glia-like Cells

Passage	Nondividers on day 10 (%)	
	Experiment A[a]	Experiment B[b]
6	2	
11	5	18
16	8	
21		32
22	12	
30	30	52
40		73

[a]Westermark (1978). Growth control in miniclones of human glial cells.
[b]Blomquist et al. (1980).

VII. Conclusions

In all systems that have been thoroughly studied a definite relation has been found between cell generation and aging. This is best understood at the population level. With only a few exceptions populations of somatic cells will gradually approach a stage where no further division is possible. The speed at which this process takes place seems to be predominantly regulated by the number of cell doublings. This raises important questions of a general cell biological nature. Are there, in fact, any true stem cells in an adult organism? If so, why are they impossible to maintain in serial transplantation *in vivo* or in serial transfer *in vitro*? The old, recently revived suggestion (Kirkwood, 1977) that somatic death is of evolutionary advantage may obtain its necessary cellular foundation from the concept that true, immortal, noncommitted stem cells do not exist in somatic tissues. It then becomes extremely important to understand how certain animal cells may remain stem cell-like and whether there is a common denominator for this class of cells. The problem can be rephrased into how the initiation of the cell cycle is controlled. One idea is that the S phase cannot proceed and/or give rise to daughter DNA strands consistent with cell viability if too many mutations have accumulated. This is normally prevented by the elaborate and effective DNA repair mechanism existing in all living systems. Senescence could then depend on a (programmed?) abolition of DNA repair. Some support for this may be derived from the observations that at least certain aspects of DNA repair seem to diminish in effectiveness in late-passage cells (Mattern and Cerutti, 1975). However, a powerful argument against this idea is that fibroblasts from xeroderma pigmentosum—a disease with a known DNA repair defect—do not seem to age prematurely *in vitro* (Goldstein, 1971). It is important to investigate this possibility further by careful comparisons between transformed immortal cells and nontransformed mortal cells.

The dogma that the DNA of all somatic cells possesses identical genetic information has recently been challenged by observations of differentiation of B lymphocytes. Evidently DNA coding for immunoglobulin chains is transferred from one site to another during embryogenesis (Hozumi and Tonegawa, 1976; Bernard *et al.*, 1978). The immunoglobulin-producing cells are thus genetically different from other somatic cells at least with respect to the ordering of their immunoglobulin genes. It is not excluded that other examples of similar nature may be discovered and that somatic differentiation is irreversible in the sense that new genotypes are created. Of course, this concept is strongly contradicted by Gurdon's finding that a somatic differentiated frog nucleus contains the entire program for the creation of a new individual (Gurdon and Woodland, 1968). A crucial experiment is whether the same results can be obtained with frog lymphocytes (if they can be shown to translocate DNA in the same fashion as mice seem to do).

Sexual DNA is obviously immortal and virtually identical over many generations. The reason for mortality not being operative in these cells is not known. The molecular basis for this mechanism has not been looked into. Meiosis involves extensive exchanges of DNA between different chromosomes. It may be suggested that this is essential not only for the creation of genetic variability but more importantly for ordering and repairing DNA to a "virgin" state where any possible somatic programs for terminal differentiation and/or commitment to reproductive standstill are erased. There may be a major difference between sexual and somatic DNA. The developmental biological problem of how and when in the early embryo the decision is made as to whether a cell will become sexual or somatic does not seem to have attracted much interest but could be of fundamental importance for the understanding of aging.

Cancer cells are also immortal. It should be recalled that they are not necessarily stem cells in the sense that they are undifferentiated. For instance, a myeloma line *in vitro* is composed of cells all of which synthesize immunoglobulin (Nilsson, 1971). Their stem cell character is instead functional, because they produce uncommitted cell copies capable of infinite multiplication.

Similarities between sexual and malignant DNA reproduction are difficult to see. Their capacity for endless rounds of reduplication may be entirely unrelated. It has often been observed (Zetterberg, 1970) that in cancer cells division tends to be very irregular, with unequal DNA distribution between the daughter cells and the continuous production of new genotypes. This could possibly reflect the loss of a mechanism by which normal somatic cells maintain and execute their differentiation–senescence program. Superficially it is reminiscent of the extensive recombination seen in meiosis, but this may be entirely fortuitous.

As with so many fundamental problems in biology the relation between cell generation and aging in the nervous system and elsewhere has turned out to be exceedingly complex. New methods and concepts are certainly required to elucidate this vexing problem.

References

Bell, E., Marek, L. F., Levinstone, D. S., Merrill, C., Sher, S., Young, I. T., and Eden, M. (1978). Loss of division potential *in vitro*. Aging or differentiation? *Science* **202**, 1158–1163.

Bernard, O., Hozumi, N., and Tonegawa, S. (1978). Sequences of mouse immunoglobulin light chain genes before and after somatic changes. *Cell* 15, 1133–1144.

Bignami, A., Eng, L. F., Dahl, D., and Uyeda, C. T. (1972). Localization of the glial fibrillary acidic protein in astrocytes by immunofluorescence. *Brain Res.* 43, 429–435.

Blomquist, E., Arro, E., Brunk, U. T., and Westermark, B. (1978). Plasma membrane motility of cultured human glia cells in phase II and phase III. *Acta Pathol. Microbiol. Scand. Sect. A* 86, 257–263.

Blomquist, E., Westermark, B., and Pontén, J. (1980). Aging of human glial cells in culture: Increase in the fraction of non-dividers as demonstrated by a mini-cloning technique. *Mechanism Aging Dev.* 12, 172–182.

Brody, H., and Vijayashankar, N. (1977). Anatomical changes in the nervous system. *In* "Handbook of the Biology of Aging" (D. Finch and L Hayflick, eds.), pp. 241–261. Van Nostrand-Reinhold, Princeton, New Jersey.

Brunk, U., Ericsson, J., Pontén, J., and Westermark, B. (1971). *In vitro* differentiation and specification of cell surfaces in human glia-like cells. *Acta Pathol. Microbiol. Scand., Sect. A* 79:309–312.

Brunk, U., Ericsson, J., Pontén, J., and Westermark, B. (1973). Residual bodies and "aging" in cultured human glia cells. *Exp. Cell Res.* 79, 1–14.

Cristofalo, V. S., and Scharf, B. B. (1973). Cellular senescence and DNA synthesis. Thymidine incorporation as a measure of population age in human diploid cells. *Exp. Cell Res.* 76, 419–427.

Daniel, C. W. (1977). Cell longevity *in vivo*. *In* "Handbook of the Biology of Aging" (C. Finch and L. Hayflick, eds.), pp. 122–154. Van Nostrand-Reinhold, Princeton, New Jersey.

Gimbrone, M. A., Jr., and Cotran, R. S. (1975). Human vascular smooth muscle in culture: Growth and ultrastructure. *Lab. Invest.* 33, 16–27.

Goldstein, S. (1971). The role of DNA repair in aging of cultured fibroblasts. *Proc. Soc. Exp. Biol. Med.* 137, 730.

Gurdon, J. B., and Woodland, H. R. (1968). The cytoplasmic control of nuclear activity in animal development. *Biol. Rev. Cambridge Philos. Soc.* 43, 233–267.

Haugen, Å (1978). Chemical carcinogenesis and ultrastructure of fetal brain cells in culture. Dissertation, University of Bergen, Norway.

Haugen, Å, and Laerum, O. D. (1978). Induced glial differentiation of fetal and brain cells in culture: An ultrastructural study. *Brain Res.* 150, 225–238.

Hayflick, L. (1965). The limited *in vitro* lifetime of human diploid cell strains. *Exp. Cell Res.* 37, 614–636.

Hayflick, L. (1966). Senescence and cultured cells. *Perspect. Exp. Gerontol.* 14, 195–211.

Hayflick, L. (1977). The cellular basis for biological aging. *In* "Handbook of Biology Aging" (C. Finch and L. Hayflick, eds.), pp. 159–179. Van Nostrand-Reinhold, Princeton, New Jersey.

Hayflick, L., and Moorhead, P. (1961). The serial cultivation of human diploid cell strains. *Exp. Cell Res.* 25, 586–621.

Holliday, R., and Tarrant, G. M. (1972). Altered enzymes in aging human fibroblasts. *Nature (London)* 238, 26–30.

Holliday, R., Huschtscha, L. T., Tarrant, G. M., and Kirkwood, R. B. L. (1977). Testing the commitment theory of cellular aging. *Science* 198, 366–372.

Hozumi, N., and Tonegawa, S. (1976). Evidence for somatic rearrangement of immunoglobulin genes coding for variable and constant regions. *Proc. Natl. Acad. Sci. U.S.A.* 73, 3628–3632.

Kirkwood, T. B. L. (1977). Evolution of aging. *Nature (London)* 270, 301–304.

Kirkwood, T. B. L., and Holliday, R. (1975). Commitment to senescence: A model for the finite and the infinite growth of diploid and transformed human fibroblasts in culture. *J. Theor. Biol.* **53**, 481–496.

Lindgren, A., and Westermark, B. (1977). Reset of the pre-replicative phase of human glia cells in culture. *Exp. Cell Res.* **106**, 89–93.

Linn, S., Kairis, M., and Holliday, R. (1976). Decreased fidelity of DNA polymerase activity isolated from aging human fibroblasts. (DNA nucleotidyl transferase/error-prone replication/cell culture senescence.) *Proc. Natl. Acad. Sci. U.S.A.* **73**, 2818–2822.

Lumsden, C. E. (1959). Tissue culture in relation to tumours of the nervous system. *In* "Pathology of the Nervous System" (D. S. Russel and L. J. Rubinstein, eds.), pp. 272–309. Arnold, London.

Macieira-Coelho, A. (1966). Action of cortisone on human fibroblasts *in vitro*. *Experientia* **22**, 390–391.

Macieira-Coelho, A. (1970). The decreased growth potential *in vitro* of human fibroblasts of adult origin. *In* "Aging in Cell and Tissue Culture" (B. T. Holecková and V. S. Cristofalo, eds.), pp. 121–132. Plenum, New York.

Macieira-Coelho, A. (1973). Aging and cell division. *Front. Matrix Biol.* **1**, 4–77.

Macieira-Coelho, A., Pontén, J., and Philipson, L. (1966). The division cycle and RNA synthesis in diploid human cells at different passage levels *in vitro*. *Exp. Cell Res.* **42**, 673–684.

Mattern, M. R., and Cerutti, P. A. (1975). Age dependent excision repair of damaged thymidine from gamma-irradiated DNA by isolated nuclei from fibroblasts. *Nature (London)* **254**, 450–452.

Miller, R. C., Nichols, W. W., Pottash, J., and Aronson, M. M. (1977). *In vitro* aging. Cytogenetic comparison of diploid human fibroblast and epithelioid cell lines. *Exp. Cell Res.* **110**, 63–73.

Nilsson, K. (1971). Characteristics of established myeloma and lymphoblastoid cell lines derived from an E myeloma patient: A comparative study. *Int. J. Cancer* **7**, 380–396.

Nilsson, K., and Pontén, J. (1975). Classification and biological nature of established human hematopoietic cell lines. *Int. J. Cancer* **15**, 321–341.

Orgel, L. E. (1970). The maintenance of the accuracy of protein synthesis and its relevance to aging. A correction. *Proc. Natl. Acad. Sci. U.S.A.* **67**, 1476.

Pendergrass, W. R., Martin, G. M., and Bornstein, P. (1976). Evidence contrary to the protein error hypothesis for *in vitro* senescence. *J. Cell. Physiol.* **87**, 3–14.

Pontén, J. (1971). "Spontaneous and Virus-Induced Transformation in Cell Culture," Virol. Monogr. No. 8. Springer-Verlag, Berlin and New York.

Pontén, J. (1973). Aging properties of human glia. *INSERM* **27**, 53–64.

Pontén, J. (1975). Neoplastic human glia cells in culture. *In* "Human Tumor Cells *in Vitro*" (J. Fogh, ed.), Vol. 7, pp. 175–206. Plenum, New York.

Pontén, J. (1977). Abnormal cell growth (neoplasia) and aging. *In* "Handbook of Biology Aging" (C. Finch and L. Hayflick, eds.), pp. 536–560. Van Nostrand-Reinhold, Princeton, New Jersey.

Pontén, J., and Macintyre, E. (1968). Long-term culture of normal and neoplastic human glia. *Acta Pathol Microbiol. Scand., Sect. A* **74**, 465–486.

Pontén, J., Westermark, B., and Hugosson, R. (1969). Regulation of proliferation and movement of human glia-like cells in culture. *Exp. Cell Res.* **58**, 393–400.

Ross, R. (1971). The smooth muscle cell. II. Growth of smooth muscle in culture and formation of elastic fibers. *J. Cell Biol.* **50**, 172–186.

Schachner, M., Schoonmaker, G., and Hynes, R. O. (1978). Cellular and subcellular localization of LETS protein in the nervous system. *Brain Res.* **158**, 149–158.

Shall, S., and Stein, W. D. (1979). A mortalization theory for the control of cell proliferation and for the origin of immortal cell lines. *J. Theor. Biol.* **76**, 219–231.

Vaheri, A., Ruoslahti, E., Westermark, B., and Pontén, J. (1976). A common cell type specific surface antigen in cultured human glial cells and fibroblasts: Loss in malignant cells. *J. Exp. Med.* **143**, 64–72.

Westermark, B. (1971). Proliferation control of cultivated human glia-like cells under "steady state" conditions. *Exp. Cell Res.* **69**, 259–264.

Westermark, B. (1973). Growth control of normal and neoplastic human glia-like cells in culture. Dissertation, Acta Univ. Upsaliensis.

Westermark, B. (1978). Growth control in miniclones of human glial cells. *Exp. Cell Res.* **111**, 295–299.

Zetterberg, A. (1970). Nuclear and cytoplasmic growth during interphase in mammal cells. *Adv. Cell Biol.* **1**, 211–232.

ADVANCES IN CELLULAR NEUROBIOLOGY, VOLUME 1

AGE-RELATED CHANGES IN NEURONAL AND GLIAL ENZYME ACTIVITIES[1]

ANTONIA VERNADAKIS AND
ELLEN BRAGG ARNOLD

Department of Psychiatry and Pharmacology and
University of Colorado School of Medicine
Denver, Colorado

[1] The preparation of this review was supported by a USPHS Training Grant T32 07072 and a Research Scientist Career Development Award from the National Institute of Mental Health to A. Vernadakis.

I. General Introduction

Numerous studies have been devoted to the biochemical development and aging of the central nervous system (CNS), and this work has been extensively reviewed (see reviews by Timiras *et al.,* 1968; Timiras, 1972; Timiras and Vernadakis, 1972). In this chapter we will consider CNS development and, wherever possible, aging at a cellular level. We will discuss only one aspect of cellular function, namely, changes in the activities of neural enzymes. The CNS contains numerous types of neuronal cells as well as glial cells, which play an important role in the integration of neural activity. Previous studies have demonstrated that various CNS areas attain maturity at different times and that aging is characterized by variable patterns in individual CNS areas. It might be inferred, therefore, that such regional differences reflect changes in the proportions of specific cell types constituting individual brain regions.

Because certain enzymes are known to be involved in specific biochemical processes unique to either neurons or glial cells, they can be used as indices for the presence of these cell types. Therefore, in our discussion we have considered neural enzymes from two points of view: as functional cellular entities and as markers for specific cell types.

II. Neuron-Specific Enzymes

A. Choline Acetyltransferase

The enzymes choline acetyltransferase (CAT) and acetylcholinesterase (AChE) are responsible for the synthesis and degradation, respectively, of the putative excitatory neurotransmitter acetylcholine (ACh). Although the presence of either enzyme is invariably associated with the occurrence of ACh, CAT probably represents a more reliable index of functional maturation of cholinergic neurons than AChE (Burt, 1973; Vernadakis, 1973a; Singh and McGeer, 1977). AChE is more ubiquitously distributed in tissues than CAT; for example, it is found in both the erythrocyte and the noninnervated placenta (Vijayan, 1977), where its function is presumably unrelated to neurotransmission.

1. AGE-RELATED CHANGES: *In Vivo* STUDIES

a. Mammalian Brain. The distribution and maturational profile of CAT activity in the CNS have been described for a number of mammalian and avian species. One of the earliest studies documenting age-related changes in CAT activity was reported by Hebb (1956). She found that, in both the rabbit and guinea pig, activity in the cerebellum reached a peak at or shortly before birth, after which it declined throughout further development. In the cerebrum, on the other hand, activity continued to increase throughout early postnatal life and did not peak until 100 days after birth in the rabbit, and 120 days after birth in the guinea pig. A similar pattern was described by McCaman and Aprison (1964), who found CAT activity in the cerebral cortex of the immature rabbit brain to be very low immediately after birth and to increase rapidly during early postnatal life, reaching adult levels of activity at 1 month of age.

Valcana and her associates (1969) were among the first to examine the development of CAT activity in discrete areas of rat brain. During early postnatal life, activity increased significantly with age in the cerebral cortex and hypothalamus, whereas no such change was observed in the cerebellum. Likewise, Coyle and Yamamura (1976) found that CAT activity in whole rat brain remained low during the first week after birth and subsequently exhibited a 10-fold increase over a 3-week period. In a number of discrete brain regions examined (cortex, hypothalamus, medulla–pons, and midbrain–thalamus), activity was also found to increase at the end of the first postnatal week. By the end of the fourth postnatal week, the medulla–pons region had achieved adult levels of activity, whereas other regions had attained 70–80% of adult levels and continued to increase in activity during subsequent weeks. In contrast, the specific activity of cerebellar CAT remained constant throughout development. Singh and McGeer (1977) have reported similar findings in the developing rat brain. In their studies, CAT activity in the cerebral cortex and caudate–putamen was low immediately after birth and began to increase at 10–15 days thereafter. It increased 3- to 4-fold in these areas from 15 to 25 days, at which time adult levels were reached.

The developmental decline in rat cerebellar CAT activity has provoked considerable study. It has been proposed (Valcana, 1971) that the increased activity of CAT in the cerebral cortex may be associated with the maturation of synaptic contacts between cholinergic nerve endings in the reticular activating system and cortical neurons, and that the decreased activity of CAT in the cerebellum may reflect the enhanced development of noncholinergic cerebellar elements. A more detailed investigation of rat cerebellar cholinergic development was subsequently reported (Valcana et al., 1974a). In this study, the age-related decrease in the specific activity of CAT in cerebellar homogenates was confirmed, however, total cerebellar CAT activity (i.e., activity per cerebellum, rather than activity per unit of wet tissue weight) showed a progressive increase with age. This trend was also observed for total and specific activity in certain cerebellar subcellular fractions (soluble, crude mitochondrial, and microsomal). In the nuclear fraction, total activity exhibited a gradual age-related increase, although no significant developmental change in specific activity was observed.

Valcana (1971) has also investigated the distribution of CAT activity in subcellular fractions of cerebral hemispheres obtained from young (22-day-old) rats. Following fractionation of brain tissue homogenates, activity was found to be highest in the mitochondrial and microsomal fractions, with little activity in the nuclear fraction. Further subfractionation of the mitochondrial fraction showed activity to be associated with the synaptosomal fraction. This observation was in accordance with earlier reports (Tucek, 1967; Hebb and Morris, 1969; Whittaker, 1969). The question of subcellular localization of CAT activity during brain maturation was also addressed by Singh and McGeer (1977), who described an age-related sensitivity of CAT to activation by sodium chloride and Triton X-100. Enzyme activity in brain homogenates from adult rats was more strongly stimulated, by either sodium chloride or Triton X-100, than activity in neonatal brain homogenates. These authors did not speculate as to whether the differential response to salt activation in adult and neonatal brain reflected a change in enzymatic form or a change in subcellular distribution during the course of development. It is known, however, that detergents increase the solubilization of CAT from cellular particulate fractions (Fonnum, 1966) and can thus lead to enhanced CAT activity in tissue homogenates. These authors therefore suggested that the increased sensitivity of CAT to detergent activation during maturation reflected a progressive change in its intracellular localization in brain. Since the soluble form of CAT is thought to be derived from cholinergic cell bodies, dendrites, and axons, and the particulate form from synaptosomes (Fonnum and Malthe-Sorenssen, 1972), neonatal rat brain CAT may be predominantly soluble and of cell body origin, and the more detergent-sensitive adult CAT may be predominantly particulate and of synaptosomal origin (Singh and McGeer, 1977). This hypothesis, they point out, is supported by the fact that postnatal CAT development in the rat closely parallels the postnatal development of cholinergic synapses in the rat neostriatum (Hattori and McGeer, 1973).

Changes in CAT activity have also been studied in bulk-separated cellular fractions of developing rat brain (Nagata et al., 1976). Fractions enriched in neuronal cell bodies and glial cells were prepared according to the method of Nagata et al. (1974) and subjected to further subcellular purification according to the method of Whittaker and Barker (1972). The resulting purified neuronal cell body fraction exhibited CAT activity which increased rapidly during early development (postnatal days 5–20), at which time adult levels of activity were attained. CAT activity in the purified glial cell fraction followed the same time course and, by postnatal day 20, was approximately one-half of that in the purified neuronal fraction. As these authors suggest, the occurrence of CAT in the purified glial cell fraction may reflect residual synaptosomal contamination or signify that some CAT activity is localized in glial cells.

Postmaturational changes in CAT activity are less well documented than those occurring in early development. Although CAT activity declines in the spinal cord of aged rats, levels of CAT activity in the cerebral cortex, cerebellum, and hypothalamus do not differ significantly between young and aged rats (Valcana and Timiras, 1969). A postmaturity decline in CAT activity in the rat caudate nucleus has been observed (McGeer et al., 1971). In the mouse, CAT activity in the cerebellum and hippocampus was studied during the postmaturational period of life (3–24 months of age) (Vijayan, 1977). Specific CAT activity in the cerebellum did not change significantly during this period, but specific CAT activity in the hippocampus showed a progressive decline with age. By the twenty-fourth month of life, CAT activity in the hippocampus was 56% of the specific activity obtained in the 8-month-old rat. These findings have been interpreted, in light of the fact that CAT activity in the hippocampus is predominantly localized to presynaptic terminals of afferent septohippocampal fibers (Silver, 1974), to suggest that the decreased specific activity of CAT in the senescent hippocampus may reflect loss of the enzyme from septohippocampal axons, loss of the axons themselves, or dilution of CAT activity caused by glial or other nonneuronal cell proliferation (Vijayan, 1977).

b. Avian Brain. The first study of developmental changes in CAT activity in the avian brain was reported by Burdick and Strittmatter (1965). They found that CAT activity in whole brain exhibited a slow, gradual increase between embryonic days 8 and 16, followed by a more rapid increase until the second day after hatching. The total developmental increase in CAT activity was sevenfold, and the peak level attained by the second day was maintained throughout the first 3 weeks of life. Subsequently, Marchisio and Giacobini (1969) analyzed regional variations in the development of CAT activity in the CNS of the chick embryo. They characterized the maturational profile of CAT activity, from day 6 of embryogenesis through 60 days after hatching, in cerebral hemispheres, optic lobes, cerebellum, midbrain, medulla, and cervical spinal cord. CAT activity appeared throughout the brain in a sequential fashion. In the cerebral hemispheres and optic lobes, it was detectable as early as 6 days of incubation, and it appeared at 8 days in the medulla and at 11 days in both the midbrain and cerebellum. Until 16 days of incubation, CAT activity increased slightly in all regions except the cerebral hemispheres. In the optic lobes, the rate of increase in CAT activity during this period was much greater than in any other brain region. From day 16 until hatching, it increased markedly in all brain regions studied. At hatching, the highest level of CAT activity was found in the optic lobes. Immediately after hatching, a significant decline in CAT activity occurred in all regions other than the optic lobes; this fall took place during the first 24 hr and was followed by a recovery of CAT activity, generally within the first few days of life. This period of decline was prolonged in the medulla, however, lasting through day 15 and recovering shortly thereafter. Although the precise mechanisms underlying the posthatching decline and subsequent recovery of CAT activity are not known, these events undoubtedly reflect an adaptive response to new environmental conditions (Giacobini and Filogamo, 1973). After recovery of activity, progressive increases were seen in all brain areas throughout the remainder of the time course examined (to 60 days after hatching). CAT activity in the optic lobes increased most rapidly and achieved the highest levels, whereas CAT activity in all other regions was not, by the end of the second month of life, markedly different from that measured at hatching. In the same study, Marchisio and Giacobini (1969) were unable to establish a correlation between the maturation of CAT activity and the onset of spontaneous and reflex electrical activity in the developing chick brain. This observation does not support the concept of a close relationship between the development of cholinergic systems in the brain and the development of electrical function, at least in the earliest phases of development (Giacobini and Filogamo, 1973).

Our laboratory has investigated alterations in parameters of cholinergic neurotransmission during maturation and aging of the avian brain (Vernadakis, 1973a). This study employed chick embryos, young chicks, and chickens up to 3 yr of age; brain areas examined included the optic lobes, cerebellum, cerebral hemispheres, and diencephalon–midbrain. In all these CNS structures, CAT activity increased gradually throughout embryonic development (Fig. 1). In the cerebral hemispheres, it increased until 3 months after hatching and then declined, falling to embryonic levels by 3 yr of age. In the cerebellum, CAT activity peaked at 6 weeks after hatching and was thereafter maintained at that level. In the diecephalon–midbrain, it increased markedly until 12 months after hatching and then declined, although the level of activity at 3 yr of age was still twice as high as that measured at hatching. Finally, CAT activity in the optic lobes increased gradually throughout the entire 3-yr period.

2. AGE-RELATED CHANGES: *In Vitro* STUDIES

Developmental aspects of cholinergic functions have also been studied in a variety of cell and tissue culture systems. Cultured explants of brain tissue

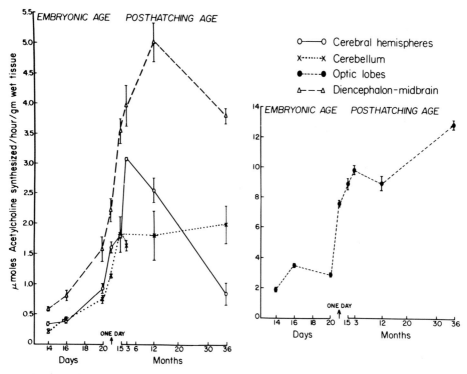

Fig. 1. Changes in CAT activity expressed as micromoles of ACh synthesized per hour per gram wet tissue in four CNS structures of chicks during embryogenesis (days) and posthatching (months). Points represent mean plus or minus standard error. (From Vernadakis, 1973a.)

have proved valuable in investigating morphological and electrophysiological parameters of brain maturation; they have been less often used for biochemical studies. One of the first detailed accounts correlating the morphological and biochemical development of explanted embryonic brain tissue was provided by Kim and Tunnicliff (1974). Portions of chick embryo cerebrum, obtained at 10–12 days of embryonic age, were maintained in culture for up to 21 days. Morphological maturation of the explanted tissue was followed by light and electron microscopy and was characterized by the appearance of well-differentiated neurons and myelinated axons and by an increase in the number and complexity of synapses. A close temporal relationship was observed between these morphological changes and an increase in the specific activity of several enzymes related to neurotransmission, including CAT. CAT activity reached maximal levels at 2 weeks of age in culture; the total increase in activity over the 21-day culture period was sixfold.

Monolayer cell cultures derived from mechanically or enzymatically dissociated embryonic neural tissue have been more extensively used to study biochemical development *in vitro*. The fact that such cultures can only be maintained until confluency is reached, or shortly thereafter, limits the time span over which development of their component cells can be studied. Nonetheless, dissociated cell culture systems provide a more abundant source of cultured neural cells than explant cultures, and a more well-defined population of individual cell types. The development of CAT activity has been studied in monolayer cultures of newborn mouse brain by Wilson *et al.* (1972). Between days 3 and 13 of cultivation, CAT activity increased 48-fold. Although it continued to increase throughout the entire culture period, CAT activity by day 30 was 3-fold lower than CAT activity in freshly dissociated newborn mouse brain cells. Shapiro and Schrier (1973) have described the development of CAT activity in dissociated cultures of fetal rat brain. In their culture system, the magnitude of increase in CAT activity was dependent upon the nature of the growth medium used, being higher in cultures grown in Ham's F-12 medium than in cultures grown in Dulbecco's modified Eagle's medium. Serial subculture of dissociated fetal rat brain cells did not result in progressive dilution of CAT activity with each subculture, suggesting that the cells in this culture system that produce CAT may be proliferative (Schrier, 1973). In a subsequent study (Schrier and Shapiro, 1974), the specific activity of CAT was found to be increased by the addition of 5-fluoro-2'-deoxyuridine, an inhibitor of cell division, an observation suggesting that cell division regulates the activity of CAT in some manner. Likewise, Wilson *et al.* (1972) found that CAT activity in monolayer cultures of newborn mouse brain cells continued to increase during the postconfluent phase of growth, when cell division was markedly restricted.

The development of CAT activity in dissociated cell cultures of chick embryo brain has been studied by several groups (Werner *et al.,* 1971; Peterson *et al.,* 1973; Vernadakis *et al.,* 1978). In cultures of 7-day-old chick embryo brain cells, activity increased 45-fold over a 20-day cultivation period (Werner *et al.,* 1971). In spite of this pronounced increase and although the pattern of increase in specific CAT activity *in vitro* closely paralleled that seen *in vivo,* the maximum level of activity attained in cultured dissociated brain cells was only 25% of that found in the chick brain at hatching (Werner *et al.,* 1971).

Peterson *et al.* (1973) confirmed the observations of Werner *et al.* that the specific activity of CAT in cultured chick embryo brain cells was markedly lower than that found *in vivo*. These investigators also studied CAT activity in isolated cultured neurons during development in culture. A suspension of dissociated embryonic brain cells was plated on glass, rather than the usual plastic culture vessels. Because they adhere poorly to glass surfaces, the neuronal cells could be readily separated from other cell types and either assayed or replated under suitable culture conditions. Peterson *et al.* found that neurons which aggregated and subsequently extended cellular processes demonstrated an increase in CAT activity over a 3- to 4-day cultivation period, but that the specific activity of CAT in these cells did not approximate *in vivo* levels.

We have also studied the maturational profile of CAT activity in dissociated cultures of chick embryo brain (Vernadakis *et al.*, 1978). Cerebral hemispheres from 8-day-old chick embryos were dissociated through a nylon mesh (73 μm pore size), and the dispersed cells were plated in Falcon plastic flasks at a density of 3 × 10⁶ cells/25-cm² flask. The cells were suspended in 4 ml of Dulbecco's modified Eagle's medium supplemented with 20% fetal bovine serum (Reheis). In this culture system, the dispersed cells formed aggregates within 24 hr and became attached to the surface of the flask. By day 5 in culture, neurites were observed to extend from the cellular aggregates. During the first 10 days in culture, neurons appeared to predominate and glial cells were not observed; after 15 days in culture, glial cells predominated and continued to proliferate throughout the life span of the cultures (approximately 4 weeks).

In these cultures, the time course of development of CAT activity was biphasic, showing an initial peak at day 10 and a later peak between days 25 and 28 in culture (Fig. 2). The early increase in activity may reflect the progressive differentiation of cholinergic neuroblasts in culture, and the later increase in activity may be related to the continued proliferation of glial cells in older cultures. There is evidence that contact with glial cells enhances and maintains the differentiation of cultured neurons. For example, Monard *et al.* (1973) have demonstrated the release of a macromolecular factor by cultured glial cells that can induce morphological differentiation of neuroblastoma cells, and Murphy *et al.* (1977) have reported that cultured glial cells secrete a factor that is biologically and immunologically similar to nerve growth factor. In addition to possible stimulatory influences on cholinergic neurons from surrounding glial cells, the possibility that glial or other nonneuronal cells contain CAT might be considered. In this regard, the work of Schrier, as discussed previously, suggested that CAT activity in dissociated fetal rat brain cell cultures was attributable to a proliferative population of cells (Schrier, 1973). Normal (i.e., nonneoplastic) neurons are unable to divide in culture; therefore, if CAT activity is localized, in part, to dividing cells, glial cells represent possible candidates for such a site.

Seeds (1973) has noted that monolayer brain cell cultures possess certain disadvantages that tend to limit their usefulness in addressing developmental problems. Chief among these limitations are the facts that developmental

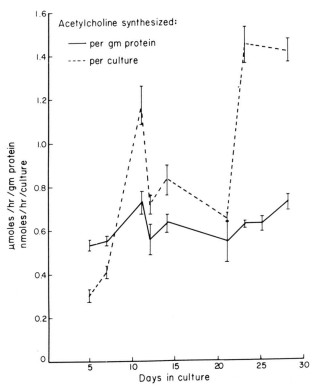

FIG. 2.   Changes in CAT activity in 8-day-old chick embryo dissociated brain cells with days in culture. Points represent mean plus or minus the standard error. (From Vernadakis *et al.*, 1978.)

changes in the specific activity of neural enzymes, although occurring in such cultures, rarely surpasses the specific activity of the undissociated fetal tissue (Rosenberg, 1973) and that certain biochemical properties of neural cells may be lost altogether in monolayer cultures (Seeds, 1973). Seeds has suggested, on the basis of such findings, that three-dimensional cellular interactions, which are destroyed by tissue dissociation and are not reestablished under monolayer culture conditions, may be required for normal differentiation to proceed. Using aggregation culture methods, which allow dissociated cells to regain specific cell–cell contacts and establish histotypic patterns characteristic of the original tissue, Seeds (1971) has described the development of CAT activity in reaggregated dissociated fetal mouse brain cells. Fetal mouse brain tissue (16–18 days of gestation) was mechanically and enzymatically dissociated; the resulting suspension was maintained under culture conditions that facilitated the reaggregation of cells. During a 20-day cultivation period, the specific activity of CAT increased 20-fold to a level that was 70% (Seeds, 1971) to 90% (Seeds, 1973) of the specific activity found in adult mouse brain. When dissociated brain cells from 17-day-old fetal mice were cultured as monolayers, rather than in an aggregate system, the specific activity of CAT in the monolayer cultures decreased over an 11-day cultivation period (Seeds, 1971).

The reaggregated cell culture system has also been used by Ramirez, who characterized the development of CAT activity in chick embryo brain (Ramirez, 1977a), optic tectum, and retina (Ramirez, 1977b). In aggregate cultures derived from 7-day-old chick embryo brain, specific activity increased 32-fold over a 30-day cultivation period to a level which was 60% of that found in chick brain *in vivo* on the day of hatching. Of additional interest is the observation that CAT activity in freshly dissociated chick embryo brain cells was only 15% of that in undissociated embryonic brain tissue, a finding that suggests a possible correlation between cell–cell contact and expression of CAT activity (Ramirez, 1977a).

B. Acetylcholinesterase

Youngstrom (1938) was the first to demonstrate that the initiation of neuromuscular function in amphibian embryos was associated with a marked increase in the activity of AChE in the whole embryo. Later studies reported a correlation between increases in AChE activity in the developing CNS and the initiation of nervous function, as evidenced by the development of muscular activity in the whole embryo in both amphibian (Boell and Shen, 1950) and mammalian (Nachmansohn, 1940; Metzler and Humm, 1951) species. The initiation of electrical activity in the chick embryo brain is likewise associated with elevated AChE activity in the brain (Burdick and Strittmatter, 1965). Biochemical and histochemical analyses of AChE activity in the developing CNS have further confirmed the correlation between markedly increased activity during neurogenesis and functional maturation of the brain (Karczmar *et al.*, 1973). Because of this often-documented association with the onset of CNS function, AChE activity has been a parameter of numerous developmental studies.

Specific AChE activity is considered characteristic of neuronal differentiation, and its presence is thought to represent a marker for neuronal cells (Wilson *et al.*, 1972; Seeds, 1973, 1975).[2] As such, AchE activity in cultured cells is often used as an index of neuronal maturation *in vitro*. However,

---

[2] The nonspecific cholinesterase, butyrylcholinesterase, which is treated separately (Section III, D), has in contrast a mainly nonneuronal localization.

studies in our laboratory have shown that AChE activity can be detected in certain clonal lines of glial cells (C6) (Vernadakis *et al.*, 1976; Vernadakis and Nidess, 1976). It seems likely that the presence of AChE in this system is related more to general neural growth processes than to neurotransmissional processes. As will be discussed, evidence suggests that substances such as ACh and certain biogenic amines, which function as neurotransmitters in the mature animal, may at times serve in a different capacity as neural growth factors during embryonic life.

1. Age-Related Changes: *In Vivo* Studies

a. Mammalian Brain. Maletta *et al.* (1967a) studied the maturation of AChE activity in several areas of the rat brain, from day 18 of gestation through day 44 of postnatal life. During this period, specific AChE activity increased with age in all areas studied (cerebellum, sensory-motor cortex, hypothalamus, and brain stem). By postnatal day 44, the specific activity of AChE was highest in the sensory-motor cortex, followed, in descending order, by the brain stem, hypothalamus, and cerebellum. AChE activity in the sensory-motor cortex reached a peak on day 44, whereas in all other brain regions it peaked by day 23.

Valcana (1971) has also described the maturational profile of rat brain AChE activity. She measured changes in enzymatic activity in the cerebellum and cerebral hemispheres between 11 and 29 days of postnatal life and found that AChE activity increased markedly during this period in both brain regions. In addition, the subcellular distribution of AChE activity in cerebral hemispheres of 22-day-old rats was studied. Among the primary subfractions of cortical homogenates, AChE activity was high in both the mitochondrial and microsomal fractions, the mitochondrial fraction having the higher activity. Among mitochondrial subfractions, AChE activity was highest in the synaptosomal fraction. This finding agreed with earlier observations on the subcellular distribution of AChE (De Robertis and Rodriguez De Lores Arnaiz, 1969; Whittaker, 1969).

A study of AChE activity in rat cerebral hemispheres and cerebellum, which covered a major portion of the life span, was reported by Kaur and Kanungo (1970). Animals at 6, 22, 52, and 96 weeks of age were studied; these ages represented an immature phase, an adult phase, an advanced reproductive phase, and a postreproductive (senescent) phase. In both brain areas, AChE activity was highest at 22 weeks of age. Enzymatic activity in the cerebral hemispheres declined sharply between 22 and 52 weeks of age and declined still further during senescence. The level of cerebellar AChE activity at 52 weeks of age was not significantly different than that found at 22 weeks of age, but a significant fall in specific activity occurred by 96 weeks of age. This study also described the effects of 5-hydroxytrypt-amine (5-HT, serotonin), epinephrine (Epi), norepinephrine (NE), γ-aminobutyric acid (GABA), and glutamic acid on AChE activity in an *in vitro* assay system. At all ages studied, AChE activity was inhibited by 5-HT, Epi, and NE, and stimulated by GABA. In neural tissues from aged rats, the inhibitory effects of 5-HT and Epi were diminished, and those of NE were enhanced. The stimulatory effects of GABA were decreased in senescence. Glutamic acid stimulated AChE activity in cerebral hemis-pheres at all ages, but stimulated cerebellar AChE only in young animals. This group of investigators subsequently reported a somewhat different profile of AChE development in rat brain (Moudgil and Kanungo, 1973). In this later study, AChE activity was highest in the cerebral hemispheres of immature animals (9 weeks of age), declined by 50% in adult animals (29 weeks of age), and declined still further in senescent animals (65 weeks of age). No such age-related decrease in cerebellar AChE activity was seen. At 9 weeks of age, AChE activity in the cerebral hemispheres was 2.5-fold higher than cerebellar AChE activity. With advancing age, the decline in cerebral hemisphere activity, coupled with the maintenance of a constant level of activity in the cerebellum, resulted in no significant difference be-tween these two brain regions by the sixty-fifth week of life.

Valcana and her associates (1974a) reported a study of cerebellar cholin-ergic development from birth to maturity in the rat. By examining the developmental patterns of both AChE and CAT, they sought to confirm the hypothesis that, in the cerebellum, AChE activity increases and CAT activity decreases with age, in contrast to the characteristic developmental pattern of other brain regions in which AChE and CAT show parallel changes with age (Silver, 1967). Animals were studied from day 1 to day 100 of postnatal life. On day 1, AChE activity was much higher than that of CAT. During the first 3 weeks of life, CAT activity decreased, whereas AChE activity increased, with maximal AChE activity consistently cor-responding to minimal CAT activity, and vice versa. Beyond the first month of life, AChE activity gradually but steadily declined, and CAT ac-tivity remained relatively constant. When these data were expressed as total rather than specific activities, however, the developmental pattern of activ-ity in the cerebellum showed an overall marked increase for both enzymes. It was suggested that the inverse relationship between the maturational pro-files of cerebellar AChE and CAT activity reflected changing equilibria be-tween the rates of development of the state of activity of cholinergic versus noncholinergic elements (Valcana *et al.*, 1974a). The divergent patterns of AChE and CAT activity, as well as the much higher relative specific activ-ity of AChE, taken together with the fact that total ACh content decreases in the developing cerebellum (Valcana *et al.*, 1974b), suggested that the rate of ACh catabolism exceeded that of ACh synthesis during early development and that the marked increase in AChE activity and concomitant decrease in ACh content were manifestations of this active metabolic process.

In contrast to these studies of cerebellar cholinergic development, which extended from birth to maturity, a series of subfractionation studies was also carried out at selected ages (9, 22, and 60 days of postnatal life) considered to represent critical and significant periods of cerebellar maturation (Valcana *et al.*, 1974a). The developmental pattern of total AChE activity in subcellular fractions of cerebellar homogenates resembled that described above. Enzymatic activity in the nuclear (P_1), soluble, and microsomal (P_3) fractions reached maximal levels at 22 days of age. The only cerebellar subfraction that showed a persistent developmental increase in both AChE and CAT activity was the nuclear (P_1) fraction, which is known to contain mossy fiber nerve endings (Israël and Whittaker, 1965). These findings were interpreted to suggest that the age-related increase in total cholinergic enzyme activity was fundamentally related to an increase in mossy fiber innervation to the cerebellum. Although the specific (as opposed to the total) activity of AChE in cerebellar homogenates did not show significant developmental changes, the specific AChE activity in each subcellular fraction showed a distinct developmental pattern. This activity showed no significant changes in the soluble fraction, an increase in the nuclear (P_1) fraction, and a progressive decrease in the microsomal (P_3) and crude mitochondrial (P_2) fractions, as well as in the subcomponents of the crude mitochondrial fraction (myelin, synaptosomes, and free mitochondria).

The solubilization and physicochemical characterization of rat brain AChE was reported by Rieger and Vigny (1976), who studied the development and progressive maturation of diverse molecular forms of AChE. Two active forms were found, one easily solubilized and having a sedimentation coefficient of 5 S (ES), and the other more resistant to solubilization and having a sedimentation coefficient of 10 S (HS). The relative proportion of HS to ES increased during development. The absolute amount of the HS form also increased sevenfold during the first postnatal month. In contrast, the ES form was already present at adult levels at birth and was not subject to maturational changes. The relative proportion of these two molecular forms remained constant throughout adult life: 90% HS and 10% ES. Finally, it was determined that neither form of rat brain AChE corresponded to the EP form of AChE, which is characteristic of the muscle end plate region.

Butcher and Hodge (1976) employed histochemical methods to follow the postnatal development of AChE in the caudate–putamen and substantia nigra of rats from 3 to 90 days of age. In previous studies, this group had shown that dopaminergic neurons in the pars compacta of the substantia nigra contained AChE, possibly for the purpose of inactivating ACh released from afferent cholinergic nerve endings synapsing upon dendrites in the pars compacta, which in turn project into the pars reticulata (Butcher and Bilezikjian, 1975; Butcher *et al.*, 1975a,b). In this report, AChE-containing filaments were observed in the pars reticulata during development, and their origin was speculated to have been AChE-containing cell bodies in the pars compacta (Butcher and Hodge, 1976).

AChE activity in the cerebellum and hippocampus of the mouse during senescence has been studied by Vijayan (1977). It had previously been determined that AChE levels in whole mouse brain declined during senescence (Ordy and Schjeide, 1973) and that AChE activity decreased with age in the rat cerebral hemispheres (Moudgil and Kanungo, 1973) and cerebral cortex (Hollander and Barrows, 1968). Vijayan found that the specific activity of AChE did not vary significantly as a function of age (3, 8, 16, and 24 months) in either the cerebellum or hippocampus.

b. Avian Brain. In the embryonic chick brain, the maturational profiles of a number of biochemical components of the cholinergic system were described by Burdick and Strittmatter (1965). They found that AChE activity increased slowly between 6 and 16 days of embryonic age; the rate of increase was then accelerated until the maximum level of activity was reached 7 days after hatching. The total increase over this period was 16-fold. The period of rapid increase in AChE activity corresponds to a time when electrical function is initiated in the major portions of the chick brain (Boell and Shen, 1950).

In order to further elucidate the relationship of AChE activity to functional brain development, AChE activity was investigated in the optic lobes of chicks (Vernadakis and Burkhalter, 1967). In this study, activity was assayed in the optic lobes of chick embryos from 5–20 days of embryogenesis, hatched and unhatched chicks at 20–21 days of embryogenesis, and young chicks from day 1 through day 90 after hatching. AChE activity increased gradually between 5 and 18 days of embryogenesis and more rapidly between 18 and 20 days. Enzymatic activity reached a peak on day 1 after hatching and remained constant thereafter. Activity was much higher in the optic lobes of hatched chicks than in those of age-matched unhatched chicks. It was inferred that an increase in AChE activity was associated with accelerated functional development of neural units in the optic lobes during hatching. The importance of increased sensory input, which occurs at the time of hatching as a consequence of abrupt exposure to light, must also be considered. Other studies, in developing rats, have shown that light stimulation increases AChE activity in some brain areas (Kling *et al.*, 1965).

We have examined the development and maturation of AChE activity in four discrete areas of the chick brain over the entire life span (14 days of embryonic age through 36 months after hatching) (Vernadakis, 1973a). In the cerebral hemispheres and cerebellum, the developmental pattern of AChE activity was similar (Fig. 3). AChE activity increased gradually throughout embryonic life and exhibited a fall 1 day after hatching. The level of activity recovered shortly and increased rapidly during the first 3 months of postnatal life. There was then a decline in AChE activity until 20 months of age and finally a period of rapid increase which persisted throughout the third year of life. In the optic lobes and diencephalon–midbrain, AChE activity increased rapidly during embryonic life. As in the cerebral hemispheres and cerebellum, a fall in activity was seen immediately after hatching but, in contrast to AChE activity in the former structures, AChE activity in the optic lobes and diencephalon–midbrain did not surpass prehatching levels at any time during the 3-yr period. AChE activity remained fairly constant, whereas activity in the diencephalon–midbrain fluctuated throughout postnatal life. This study also demonstrated, during senescence, an inverse relationship between the activities of AChE and CAT in the cerebral hemispheres, AChE activity being high and CAT ac-

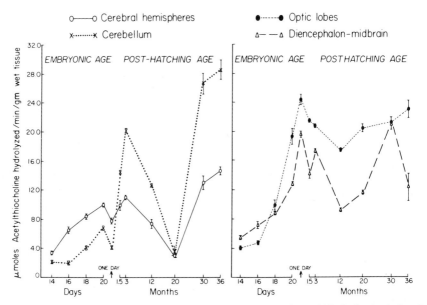

FIG. 3. Changes in AChE activity expressed as micromoles of acetylthiocholine hydrolyzed per minute per gram wet tissue in four CNS structures of chicks during embryogenesis (days) and posthatching (months). Points with vertical lines represent means and standard errors. (From Vernadakis, 1973a.)

tivity low (Figs. 1 and 3). Because it is now generally accepted that the presence of CAT is a more conclusive index for the intracellular presence of ACh, the low levels of CAT in aging cerebral hemispheres may suggest that cholinergic neurons are profoundly decreased in activity and/or number. The marked increase in AChE activity during aging, particularly in the cerebral hemispheres, may represent the participation of ACh and AChE in functions unrelated to neurotransmissional processes.

Developmental patterns of AChE and CAT in the cerebellum were similar to those in the cerebral hemisphere, however, changes during aging were not as pronounced. Cholinergic neurons in the cerebellum, in contrast to those in the cerebral hemispheres, may therefore remain more active during senescence (Vernadakis, 1973a).

As demonstrated for the rat brain (Rieger and Vigny, 1976), AChE in the chick brain exists throughout development in multiple forms. Using techniques of isoelectric focusing, Atherton (1973) followed changes in developmental patterns of AChE isozymes in the chick cerebellum from 10 to 14 days of embryonic age, a period of rapid cerebellar differentiation temporally associated with the onset of body movements in the embryo. Three AChE isozymes were found in the 10-day-old cerebellum, and five were found in the 14-day-old cerebellum. These studies led to the interpretation that the shift from the 10-day-old isozyme complex to the 14-day-old isozyme complex constituted a differentiating step that was under genetic control (Atherton, 1973).

Developmental variations in AChE activity and in molecular forms of AChE in the chick brain were studied by Marchand et al. (1977). Two regions of the chick brain were studied: the optic tectum and the forebrain hemispheres. Over a period from 10 days of embryonic age to day 15 of postnatal life, AChE activity increased fivefold in the whole brain, fourfold in the forebrain hemispheres, and eightfold in the optic tectum. Although the optic tectum represented only 20% of the total brain weight by 15 days of age, it accounted for 40% of the total brain AChE activity. Three molecular forms of AChE, distinguished on the basis of their sedimentation coefficients, have been identified in the chick brain: a heavy form (11 S), a light form (6 S), and a minor component (4 S). In the adult chick brain, the 11 S form represents most of the total brain AChE activity (Vigny et al., 1976). Marchand et al. (1977) found that the specific activity of the lighter components (6 and 4 S) remained virtually constant and equal in both the optic tectum and forebrain hemispheres during the period studied, whereas the specific activity of the 11 S form increased with age. In both regions, increases in the specific activity of the 11 S form constituted the major part of the total increase in AChE activity. Accumulation of the 11 S form in the optic tectum was three times higher than that in the hemispheres. These authors have interpreted this finding to suggest that a high level of AChE activity in the optic tectum reflects a high density of cholinergic synapses and may be related to the predominance of the visual system in birds.

Marchand *et al.* (1977) found significant quantitative differences between AChE activity in chick and rat brain (Rieger and Vigny, 1976). The specific activity of AChE at birth is 10-fold lower in the rat than in the chick, while adult levels of rat brain AChE are only 40% of those in the 15-day-old chick brain. These two species have in common, however, the fact that the specific activity of the light molecular forms of AChE does not change during development, whereas the heavier molecular forms accumulate and appear to be associated with maturational processes (Rieger and Vigny, 1976; Marchand *et al.*, 1977).

2. AGE-RELATED CHANGES: *In Vitro* STUDIES

a. Mammalian Systems. Seeds (1971) has investigated the development of AChE activity in reaggregated cell cultures of fetal mouse brain. In cultures prepared from 17-day-old fetal mice, AChE activity increased 10-fold over a 3-week period and attained a level equivalent to 40% of that found in adult mouse brain. The pattern of increase resembled that seen *in vivo,* suggesting that cellular or intracellular mechanisms responsible for controlling the time course for the expression of differentiation are retained in this *in vitro* system. The reestablishment of normal cell–cell contact appears to be crucial in this regard; when fetal mouse brain cells were removed and cultured in a monolayer system, rather than in aggregates, the specific activity of AChE did not increase over an 11-day cultivation period. The specific activity in these cultures was slightly lower than that in freshly dissociated fetal brain cells. The disparity between the maturational profiles of AChE activity in these two culture systems strongly implicates the importance of proper spatial orientation for the acquisition of biochemical characteristics of differentiation (Seeds, 1971, 1973). Similar findings were reported for the development of AChE activity in trypsin-dissociated newborn mouse brain cells maintained in a monolayer culture system (Wilson *et al.,* 1972). In these cells, AChE activity declined by 90% during the first 5 days of cultivation and remained at low levels over a 30-day period.

AChE activity has also been studied in clonal cell lines, such as the C-1300 mouse neuroblastoma, a neoplastic line composed of neuroblast cells. Although such cell lines cannot be used to study normal developmental processes, they provide a system well-suited to the study of certain aspects of neuronal differentiation, particularly in regard to the differentiation and regulation of enzymes involved in neurotransmission (Rosenberg, 1973). For example, Blume *et al.* (1970) have used the C-1300 line to show that AChE activity in neuroblastoma cells is subject to regulation, and that the regulatory mechanism involved is inversely related to the rate of cell division. In this study, the specific activity of AChE increased 25-fold when cell division was restricted by decreasing the concentration of serum in the growth medium.

b. Avian Systems. Kim and Tunnicliff (1974) have described the morphological and biochemical maturation of 10- to 12-day-old chick embryo cerebrum in explant culture. Over a 3-week period, AChE activity increased twofold; in addition, a close temporal relation between synaptogenesis and the development of activity in the explants was observed.

The development of AChE activity in dissociated cell cultures of 7-day-old chick embryo brain was studied by Werner *et al.* (1971), who found that AChE activity increased threefold by the fourth day in culture but then declined to basal levels throughout the remainder of the 20-day cultivation period. Because of the possibility that changes in AChE activity were being obscured by the presence of a rapidly proliferating population of nonneural cells, this group subsequently studied the development of AChE activity in dissociated cultured cells which aggregated and extended processes, as distinguished from the underlying flat cells. Many of the former group of cells possessed the morphological, histological, and biochemical attributes of neurons and were designated as such. Following physical separation of the two cell types, the neuronal cells were shown to contain AChE activity, as demonstrated by both histochemical and biochemical techniques. AChE activity in the cultured neurons increased fivefold over an 8-day period (Peterson *et al.*, 1973).

Studies from this laboratory demonstrated that dissociated cell cultures obtained from 6- or 10-day-old chick embryo cerebral hemispheres exhibited age-related differences in their patterns of growth and enzyme development (Vernadakis and Arnold, 1979). In cultures of cerebral hemispheres from 6-day-old chick embryos, AChE activity declined by 65% from initially high levels over a 2-week time period (day 3 through day 18 in culture). In similar cultures from 10-day-old embryos, AChE activity on day 3 in culture was approximately twice as high as in cultures obtained from younger embryos, and the subsequent decline in ezymatic activity was more rapid and more pronounced, falling by 90% on day 11 in culture. These observations correspond to those of others who have found a decline in AChE activity during culturing of dissociated brain cell cultures from chick embryos (Werner *et al.*, 1971; Ebel *et al.*, 1974), newborn mice (Wilson *et al.*, 1972), and fetal rats (Shapiro and Schrier, 1973). Ebel *et al.*

(1974) attributed the fall in AChE activity in their cultures to a progressive proliferation of astroblasts and mesenchymal cells in which AChE activity was negligible. The possibility that a decrease in AChE activity reflects dilution by proliferating cells not containing AChE is also suggested by the experiments of Werner *et al.* (1971), where the addition of 5-fluorouracil prevented overgrowth of nonneuronal cells and significantly elevated AChE activity in comparison to that in untreated cultures. In neuroblastoma cell cultures, AChE decreases with increasing cell density (Vernadakis *et al.*, 1976). As previously discussed, Blume *et al.* (1970) reported that, although the specific activity of AChE did not change appreciably during the period of rapid cell division, the specific activity during the stationary phase of growth increased by 25-fold. It has therefore been suggested that AChE responds to a regulatory mechanism coupled to the rate of cell division.

3. The Role of Acetylcholine in Neural Growth

The suggestion that ACh, as well as a number of other neurotransmitters, is involved in basic cellular processes affecting early growth has developed chiefly from the work of Buznikov (1971; Buznikov *et al.*, 1964, 1968, 1970, 1972) and Gustafson (1969; Gustafson and Toneby, 1970). These groups have gathered evidence that certain neurotransmitters are directly involved in the major developmental processes of cell division and morphogenetic movements of cells and that their involvement occurs at a time prior to the appearance of neurons in the embryo. The biochemical bases for these actions are not known, although it has been suggested that neurotransmitters may regulate the synthesis of cyclic nucleotides which, in turn, can guide morphogenetic movements by controlling cellular motility or cellular adhesiveness (McMahon, 1974). Another suggestion is that neurotransmitter substances may be involved in nucleic acid synthesis. Prives and Quastel (1969) found that ACh increased the rate of incorporation of uridine into RNA in rat brain slices, and Kasa *et al.* (1966) hypothesized the involvement of ACh in stimulating protein synthesis within developing nerve cells. Thus, during early neural growth, ACh may be involved in the initiation of neuroblast synthetic processes (Vernadakis and Gibson, 1974).

With regard to the participation of ACh in neural growth processes, there is ample evidence that components of the cholinergic system appear very early in a number of developing systems. Filogamo has detected AChE activity in the perikarya of neurons migrating toward the surface layer of the chick optic tectum as early as the third day of embryogenesis (Giacobini and Filogamo, 1973). AChE activity has been found in the perikarya of the anterior horn cells of the chick spinal cord on the fourth day of embryogenesis (Bonichon, 1958), in embryonic rabbit dorsal root neuroblasts (Tennyson and Brzin, 1970), in the chick optic lobe on the fifth day of embryogenesis (Vernadakis and Burkhalter, 1967), and in the fetal rat spinal cord at 8 days of gestation (Maletta *et al.*, 1967b). Electron microscopic studies have shown that AChE initially appears in the nuclear envelope of the neuroblast, later in the endoplasmic reticulum, and finally at the synapse (Duffy *et al.*, 1967; Tennyson and Brzin, 1970).

Filogamo and Marchisio (1971) have presented evidence that ACh, AChE, and CAT are found in neuroblasts and that these neuroblasts do not necessarily give rise to cholinergic neurons as development proceeds. For example, the presence of AChE can be demonstrated in chick and rabbit olfactory neuroblasts that do not correspond to cholinergic neurons in adult animals. A transient positive reaction for AChE is found in embryonic ganglia and amacrine cells of chick embryo retina, but a positive reaction is lacking in the mature cells. High activities of CAT and AChE are found in spinal ganglion neuroblasts, the majority of which give rise to noncholinergic neurons. These authors speculated that the ACh system may be involved in early development through mechanisms not yet adapted to synaptic transmission. The suggestion that early increases in AChE activity reflect growth and differentiation of the neuroblast and its processes, rather than functional cholinergic transmission, has been advanced by several other investigators (Burt, 1968; McMahon, 1974; Butcher and Hodge, 1976).

C. Tyrosine Hydroxylase

Tyrosine hydroxylase (TH), the enzyme that catalyzes the hydroxylation of tyrosine to dihydroxyphenylalanine (dopa), represents the rate-limiting step in catecholamine biosynthesis. TH activity is found in brain tissue, adrenergic nervous tissue, and the adrenal medulla and is believed to be localized within adrenergic nerves (for review, see Weiner, 1970). Weiner and his associates have demonstrated, in peripheral adrenergic nerve preparations, that increased neural activity is associated with an enhanced synthesis of NE from tyrosine and that this increased synthesis can be partially or completely blocked by the addition of NE to the *in vitro* system. These studies led to the proposal that end product feedback inhibition represents an important mechanism for the intrinsic regulation of TH activity (Alousi and Weiner, 1966; Weiner and Rabadjija, 1968; Weiner, 1970). It has been further suggested that the basis of this neural activation-induced increase in NE synthesis may be related to a progressive depletion of amine transmitter from the nerve ending, which serves to release TH from its normal (i.e., nonstimulated) state of product feedback inhibition (Weiner *et al.,* 1977).

1. AGE-RELATED CHANGES: *In Vivo* STUDIES

Coyle and Axelrod (1972) have studied the development of TH activity in various regions and subcellular fractions of rat brain. TH activity was detected in the fetal rat brain at 15 days of gestation and increased 4-fold between days 15 and 17. This 2-day period constituted the greatest increase in specific activity throughout the life span; from birth to adulthood, the specific activity of TH increased only 2.5-fold. Overall, between day 15 of gestation and adulthood, TH activity exhibited a 15-fold increase in specific activity. Although the rate of increase in TH activity was greatest during gestation, almost 95% of the total TH activity in the brain appeared after birth. In this same study, examination of specific brain regions showed that the regions of the brain containing predominantly catecholaminergic neurons exhibited the most pronounced developmental increases in TH activity. The largest increases in specific activity occurred in the cerebral cortex, cerebellum, and striatum, areas that contain catecholaminergic nerve terminals; much smaller increases in specific activity took place in the pons–medulla and midbrain–hypothalamus, areas that contain chiefly cell bodies of catecholaminergic neurons. As brain maturation proceeds, this shift in distribution of TH activity to regions containing nerve terminals was accompanied by a change in the subcellular distribution of enzyme activity. In fetal brain, TH activity was found predominantly in the soluble fraction of brain homogenates. In adult brain, the greatest proportion of activity was found in the synaptosome-enriched P_2 fraction. Correlation of these biochemical findings with histochemical observations led Coyle and Axelrod to conclude that changes in the levels of activity of TH, and perhaps other enzymes involved in catecholamine synthesis, probably reflect stages of differentiation of central aminergic neurons.

Studies by Algeri *et al.* (1977) confirmed this pattern of TH development. These investigators found that TH activity was undetectable in fetal rat brain during the first 15 days of gestation but developed rapidly thereafter, reaching 33% of the adult level of activity in the 17-day-old fetal brain and 67% of the adult value at birth (day 21). They extended their observations of TH maturation to include the senescent rat (30 months of age). TH activity declined significantly in the striatum and diencephalon of aged rats and, although the changes were not significant, was also lowered in the cerebral cortex and brain stem.

Lydiard and Sparber (1974) studied TH development in the embryonic chick brain. They found the greatest increase in specific activity (8.5-fold) to occur between days 15 and 20 of gestation. Between day 20 of gestation and day 29 after hatching, TH activity increased only 40%. A critical period for the development of TH activity in the chick brain was found to exist between 10 and 20 days of development. Prenatal administration of reserpine or α-methyl-p-tyrosine during this period resulted in an increase in brain TH activity, which persisted during the first month of postnatal life.

We have reported a similar maturational profile of TH activity in the cerebral hemispheres of the developing chick embryo (Arnold and Vernadakis, 1979). TH activity was determined as described by Waymire *et al.* (1971). This procedure involves the recovery and measurement of [*carboxyl*-^{14}C]dopa formed from [*carboxyl*-^{14}C]tyrosine. Enzymatic activity was expressed as picomoles of [^{14}C]carbon dioxide formed per milligram of protein per hour. TH activity in the cerebral hemispheres was undetectable prior to 14 days of embryonic age (Fig. 4). From this time until immediately prior to hatching (day 21), TH activity increased fivefold. The chick cerebral hemisphere (accessory hyperstriatum) is composed chiefly of stellate neurons arrayed without layering or other structural subdivisions (Jones and Levi-Montalcini, 1958; Corner *et al.*, 1974). Corner *et al.* (1977) have reported that the first identifiable synapses in this region are found at 10 days of embryonic age. As maturation proceeds, there is a gradual increase in synaptogenesis and, shortly before hatching, an explosive formation of new synapses. This latter phase of synaptogenesis coincides with a period of rapid dendritic maturation (Corner *et al.*, 1977). The develop-

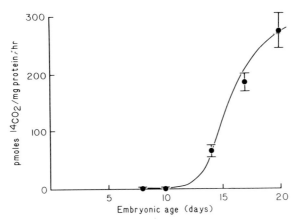

FIG. 4. TH activity in cerebral hemisphere tissue from chick embryos, 8–20 days of embryonic age. Activity is expressed as picomoles of [^{14}C]carbon dioxide formed per milligram of protein per hour. Each point represents the mean of four determinations plus or minus the standard error. (From Arnold and Vernadakis, 1979.)

mental profile of hyperstriatal TH activity described above closely parallels the time course of synaptogenesis in the embryonic chick brain.

2. AGE-RELATED CHANGES: *In Vitro* STUDIES

We have described the development of TH activity in dissociated cultures of chick embryo brain (Arnold and Vernadakis, 1979). In studying factors involved in the maturation of neurotransmission processes, dissociated cell culture systems offer a distinct advantage in that they allow one to examine properties of a heterogeneous population of neural cells derived from normal embryonic tissue. Undifferentiated, dissociated brain cells obtained from chick embryos can differentiate, in culture, into morphologically mature neurons and glial cells (Sensenbrenner *et al.*, 1972). The development of enzyme activities characteristic of neuronal differentiation, such as TH, provides an index of neuronal maturation in such cultures. Cultures of cerebral hemispheres obtained from 8-day-old chick embryos exhibited very low levels of TH activity prior to 14 days in culture. Thereafter, TH activity increased steadily throughout the third and fourth weeks in culture. Over this 2-week period, TH activity increased 10-fold (Fig. 5). Thus, TH activity in cultures of embryonic chick brain cells exhibits a pronounced maturational profile, with a critical period occurring at approximately 2 weeks of age in culture. This pattern resembles that observed *in vivo*, where TH activity increases markedly during the week immediately prior to hatching.

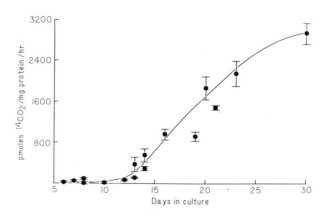

FIG. 5. TH activity in dissociated cerebral hemisphere cell cultures from 8-day-old chick embryos. Activity is expressed as picomoles of [^{14}C]carbon dioxide formed per milligram of protein per hour. Each point represents the mean of values obtained from six or more cultures plus or minus the standard error. (From Arnold and Vernadakis, 1979.)

3. THE RELATION OF TH ACTIVITY TO NEURONAL
   DIFFERENTIATION

The expression of TH activity is a characteristic feature of biochemical differentiation in mature neurons. This relationship has been extensively studied in mouse neuroblastoma cells (for review, see Prasad, 1975). TH activity in these cells is normally very low but can be enhanced by a variety of agents—most notably by adenosine 3′,5′-cyclic monophosphate (cyclic AMP) analogs or agents that elevate intracellular levels of cyclic AMP, such as prostaglandin E_1, R020–1724, and papaverine (Prasad, 1975). Furthermore, this response can be modulated by the inclusion of various types of serum in the neuroblastoma growth medium (Arnold *et al.*, 1978; Prasad *et al.*, 1979). Biochemical differentiation of neuroblastoma cells, as evidenced by increases in the activities of TH or other neuron-specific enzymes, increases in intracellular levels of cyclic AMP, or increased adenyl cyclase activity, is frequently associated with morphological differentiation. However, Prasad has shown that morphological differentiation (i.e., neurite extension) in neuroblastoma cells can occur in the absence of increased activities of TH, CAT, AChE, or adenyl cyclase. Conversely, these biochemically differentiated functions can be expressed in the absence of neurite formation (Prasad, 1975). This observation was confirmed in a later study which showed that, whereas elevated cyclic AMP levels in neuroblastoma cells were invariably associated with changes in both TH activity and morphology, changes in TH activity alone were not necessarily accompanied by enhanced morphological differentiation (Arnold *et al.*, 1978). These findings suggest that biochemical and morphological aspects of differentiation in neuroblastoma cells are subject to more than one mode of regulation and illustrate the applicability of the neuroblastoma cell culture system to study of the regulation of TH activity and other biochemically differentiated functions in neuronal cells.

D. *Glutamic Acid Decarboxylase and γ-Aminobutyric Acid Transaminase*

The major inhibitory neurotransmitter in the CNS is GABA (Baxter, 1969; Curtis and Johnston, 1974). The synthetic and degradative enzymes responsible for its metabolism, L-glutamic acid decarboxylase (GAD) and GABA transaminase (GABA-T), are found chiefly in the gray matter of the CNS; GAD in particular is considered a neuronal enzyme (Seeds, 1973). Because GABA synthesis appears to be involved in the regulation of cerebral excitability (Tapia, 1974; Tapia *et al.*, 1975b) and because heightened cerebral excitability is a characteristic feature of brain development in a number of species (Purpura, 1969; Tapia *et al.*, 1975a), the role of GAD activity during brain maturation may be of importance.

In the newborn rat brain, the specific activity of GAD is low, representing only 10% of adult levels (Coyle and Enna 1976). Similar observations of a developmental increase in activity have been made by Ossola *et al.* (1978), who reported a fourfold increase in rat cerebral hemisphere GAD activity during the first 3 weeks of life and a lesser increase in cerebellar GAD activity during the same period. In mouse brain, activity increased nearly 30-fold during this 3-week period (Seeds, 1971).

The developing GABA system has also been studied in the embryonic chick cerebellum (Kuriyama *et al.*, 1968) and tectum (McGeer *et al.*, 1974; McGeer and Maler, 1975). McGeer and associates have correlated increases in GAD activity with increases in [^{14}C]GABA accumulation by embryonic chick tectal tissue and have found the rate of increase in these two parameters to be parallel as well as age-related. The correlation between GAD activity and GABA accumulation was especially pronounced from the eighth to the twentieth embryonic day, suggesting that both parameters reflect a period of synaptic development in the chick tectum. A similar inference has been drawn for changes in GAD activity in the developing chick cerebellum (Kuriyama *et al.*, 1968).

Based on observations of differences in the biochemical properties of GAD in the brains of newborn and adult mice, Tapia and Meza-Ruiz have postulated the existence of two distinct enzymatic forms. Relative to the adult form of the enzyme, newborn GAD is less stable to a variety of treatments and has a higher affinity for the cofactor pyridoxal 5′-phosphate. This age-related distinction was not found for other amino acid decarboxylases (Tapia and Meza-Ruiz, 1975). Further studies demonstrated that these changes in GAD activity occurred progressively during postnatal development and that the immature form of the enzyme appeared to be replaced by the adult form during maturation of the brain. The changes were most marked between days 10 and 15 of postnatal life, an age that coincides with pronounced histological, electroencephalographic, and behavioral maturation of the mouse brain (Tapia and Meza-Ruiz, 1976).

The cellular localization of GAD during development has been described by Nagata *et al.* (1976), who studied changes in enzymatic activity in neuronal cell body-enriched and glial cell-enriched fractions of developing rat brain. Purified neuronal fractions exhibited high levels of GAD activity, which increased rapidly during the first month of postnatal life. Activity in purified glial fractions was low and increased only slightly during development.

The possibility has been raised that, in the developing animal, GABA levels may also be influenced at the level of transport or degradation. Ossola *et al.* (1978) reported that levels of GABA in the rat brain remained constant from birth to adulthood, in spite of increases in GAD activity throughout the same period. The specific activity of GABA-T increased during brain maturation in a fashion that suggested the presence of an effector of GABA-T activity. These investigators surmised that modulation of the levels of such an effector in turn modified GABA-T activity during growth and thereby contributed to the regulation of GABA levels in the developing rat brain.

Studies using culture systems have shown a developmental pattern of GAD similar to that seen *in vivo*. In aggregate cell cultures the enzymatic activity increased fivefold during a 3-week period in culture and attained a level of specific activity one-third of that found in adult mouse brain (Seeds, 1971). However, dissociated fetal mouse brain cells maintained in monolayer rather than aggregate culture did not exhibit developmental increases in GAD activity. After 11 days *in vitro*, GAD activity in such cultures declined by 40% (Seeds, 1971). Developmental increases in GAD activity have been found when newborn mouse brain cells were cultured for a longer period of time in a monolayer system (Wilson *et al.*, 1972). Brain tissue from newborn BALB/c mice was enzymatically dissociated; the resultant single-cell suspension was plated in petri dishes, and the cultures were maintained for up to 30 days. Under these conditions, GAD activity remained low throughout the period of rapid cell division (approximately 2 weeks in culture) and rose rapidly after cells reached confluency (from 2 to 4 weeks in culture). During this period, GAD activity increased approximately threefold. In contrast to these observations of increased GAD activity in developing mouse brain cell cultures, Kim and Tunnicliff (1974) have observed a 70% decrease in GAD activity in explants of chick embryo cerebrum maintained in culture for 3 weeks.

E. Tryptophan Hydroxylase and 5-Hydroxytryptophan Decarboxylase

Biochemical and morphological data indicate that, in the brain, the putative neurotransmitter 5-HT is localized to specific tracts of nerve cells, chiefly, the raphe nuclei of the lower pons and upper brain stem. The endogenous amino acid precursor, tryptophan, is converted by tryptophan hydroxylase (TPH) to 5-hydroxytryptophan (5-HTP), which is in turn metabolized to 5-HT by 5-HTP decarboxylase. This enzyme, which may be the same enzyme involved in the decarboxylation of the catecholamine precursor dopa to dopamine, is localized to nerve terminals. In the rat brain, this decarboxylase activity is highest in the caudate nucleus, hypothalamus, and midbrain.

Bennett and Giarman (1965) studied the pattern of appearance of total and particulate-bound 5-HT, the activity of 5-HTP decarboxylase, and the *in vivo* conversion of tryptophan to 5-HTP in the brains of fetal, neonatal, and adult rats. In both immature and adult animals, 80-90% of the total 5-HT content was bound to the particulate fraction of brain tissue homogenates. The major increase in 5-HT content occurred shortly after birth; newborn levels of 5-HT were 38% of adult levels. In contrast, 5-HTP decarboxylase activity developed primarily during the prenatal period; newborn levels of enzyme activity were 68% of adult levels. These authors observed that the capacity of immature rat brain to hydroxylate tryptophan to 5-HTP was reduced relative to that of adult rat brain, while the capacity to synthesize 5-HT from 5-HTP was similar in both immature and adult rat brain. They concluded that the level of 5-HT in the developing brain was regulated by the activity of TPH and that lowered TPH activity was responsible for the relatively low levels of 5-HT found in the newborn rat brain (Bennett and Giarman, 1965).

Regional changes in the levels of 5-HT and its associated biosynthetic enzymes were studied in various regions of the developing chick brain by Eiduson (1966). In this study, Eiduson found the concentration of 5-HTP in embryonic brain to be a limiting factor, apparently because of reduced TPH activity. From day 14 of embryonic development to day 6 after hatching, the activity of L-amino acid decarboxylase (5-HTP decarboxylase) increased gradually in the optic lobes and cerebral hemispheres. In the cerebellum, no enzymatic activity was detected prior to 6 days after hatching. Following 18 days of embryogenesis, increased 5-HTP decarboxylase activity and subsequent increases in 5-HT content occurred. The observation of a long-lasting (30 days after hatching) decline in 5-HT levels in the chick cerebral hemispheres, following alteration of embryonic 5-HT levels by prenatal administration of exogenous 5-HTP, suggested that end product inhibition or repression of TPH might be an important mechanism for the regulation of 5-HT biosynthesis in the developing brain (Eiduson, 1966).

The development of the serotonergic system in various regions of the mouse brain was studied by Baker and Hoff (1972). In this study, whole-brain levels of 5-HT in newborn mice were 44% of adult levels, which were attained at 6 weeks of age. The 5-HT system matured most rapidly in the medulla–pons, followed by the mesencephalon–diencephalon, cerebral hemispheres, and cerebellum. The level of 5-HT in whole brain increased rapidly following the first postnatal week. This observation correlates with data reported by Schmidt and Sanders-Bush (1971), who found TPH activity in the rat brain to be minimal prior to birth and during the first postnatal week but to increase rapidly during the next three postnatal weeks. Thus, numerous data support the hypothesis that TPH activity represents the developmentally limiting step in indoleamine biosynthesis (Bennett and Giarman, 1965; Schmidt and Sanders-Bush, 1971; Baker and Hoff, 1972).

An extracerebral influence on the maturation of 5-HT levels in the brain has been suggested by the work of Kellogg and Lundborg (1971), who found that the ontogenetic development of peripheral L-amino acid decarboxylase markedly affected the production of tritiated catecholamines in the immature rat brain following systemic administration of the tritiated catecholamine precursor [³H]dopa. In a subsequent study, they measured the *in vivo* uptake and utilization of [³H]5-HTP in the cerebral hemispheres, diencephalon, and brain stem of rats during early postnatal development and found that the capacity of peripheral decarboxylase increased markedly with age and thus influenced the amount of peripherally administered 5-HT precursor reaching the brain (Kellogg and Lundborg, 1972).

III. Glia-Specific Enzymes

A. Glutamine Synthetase

Elucidating the metabolism of amino acids and proteins during growth and differentiation has constituted a central problem in developmental physiology. In 1954, Rudnick *et al.* reported a general developmental survey of glutamotransferase and glutamine synthetase activity in the chick embryo. These enzymatic activities, which may be aspects of a single enzymatic process, are concerned with the metabolism of glutamic acid and glutamine. The specific activity of glutamotransferase was minimal in most embryonic tissues and membranes. The exceptions were yolk sac epithelium, during the first two-thirds of incubation; liver, as soon as it became a morphologically distinct organ; and brain, during the last third of incubation (chick embryonic incubation is 21 days). In general, no striking differences among the brain areas were observed, although the cerebral hemispheres (both pallium and neostriatum) and the spinal cord deviated slightly but significantly in enzymatic development from other brain regions (Rudnick and Waelsch, 1955). The retina, on the other hand, was found to have a quite unrelated and rather spectacular enzymatic history. The specific activity of transferase was minimal in the retina until day 17 of incubation. It then rose markedly, increasing 20-fold on day 17 and 50-fold on day 18. This increase continued after hatching. It was apparent from this study that the development of transferase and synthetase occurred relatively late in the development of the chick nervous system. Schroeder had suggested in 1911 that these two enzymatic activities may be associated with the maturation of a number of cell types, of interstitial material, or of myelin, which makes its gradual appearance during this period. The time at which the dramatic rise in transferase activity occurs in the retina strongly suggests a correlation with visual maturation. By 17 days of incubation, the retina has attained its full complement of cells, and their arrangement in layers and preliminary differentiation is complex (Weysse and Bugess, 1906). Retinal ganglionic and bipolar cells have attained their definitive size and alignment, and in the visual cells the carotenoid filter pigments have aggregated into droplets (Wald and Zussman, 1938; Coulombre, 1955). The final steps in differentiation of the rods and cones take place at about 17 days. Rudnick and Waelsch (1955) suggested that the high transferase and synthetase activities found in the retina may subserve a much different metabolic role in this tissue than in the brain.

Jolley and Labby (1960) described changes in the activity of enzymes concerned with glutamic acid metabolism in fetal monkey and pig. They investigated glutamotransferase (GTF), glutamine dehydrogenase (GDH), and glutamic-oxaloacetic transaminase (GOT). They found that all three enzymes increased in specific activity during the fetal period, with GOT showing the smallest relative increase. GTF activity increased dramatically during the last part of intrauterine life, especially in the fetal pig. Berl (1966) studied the distribution of glutamine synthetase in kitten brain during development. In the neocortex, the enzyme showed a four- to fivefold increase between the first 2 days and the ninth week of life. In the hippocampus, part of the archicortex, the increase in enzyme activity was more gradual and reached values that were half of those in the neocortex at 8 weeks of age. Similar changes occurred in the cerebellum. In the mesodiencephalon and brain stem, the level of enzymatic activity was somewhat higher in neonatal animals, and its rate of increase was less than in the other brain areas. Neurons in the hippocampus appear to be 2 weeks more advanced in structure and physiological properties than those of the neocortex (Purpura, 1964). In contrast, Purkinje cells of the cerebellum are much less developed than the major cell types of the neocortex (Purpura et al., 1964). It is of interest that, in these three areas, the level of enzyme activity is uniformly low shortly after birth. Therefore, glutamine synthetase activity does not appear to bear any relationship to early initial maturation of these neuronal cell types.

Since glutamine synthetase, which catalyzes the amidation of glutamate to glutamine, is a key enzyme of the small glutamate compartment, the exact localization of this enzyme could provide strong evidence for the site of this metabolic compartment. Glutamine synthetase is present chiefly in the microsomal fraction of brain (Sellinger and deBalbian Verster, 1962; Salganicoff and De Robertis, 1965), although a small amount has also been found in the synaptosomal fraction (De Robertis et al., 1967; Bradford and Thomas, 1969). Attempts to determine the localization of glutamine synthetase more precisely by bulk cell isolation techniques have been unsuccessful (Hamberger et al., 1976). Martinez-Hernandez et al. (1977), using immunohistochemical techniques specific for glutamine synthetase, found that the enzyme was localized in rat brain glial cells. The enzyme-labeled antibody method disclosed that all positive cells were glial cells, but it could not be determined whether these positive cells were astrocytes, oligodendroglia, or both. The same investigators subsequently reported that the enzyme-labeled glial cells were, in fact, astrocytes (Norenberg and Martinez-Hernandez, 1979). This finding substantiates the hypothesis that glial cells participate actively in glutamic acid and GABA metabolism (Hamberger et al., 1976; Henn and Hamberger, 1971; Schrier and Thompson, 1974; Ehinger and Falck, 1971; Schon and Kelly, 1974; Hutchison et al., 1974).

Using immunohistochemistry, Riepe and Norenberg (1977, 1978) studied the distribution of glutamine synthetase in the rat retina from birth to adulthood. Glutamine synthetase was present in the pigment epithelium in the newborn rat and, over the first few days of life, was demonstrated around blood vessels and within small glial cells. The enzyme was first detected in perikarya and processes of Muller cells on day 5. The normal adult glutamine synthetase pattern was acquired by day 12, except for persistence of glutamine synthetase in the pigment epithelium. These data show that the dramatic rise in retinal glutamine synthetase demonstrated biochemically by others, as discussed above, occurs exclusively in Muller cells. Analogous to this finding is the localization of glutamine synthetase to glial cells in rat brain (Martinez-Hernandez et al., 1977). Perhaps an integral role can now be assigned to the Muller cell in the GABA–glutamate–glutamine cycle proposed to occur in the retina (Starr, 1974). According to this concept, Muller cells actively accumulate neuronally released glutamate and GABA, thereby inactivating and conserving these amino acids which are subsequently converted to glutamine by glutamine synthetase. Glutamine, which is freely diffusible, then diffuses from the Muller cell and is thus made available to neurons for the formation of glutamate and GABA.

Extensive work has been focused on the regulation and function of glutamine synthetase in the retina. Moscona and associates, in a series of studies (see Moscona, 1967), have shown that the characteristic sharp rise in retinal glutamine synthetase activity discussed above can be induced precociously, days ahead of the normal schedule, by the application of "inducers" such as fetal calf serum, which includes hormones of known chemical identity, to embryos or to organ cultures of retinal tissue. The acceleration of the glutamine synthetase schedule is followed by enhancement of morphological features of differentiation in the retina. Both the precocious and the normally timed rises in glutamine synthetase activity require continuous protein synthesis; they are, in all likelihood, due to the accumulation of new enzyme protein and are thought to represent, in part, an increased rate of template activity. The onset of this process is under genetic control, in that synthesis of new RNA is required for the initiation of the sharp rise in glutamine synthetase activity; however, glutamine synthetase activity increases independently of new RNA synthesis.

More recently, Moscona and Wiens (1975) reported that proflavine (3,6-diaminoacridine) could be used as an effective molecular probe for analysis of tissue-specific gene expression in embryonic neural retina cells. The induction of glutamine synthetase by hydrocortisone was differentially and reversibly blocked by proflavine. Concentrations of proflavine that totally inhibited the synthesis of enzymatic and immunologically reactive glutamine synthetase only minimally reduced total RNA and protein synthesis. Proflavine did not hinder the uptake of the steroid inducer into retinal cells or its binding to cytoplasmic receptors; nor did it directly prevent the translation of glutamine synthetase mRNA. The authors suggested that proflavine hindered, reversibly and differentially, processes necessary for the provision of functional transcripts for glutamine synthetase synthesis, and that proflavine functioned in this case as a nonmutagenic, differential modifier of gene expression.

B. Glycerophosphate Dehydrogenase

GPDH is a soluble cytoplasmic enzyme that is NAD^+-dependent. GPDH catalyzes the conversion of dihydroxyacetone phosphate to glycerophosphate. The reaction is reversible, but its equilibrium favors glycerophosphate formation. Mitochondrial GPDH, a very different protein, is a flavoprotein that converts glycerophosphate to dihydroxyacetone phosphate. This reaction is irreversible and requires oxygen. These two enzymes together form the glycerophosphate cycle, whose net result is the generation of NAD^+ from NADH. NAD^+ is required for hydrolysis. Another function of cytoplasmic GPDH is to provide precursors for phospholipids, since glycerophosphate can be converted to phosphatic acid. It should be mentioned that the brain has no glycerokinase and therefore cannot make glycerophosphate from glycerol. Mitochondrial GPDH is not under hormonal control in the brain (de Vellis and Inglish, 1968).

In the developing rat brain, most enzymes increase two- to threefold during the first three postnatal weeks (de Vellis and Clemente, 1970). On the other hand, a few enzymes, such as isocitrate dehydrogenase, decrease in activity. Following the critical early postnatal period there is a second period of enzyme development which lasts until 40 days of age, during which time enzymes such as GPDH (de Vellis and Inglish, 1969), S-100, a brain-specific protein (Herschman et al., 1971), and carbonic anhydrase (Millichap, 1957) are induced. This second period of postnatal enzyme development follows a period of glial proliferation, which suggests that these three proteins may originate from glial cells. Carbonic anhydrase is present in glial cells of Deiters' nucleus but not in neurons (Giacobini, 1964). S-100 protein is also thought to be of glial cell origin, because its concentration in the optic nerve rises during the degeneration of axons of the optic nerve (Perez et al., 1970) and it is characteristically high in glial tumors and glial cell lines (Benda et al., 1968). Similarly, GPDH concentration increases during degeneration of the optic lobe (McCaman and Robins, 1959) and is inducible in the same cell line as S-100 protein (de Vellis and Inglish, 1969).

The concentration of GPDH in the rat brain is controlled by glucocorticoids (de Vellis and Inglish, 1968). Hypophysectomy of 20-day-old rats halted the developmental increase in GPDH, whereas several other enzymes, and brain protein content, remained unchanged, indicating that cortisol regulates GPDH in a specific fashion.

Both in vivo (rat brain) and in vitro (C6 glial cells) the presence of cortisol is required for both the induction of GPDH and its maintenance. After the hormone enters the cell by diffusion, cortisol binds to a receptor protein (cytoplasmic receptor), and the complex is apparently translocated to the nucleus where it then binds to an acceptor site (de Vellis, 1973). Inhibition of nuclear uptake and binding of the hormone occur rapidly (within a matter of minutes) and precede the requirement for the de novo synthesis of RNA and protein specific to GPDH.

Glial cell proliferation in the rat brain occurs mainly during the first 12 postnatal days. It is not known whether the development of GPDH is limited to factors other than the appearance of glial cells and their growth. Regional differences within the brain have been found for the induction of GPDH by cortisol, with the cerebellum and brain stem showing higher levels of induction than the cerebral hemispheres. The adult level of GPDH activity is reached by 35 days of age (de Vellis et al., 1967). The delayed responsiveness in the cerebrum is difficult to explain, although it is known that glial cells appear later in the cerebrum than in the brain stem. However, the cerebellum is also quite immature, yet responds to the administration of cortisol. The role of other factors cannot be easily ascertained because of the difficulty of obtaining intact pure glial cells. A regional study of the concentration of cortisol receptors in glial cells during development is essential in order to understand fully the observed regional differences. Another approach is the use of clonal cultures of glial cells, which provide a model system allowing quantitative biochemical measurements. It is hoped that such systems also reflect some aspects of the activity of glial cells *in vivo*.

The hormonal specificity of GPDH induction established *in vivo* was further confirmed in rat glioma (RG) C6 glial cells. Thyroid hormones, insulin, growth hormone, estradiol, testosterone, progesterone, and catecholamines cannot induce GPDH. Furthermore, cortisol induces GPDH in cells maintained in serum-free medium (de Vellis *et al.*, 1971b). Therefore, *in vivo* as well as *in vitro*, the presence of a glucocorticoid appears to constitute the only hormonal requirement for GPDH induction. Unlike lactic acid dehydrogenase (LDH), GPDH is not inducible by dibutyryl cyclic AMP. This is in agreement with the fact that cortisol does not cause a rise in cyclic AMP in mouse glioma (MG) C6 cells. In addition, both NE and cortisol result in GPDH induction identical to that obtained with cortisol alone, suggesting that cyclic AMP does not have even a secondary effect on GPDH induction.

S-100 protein accumulation in RG C6 cells begins only after cultures have reached confluency (Benda *et al.*, 1968; Pfeiffer *et al.*, 1970). Unlike S-100 protein accumulation, GPDH induction was observed in cultures during logarithmic growth as well as in the later stationary phase. However, clones 11D and 2B, which displayed good levels of induction, induced GPDH maximally 8–10 days after passage. Inducibility is very low in proliferating cells and in old cells (38 days of age in culture) (de Vellis and Inglish, 1969; de Vellis *et al.*, 1971a).

The lowered inducibility of GPDH in rapidly dividing cells (generation time 20 hr) and in old cells could be due to a reduction in the overall rate of synthesis of RNA and/or protein, or to a decrease in the nuclear uptake and binding of the hormone into the nucleus (or a combination of these factors), as compared to early stationary-phase cells. Nuclear retention of cortisol varied with the age of the culture. Nuclear retention was maximum at 10 days of age, when inducibility was maximal (de Vellis, 1973). The lowered inducibility observed in proliferating and stationary cells correlates with the reduction in nuclear retention of cortisol. These findings suggest that a lower hormonal receptor activity is responsible for the reduced inducibility of GPDH in proliferating and old cells. This conclusion points out the importance of studying the regulation of the cortisol receptor of glial cells *in vivo* during development and aging. Such studies might explain the observed regional differences in induction of GPDH during early postnatal development.

C. 2′,3′-Cyclic-Nucleotide 3′-Phosphohydrolase

The presence of an enzyme capable of hydrolyzing ribonucleoside 2′,3′-cyclic phosphates, but which did not hydrolyze internucleotide bonds, was first reported in the CNS in 1962 by Drummond *et al.* This enzyme, 2′,3′-cyclic-nucleotide 3′-phosphohydrolase (CNP), is bound to membranes (Pfeiffer, 1973) and has proven to be a very useful marker for myelination.

Kurihara and Tsukada (1967) characterized the regional and subcellular distribution of this enzyme in the CNS of several vertebrate species. The most notable observation was that cerebral white matter (corpus callosum or the superior longitudinal bundle) of rabbit, ox, and dog was approximately 10 times as active as the corresponding cerebral cortex. The brain tissues of rat and mouse were less enzymatically active than those of rabbit, ox, and dog. The midbrain, medulla oblongata, and spinal cord of hen and frog contained considerable levels of activity, although their cerebral cortices showed extremely low activity. In particular, the frog spinal cord was most active among the spinal cords of various species, and its activity corresponded to that of the cortical white matter of rabbit, ox, or dog. The regional distribution of CNP was paralleled by the distribution of myelinated fibers; the corpus callosum was approximately 10 times as active as the cerebral cortex. After fractionation of cerebral cortex homogenates, the greatest proportion of enzyme activity was recovered from the myelin-rich fractions. It seems reasonable to conclude, therefore, that CNS phosphohydrolase is localized in the myelin sheath or intimately associated structures.

Developmental changes in CNP activity in the chick brain and spinal cord have been reported by Kurihara and Tsukada (1968). The periods of greatest increase in enzyme activity occurred between 18 days of incubation and 3 days after hatching in the whole brain, and between 18 and 21 days of incubation in the spinal cord. Since these periods correspond to those of active myelination in the chick brain and spinal cord, the possibility that CNP activity can serve as a marker for the formation of myelin sheaths is further supported.

Studies by Toews and Horrocks (1976) showed that CNP activity in human cerebral matter samples was very low until approximately 2 months of age, after which it increased rapidly, approximating adult levels by 2 yr of age. CNP activity in adult (15–60 yr) cerebral white matter was 8.1 ± 1.0 μmol/min/mg protein. CNP activity in cerebral gray matter was initially very low and showed only a small increase during development, to adult values of approximately 1.4 μmol/min/mg protein. In the spinal cord, adult values (3.7 ± 0.56 μmol/min/mg protein) were found shortly after birth. These findings are consistent with the hypothesis that CNP is an intrinsic myelin component in human CNS myelin. The marked increase in CNP activity in white matter coincides with a period of rapid myelin deposition.

Zanetta et al. (1972) reported the occurrence of CNP in a glial cell line derived from a chemically induced rat brain tumor. This C6 clone was first developed by Benda et al. (1968). Since the oligodendrocyte plasma membrane is the site that gives rise to brain myelin (Bunge et al., 1962), it might be expected that the enzyme would be found in the oligodendrocyte plasma membrane. It is perhaps significant that these glial cells are known to contain the protein S-100 (Benda et al., 1968; Benda, 1968b) which has been claimed, on the basis of immunological studies, to be found in oligodendroglia and only rarely in astrocytoma cells (Benda, 1968a). Further evidence that these C6 glial cells, at least at some stage of their life span, have oligodendroglial properties has been provided by the studies of de Vellis and associates, using GPDH activity as a glial cell marker, as discussed earlier.

D. Butyrylcholinesterase (EC 3.1.1.8)

In contrast to AChE, the enzyme that specifically catalyzes the hydrolysis of ACh, there exist other, nonspecific cholinesterases (pseudocholinesterase, butyrylcholinesterase, BuChE) that hydrolyze other esters in addition to choline esters (Giacobini, 1959). One of the earliest suggestions concerning the localization of BuChE in neural tissue was made by Koelle (1951). Using histochemical methods, he reported the occurrence of BuChE in glial cells of the spinal and sympathetic ganglia, Schwann cells of myelinated nerves, fibrous astrocytes, and CNS vascular epithelium (Koelle, 1951, 1955). Koelle's results were confirmed by other investigators who found BuChE activity in CNS tumors of glial origin (primarily astrocytomas) to be much higher than in tumors of nonglial origin, such as neuromas, meningiomas, and medulloblastomas (Youngstrom et al., 1941; Bülbring et al., 1953; Cavanagh et al., 1954).

The proposed extraneuronal localization of BuChE was also supported by degeneration studies which showed that BuChE activity in neural tissues was virtually unaffected following complete destruction of nerve fibers (Sawyer, 1946; Cavanagh *et al.*, 1954). From these experiments it was concluded that BuChE was primarily a component of the nerve sheath, probably localized to the Schwann cells. It has also been suggested that BuChE may have a dual origin: in oligodendrocytes and in microglial histiocytes (Lumsden, 1957). Finally, Giacobini found, using three different neural cell preparations (Deiters' cell–oligodendrocyte, anterior horn cell–astrocyte, spinal ganglion cell–oligodendrocyte, and sympathetic ganglion cell–oligodendrocyte), that neither oligodendrocytes nor astrocytes showed AChE activity, while levels of BuChE activity found in glial cells were consistently higher than those found in neuronal cells (Giacobini, 1964).

The development and maturation of BuChE activity throughout the life span has been studied in four discrete regions of the avian brain (Vernadakis, 1973b). In the cerebral hemispheres and cerebellum, BuChE activity rose gradually during embryonic life and sharply immediately after hatching; the posthatching increase continued for 3 months. BuChE activity in these areas declined during the first year of life and then began a second period of increase, which continued throughout the second and third years of life. In the optic lobes and diencephalon–midbrain, BuChE activity followed a similar maturational pattern during embryonic and early posthatching life. The decline in activity after the third month was not so pronounced as in the cerebral hemispheres and cerebellum. The two periods of most rapid increase in BuChE activity (between 18 days of embryogenesis and 3 months after hatching, and during the latter third of the life span) are believed to reflect marked glial cell proliferation during these periods (Vernadakis, 1973b).

IV. Enzymes Associated with Specific Cellular Processes

A. Adenosinetriphosphatase

The enzyme system Na^+,K^+-ATPase was first proposed to be responsible for the coupled active transport of Na^+ and K^+ across the cytoplasmic membrane by Skou (1965). Very high ATPase activity has been found in cerebral gray matter (Bonting *et al.*, 1961), and severe neurological disturbances have been shown to result from the inhibition of ATPase by ouabain *in vivo* (Bignami and Palladini, 1966). Histochemical demonstration of ATPase in intact gray matter has led to the postulation that higher Na^+,K^+-ATPase activity exists in neurons than in glial cells (Stahl and Broderson, 1975), but biochemical measurements in isolated cell preparations have repeatedly shown that the activity of Na^+,K^+-ATPase is higher in glial than in neuronal preparations, and that a considerable postnatal increase in activity occurs in glial cells (Cummins and Hydén, 1962; Henn *et al.*, 1972; Medzihradszky *et al.*, 1972; Nagata *et al.*, 1974; Latzkovitz, 1978). A general trend seems to be that the highest activity is found in bulk-prepared glial cells, intermediate activity is found in glial cells in primary cultures, and the lowest activity is found in glial cells lines (see review by Hertz, 1977). According to the data reviewed by Hertz (1978), the sum of the Mg^{2+}-ATPase and the Na^+,K^+-ATPase activities in both bulk-prepared glial cells and astrocytes in primary cultures approximates the activity of the brain cortex. However, Na^+,K^+-ATPase activity is low in cultured cells. It is not clear whether this phenomenon is related to the lack of enzyme induction (by high potassium ion concentrations) during the culture period. Such findings suggest, however, that cell culture systems may be useful models for studying the regulatory mechanisms involved in the expression of Na^+,K^+-ATPase activity.

The developmental pattern of this enzyme is in good agreement with data reported for the onset of electrical activity in the spinal cord (Provine *et al.*, 1970) and with the pattern of embryonic mobility in the chick embryo (Hamburger, 1970). The sharp rise in NA^+,K^+-ATPase activity seen from day 18 coincides with the onset of a more complex electrical activity, in the lower two-thirds of the spinal cord, consisting of polyneuronal burst discharges (Provine *et al.*, 1970). Samson *et al.* (1964), investigating Na^+,K^+-ATPase activity during postnatal maturation of the rat brain, found a rapid increase to occur between days 10 and 21 after birth. In chick cerebral hemispheres, a rapid increase in Na^+,K^+-ATPase activity occurred at 19 days of incubation, and at hatching the enzymatic activity per unit of weight reached 80% of adult values (Bignami *et al.*, 1966). The increase in Na^+,K^+-ATPase activity in the chick brain coincides with a period of EEG and histological maturation (Sharma *et al.*, 1964). After hatching, Na^+, K^+-ATPase activity increased above the adult level between days 7 and 30. During this period, the cerebral hemispheres grow more rapidly than during any other period of postnatal development, almost doubling their weight (Latimer, 1925).

Alfei and Venturini (1972) further explored Na^+,K^+- and Mg^{2+}-ATPase activities in the developing lumbosacral region of chick spinal cord in an attempt to establish correlations among enzyme concentration, the pattern of embryonic mobility, and the appearance of electrical activity in this particular region of the nervous system. ATPase activity was already detectable on day 8, and there was a slight rise until day 16, after which time the rate of increase was much sharper from day 18 until postnatal day 10. At hatching, Na^+,K^+-ATPase activity per unit of weight was 50% of the adult values; Mg^{2+}-ATPase activity remained consistently higher than that of Na^+,K^+-ATPase until hatching. Na^+,K^+-ATPase activity in the spinal cord was detectable 2 days earlier than in the cerebral hemispheres.

In a series of studies designed to investigate the effects of neonatal hypothyroidism on ionic composition of the developing rat brain, Valcana (1974) correlated changes in ionic composition with changes in Na^+,K^+- and Mg^{2+}-ATPase. During normal maturation, Cl^-, Na^+, and Ca^{2+} concentrations in plasma increased, whereas K^+ concentration decreased (Valcana, 1974). No changes occurred in Mg^{2+} concentration, in agreement with Silverman and Gardner (1954), who found no differences in total plasma Mg^{2+} between children and adults. The changes in Na^+, Cl^-, and K^+ contents in plasma correspond to those previously observed by Vernadakis and Woodbury (1962). In the cerebral cortex, K^+ and Mg^{2+} contents increased and Na^+ and Cl^- contents decreased with development. Ca^{2+} levels in the cerebral cortex, however, showed no definite developmental pattern. When changes in ATPase activities were correlated with developmental changes in Na^+, K^+, and Cl^- contents, the specific activity of Mg^{2+}-ATPase and Na^+,K^+-ATPase increased with age and the latter showed a significant increase during that period of development characterized by marked redistribution of Na^+ and K^+ contents.

The decrease in Na^+ concentration and the increase in K^+ concentration between postnatal days 10 and 22 implies an exchange of these ions during neonatal development. This ionic redistribution in the cerebral cortex has been previously ascribed to acid–base changes that may occur during maturation; according to Woodbury and his co-workers (1958), for example, K^+ concentration is regulated by the total anion content of the cell. This may vary considerably, as it is dependent on cellular pH as well as on the metabolic processes affecting the concentrations of free acidic and basic amino acids and on the cellular concentration of bicarbonate. In addition, K^+ concentration depends on the coupling of potassium movement to active sodium transport. The results of Valcana presented here suggest that Na^+ and K^+ redistribution may occur subsequent to the onset of active transport, inasmuch as it occurs concomitantly with a marked increase in the Na^+,K^+-ATPase activity in the brain and other tissues in which it is found (Deul and McIlwain, 1961; Skou, 1962, 1965; Bonting et al., 1961).

B. Enzymes Involved in Cyclic Nucleotide Metabolism

The biological regulatory processes that control the growth, differentiation, and hormonal responses of cells are poorly understood. It is known, however, that many regulatory events are affected by the intracellular concentrations of the cyclic nucleotide cyclic AMP. A second cyclic nucleotide, guanosine $3',5'$-monophosphate (cyclic GMP), is also present in all living systems (Goldberg et al., 1973). It has been proposed that cellular regulatory processes may be influenced by the interaction of cyclic AMP and cyclic GMP, a proposal that has generated considerable interest among biologists. We will herein review only briefly the enzymes involved in catalyzing cyclic nucleotides, their age-related changes, and the implications thereof.

1. ADENYL CYCLASE

The enzymatic conversion of adenosine triphosphate (ATP) to cyclic AMP was first described in detail by Sutherland and Rall and their co-workers (Sutherland et al., 1962; Rall and Sutherland, 1962; Murad et al., 1962; Klainer et al., 1962). It was shown that the enzyme adenyl cyclase required ATP and Mg^{2+}, was particulate-bound, and was present in tissues from at least four phyla. In every case tested, enzyme activity was stimulated by sodium fluoride and was stimulated in a tissue-specific manner by different hormones. Their suggestion that the action of many hormones might be mediated by cyclic AMP, as a result of stimulation of adenyl cyclase activity, has been confirmed and extended by a myriad of subsequent studies (for reviews, see Robison et al., 1971; Greengard and Robison, 1972).

Cyclic AMP appears to be present in all animal species, as well as in at least 10 strains of bacteria (Ide, 1971). In mammalian cells, the synthesis of cyclic AMP is regulated by hormonal influences on adenyl cyclase activity, the hormone being produced in a cell type distinct from that which it affects.

There has long been indirect evidence that the adenyl cyclase system is composed of at least two components, a receptor or regulatory component and a catalytic component. As a working hypothesis, Robison et al (1967) proposed a model that assumed the existence of two subunits with a specific directional orientation. Hormone interaction with the receptor unit causes an alteration in the receptor, which in turn alters the catalytic unit in a way that leads to increased enzyme activity. An alternate model has been suggested by others (Hechter and Halkerston, 1964; Hechter, 1965; Birnbaumer et al., 1972; Rodbell, 1971; for a detailed review, see Perkins, 1973).

One of the most interesting properties of adenyl cyclase is its responsiveness to hormones. Studies on the regulation of the cyclic AMP content of brain slices indicate a strikingly complex picture, apparently involving multivalent regulation of adenyl cyclase activity. These studies have been frequently reviewed (Rall and Sattin, 1970; Shimizu *et al.,* 1970; Krishna *et al.,* 1970; Rall and Gilman, 1970). Briefly, the level of cyclic AMP in brain slices is increased by catecholamines, 5-HT, histamine, and adenosine. Potassium, veratridine, butrachotoxin, and ouabain apparently act indirectly by causing the release of adenosine.

Rosen and Rosen (1968) first observed that the expression of hormone sensitivity of adenyl cyclase could occur at a stage of development later than the expression of catalytic activity. Whether such an observation can be attributed to the absence of the receptor component or to a lack of the coupling component is not known. Schmidt *et al.* (1970) have reported a distinct difference in the time of development of the responsiveness of adenylate cyclase of brain slices to catecholamines and the appearance of basal catalytic activity. Perkins and his associates (1971) have carried out similar studies using slices of rat cerebrum and have suggested that the catecholamine receptor may be present in the developing rat cerebral cortex before it is coupled to the catalytic component. In slices from 10-day-old cerebrum, cyclic AMP was not increased by NE, however, in the presence of adenosine, NE markedly increased cyclic AMP content above the level observed in the presence of adenosine alone. At days 11–12 there is an abrupt increase in the responsiveness of the slices to NE alone, but the synergistic effect of NE and adenosine in combination is still not observed. These observations suggest that adrenergic receptors are present in tissue prior to day 11. The fact that NE alone has no effect suggests that either the receptor does not bind to the amine in the absence of adenosine or that the coupling function is not fully operative in the absence of adenosine.

The use of tumor cells has provided some insights into the organization of the adenyl cyclase system and has suggested a means whereby this organization might be manipulated experimentally. Perkins and his associates (1971; Clark and Perkins, 1971, 1972; Perkins, 1973) have used a spectrum of clonal cell lines with selective deletions of different components of the adenyl cyclase system. They found that each cell type responded differentially to each of four effectors used. For example, 118132 cells have apparently lost their sensitivity to histamine but retain their responsiveness to catecholamines, adenosine, and prostaglandin E_1. In no case does there appear to be a change with respect to only one effector; quantitatively, alterations occur in the response to each of the agonists.

Studies by von Hungen *et al.* (1974) reveal the presence of adenyl cyclase systems responsive to NE, dopamine, and 5-HT in preparations obtained from different regions of the rat brain. These adenylate cyclase systems exhibit diverse patterns of maturation during development. At birth, adenyl cyclase activity was much higher in the hindbrain–medulla preparations than in the cerebellum, cerebral cortex, or subcortex (including midbrain, corpus striatum, hypothalamus, and hippocampus). Adenyl cyclase activity increased during early development in preparations from all areas of the brain. Maximal levels were reached at 14 days of age or later. Adenyl cyclase systems in the cerebral cortex and subcortical preparations were activated by NE and dopamine throughout development. 5-HT also stimulated adenyl cyclase activity from young animals but was much less effective in comparable fractions from adult rats. The response to dopamine was diminished with age in cerebral cortical preparations but not in subcortical fractions. The response to NE increased in both brain regions during early development. Adenyl cyclase from the cerebellum and hindbrain–medulla areas exhibited relatively poor responses to the biogenic amines. These authors suggest that the differences in the responsiveness of adenyl cyclase to hormones in various brain regions with development may be partly attributed to glial cells (cf. chapter by van Calker and Hamprecht, this volume). The proportion of neurons to glial cells in brain regions is known to differ (Brizzee et al., 1964). Recent investigations have shown that adenyl cyclase systems in particular preparations from neuron-enriched and glial-enriched fractions isolated from the brain of the young rat possess different sensitivities to biogenic amines (Palmer, 1973). Moreover, homogenates of cultured astroglioma cells derived from rat brain contain adenyl cyclase systems which are responsive to both NE and dopamine but not to 5-HT (Jard et al., 1972).

2. GUANYL CYCLASE

Guanyl cyclase, the enzyme that catalyzes the formation of cyclic GMP to guanosine triphosphate (GTP), has been detected in all mammalian tissues examined to date (Clark and Bernlohr, 1972). One unique property of guanyl cyclase that appears to distinguish it from adenyl cyclase is its apparent subcellular distribution. Unlike mammalian adenyl cyclase, which is considered totally particulate, the major portion of guanyl cyclase activity appears to reside in the soluble fraction after homogenization of most tissues (White et al., 1969; White and Aurbach, 1969; Hardman and Sutherland, 1969; Schultz et al., 1969). Reports that guanyl cyclase appears to be a cytoplasmic enzyme in some tissues lend support to the suggestion that hormones such as steroids and thyroxine, which enter the cell, may be the agents responsible for regulating its activity.

Goridis et al. (1974) studied the activity of guanyl cyclase in dissociated chick embryo brain cell cultures containing varying ratios of neurons and glial cells. The cultures containing neurons in substantial numbers had consistently higher levels of guanyl cyclase activity than those consisting mainly of glial cells. Moreover, no guanyl cyclase activity could be found in cultures composed of pure glial or meningeal cells. In order to compare these results with guanyl cyclase development *in vivo*, guanyl cyclase activity in chick embryo cerebral hemispheres was determined at various times after fertilization. There was a pronounced increase in guanyl cyclase activity between days 12 and 16, after which time enzymatic activity decreased markedly. Since glial cells actively proliferate after 16 days of incubation in the embryonic chick brain (Hanaway, 1967; Vernadakis, 1973b), it is possible that the decrease in guanyl cyclase activity after 16 days reflects dilution by glial cells.

References

Alfei, L., and Venturini, B. (1972). ATPase activity in the developing chick spinal cord. *Brain Res.* **43**, 314–316.

Algeri, S., Bonati, M., Brunello, N., and Ponzio, F. (1977). Dihydropteridine reductase and tyrosine hydroxylase activities in rat brain during development and senescence: A comparative study. *Brain Res.* **132**, 569–574.

Alousi, A., and Weiner, N. (1966). The regulation of norepinephrine synthesis in sympathetic nerves: Effect of nerve stimulation, cocaine, and catecholamine-releasing agents. *Proc. Natl. Acad. Sci. U.S.A.* **56**, 1491–1496.

Arnold, E. B., and Vernadakis, A. (1979). Development of tyrosine hydroxylase activity in dissociated cerebral cell cultures. *Dev. Neurosci.* **2**, 46–50.

Arnold, E. B., Spuhler, K., Vernadakis, A., and Prasad, K. N. (1978). Effects of different sera on tyrosine hydroxylase activity in neuroblastoma cells. *Fed. Proc., Fed. Am. Soc. Exp. Biol.* **37**, 345.

Atherton, R. W. (1973). Ontogenetic differences between acetyl- and butyrylcholinesterase isozymes in the chick embryo cerebellum. *Experientia* **29**, 1069–1070.

Baker, P. C., and Hoff, K. M. (1972). Maturation of 5-hydroxytryptamine levels in various brain regions of the mouse from 1 day postpartum to adulthood. *J. Neurochem.* **19**, 2011–2015.

Baxter, C. F. (1969). The nature of gamma-aminobutyric acid. *In* "Handbook of Neurochemistry" (A. Lajtha, ed.), Vol. 3, pp. 289–353. Plenum, New York.

Benda, P. (1968a). Protéine S-100 et cellules gliales du rat *Rev. Neurol.* **118**, 364–367.

Benda, P. (1968b). Protéine S-100 et tumeurs cérébrales humaines. *Rev. Neurol.* **118**, 368–372.

Benda, P., Lightbody, L., Sato, G., Levine, L., and Sweet, W. (1968). Differentiated rat glial cell strain in tissue culture. *Science* **161**, 370–371.

Bennett, D. S., and Giarman, N. J. (1965). Schedule of appearance of 5-hydroxytryptamine (serotonin) and associated enzymes in the developing rat brain. *J. Neurochem.* **12**, 911–918.

Berl, S. (1966). Glutamine synthetase: Determination and its distribution in brain during development. *Biochemistry* **5**, 916–922.

Bignami, A., and Palladini, G. (1966). Experimentally produced cerebral status spongiosus and continuous pseudorhythmic electroencephalographic changes with a membrane-ATPase inhibitor in rat. *Nature (London)* **209**, 413–414.

Bignami, A., Palladini, G., and Venturini, G. (1966). Sodium-potassium adenosine triphosphatase in the developing chick brain. *Brain Res.* **3**, 207–209.

Birnbaumer, L., Pol, S. L., Rodbell, M., and Sundby, F. (1972). The glucagon-sensitive adenylate cyclase system in plasma membranes of rat liver. VII. Hormonal stimulation: Reversibility and dependence on concentration of free hormone. *J. Biol. Chem.* **247**, 2038–2043.

Blume, A., Gilbert, F., Wilson, S., Farber, J., Rosenberg, R., and Nirenberg, M. (1970). Regulation of acetylcholinesterase in neuroblastoma cells. *Proc. Natl. Acad. Sci. U.S.A.* **67**, 786–792.

Boell, E. J., and Shen, S. C. (1950). Development of cholinesterase in the central nervous system of *Amblystoma punctatum*. *J. Exp. Zool.* **113**, 583–599.

Bonichon, A. (1958). L'acetylcholinesterase dans la cellule et la fibre nerveuse au cours du développement. I. Différenciation biochimique précoce du neuroblaste. *Ann. Histochim.* **3**, 85–93.

Bonting, S. L., Simon, K. A., and Hawkins, N. M. (1961). Studies on sodium-potassium-activated adenosine triphosphatase. IV. Correlation with cation transport sensitive to cardiac glycosides. *Arch. Biochem. Biophys.* **95**, 416–423.

Bradford, H. F., and Thomas, A. J. (1969). Metabolism of glucose and glutamate by synaptosomes from mammalian cerebral cortex. *J. Neurochem.* **16**, 1495–1504.

Brizzee, K. R., Vogt, J., and Kharetchko, X. (1964). Postnatal changes in glial neuron index with a comparison of methods of cell enumeration in the white rat. *Prog. Brain Res.* **4**, 136–149.

Bülbring, E., Philpot, F. J., and Bosanquet, F. D. (1953). Amine oxidase, pressor amines and cholinesterase in brain tumours. *Lancet* **1**, 865.

Bunge, M. B., Bunge, R. P., and Pappas, G. D. (1962). Electron microscopic demonstration of connections between glial and myelin sheaths in the developing mammalian central nervous system. *J. Cell Biol.* **12**, 448–453.

Burdick, C. J., and Strittmatter, C. F. (1965). Appearance of biochemical components related to acetylcholine metabolism during the embryonic development of chick brain. *Arch. Biochem. Biophys.* **109**, 293–301.

Burt, A. M. (1968). Acetylcholinesterase and choline acetyltransferase activity in the developing chick spinal cord. *J. Exp. Zool.* **169**, 107–112.

Burt, A. M. (1973). Choline acetyltransferase and neuronal maturation. *Prog. Brain Res.* **40**, 245–252.

Butcher, L. L., and Bilezikjian, L. (1975). Acetylcholinesterase-containing neurons in the neostriatum and substantia nigra revealed after punctate intracerebral injection of diisopropylfluorophosphate. *Eur. J. Pharmacol.* **34**, 115–125.

Butcher, L. L., and Hodge, G. K. (1976). Postnatal development of acetylcholinesterase in the caudate-putamen nucleus and substantia nigra of rats. *Brain Res.* **106**, 223–240.

Butcher, L. L., Talbot, K., and Bilezikjian, L. (1975a). Localization of acetylcholinesterase within dopamine-containing neurons in the zona compacta of the substantia nigra. *Proc. West. Pharmacol. Soc.* **18**, 256–259.

Butcher, L. L., Talbot, K., and Bilezikjian, L. (1975b). Acetylcholinesterase neurons in dopamine-containing regions of the brain. *J. Neural Transm.* **37**, 127–153.

Buznikov, G. A. (1971). The role of mediators of the nervous system in ontogenesis. *Ontogenesis* **2**, 5–13.

Buznikov, G. A., Chudakova, I. V., and Zvezdina, N. D. (1964). The role of neurohumors in early embryogenesis. I. Serotonin content of developing embryos of sea urchin and leech. *J. Embryol. Exp. Morphol.* **12**, 563–573.

Buznikov, G. A., Chudakova, I. V., Berdysheva, L. V., and Vyazmina, N. M. (1968). The role of neurohumors in early embryogenesis. II. Acetylcholine and catecholamine content in developing embryos of sea urchin. *J. Embryol. Exp. Morphol.* **20**, 119-128.

Buznikov, G. A., Kost, A. N., Kucherova, N. F., Mundzhoyan, A. L., Suvorov, N. N., and Berdysheva, L. V. (1970). The role of neurohumors in early embryogenesis. III. Pharmacological analysis of the role of neurohumors in cleavage divisions. *J. Embryol. Exp. Morphol.* **23**, 549-569.

Buznikov, G. A., Sakharova, A. V., Manukhin, B. N., and Markova, L. N. (1972). The role of neurohumors in early embryogenesis. IV. Fluorometric and histochemical study of serotonin in cleaving eggs and larvae of sea urchins. *J. Embryol. Exp. Morphol.* **27**, 339-351.

Cavanagh, J. B., Thompson, R. H. S., and Webster, G. R. (1954). The localization of pseudocholinesterase activity in nervous tissue. *J. Exp. Physiol. Cogn. Med. Sci.* **39**, 185-197.

Clark, R. B., and Perkins, J. P. (1971). Regulation of adenosine $3',5'$-cyclic monophosphate concentration in cultured human astrocytoma cells by catecholamines and histamine. *Proc. Natl. Acad. Sci. U.S.A.* **68**, 2757-2760.

Clark, R. B., and Perkins, J. P. (1972). The effect of adenosine on the formation of cyclic AMP (cAMP) in cultured human astrocytoma cells. *Fed. Proc., Fed. Am. Soc. Exp. Biol.* **31**, 513.

Clark, V. L., and Bernlohr, R. W. (1972). Guanyl cyclase of *Bacillus licheniforms. Biochem. Biophys. Res. Commun.* **46**, 1570-1576.

Corner, M. A., Smith, J., and Romijn, H. J. (1974). Maturation of cerebral bioelectric activity in the chick embryo in relation to morphological and biochemical factors. *In* "Ontogenesis of the Brain" (L. Jilek and F. Trojan, eds.), pp. 31-42. Charles Univ. Press, Prague.

Corner, M. A., Romijn, H. J., and Richter, A. P. J. (1977). Synaptogenesis in the cerebral hemisphere (accessory hyperstriatum) of the chick embryo. *Neurosci. Lett.* **4**, 15-19.

Coulombre, A. J. (1955). Correlations of structural and biochemical changes in the developing retina of the chick. *Am. J. Anat.* **96**, 153-190.

Coyle, J. T., and Axelrod, J. (1972). Tyrosine hydroxylase in rat brain: Developmental characteristics. *J. Neurochem.* **19**, 1117-1123.

Coyle, J. T., and Enna, S. J. (1976). Neurochemical aspects of the ontogenesis of gabanergic neurons in the rat brain. *Brain Res.* **111**, 119-133.

Coyle, J. T., and Yamamura, H. I. (1976). Neurochemical aspects of the ontogenesis of cholinergic neurons in the rat brain. *Brain Res.* **118**, 429-440.

Cummins, J., and Hydén, H. (1962). Adenosine triphosphate levels and adenosine triphosphatases in neurons, glia, and neuronal membranes of the vestibular nucleus. *Biochim. Biophys. Acta* **60**, 271-283.

Curtis, D. R., and Johnston, G. A. R. (1974). Amino acid transmitters in the mammalian central nervous system. *Ergeb. Physiol. Biol. Chem. Exp. Pharmakol.* **69**, 97-188.

De Robertis, E., and Rodriguez De Lores Arnaiz, G. (1969). Structural components of the synaptic region. *In* "Handbook of Neurochemistry" (A. Lajtha, ed.), Vol 2, pp. 365-392. Plenum, New York.

De Robertis, E., Sellinger, O. Z., Rodriguez De Lores Arnaiz, G., Alberici, M., and Zieher, L. M. (1967). Nerve endings in methionine sulphoximine convulsant rats, a neurochemical and ultrastructural study. *J. Neurochem.* **14**, 81-89.

Deul, D. H., and McIlwain, H. (1961). Activation and inhibition of adenosine triphosphatases of subcellular particles from the brain. *J. Neurochem.* **8**, 246-256.

de Vellis, J. (1973). Mechanisms of enzymatic differentiation in the brain and in cultured cells. *In* "Development and Aging in the Nervous System" (M. Rockstein and M. L. Sussman, eds.), pp. 171-198. Academic Press, New York.

de Vellis, J., and Clemente, C. D. (1970). Neural cell differentiation. *In* "Cell Differentiation" (O. A. Schjeide and J. de Vellis, eds.), pp. 529–574. Van Nostrand-Reinhold, Princeton, New Jersey.

de Vellis, J., and Inglish, D. (1968). Hormonal control of gycerolphosphate dehydrogenase in the rat brain. *J. Neurochem.* **15**, 1061–1070.

de Vellis, J., and Inglish, D. (1969). Glycerolphosphate dehydrogenase induction in a cloned glial cell culture by glucocorticoids. *Int. Meet. Soc. Neurochem., 2nd, 1969* pp. 151–152.

de Vellis, J., Schjeide, O. A., and Clemente, C. D. (1967). Protein synsthesis and enzymic patterns in the developing brain following head x-irradiation of newborn rats. *J. Neurochem.* **14**, 499–511.

de Vellis, J., Inglish, D., and Galey, F. (1971a). Effects of cortisol and epinephrine on glial cells in culture. *In* "Cellular Aspects of Growth and Differentiation in Nervous Tissue" (D. Pease, ed.), pp. 23–32. Univ. of California Press, Berkeley and Los Angeles.

de Vellis, J., Inglish, D., Cole, R., and Molson, J. (1971b). Effects of hormones on the differentiation of cloned lines of neurons and glial cells. *In* "Influence of Hormones on the Nervous System" (D. H. Ford, ed.), pp. 25–39. Karger, Basel.

Drummond, G. I., Iger, N. T., and Keith, J. (1962). Hydrolysis of ribonucleoside 2′,3′-cyclic phosphates by a diesterase from brain. *J. Biol. Chem.* **237**, 3535–3539.

Duffy, P. E., Tennyson, V. M., and Brzin, M. (1967). Cholinesterase in adult and embryonic hypothalamus. *Arch. Neurol. (Chicago)* **16**, 385–403.

Ebel, A., Massarelli, R., Sensenbrenner, M., and Mandel, P. (1974). Choline acetyltransferase and acetylcholinesterase activities in chicken brain hemispheres *in vivo* and in cell culture. *Brain Res.* **76**, 461–472.

Ehinger, B., and Falck, B. (1971). Autoradiography of some suspected neurotransmitter substances: GABA, glycine, glutamic acid, histamine, dopamine, and L-dopa. *Brain Res.* **33**, 157–172.

Eiduson, S. (1966). 5-Hydroxytryptamine in the developing chick brain: Its normal and altered development and possible control by end-product repression. *J. Neurochem.* **13**, 923–932.

Filogamo, G., and Marchisio, P. C. (1971). Acetylcholine system and neural development. *Neurosci. Res.* **4**, 29–64.

Fonnum, F. (1966). A radiochemical method for the estimation of choline acetyltransferase. *Biochem. J.* **100**, 479–484.

Fonnum, F., and Malthe-Sorenssen, D. (1972). Molecular properties of choline acetyltransferase and their importance for the compartmentation of acetylcholine synthesis. *Prog. Brain Res.* **36**, 13–27.

Giacobini, E. (1959). The distribution and localization of cholinesterases in nerve cells. *Acta Physiol. Scand.* **45**, 1–45.

Giacobini, E. (1964). Metabolic relations between glia and neurons studied in single cells. *In* "Morphological and Biochemical Correlates of Neural Activity" (M. M. Cohen and R. S. Snider, eds.), pp. 15–38. Harper, New York.

Giacobini, G., and Filogamo, G. (1973). Changes in the enzymes for the metabolism of acetylcholine during development of the central nervous system. *In* "Central Nervous System—Studies on Metabolic Regulation and Function" (E. Genazzani and H. Herken, eds.), pp. 153–157. Springer-Verlag, Berlin and New York.

Goldberg, N. D., O'Dea, R. F., and Haddox, M. K. (1973). Cyclic GMP. *Adv. Cyclic Nucleotide Res.* **3**, 155–223.

Goridis, C., Massarelli, R., Sensenbrenner, M., and Mandel, P. (1974). Guanyl cyclase in chick embryo brain cell cultures: Evidence of neuronal localization. *J. Neurochem.* **23**, 135–138.

Greengard, P., and Robison, G. A., eds. (1972). "Advances in Cyclic Nucleotide Research," Vol. 1. Raven Press, New York.

Gustafson, T. (1969). Cell recognition and cell contacts during sea urchin development. In "Cellular Recognition" (R. T. Smith and R. A. Good, eds.), pp. 47–60. Appleton, New York.

Gustafson, T., and Toneby, M. (1970). On the role of serotonin and acetylcholine in sea urchin morphogenesis. Exp. Cell Res. 62, 102.

Hamberger, A., Nyström, B., Sellström, A., and Woiler, C. T. (1976). In "Transport Phenomena in the Nervous System (G. Levi, L. Battistin, and A. Lajtha, eds.), p. 221. Plenum, New York.

Hamburger, V. (1970). Embryonic motility in vertebrates. In "The Neurosciences, Second Study Program" (F. O. Schmitt, ed.), pp. 141–151. Rockefeller Univ. Press, New York.

Hanaway, J. (1967). Formation and differentiation of the external layer of the chick cerebellum. J. Comp. Neurol. 131, 1–14.

Hardman, J. G., and Sutherland, E. W. (1969). Guanyl cyclase, an enzyme catalyzing the formation of guanosine 3′,5′-monophosphate from guanosine triphosphate. J. Biol. Chem. 244, 6363–6370.

Hattori, T., and McGeer, P. L. (1973). Synaptogenesis in the corpus striatum of infant rat. Exp. Neurol. 38, 70–79.

Hebb, C., and Morris, D. (1969). Identification of acetylcholine and its metabolism in nervous tissue. In "The Structure and Function of Nervous Tissue" (G. H. Bourne, ed.), Vol 3, pp. 25–60. Academic Press, New York.

Hebb, C. O. (1956). Choline acetylase in the developing nervous system of the rabbit and guinea pig. J. Physiol. (London) 133, 566–570.

Hechter, O. (1965). Hormone action at the cell membrane. In "Mechanisms of Hormone Action" (P. Karlson, ed.), pp. 61–93. Academic Press, New York.

Hechter, O., and Halkerston, I. D. K. (1964). On the action of mammalian hormones. In "The Hormones" (G. Pinkus, K. V. Thimann, and F. B. Astwood, eds.), pp. 697–825. Academic Press, New York.

Henn, F. A., and Hamberger, A. (1971). Glial cell function: Uptake of transmitter substances. Proc. Natl. Acad. Sci. U.S.A. 68, 2686–2690.

Henn, F. A., Halijamae, H., and Hamberger, A. (1972). Glial cell function: Active control of extracellular K^+ concentration. Brain Res. 43, 437–443.

Herschman, H. R., Levine, L., and de Vellis, J. (1971). Appearance of a brain-specific antigen (S-100 protein) in the developing rat brain. J. Neurochem. 18, 629–633.

Hertz, L. (1978). Biochemistry of glial cells. In "Cell, Tissue and Organ Cultures in Neurobiology" (S. Fedoroff and L. Hertz, eds.), pp. 39–71. Academic Press, New York.

Hollander, J., and Barrows, C. H. (1968). Enzymatic studies in senescent rodent brains. J. Geront. 23, 174–186.

Hutchison, H. T., Werrbach, K., Vance, C., and Haber, B. (1974). Uptake of neurotransmitters by clonal lines of astrocytoma and neuroblastoma in culture. I. Transport of γ-aminobutyric acid. Brain Res. 66, 265–274.

Ide, M. (1971). Adenyl cyclase of bacteria. Arch. Biochem. Biophys. 144, 262–268.

Israël, M. and Whittaker, V. P. (1965). The isolation of mossy fibre endings from the granular layer of the cerebellar cortex. Experientia 21, 325–326.

Jard, S., Premont, J., and Benda, P. (1972). Adenylate cyclase, phosphodiesterases and protein kinase of rat glial cells in culture. FEBS Lett. 26, 344–348.

Jolley, R. L., and Labby, D. H. (1960). Development of brain enzymes concerned with glutamic acid metabolism in fetal monkey and pig. Arch. Biochem. Biophys. 90, 122–124.

Jones, A. W., and Levi-Montalcini, R. (1958). Patterns of differentiation of the nerve centers and fiber tracts of the avian cerebral hemispheres. *Arch. Ital. Biol.* **96**, 231–284.

Karczmar, A. G., Srinivasan, R., and Bernsohn, J. (1973). Cholinergic function in the developing fetus. In "Fetal Pharmacology" (L. Boreus, ed.), pp. 127–177. Raven Press, New York.

Kasa, P., Csillik, R., Joo, F., and Kynihar, E. (1966). Histochemical and ultrastructural alterations in the isolated archicerebellum of the rat. *J. Neurochem.* **13**, 173–178.

Kaur, G., and Kanungo, M. S. (1970). Alterations in the activity and regulation of cholinesterase of the nervous tissue of rats of various ages. *Indian J. Biochem.* **7**, 122–125.

Kellogg, C., and Lundborg, P. (1971). Production of [³H]catecholamines in the brain following the peripheral administration of [³H]DOPA during pre- and postnatal development. *Brain Res.* **36**, 333–342.

Kellogg, C., and Lundborg, P. (1972). Uptake and utilization of [³H]5-hydroxytryptophan by brain tissue during development. *Neuropharmacology* **11**, 363–372.

Kim, S. U., and Tunnicliff, G. (1974). Morphological and biochemical development of chick cerebrum cultured *in vitro. Exp Neurol.* **43**, 515–526

Klainer, L. M., Chi, Y. M., Friedberg, S. L., Rall, T. W., and Sutherland, E. W. (1962). Adenyl cyclase. IV. The effects of neurohormones on the formation of adenosine 3′,5′-phosphate by preparations from brain and other tissue. *J. Biol. Chem.* **237**, 1239–1243.

Kling, A., Finer, S., and Nair, V. (1965). Effects of early handling and light stimulation on the acetylcholinesterase activity of the developing rat brain. *Int. J. Neuropharmacol.* **4**, 353–357.

Koelle, G. B. (1951). Elimination of enzymatic diffusion artifacts in the histochemical localization of cholinesterases and a survey of their cellular distributions. *J. Pharmacol. Exp. Ther.* **103**, 153–171.

Koelle, G. B. (1955). Cholinesterases of the central nervous system. J. Neuropathol. Exp. *Neurol.* **14**, 23–27.

Krishna, G., Forn, J., Voigt, K., Paul, M., and Gessa, G. L. (1970). Dynamic aspects of neurohormonal control of cyclic 3′,5′-AMP synthesis in brain. *Adv. Biochem. Psychopharmacol.* **3**, 155–173.

Kurihara, T., and Tsukada, Y. (1967). The regional and subcellular distribution of 2′,3′-cyclic nucleotide 3′-phosphohydrolase in the central nervous system. *J. Neurochem.* **14**, 1167–1174.

Kurihara, T., and Tsukada, Y. (1968). 2′,3′-cyclic nucleotide 3′-phosphohydrolase in the developing chick brain and spinal cord. *J. Neurochem.* **15**, 827–832.

Kuriyama, K., Sisken, B., Ito, J., Simonsen, G., Haber, B., and Roberts, E. (1968). The γ-aminobutyric acid system in the developing chick embryo cerebellum. *Brain Res.* **11**, 132–152.

Latimer, H. B. (1925). The postnatal growth of the central nervous system of the chicken. *J. Comp. Neurol.* **38**, 251–297.

Latzkovitz, L. (1978). Neuronal-glial interactions in potassium transport. In "Dynamic Properties of Glial Cells" (E. Schoffenick, G. Franck, L. Hertz, and D. B. Tower, eds.), pp. 327–336. Pergamon, Oxford.

Lumsden, C. E. (1957). The problem of correlation of quantitative methods and tissue morphology in the central nervous system (the distribution of cholinesterase). In "Metabolism of the Nervous System" (D. Richter, ed.), pp. 91–100. Pergamon, Oxford.

Lydiard, R. B., and Sparber, S. B. (1974). Evidence for a critical period for postnatal elevation of brain tyrosine hydroxylase activity resulting from reserpine administration during embryonic development. *J. Pharmacol. Exp. Ther.* **189**, 370–379.

McCaman, R. E., and Aprison, M. H. (1964). The synthetic and catabolic enzyme systems for acetylcholine and serotonin in several discrete areas of the developing rabbit brain. *Prog. Brain Res.* **9**, 220–233.

McCaman, R. E., and Robins, E. (1959). Quantitative biochemical studies of Wallerian degeneration in the peripheral and central nervous systems. II. Twelve enzymes. *J. Neurochem.* **5**, 32–42.

McGeer, E. G., and Maler, L. (1975). GABA accumulation by embryonic chick tectum. *Dev. Biol.* **47**, 464–465.

McGeer, E. G., Fibiger, H. C., McGeer, P. L., and Wickson, V. (1971). Aging and brain enzymes. *Exp. Gerontol.* **6**, 391–396.

McGeer, E. G., Maler, L., and Fitzsimmons, R. C. (1974). Comparative enzymatic development in chick embryonic brain areas. *Dev. Biol.* **38**, 165–174.

McMahon, D. (1974). Chemical messengers in development: A hypothesis. *Science* **185**, 1012–1021.

Maletta, G. J., Vernadakis, A., and Timiras, P. S. (1967a). Acetylcholinesterase activity and protein content of brain and spinal cord in developing rats after prenatal x-irradiation. *J. Neurochem.* **14**, 647–652.

Maletta, G. J., Vernadakis, A., and Timiras, P. S. (1967b). Pre- and postnatal development of the spinal cord: Increased acetylcholinesterase activity. *Proc. Soc. Exp. Biol. Med.* **121**, 1210–1211.

Marchand, A., Chapouthier, G., and Massoulie, J. (1977). Developmental aspects of acetylcholinesterase activity in chick brain. *FEBS Lett.* **78**, 233–236.

Marchisio, P. C., and Giacobini, G. (1969). Choline acetyltransferase activity in the central nervous system of the developing chick. *Brain Res.* **15**, 301–304.

Martinez-Hernandez, A., Bell, K. P., and Norenberg, M. D. (1977). Glutamine synthetase: Glial localization in brain. *Science* **195**, 1356–1358.

Medzihradszky, F., Sellinger, O. Z., Nandhasvl, P. S., and Santiago, J. C. (1972). ATPase activity in glial cells and in neuronal perikarya of rat cerebral cortex during early postnatal development. *J. Neurochem.* **19**, 543–545.

Metzler, C. J., and Humm, D. G. (1951). The determination of cholinesterase activity in whole brains of developing rats. *Science* **113**, 382–383.

Millichap, J. G. (1957). Development of seizure patterns in newborn animals. Significance of brain carbonic anhydrase. *Proc. Soc. Exp. Biol. Med.* **96**, 125–129.

Monard, D., Solomon, F., Rentsch, M., and Gysin, R. (1973). Glia-induced morphological differentiation in neuroblastoma cells. *Proc. Natl. Acad. Sci. U.S.A.* **70**, 1894–1897.

Moscona, A. A. (1967). Induction of retinal glutamine synthetase in the embryo and in culture. *Accad. Naz. Lincei* **104**, 237–256.

Moscona, A. A., and Wiens, A. W. (1975). Proflavine as a differential probe of gene expression: Inhibition of glutamine synthetase induction in embryonic retina. *Dev. Neurobiol.* **44**, 33–45.

Moudgil, V. K., and Kanungo, M. S. (1973). Effect of age of the rat on induction of acetylcholinesterase of the brain by 17β-estradiol. *Biochim. Biophys. Acta* **329**, 211–220.

Murad, F., Chi, M., Rall, T. W., and Sutherland, E. W. (1962). Adenyl cyclase. III. The effect of catecholamines and choline esters on the formation of adenosine 3′,5′-phosphate by preparations from cardiac muscle and liver. *J. Biol. Chem.* **237**, 1233–1238.

Murphy, R. A., Oger, J., Saide, J. D., Blanchard, M. H., Aranson, G. W., Hogan, C., Pantazis, N. J., and Young, M. (1977). Secretion of nerve growth factor by central nervous system glioma cells in culture. *J. Cell Biol.* **72**, 769–773.

Nachmansohn, D. (1940). Choline esterase in brain and spinal cord of sheep embryos. *J. Neurophysiol.* **3**, 396–402.

Nagata, Y., Mikoshiba, K., and Tsukada, Y. (1974). Neuronal cell body-enriched and glial cell-enriched fractions from young and adult rat brains: Preparation and morphological and biochemical properties. *J. Neurochem.* **22**, 493–503.

Nagata, Y., Nanba, T., and Ando, M. (1976). Changes in some enzymic activities of separated neuronal and glial cell-enriched fractions from rat brains during development. *Neurochem. Res.* **1**, 299–312.

Norenberg, M. D., and Martinez-Hernandez, A. (1979). Fine structural localization of glutamine synthetase in astrocytes of rat brain. *Brain Res.* **161**, 303–310.

Ordy, J. M., and Schjeide, O. A. (1973). Univariate and multivariate models for evaluating long-term changes in neurobiological development maturity and aging. *Prog. Brain Res.* **40**, 25–52.

Ossola, L., Maitre, M., Blindermann, J. M., and Mandel, P. (1978). Some aspects of GABA level regulation in developing rat brain. *In* "Maturation of Neurotransmission" (A. Vernadakis, E. Giacobini, and G. Filogamo, eds.), pp. 83–90. Karger, Basel.

Palmer, G. C. (1973). Adenyl cyclase in neuronal and glial-enriched fractions from rat and rabbit brain. *Res. Commun. Chem. Pathol. Pharmacol.* **5**, 603–613.

Perez, V. J., Olney, J., Cicero, T. J., Moore, B. W., and Bahn, B. A. (1970). Wallerian degeneration in rabbit optic nerve: Cellular localization in the central nervous system of the S-100 and 14-3-2 proteins. *J. Neurochem.* **17**, 511–519.

Perkins, J. P. (1973). Adenyl cyclase. *Adv. Cyclic Nucleotide Res.* **3**, 1–61.

Perkins, J. P., MacIntyre, E. H., Riley, W. D., and Clark, R. B. (1971). Adenyl cyclase, phosphodiesterase and cyclic AMP dependent protein kinase of malignant glial cells in culture. *Life Sci.* **10**, 1069–1080.

Peterson, G. R., Webster, G. W., and Shuster, L. (1973). Characteristics of choline acetyltransferase and cholinesterase in two types of cultured cells from embryonic chick brain. *Dev. Biol.* **34**, 119–134.

Pfeiffer, S. E. (1973). Clonal lines of glial cells. *In* "Tissue Culture of the Nervous System" (G. Sato, ed.), pp. 203–210. Plenum, New York.

Pfeiffer, S. E., Herschman, J. R., Lightbody, J., and Sato, G. (1970). Synthesis by a clonal line of rat glial cells of a protein unique to the nervous system. *J. Cell. Physiol.* **75**, 329–339.

Prasad, K. N. (1975). Differentiation of neuroblastoma cells in culture. *Biol. Rev. Cambridge Philos. Soc.* **50**, 129–165.

Prasad, K. N., Spuhler, K., Arnold, E. B., and Vernadakis, A. (1979). Modification of response of mouse neuroblastoma cells in culture by serum type. *In Vitro* **15**, 807–812.

Prives, C., and Quastel, J. H. (1969). Effects of cerebral stimulation on the biosynthesizing *in vitro* of nucleotides and RNA in brain. *Nature (London)* **221**, 1053.

Provine, R. R., Sharma, S. C., Sandel, T. T., and Hamburger, V. (1970). Electrical activity in the spinal cord of the chick embryo *in situ*. *Proc. Natl. Acad. Sci. U.S.A.* **65**, 508–515.

Purpura, D. P. (1964). Relationship of seizure susceptibility to morphologic and physiologic properties of normal and abnormal immature cortex. *In* "Neurological and Electroencephalographic Correlative Studies in Infancy" (P. Kellaway, ed.), pp. 117–154. Grune & Stratton, New York.

Purpura, D. P. (1969). Stability and seizure susceptibility of immature brain. *In* "Basic Mechanisms of the Epilepsies" (H. H. Jasper, A. A. Ward, and A. Pope, eds.), pp. 481–505. Little, Brown, Boston, Massachusetts.

Purpura, D. P., Shofer, R. J., Housepian, E. M., and Noback, C. R. (1964). Comparative ontogenesis of structure function relations in cerebral and cerebellar cortex. *Prog. Brain Res.* **2**, 187–221.

Rall, T. W., and Gilman, A. G. (1970). Formation and disposition of cyclic AMP in nervous system tissue. *Neurosci. Res. Program, Bull.* **8**, 239–266.

Rall, T. W., and Sattin, A. (1970). Factors influencing the accumulation of cyclic AMP in brain tissue. *Adv. Biochem. Psychopharmacol.* **3**, 113–133.

Rall, T. W., and Sutherland, E. W. (1962). Enzymes concerned with interconversion of liver phosphorylases. *In* "Methods in Enzymology" (T. W. Rall and E. H. Sutherland, eds.), Vol. 5, pp. 377–394. Academic Press, New York.

Ramirez, G. (1977a). Cholinergic development in chick brain reaggregated cell cultures. *Neurochem. Res.* **2**, 417–425.

Ramirez, G. (1977b). Cholinergic development in chick optic tectum and retina reaggregated cell cultures. *Neurochem. Res.* **2**, 427–438.

Rieger, S., and Vigny, M. (1976). Solubilization and physical chemical characterization of rat brain acetylcholinesterase: Development maturation of its molecular forms. *J. Neurochem.* **27**, 121–129.

Riepe, R. E., and Norenberg, M. D. (1977). Müller cell localization of glutamine synthetase in rat retina. *Nature (London)* **268**, 654–655.

Riepe, R. E., and Norenberg, M. D. (1978). Glutamine synthetase in the developing rat retina: An immunohistochemical study. *Exp. Eye Res.* **27**, 435–444.

Robison, G. A., Butcher, R. W., and Sutherland, E. W. (1967). Adenyl cyclase as an adrenergic receptor. *Ann. N.Y. Acad. Sci.* **139**, 703–723.

Robison, G. A., Butcher, R. W., and Sutherland, E. W. (1971). "Cyclic AMP." Academic Press, New York.

Rodbell, M. (1971). Hormones, receptors and adenyl cyclase activity in mammalian cells. *In* "Colloquium on the Role of Adenyl Cyclase and Cyclic 3′,5′-AMP in biological Systems" (P. Condliffe and M. Rodbell, eds.), pp. 88–95. Fogarty Int. Cent., U.S. Gov. Printing Office, Washington, D.C.

Rosen, O. M., and Rosen, S. M. (1968). The effect of catecholamines of the adenyl cyclase of frog and tadpole hemolysates. *Biochem. Biophys. Res. Commun.* **31**, 82–91.

Rosenberg, R. N. (1973). Regulation of neuronal enzymes in cell culture. *In* "Tissue Culture of the Nervous System" (G. Sato, ed.), pp. 107–134. Plenum, New York.

Rudnick, D., and Waelsch, H. (1955). Development of glutamotransferase and glutamine synthetase in the nervous system of the chick. *J. Exp. Zool.* **129**, 309–326.

Rudnick, D., Mela, P., and Waelsch, H. (1954). Enzymes of glutamine metabolism in the developing chick embryo. *J. Exp. Zool.* **126**, 297–321.

Salganicoff, L., and De Robertis, E. (1965). Subcellular distribution of the enzymes of the glutamic acid, glutamine and γ-aminobutyric acid cycles in rat brain. *J. Neurochem.* **12**, 287–309.

Samson, F. E., Dick, H. C., and Balfour, W. M. (1964). Na$^+$-K$^+$-stimulated ATPase in brain during neonatal maturation. *Life Sci.* **3**, 511–515.

Sawyer, C. H. (1946). Cholinesterases in degenerating and regenerating peripheral nerves. *Am. J. Physiol.* **146**, 246–253.

Schmidt, M. J., and Sanders-Bush, E. (1971). Tryptophan hydroxylase activity in developing rat brain. *J. Neurochem.* **18**, 2549–2551.

Schmidt, M. J., Palmer, E. C., Dettbarn, W. D., and Robinson, G. A. (1970). Cyclic AMP and adenyl cyclase in the developing rat brain. *Dev. Psychobiol.* **3**, 53–67.

Schon, F., and Kelly, J. S. (1974). Autoradiographic localization of [^{3}H]GABA and [^{3}H]glutamate over satellite glial cells. *Brain Res.* **66**, 275–288.

Schrier, B. K. (1973). Surface culture of fetal mammalian brain cells: Effect of subculture on morphology and choline acetyltransferase activity. *J. Neurobiol.* **4**, 117–124.

Schrier, B. K., and Shapiro, D. L. (1974). Effects of fluorodeoxyuridine on growth and choline acetyltransferase activity in fetal rat brain cells in surface culture. *J. Neurobiol.* **5**, 151–159.

Schrier, B. K., and Thompson, E. J. (1974). On the role of glial cells in the mammalian nervous system. Uptake, excretion, and metabolism of putative neurotransmitters by cultured glial tumor cells. *J. Biol. Chem.* **249**, 1769–1780.

Schroeder, K. (1911). Der faserverlauf im vorderhirn des hühnes. *J. Psychol. v. Neurol.* **18**, 115–173.

Schultz, G., Bohme, E., and Munske, K. (1969). Guanyl cyclase. Determination of enzyme activity. *Life Sci.* **8**, 1323–1332.

Seeds, N. W. (1971). Biochemical differentiation in reaggregating brain cell culture. *Proc. Natl. Acad. Sci. U.S.A.* **68**, 1858–1861.

Seeds, N. W. (1973). Differentiation of aggregating brain cell cultures. *In* "Tissue Culture of the Nervous System" (G. Sato, ed.), pp. 35–53. Plenum, New York.

Seeds, N. W. (1975). Expression of differentiated activities in reaggregated brain cell cultures. *J. Biol. Chem.* **250**, 5455–5458.

Sellinger, O. Z., and deBalbian Verster, J. (1962) Glutamine synthetase of rat cerebral cortex: intracellular distribution and structural latency. *J. Biol. Chem.* **237**, 2836–2844.

Sensenbrenner, M., Springer, N., Booher, J., and Mandel, P. (1972). Histochemical studies during the differentiation of dissociated nerve cells cultivated in the presence of brain extracts. *Neurobiology* **2**, 49–60.

Shapiro, D. L., and Schrier, B. K. (1973). Cell cultures of fetal rat brain: Growth and marker enzyme development. *Exp. Cell Res.* **77**, 239–247.

Sharma, K. N., Dua, S., Singh, B., and Anand, B. K. (1964). Electroontogenesis of cerebral and cardiac activities of the chick embryo. *Electroencephalogr. Clin. Neurophysiol.* **16**, 503–509.

Shimizu, H., Creveling, C. R., and Daly, J. W. (1970). Effect of membrane depolarization and biogenic amines on the formation of cyclic AMP in incubated slices. *Adv. Biochem. Psychopharmacol.* **3**, 135–154.

Silver, A. (1967). Cholinesterases of the central nervous system with special reference to the cerebellum. *Int. Rev. Neurobiol.* **10**, 57–109.

Silver, A. (1974). "The Biology of Cholinesterases." Elsevier, Amsterdam.

Silverman, S. H., and Gardner, L. I. (1954). Ultrafiltration studies on serum magnesium. *N. Engl. J. Med.* **250**, 938.

Singh, V. K., and McGeer, E. G. (1977). Choline acetyltransferase in developing rat brain and spinal cord. *Brain Res.* **127**, 159–163.

Skou, J. C. (1962). Preparation from mammalian brain and kidney of the enzyme system involved in active transport of Na^+ and K^+. *Biochim. Biophys. Acta* **58**, 314.

Skou, J. C. (1965). Enzymatic basis for active transport of Na^+ and K^+ across cell membrane. *Physiol. Rev.* **45**, 596–617.

Stahl, W. L., and Broderson, S. H. (1975). Localization of Na^+,K^+-ATPase in brain. *Fed. Proc., Fed. Am. Soc. Exp. Biol.* **35**, 1260–1265.

Starr, M. S. (1974). Evidence for the compartmentation of glutamate metabolism in isolated rat retina. *J. Neurochem.* **23**, 337–344.

Sutherland, E. W., Rall, T. W., and Menon, T. (1962). Adenyl cyclase. I. Distribution, preparation, properties. *J. Biol. Chem.* **237**, 1220–1227.

Tapia, R. (1974). The role of γ-aminobutyric acid metabolism in the regulation of cerebral excitability. *In* "Neurohumoral Coding of Brain Function" (R. R. Drucker-Colin and R. D. Myers, eds.), pp. 3–26. Plenum, New York.

Tapia, R., and Meza-Ruiz, G. (1975). Differences in some properties of newborn and adult brain glutamate decarboxylase. *J. Neurobiol.* **6**, 171–181.

Tapia, R., and Meza-Ruiz, G. (1976). Changes in some properties of glutamate decarboxylase activity during the maturation of the brain. *Neurochem. Res.* **1**, 133–140.

Tapia, R., Pasante-Morales, H., Taborda, E., and Perez de la Mora, M. (1975a). Seizure susceptibility in the developing mouse and its relationship to glutamate decarboxylase and pyridoxal phosphate in brain. *J. Neurobiol.* **6**, 159–170.

Tapia, R., Sandoval, M. E., and Contreras, P. (1975b). Evidence for a role of glutamate decarboxylase activity as a regulatory mechanism of cerebral excitability. *J. Neurochem.* **24**, 1283–1285.

Tennyson, V. M., and Brzin, M. (1970). The appearance of acetylcholinesterase in the dorsal root neuroblast of the rabbit embryo. A study by electron microscope cytochemistry and microgasometric analysis with the magnetic diver. *J. Cell Biol.* **46**, 64–80.

Timiras, P. S. (1972). Development and plasticity of the nervous system. *In* "Developmental Physiology and Aging" (P. S. Timiras, ed.), pp. 129–165. Macmillan, New York.

Timiras, P. S., and Vernadakis, A. (1972). Structural, biochemical and functional aging of the nervous system. *In* "Developmental Physiology and Aging" (P. S. Timiras, ed.), pp. 502–526. Macmillan, New York.

Timiras, P. S., Vernadakis, A., and Sherwood, N. (1968). Development of plasticity of the nervous system. *In* "Biology of Gestation" (N. Assali, ed.), Vol. 2, pp. 261–319. Academic Press, New York.

Toews, A. D., and Horrocks, L. A. (1976). Developmental and aging changes in protein concentration and 2′,3′-cyclic nucleoside monophosphate phosphodiesterase activity (EC 3.1.4.16) in human cerebral white and gray matter and spinal cord. *J. Neurochem.* **27**, 545–550.

Tucek, S. (1967). Subcellular distribution of acetyl-coA synthetase, ATP citrate lyase, citrate synthase, choline acetyltransferase, fumarate hydratase and lactate dehydrogenase in mammalian brain tissue. *J. Neurochem.* **14**, 531–545.

Valcana, T. (1971). Effect of neonatal hypothyroidism on the development of acetylcholinesterase and choline acetyltransferase activities in the rat brain. *In* "Influence of Hormones on the Nervous System" (D. H. Ford, ed.), pp. 174–184. Karger, Basel.

Valcana, T. (1974). Developmental changes in ionic composition of the brain in hypo- and hyperthyrodism. *In* "Drugs and the Developing Brain" (A. Vernadakis and N. Weiner, eds.), pp. 289–304. Plenum, New York.

Valcana, T., and Timiras, P. S. (1969). Choline acetyltransferase activity in various brain areas of aging rats. *Proc. Int. Congr. Gerontol., 8th, 1969* Vol. 2, p. 24.

Valcana, T., Vernadakis, A., and Timiras, P. S. (1969). Effects of neonatal X-radiation on choline acetyltransferase activity in various areas of the developing central nervous system. *In* "Radiation Biology of the Fetal and Juvenile Mammal" (M. R. Skov and D. D. Mahlum, eds.), pp. 887–898. U.S. At. Energy Comm., Washington, D.C.

Valcana, T., Liao, C., and Timiras, P. S. (1974a). Effects of X-radiation on the subcellular distribution of cholinergic enzymes in the developing rat cerebellum. *Brain Res.* **73**, 105–120.

Valcana, T., Liao, C., and Timiras, P. S. (1974b).Effects of X-radiation on development of cholinergic system of rat brain. 2. Investigation of alterations in acetylcholine content. *Environ. Physiol. & Biochem.* **4**, 58–63.

Vernadakis, A. (1973a). Comparative studies of neurotransmitter substances in the maturing and aging central nervous system of the chicken. *Prog. Brain Res.* **40**, 231–243.

Vernadakis, A. (1973b). Changes in nucleic acid content and butyrylcholinesterase activity in CNS structures during the life span of the chicken. *J. Gerontol.* **28**, 281–286.

Vernadakis, A., and Arnold, E. B. (1979). Characterization of neural enzyme development in dissociated chick embryo brain cell cultures. *In* "NATO Advanced Study Institute, Developmental Neurobiology of Vision" pp. 433–442. Plenum, New York.

Vernadakis, A., and Burkhalter, A. (1967). Acetylcholinesterase activity in the optic lobes of chicks at hatching. *Nature (London)* **214**, 594–595.

Vernadakis, A., and Gibson, D. A. (1974). Role of neurotransmitter substances in neural growth. *In* "Perinatal Pharmacology: Problems and Priorities" (J. Dancis and J. C. Hwang, eds.), pp. 65–76. Raven Press, New York.

Vernadakis, A., and Nidess, R. (1976). Biochemical characteristics of C-6 glial cells. *Neurochem. Res.* **1**, 385–402.

Vernadakis, A., and Woodbury, D. M. (1962). Electrolyte and amino acid changes in rat brain during maturation. *Am. J. Physiol.* **203**, 748.

Vernadakis, A., Nidess, R., Timiras, M. L., and Schlesinger, R. (1976). Responsiveness of acetylcholinesterase and butyrylcholinesterase activities in neural cells to age and cell density in culture. *Exp. Cell Res.* **97**, 453–457.

Vernadakis, A., Arnold, E. B., and Hoffman, D. W. (1978). Neural tissue culture: A model for the study of the maturation of neurotransmission. *In* "Maturation of Neurotransmission" (A. Vernadakis, E. Giacobini, and G. Filogamo, eds.), pp. 160–170. Karger, Basel.

Vigny, M., De Giamberardino, L., Couraud, J. Y., Rieger, F., and Koenig, J. (1976). Molecular forms of chicken acetylcholinesterase: Effect of denervation. *FEBS Lett.* **69**, 277–280.

Vijayan, V. K. (1977). Cholinergic enzymes in the cerebellum and the hippocampus of the senescent mouse. *Exp. Gerontol.* **12**, 7–11.

Von Hungen, K., Roberts, S., and Hill, F. D. (1974). Developmental and regional variations in neurotransmitter-sensitive adenylate cyclase systems in cell-free preparations from rat brain. *J. Neurochem.* **22**, 811–819.

Wald, G., and Zussman, H. (1938). Carotenoids of the chicken retina. *J. Biol. Chem.* **122**, 449–460.

Waymire, J. C., Bjur, R., and Weiner, N. (1971). Assay of tryosine hydroxylase by coupled decarboxylation of DOPA formed from 1-^{14}C-L-tyrosine. *Anal. Biochem.* **43**, 588–600.

Weiner, N. (1970). Regulation of norepinephrine biosynthesis. *Annu. Rev. Pharmacol.* **10**, 273–290.

Weiner, N., and Rabadjija, M. (1968). The effect of nerve stimulation on the synthesis and metabolism of norepinephrine in the isolated guinea pig hypogastric nerve-vas deferens preparation. *J. Pharmacol. Exp. Ther.* **160**, 61–71.

Weiner, N., Barnes, E., and Lee, F. L. (1977). An analysis of the possible factors modulating neurotransmitter synthesis in catecholaminergic neurons consequent to nerve stimulation. *In* "Neuroregulators and Hypothesis of Psychiatric Disorders" (J. Barchas, D. Hamburg, and E. Usdin, eds.), pp. 75–87. Oxford Univ. Press, London and New York.

Werner, I., Peterson, G. R., and Shuster, L. (1971). Choline acetyltransferase and acetylcholinesterase in cultured brain cells from chick embryos. *J. Neurochem.* **18**, 141–151.

Weysse, A. W., and Burgess, W. S. (1906). Histogenesis of the retina. *Am. Nat.* **40**, 611–637.

White, A. A., and Aurbach, G. D. (1969). Detection of guanyl cyclase in mammalian tissues. *Biochim. Biophys. Acta* **191**, 686–697.

White, A. A., Aurbach, G. D., and Carlson, S. F. (1969). Identification of guanyl cyclase in mammalian tissues. *Fed. Proc., Fed. Am. Soc. Exp. Biol.* **28**, 473.

Whittaker, V. P. (1969). The synaptosome. *In* "Handbook of Neurochemistry" (A. Lajtha, ed.), pp. 327–364. Plenum, New York.

Whittaker, V. P., and Barker, L. A. (1972). The subcellular fractionation of brain tissue with special reference to the preparation of synaptosomes and their component organelles. *In* "Methods of Neurochemistry" (R. Fried, ed.), Vol. 2, pp. 1–15. Dekker, New York.

Wilson, S. H., Schrier, B. K., Farber, J. L., Thompson, E. J., Rosenberg, R. N., Blume, A.

J., and Nirenberg, M. W. (1972). Markers for gene expression in cultured cells from the nervous system. *J. Biol. Chem.* **247**, 3159–3169.

Woodbury, D. M. (1958). Effect of hormones on brain excitability and electrolytes. *Recent Prog. Horm. Res.* **10**, 65.

Youngstrom, K. A. (1938). On the relationship between cholinesterase and the development of behavior in amphibia. *J. Neurophysiol.* **1**, 357–363.

Youngstrom, K. A., Woodhall, V., and Graves, R. W. (1941). Acetylcholinesterase content of brain tumors. *Proc. Soc. Exp. Biol. Med.* **48**, 555–557.

Zanetta, J. P., Benda, P., Gombos, G., and Morgan, I. G. (1972). The presence of 2′,3′-cyclic AMP 3′-phosphohydrolase in glial cells in tissue culture. *J. Neurochem.* **19**, 881–883.

ADVANCES IN CELLULAR NEUROBIOLOGY, VOLUME 1

GLIAL FIBRILLARY ACIDIC PROTEIN (GFA) IN NORMAL NEURAL CELLS AND IN PATHOLOGICAL CONDITIONS

*AMICO BIGNAMI, DORIS DAHL, AND
DAVID C. RUEGER*

Spinal Cord Injury Research Laboratory, West Roxbury Veterans Administration
Medical Center and the Department of Neuropathology,
Harvard Medical School, Boston, Massachusetts

I. Introduction

A review of the literature on glial fibrillary acidic (GFA) protein indicates that it has found its major application as an immunohistochemical marker for astroglia, especially under conditions where identification of the cell type is difficult, that is, during development, in culture, in cellular fractions and, more recently, in human brain tumors. In our laboratory, procedures have been developed to facilitate production of the antisera, thus making immunological staining more generally available.

Another line of research concerns the biology of glial filaments and the relation of these filaments to the 100-Å filaments in neural and nonneural cells. Filaments approximately 100 Å in diameter are major cytoplasmic constituents of glial and nerve fibers. They are usually considered stable structures based on their insolubility after isolation. The solubility of GFA protein in aqueous solutions and the *in vitro* formation of filaments from purified GFA protein preparations suggest that glial filaments, like other constituents of the cytoskeleton in nonmuscle cells, namely, microtubules and actin microfilaments, may assemble and disassemble in the cytoplasm. The extensive remodeling of the glial framework during brain development and the rapid change in cell shape occurring in cultured astrocytes following serum withdrawal and the administration of cyclic AMP make this hypothesis particularly attractive. GFA protein has been included in recent reviews on brain-specific proteins (Bock, 1978; Eng and Bigbee, 1978). Proteins found to be antigenically identical to GFA protein are astroprotein (Mori and Marimoto, 1975), nervous system antigen NSA 1 (Delpech *et al.*, 1978a) and α-albumin (Löwenthal *et al.*, 1978).

II. Biochemical Properties of GFA Protein

A. *Isolation*

The original isolation of GFA protein from gliosed human brain required relatively simple procedures such as buffer extraction and ammonium sulfate precipitation in order to obtain purified preparations (Eng *et al.*, 1971). Essentially, no experimental procedure will result in the tremendous enrichment in fibrous astrocytes observed in old multiple sclerosis plaques. When normal material was used as the source, these procedures did not allow satisfactory purification, and so new methods were developed (Dahl and Bignami, 1973a, 1975).

Hydroxyapatite chromatography is the most convenient procedure for isolating the protein from normal human brain and spinal cord. However, this method cannot be used to isolate GFA protein from buffer extracts of rapidly frozen material since tubulin is a major contaminant under these conditions (Liem and Shelanski, 1978; Rueger *et al.*, 1978a). In this context, it is appropriate to comment on some remarkable properties of the hydroxyapatite procedure, which can be used to isolate most cytoskeletal proteins by varying the source. As mentioned before, tubulin is the major protein isolated from buffer extracts of rapidly frozen bovine or rat brain, whereas with autolyzed tissue (e.g., human brain obtained at autopsy) tubulin does not absorb to the gel and highly purified preparations of *degraded* GFA protein are obtained. With chicken brain the main antigenic protein isolated with the procedure is neurofilament protein (Dahl and Bignami, 1977b), and this finding has been recently confirmed by an independent group of investigators (Bennet *et al.*, 1978). When the method is applied to extracts of chicken gizzard, actin and desmin, the subunit of the 100-Å smooth muscle filaments (Lazarides and Hubbard, 1976), are isolated (D. Dahl, unpublished observations).

Nondegraded GFA protein has been recently isolated from buffer extracts of bovine brain by immunoaffinity chromatography (Rueger *et al.*, 1978a,b). Antisera were raised against *degraded* antigen isolated from human spinal cord (Section III). *Nondegraded* GFA protein is 50,000–54,000 in MW, depending on the gel system, and comigrates with tubulin on sodium dodecyl sulfate (SDS)–acrylamide gel electrophoresis in phosphate buffer. The best separation of GFA protein and the tubulin subunits was obtained using a Tris–glycine–SDS buffer system (Bryan, 1974) with gels containing 5% acrylamide and 8 M urea (Rueger *et al.*, 1978a).

B. Degradation

GFA protein isolated from human brain is heterogeneous in polypeptide composition (Dahl and Bignami, 1974, 1975). These polypeptides range in MW from 40,000 to about 54,000 on SDS–acrylamide gel electrophoresis in 0.025 M phosphate buffer; they are immunologically and biochemically related, since (1) they all cross-react with GFA protein antisera by disc immunodiffusion; (2) they contain the major amino acids of GFA protein, namely, glutamic acid, alanine, and leucine, in comparable amounts; (3) most cyanogen bromide peptides are common, including an immunologically active fragment in the myoglobin range (MW 17,200); and (4) a unique amino terminal sequence, alanine–glycine–phenylalanine, was common to degraded preparations differing in polypeptide composition.

Experiments conducted in our laboratory have conclusively shown that the heterogeneity of GFA protein is the result of degradation (Dahl and Bignami, 1975; Dahl, 1976). Our data indicate the existence of two distinctive degradative pathways for GFA protein. In one type, best demonstrated in homogenates incubated at pH 6.0 and 6.5, GFA protein is rapidly destroyed, with loss of immunological activity. The other type of degradation is observed in tissues incubated at room temperature, as a result of postmortem autolysis, and in homogenates incubated in phosphate buffer, pH 8.0, at 24° and 37° C. It presents a characteristic disc electrophoretic pattern of multiple closely spaced bands, suggesting that small fragments are cleaved from the original polypeptide chain in successive steps of degradation. The final products of this type of degradation, polypeptides in the 40,000-MW range, are remarkably resistant to tissue autolysis and are still present after many days of incubation of human spinal cord *in vitro* (Dahl and Bignami, 1976).

C. In Vitro *Assembly and Disassembly*

GFA protein is partially extracted in 0.05 M sodium phosphate buffer, pH 8.0, the remaining water-insoluble fraction being solubilized under more drastic conditions, that is, with urea, guanidine hydrochloride, and detergent (Eng *et al.,* 1971; Dahl and Bignami, 1973a). It was originally speculated that the water-soluble fraction was related to the native nonassembled subunits of GFA protein and that the fraction obtained with more drastic extractive procedures was a constituent of the glial filaments (Bignami *et al.,* 1972). The stability of brain filaments after isolation (Yen *et al.,* 1976) appeared to be consistent with this hypothesis.

Recent data from our laboratory on the assembly–disassembly at physiological pH of GFA protein isolated at low ionic strength from bovine brain suggest a different view, namely, that glial filaments may be formed and disassembled in the cytoplasm according to the needs of the cell (Rueger *et al.,* 1979). GFA protein was extracted from bovine brain with 1 mM sodium phosphate buffer, pH 8.0, and purified by immunoaffinity chromatography. Filaments assembled in 100 mM imidazole-HCl buffer, pH 6.8, and were disassembled by dialysis against 2 mM Tris-HCl, pH 8.5. They formed networks of smooth curvilinear structures approximately 100 Å in diameter and of indefinite length (Fig. 1). Although assembly could be effected in the absence of reducing agents, most of the protein did not disassemble under these conditions because of the formation of disulfide bonds. Whether disulfide bonds play a role in the stabilization of glial filaments (Huston and Bignami, 1977), as suggested for keratin

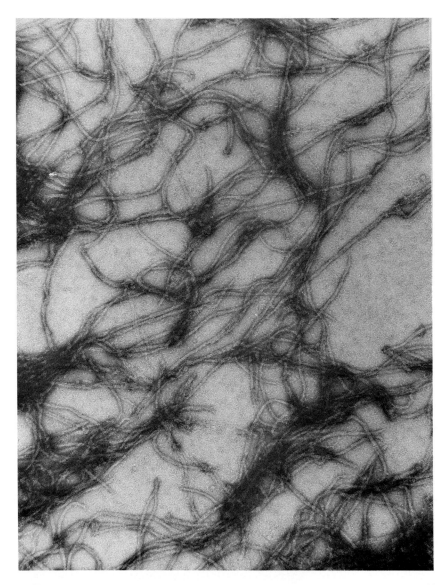

FIG. 1. *In vitro* polymerized GFA protein filaments stained with uranyl acetate on a grid. × 60,000.

tonofilaments (Sun and Green, 1978), remains to be seen. In this respect it should be noted that, as indicated by immunofluorescence studies, the astroglial framework undergoes rapid changes during development (Section V) but remains essentially unchanged over a period of months in brain scars (Section VI).

D. Relation to Neurofilament Protein and to Filament Proteins in Nonneural Cells

A controversy exists in the literature concerning the similarity of GFA and neurofilament protein. It now has been established in a number of laboratories that GFA protein, isolated glial filaments, and the ~50K "neurofilament" protein in myelin-free bovine axons are similar (Yen *et al.*, 1976; Benitz *et al.*, 1976; De Vries *et al.*, 1976; Davison and Hong, 1977; Goldman *et al.*, 1978). These data were interpreted as indicating that the building blocks of neurofilaments and glial filaments are biochemically related. According to another hypothesis, the presence of GFA-like protein in myelin-free axons is due to glial contamination. Recent data from our laboratory appear to support this hypothesis (Dahl and Bignami, 1979). Most of the 50K protein in myelin-free axons is selectively absorbed on Sepharose 4B coupled to GFA protein antisera. A smaller but still significant fraction in this MW range is not adsorbed and is probably a neurofilament component.

It should be noted that the controversy over the relation between GFA and neurofilament protein is part of the broader issue concerning the biochemical relatedness (or nonrelatedness) of different types of filaments, such as keratin tonofilaments in epidermis, desmin filaments in smooth muscle, and 100-Å filaments in tissue culture cells (see Lazarides and Balzer, 1978, for discussion). Previous immunofluorescence data on the similarity of filaments in different cells, such as neurons and endothelial cells (Blose *et al.*, 1977), should be interpreted with caution, since "nonspecific" IgG may bind to filamentous structures, suggesting the existence of nonimmunological affinities comparable to those occurring between stain and fibrils with histological methods (Bennett *et al.*, 1978; Dahl and Bignami, 1978b). Although most of the evidence derived from carefully controlled immunofluorescence studies (Bignami and Dahl, 1977; Franke *et al.*, 1978) and from peptide map analysis (Davison *et al.*, 1977) points toward the existence of biochemically separate classes of 100-Å filaments, on the basis of our experience we cannot rule out the possibility that GFA protein may cross-react to a certain extent with neurofilament antisera. GFA protein and neurofilament antisera are extremely specific with respect to their immunohistological localization (Dahl and Bignami,

1977b, 1978a). However, it is possible to absorb neurofilament antisera with highly purified GFA protein preparations. On the other hand, partially purified neurofilament preparations very effective in absorbing neurofilament antisera do not absorb GFA protein antisera with respect to their staining properties in immunofluorescence tests. Experiments pointing in the same direction have been recently published by Bock et al. (1977b). GFA protein antisera were partially absorbed with extracts of cultured human fibroblasts containing 100-Å filaments.

Recent experiments also suggest that the same cell may contain biochemically distinct types of 100-Å filaments (Paetau et al., 1979). Glioma cells in culture were stained by immunofluorescence with both GFA protein antisera and with a human autoimmune serum, demonstrating a fibrillary pattern in fibroblasts, smooth muscle, and neuroblastoma cells (these cells were not stained with GFA protein antisera). In addition, the staining pattern of the glioma cells was different with the two antisera.

III. Preparation of Antisera

The immunogenicity of GFA protein varies considerably depending on the degree of degradation. Nondegraded or partially degraded preparations are nonimmunogenic or weakly immunogenic, whereas strong antisera are obtained with degraded antigen in the 40,000-MW range (Uyeda et al., 1972; Bignami and Dahl, 1974a; Dahl and Bignami, 1976). With nondegraded or partially degraded protein immunogenicity is enhanced by treatment with SDS or by cyanogen bromide cleavage (Dahl and Bignami, 1976, 1977a).

Since many investigators are interested in GFA protein as an astroglial marker and since human spinal cord, the source of the antigen in our laboratory (Dahl and Bignami, 1976), is not readily available in biological laboratories, we now report a protocol for the production of antisera from bovine spinal cord. Bovine spinal cord is purchased at low cost from a slaughterhouse. The vertebral canal is widely opened during the processing of the carcass, and the spinal cord can be removed without difficulty.

The thick membrane covering the spinal cord (dura mater) is removed, and the spinal cord is cut into fragments. The fragments are incubated for two consecutive days at room temperature in 0.05 M sodium phosphate buffer, pH 8.0 (w/v, 1:4), with 0.1% azide as a preservative. Autolyzed spinal cord is homogenized in the incubation buffer and the pH is adjusted to 8.0 with sodium hydroxide. Following centrifugation at 48,000 g for 60 min the supernatant is added to hydroxyapatite (BioRad), stirred for 30 min (30 ml of hydroxyapatite suspension for 10 gm of tissue), and washed with 0.05 M sodium phosphate buffer, pH 8.0, until all unadsorbed protein is removed. After the initial washes, the hydroxyapatite is allowed to stand overnight. The protein is eluted with 0.1 M potassium phosphate buffer, pH 8.0. The eluates are pooled and precipitated by adding ammonium sulfate to a final concentration of 35%. The ammonium sulfate precipitate is dialyzed against water and freeze-dried. The yield of the procedure is low compared with human spinal cord (0.2–0.3 mg protein and 1–2 mg protein per gram of tissue, respectively). In addition, it was not possible to prolong the time of autolysis to allow complete degradation of GFA to a single 40,000-MW species without further decreasing the yield; most preparations obtained with the outlined procedure separated into two bands in the 40,000–45,000 MW range on SDS–polyacrylamide gel electrophoresis. For production of the antisera 2 mg of purified protein are thoroughly emulsified with Freund's complete adjuvant and injected into the footpads and at several sites subcutaneously on the back of an albino New Zealand rabbit. Three weeks later a booster of 1 mg is given subcutaneously on the back. Two rabbits injected in our laboratory with bovine spinal cord GFA protein according to this schedule produced strong antisera in the fourth week. In contrast to rabbits injected with human GFA protein, the titer decreased rapidly without frequent booster injections. It is thus important to check frequently small samples of serum after the first week in order to effect a large bleeding as soon as the antiserum has reached the desired potency for immunohistological studies (bright immunofluorescent staining of Bergmann glia in nonfixed cryostat sections of rat cerebellum at 1:80 dilutions in our laboratory). In view of the dilutions used for immunohistological studies, a single large bleeding (60 ml taken from the marginal vein of the ear) will suffice for the needs of the laboratory. The antiserum should be divided into 1-ml samples and stored frozen. Repeated freezing and defreezing of the antiserum and its dilutions should be avoided. Even at − 70°C a decrease in titer will occur after 2–3 years. The antiserum can be mailed at room temperature with 0.1% azide added as a preservative.

IV. Immunohistochemical Localization of GFA Protein in Adult CNS

As indicated by studies conducted in the rat and mouse brain with the indirect immunofluorescence method and with peroxidase-labeled GFA protein antiserum at the light and electron microscopic level, GFA protein is selectively located in astroglia (Bignami *et al.,* 1972; Bignami and Dahl, 1974a; Ludwin *et al.,* 1976; Schachner *et al.,* 1976, 1977; Delpech *et al.,* 1978a). Ependymal processes are apparently the only exception. Although it is practically impossible to distinguish ependymal processes from subependymal glia in the periventricular layer, the radially oriented processes in the rat hypothalamus can be demonstrated by immunofluorescence (Bignami and Dahl, 1976). These processes have been traced to the ependyma with the silver carbonate method in the hypothalamus of the opossum (Royce, 1971). They apparently belong to a specialized form of ependyma (tanycytes, radial glia of the third ventricle) with characteristic cytological features including a highly filamentous cytoplasm (Del Cerro and Knigge, 1979).

The distribution of astroglia demonstrated with GFA protein antisera in the rat central nervous system (CNS) and with Weigert's method in the human CNS (Weigert, 1895) is remarkably similar. In white matter astroglia form a delicate framework surrounding the myelin sheaths. The framework is reinforced around blood vessels and at the surface of the CNS where a continuous membrane is formed. Thick glial septa extend at right angles from the surface in both the optic nerve and spinal cord (Fig. 2). Bergmann radial glia extending from the Purkinje cell layer to the

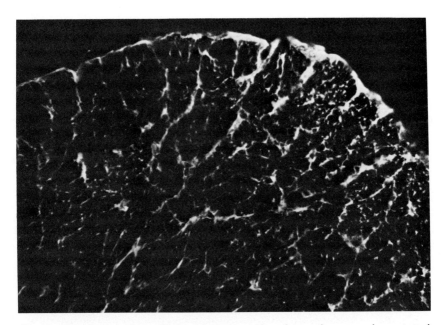

Fig. 2.   Distribution of astroglia in a transverse section of rat optic nerve as demonstrated by immunofluorescence with GFA protein antiserum. Glial processes form an external membrane from which thick septa penetrate into the nerve, separating the nerve fibers into bundles. Note the fine meshwork of fibrils within the nerve bundles. × 200.

cerebellar surface are the only processes stained in the molecular layer of the cerebellum (Fig. 3). Because of their typical appearance, they represent the most desirable structure for screening antisera for specificity to GFA protein. Typical stellate astrocytes are found in cerebral white matter, spinal cord gray matter, the granular layer of the cerebellum, thalamic nuclei, and the hippocampus (Fig. 4). In the rat cerebral isocortex astrocytes are well stained in the superficial and deep layers, whereas in the middle layers only processes abutting on blood vessels can be demonstrated by immunofluorescence. Spidery astrocytes were stained in this location with peroxidase-coupled antiserum (Ludwin *et al.*, 1976), but typical protoplasmic astrocytes of feathery appearance have been only demonstrated in the human cerebral cortex (Braak *et al.*, 1978). GFA protein has been also localized to the interstitial cells of the pineal gland (Møller *et al.*, 1978). This confirms the classification of these cells as astrocytes based on the presence of filaments. Also, Müller cells of the retina contain GFA protein. (See Section VI for the localization of GFA protein in retinal glia.)

Glial cells are selectively demonstrated in the submammalian CNS (birds, amphibians, reptiles, and fish) by immunofluorescence with GFA protein

Fɪɢ. 3.   Immunofluorescence of Bergmann glia in the rat cerebellum with GFA protein antiserum. The tangential cut of the upper folium shows that Bergmann fibers are collected in small bundles. × 280.

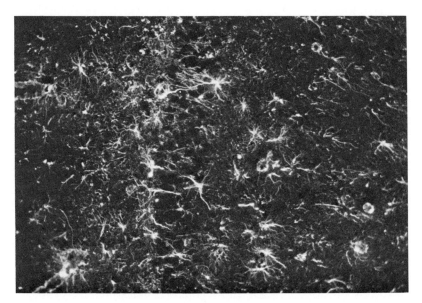

FIG. 4.   Star-shaped astrocytes demonstrated by immunofluorescence with GFA protein antiserum in the rat hippocampus. Also note the very delicate fibrils in the neuropil. × 300.

antisera raised against human antigen (Dahl and Bignami, 1973b). The radial glia characteristic of these species were stained in some locations, e.g., the optic tectum of the turtle. Brain extracts give a reaction of incomplete identity (formation of spurs) by double immunodiffusion.

V. Gliogenesis

Immunohistological studies with GFA protein antisera have provided new data in three critical areas of research on the development of neuroglia (Privat, 1975; Jacobson, 1978): (1) myelination gliosis, (2) time of appearance of Bergmann fibers in the cerebellum, and (3) the nature of radial glia.

A. Myelination Gliosis

A prominent feature of myelinating tracts is represented by the striking proliferation of glial cells, a phenomenon described as "myelination gliosis" (Roback and Scherer, 1935). Myelination gliosis, as the term implies, is a phenomenon related to the formation of myelin sheaths and oligodendroglia differentiation. The presence of myelin basic protein in oligodendroglia prior to myelin formation has been recently demonstrated (Sternberger *et al.*, 1978). Our findings indicate that in white matter a close correlation exists between the appearance of GFA-positive cells and myelination, thus suggesting that interfascicular glia (oligodendrocytes and astrocytes) differentiate from precursor cells (glioblasts) in a period of intense cell multiplication (Bignami and Dahl, 1973, 1975). In the developing mouse a peak of the specific concentration of GFA protein relative to total protein content has been found between days 10 and 14 postnatally, corresponding to a period of more extensive myelination (Bock *et al.*, 1975).

The first appearance of GFA-positive cells in white matter at the time of myelination does not allow a general statement to the effect that astroglial differentiation is a late event in ontogeny (Section V,C) but rather reflects events specifically related to the differentiation of interfascicular glia.

As evidenced by GFA expression, astroglial differentiation is not impaired in mutant mice with defective myelination (quaking and jimpy). In effect the concentration of GFA protein is increased in these mutants, probably as a result of reactive gliosis (Bignami and Dahl, 1974c; Jacque *et al.*, 1974, 1976a,b).

B. Differentiation of Bergmann Glia

Different views have been expressed concerning the development of Bergmann glia in the rodent cerebellum. According to Das and his collaborators (Das *et al.*, 1974; Das, 1976) Bergmann glia originate from the external granular layer at the end of the first week of postnatal life and then migrate through the molecular layer to reach the Purkinje cell layer on the twelfth day. According to this view Bergmann fibers spanning the whole thickness of the molecular and external granular layers are thus not present until late in postnatal development, at a time when neuronal migration from the external granular layer is well underway. According to another view (Del Cerro and Swarz, 1976), Bergmann glia are derived from the subventricular layer of the fourth ventricle and are already positioned at embryonic day 15 in the mouse and at embryonic day 17 in the rat. Although immunofluorescence studies with GFA protein antisera did not contribute to the first part of the controversy, i.e., external granular layer versus subependymal layer origin, they clearly demonstrated that in the rodent cerebellum Bergmann fibers extending through the external granular and molecular layer are already present on the fourth postnatal day (Bignami and Dahl, 1973), before neuroblasts start to migrate from the ex-

ternal granular layer. The pattern of astroglial differentiation was completely different from the one observed in white matter. In myelinating
tracts, there was a progressive increase in immunofluorescence over a
period of many days, consistent with the proliferation of glia during
myelination. In the cerebellum, Bergmann fibers suddenly become positive
over most of the cerebellum during the fourth postnatal day, and subsequent changes were mainly morphological.

The reason for paying so much attention to the development of Bergmann glia is that, according to an attractive hypothesis, they provide
guidance for neuroblasts migrating from the external granular layer (Rakic
and Sidman, 1973). While our findings are consistent with this hypothesis,
they do not provide support for the concept that failure of neuronal migration in the mutant weaver mouse is due to a lack of Bergmann fibers. Their
appearance was not delayed in this mutation (Bignami and Dahl, 1974b) in
accordance with the data of Sotelo and Changeux (1974).

C. Radial Glia

Since the classical studies of Cajal, the presence of radially oriented
fibers spanning the entire thickness of the developing fetal cerebrum has
been recognized by many investigators (reviewed by Choi and Lapham,
1978). The observation by Rakic (1971) that migrating neuroblasts are
closely apposed to radial glia led to the concept of guidance (Section V,B)
and to renewed interest in these structures. Studies aimed at determining
whether radial glia are astrocytic or multipotential (spongioblasts), based
on the presence of GFA protein, demonstrate one pitfall in assessing the
onset of differentiation by the time of appearance of a cell-specific protein.
Positive results provide the unmistakable proof of cell commitment, but
negative findings should be interpreted with caution. The cell may be fully
restricted but still not produce the specialized product or, more likely, the
concentration may be too low to be detected with the particular technique
used. With respect to radial glia, these cells were not stained by immunofluorescence in the developing rat brain (Bignami and Dahl, 1974a).
However, they became GFA-positive shortly after injury to the immature
cerebrum and thus appeared to be comparable to protoplasmic astrocytes
in the rat cerebral cortex in this respect (Section VI). The final demonstration that radial glia are biochemically differentiated astrocytes, although
immature in appearance, came from studies of human fetal brains. Radially oriented, nearly parallel immunofluorescent fibers between the ventricular zone and the external surface could be demonstrated as early as 10
weeks of age (Antanitus et al., 1976).

VI. Astroglial Response to Injury

It is generally recognized that a pathological growth of fibrillary neuroglia is the most common reaction of the brain to injury. The time course of GFA protein production following injury has been studied by immunofluorescence in two locations where astrocytes are not normally stained with this method, namely, the cerebral cortex and the retina of the rat (Bignami and Dahl, 1974d, 1976, 1979). Approximately 48 hr after a stab wound had been made, the protoplasmic astrocytes (Section IV) of the cerebral cortex became intensely immunofluorescent and assumed the characteristic stellate shape of fibrous astrocytes. A similar sequence of events occurred in the retina (Bignami and Dahl, 1979). Müller radial glia are not normally demonstrated by immunofluorescence except for their terminal expansions forming the glia limitans on the inner surface of the retina. Following retinal ischemia produced by squeezing the optic nerve behind the eye, Müller fibers became intensely immunofluorescent (Fig. 5).

In the immature brain, the changes were somewhat different. Following a stab wound to the newborn rat brain, the first cells to react were radial glia (Section V,C). One month after stabbing the reaction had become

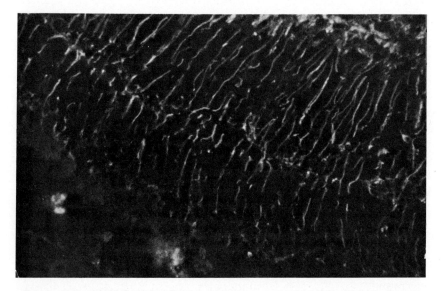

Fig. 5. Approximately 48 hr after injury (optic nerve crush) Müller fibers in the retina can be demonstrated by immunofluorescence with GFA protein antisera. (The fibers stay positive for indefinite periods.) × 200.

more severe. However, radial glia were no longer present, their place having been taken by astrocytes of mature appearance.

The controversial issue of cell division versus simple hypertrophy in astrocytes reacting to traumatic injury has been apparently solved by combining the peroxidase–antiperoxidase procedure with tritiated thymidine autoradiography (Latovit *et al.,* 1979). Following needle stabbing of the rat cerebral cortex, [³H]thymidine autoradiographic products were located in the nuclei of GFA-positive cells, thus indicating that reactive astrocytes proliferate. However, cell division does not necessarily occur in astrocytes switching to GFA protein production as a response to injury. Treatment of 6-day-old rats with hydroxyurea, an inhibitor of DNA synthesis, enriched the cerebellum in GFA-positive astrocytes on day 8. This observation allowed the development of an elegant technique for the isolation of a cell fraction enriched in viable GFA-positive astrocytes from this tissue (Cohen *et al.,* 1979). If reaction of astrocytes occurs quite early after traumatic, ischemic, and toxic injury, the same may not be true for other types of lesion. We have been following, for over 1 month, a series of rats undergoing Wallerian degeneration of the optic nerve after enucleation. Except for condensation of the glial framework probably due to the breakdown of the nerve fibers, no changes were found compared with the nonoperated site.

The astroglial reaction to injury (spinal cord transsection, stab wound) has been also investigated by immunofluorescence with GFA protein antisera in the chicken and the goldfish. Compared with the rat this reaction was extremely limited (Bignami *et al.,* 1974; Bignami and Dahl, 1976).

VII. Astroglial Marker *in Vitro*

Identification of astroglia *in vitro* has been greatly facilitated by the use of immunohistological staining for GFA protein (Raff *et al.,* 1979). *In vitro* synthesis of GFA protein has been demonstrated in primary cultures of human, rat, and mouse brain (Section VII,A), of human and chemically induced glioma (Sipe *et al.,* 1974; Vraa-Jensen *et al.,* 1976), and of a mouse experimental testicular teratoma undergoing neuroepithelial differentiation (VandenBerg *et al.,* 1975, 1976), as well as in reaggregating cultures of dissociated mouse brain cells (Section VII,B) and in glial cell lines (Section VII,C). Selective staining of astrocytes *in vitro* was first demonstrated in cryostat sections of human fetal brain explants (gestational age 12–20 weeks) which had been maintained between 8 and 39 days in culture (Antanitus *et al.,* 1975). Two types of GFA-positive cells were recognized in the outgrowth zone of the explant: broad, sheetlike epithelioid cells and cells giving rise to long processes with frequent branchings. Intermediate forms were common.

300   AMICO BIGNAMI ET AL.

A. *Dissociated Brain Cells in Primary Monolayer Culture*

   Primary monolayer cultures of dissociated brain cells have been extensively used for neurobiological studies (for a review, see Sensenbrenner,
1978). Dissociated cells from human fetus and from late fetal and newborn
rodent brain develop into a confluent monolayer of GFA-positive cells
(Fig. 6) after 10–15 days in culture(Maunoury *et al.,* 1976; Bock *et al.,*
1977a; Kozak *et al.,* 1978a,b; Manthorpe *et al.,* 1979). The concentration
of GFA protein is markedly increased in these cultures, not only compared
with newborn brain but also compared with adult brain (Bock *et al.,* 1975).
The morphology of these cells is essentially identical to that reported by
Antanitus *et al.* (1975) for the outgrowth zone of human fetal brain explants, namely, flat cells with delicate, reticulated cytoplasmic fluorescence
and cells with brightly stained processes. Flat GFA-positive cells transform
into astrocytes of typical morphology (i.e., process-bearing cells) in
response to the combined treatment of serum withdrawal and the addition
of dibutyryl cyclic AMP (Manthorpe *et al.,* 1979). These results indicate
that astrocytes have the capacity to express different morphologies in
culture and that process formation may not be used as a parameter to
assess biochemical differentiation.

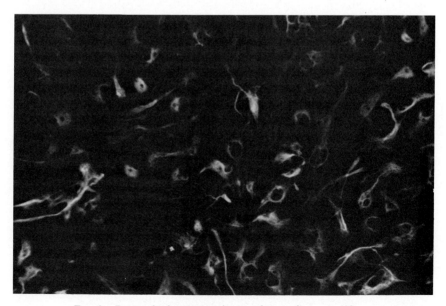

FIG. 6.   Rat cerebral astrocytes in monolayer culture. × 400.

B. Reaggregating Culture of Dissociated Brain Cells

In vitro aggregation of dissociated immature CNS cells has provided another system for the study of CNS differentiation (for a review, see Garber, 1978). GFA-positive cells first appear after 24 hr *in vitro* in reaggregates of immature mouse brain cells, and after 1 week they form a delicate framework within the aggregate and a continuous membrane on the surface (Kozak *et al.,* 1977). Whereas in reaggregates of fetal cerebral hemispheres (17 days gestation age) GFA-positive glia remained essentially unchanged for the whole period of culture (up to 55 days), a second process occurred in reaggregates of neonatal mouse cerebellum after about 10 days in culture, namely, the progressive degeneration of neurons and their gradual replacement with GFA-positive neuroglia (Kozak *et al.,* 1977, 1978a). This process could be prevented by the addition of poly-L-lysine to the medium (Kozak *et al.,* 1978b). Since immature granule neurons were the main population included in the cerebellar aggregates, these changes were interpreted as a glial reaction to granule cell degeneration, probably because granule cells were deprived of their main target for synaptic contacts since Purkinje cells were not included in the aggregates. Telencephalon aggregates exceeding 4 mm in diameter provided a convincing demonstration that glia may react *in vitro* to tissue damage, as in the intact organism (Kozak *et al.,* 1978a). The center of these large aggregates underwent necrosis, probably because of nutritional deficiencies, and glia reacted with the formation of a dense GFA-positive scar (Fig. 7).

C. Glial Cell Lines

C6 is by far the most extensively studied glial cell line (reviewed by Pfeiffer *et al.,* 1978) and thus appears to be most attractive for studies on GFA protein expression and its relation to other specific aspects of glial function. Only a few C6 cells accumulate GFA protein in monolayer and suspension culture. The concentration of the protein measured by radioimmunoassay is thus low under these conditions. The number of GFA-positive cells and the concentration of the protein are markedly increased when C6 is maintained on sponge foam matrices (organ culture). Although the concentration of GFA protein is increased in monolayer culture during the stationary stage of cell growth, it never reaches the level observed in organ culture (Bissell *et al.,* 1974, 1975; Liao *et al.,* 1978). A uniform population of GFA-positive C6 cells may also be obtained by reaggregating the cells in suspension culture without the use of supporting matrices (Bignami *et al.,* 1979). The high yield of the method may facilitate the

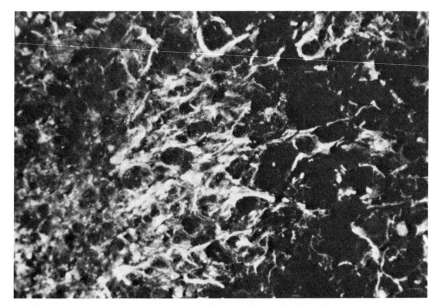

FIG. 7. Intense glial reaction surrounding a central area of necrosis in a large aggregate of dissociated fetal brain cells. Note the nonspecific fluorescein precipitates in the necrotic area (on the left). × 400.

biochemical analysis of the C6 cell line in conditions favoring the production of GFA protein.

Only two other GFA-positive cell lines have been so far reported in the literature (Bignami and Stoolmiller, 1979), G26-20 and G26-24 isolated by Sundarraj et al. (1975) from a methylcholanthrene-induced mouse glioma. As for C6 glioma, only a limited number of cells were immunofluorescent in confluent monolayers. The interest of the observation resides in the fact that both cell lines are capable of synthesizing myelin-related glycosphingolipids characteristic of oligodendroglia (Dawson et al., 1977), thus suggesting the existence of a stem cell capable of multipotential differentiation in vitro.

VIII. Diagnosis of Brain Tumors

The application of GFA protein to diagnostic and biological studies of brain tumors is fairly recent (Dittman et al., 1977; Duffy et al., 1977; Deck et al., 1978; Delpech et al., 1978b; Eng and Rubinstein, 1978; Jacque *et al.,* 1978; Manoury *et al.,* 1978; van der Meulen *et al.,* 1978; Velasco *et al.,* 1980; Vidard *et al.,* 1978; Conley, 1979; Palfreyman *et al.,* 1979). GFA protein can be localized by the peroxidase–antiperoxidase technique developed by Sternberg *et al.* (1970)[1] in formalin-fixed and paraffin-embedded tissue, an obvious advantage when dealing with human material. Although the technique has been used in pathological laboratories for only 2–3 yr, one conclusion stands on firm grounds: GFA protein is the most specific marker for gliomas. Regardless of the bizarre appearance of the cells, their location, namely, metastasis in the meninges and in distant organs, the results with neurohistological stains such as phosphotungstic acid–hematoxylin, and their electron microscopic appearance, the finding of reaction products within tumor cells provides the definitive evidence that the tumor is glial in origin. The opposite is not necessarily true, and negative results should be interpreted with caution, since poorly differentiated (anaplastic) glioma cells may not produce the protein, and in effect most studies appear to indicate an inverse relation between the concentration of GFA protein or number of GFA-positive neoplastic cells and the malignancy of the tumor. In glioblastoma multiforme, the most malignant glioma, the small, round cells show very little or no staining, while the large, bizarre, multinucleated cells are prominently stained (Velasco *et al.,* 1980).

A caveat for the inexperienced pathologist relying on the presence of a brown reaction product to make a diagnosis of glioma is that reactive astrocytes are intensely positive (Section VI). Reactive astrocytes are found at the periphery of CNS tumors, both primary and metastatic (Fig. 8), but may also be present within the tumor.

Another diagnostic problem in neuropathology is the distinction between reactive gliosis and well-differentiated astrocytomas. According to Velasco *et al.* (1978) differentiation between reactive and neoplastic astrocytes is considerably easier with immunostaining as compared to traditional methods. GFA-positive tumor cells lack the delicately branched processes of reactive astrocytes. In addition, they do not form perivascular glial membranes (van der Meulen *et al.,* 1978).

As expected, in view of its localization in a normal CNS, GFA protein is found in tumors of astrocytic lineage, but there appear to be a few exceptions. The presence of GFA protein in ependymal cells (tanycytes) has been discussed (Section IV), and it is thus not surprising that GFA-positive cells have been found in ependymomas, both *in vitro* and *in vivo* (Vraa-Jensen *et al.,* 1976; Deck *et al.,* 1978; Velasco *et al.,* 1980). More surprising is the report of typical (signet-ring) oligodendroglioma cells in grade III (malignant) tumors containing the reaction product (van der Meulen *et al.,* 1978).

[1] This method is described in some detail in the chapter by E. L. McGeer and P. L. McGeer, this volume.

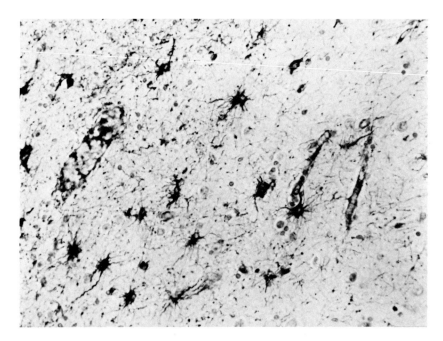

FIG. 8. Reactive astrocytes in the cerebral human cortex in proximity to a glioma. Note the delicate processes and the perivascular glial membranes. Anti-GFA treated, formalin-fixed, paraffin-imbedded section. PAP staining. × 250. (Observation by Velasco.)

Whether this indicates that the expression of differentiated functions in tumor cells is not so strictly regulated as in normal cells, or the presence in the tumor of cells of different lineage but morphologically identical, remains an open question. It should be noted, however, that in another series oligodendrogliomas were consistently negative, whereas in mixed astrocytoma–oligodendroglioma tumors a clear distinction between both components was always evident (Velasco *et al.*, 1980).

Acknowledgment

This work was supported by USPHS grant NS 13034 and by the Veterans Administration.

References

Antanitus, D. S., Choi, B. H., and Lapham, L. W. (1975). Immunofluorescence staining of astrocytes *in vitro* using antiserum to glial fibrillary acidic protein. *Brain Res.* **89,** 363–367.

Antanitus, D. S., Choi, B. H., and Lapham, L. W. (1976). The demonstration of glial fibril-
    lary acidic protein in the cerebrum of the human fetus by indirect immunofluorescence.
    Brain Res. **103**, 613–616.
Benitz, W. E., Dahl, D., Williams, L. W., and Bignami, A. (1976). The protein composition
    of glial and nerve fibers. *FEBS Lett.* **66**, 285–289.
Bennett, G. S., Fellini, S. A., Croop, J. M., Otto, J. J., Bryan, J., and Holtzer, H. (1978).
    Differences among 100-Å filament subunits from different cell types. *Proc. Natl. Acad.
    Sci. U.S.A.* **75**, 4364–4368.
Bignami, A., and Dahl, D. (1973). Differentiation of astrocytes in the cerebellar cortex and
    the pyramidal tracts of the newborn rat. An immunofluorescence study with antibodies to
    a protein specific to astrocytes. *Brain Res.* **49**, 393–402.
Bignami, A., and Dahl, D. (1974a). Astrocyte-specific protein and neuroglial differentiation.
    An immunofluorescence study with antibodies to the glial fibrillary acidic protein. *J.
    Comp. Neurol.* **153**, 27–38.
Bignami, A., and Dahl, D. (1974b). The development of Bergmann glia in mutant mice with
    cerebellar malformations: Reeler, staggerer and weaver. Immunofluorescence study with
    antibodies to the glial fibrillary acidic protein. *J. Comp. Neurol.* **155**, 219–230.
Bignami, A., and Dahl, D. (1974c). Glial fibrillary acidic protein in mutant mice with defi-
    ciency of myelination: Quaking and jimpy. *Acta Neuropathol.* **28**, 269–272.
Bignami, A., and Dahl, D. (1974d). Astrocyte-specific protein and radial glia in the cere-
    bral cortex of newborn rat. *Nature (London)* **252**, 55–56.
Bignami, A., and Dahl, D. (1975). Astroglial protein in the developing spinal cord of the
    chick embryo. *Dev. Biol.* **44**, 204–209.
Bignami, A., and Dahl, D. (1976). The astroglial response to stabbing. Immunofluorescence
    studies with antibodies to astrocyte-specific protein (GFA) in mammalian and submam-
    malian vertebrates. *Neuropathol. Appl. Neurobiol.* **2**, 99–111.
Bignami, A., and Dahl, D. (1977). Specificity of the glial fibrillary acidic protein for as-
    troglia. *J. Histochem. Cytochem.* **25**, 466–469.
Bignami, A., and Dahl, D. (1979). The radial glia of Müller in the rat retina and their response
    to injury. An immunofluorescence study with antibodies to the glial fibrillary acidic
    (GFA) protein. *Exp. Eye Res.* **28**, 63–69.
Bignami, A., and Stoolmiller, A. C. (1979). Astroglia-specific protein (GFA) in clonal cell
    lines derived from the G26 mouse glioma. *Brain Res.* **163**, 353–357.
Bignami, A., Eng. L. F., Dahl, D., and Uyeda, C. T. (1972). Localization of the glial fibril-
    lary acidic protein in astrocytes by immunofluorescence. *Brain Res.* **43**, 429–435.
Bignami, A., Forno, L., and Dahl, D. (1974). The neuroglial response to injury following
    spinal cord transection in the goldfish. *Exp. Neurol.* **44**, 60–70.
Bignami, A., Swanson, J., and Dahl, D. (1979). GFA expression in aggregating cultures of
    rat C6 glioma. *Experientia* **35**, 1170–1171.
Bissell, M.G., Rubinstein, L. J., Bignami, A., and Herman, M. M. (1974). Characteristics of
    the rat C-6 glioma maintained in organ culture systems. Production of glial fibrillary
    acidic protein in the absence of gliofibrillogenesis. *Brain Res.* **82**, 77–89.
Bissell, M. G., Eng, L. F., Herman, M. M., Bensch, K. G., and Miles, L. E. M. (1975). Quan-
    titative increase of neuroglia-specific GFA protein in rat C-6 glioma cells *in vitro*. *Nature
    (London)* **255**, 633–634.
Blose, S. H., Shelanski, M. L., and Chacko, S. (1977). Localization of bovine brain filament
    antibody on intermediate (100 Å) filaments in guinea pig vascular endothelial cells and
    chick cardiac muscle cells. *Proc. Natl. Acad. Sci. U.S.A.* **74**, 662–665.
Bock, E. (1978). Nervous system specific proteins. *J. Neurochem.* **30**, 7–14.
Bock, E., Jørgensen, O. S., Dittmann, L., and Eng, L. F. (1975). Determination of brain-

---

specific antigens in short term cultivated rat astroglial cells and in rat synaptosomes. *J. Neurochem.* **25**, 867–870.

Bock, E., Møller, M., Nissen, C., and Sensenbrenner, M. (1977a). Glial fibrillary acidic protein in primary astroglial cell cultures derived from newborn rat brain. *FEBS Lett.* **83**, 207–211.

Bock, E., Rasmussen, S., Møller, M., and Ebbesen, P. (1977b). Demonstration of a protein immunochemically related to glial fibrillary acidic protein in human fibroblasts in culture. *FEBS Lett.* **83**, 212–216.

Braak, E., Crenckhahn, D., Unsicker, K., Groschel-Stewart, U., and Dahl, D. (1978). Distribution of myosin and the glial fibrillary acidic protein (GFA protein) in rat spinal cord and in the human frontal cortex as revealed by immunofluorescence microscopy. *Cell Tissue Res.* **191**, 493–500.

Bryan, J. (1974). Biochemical properties of microtubules. *Fed. Proc., Fed. Am. Soc. Exp. Biol.* **33**, 152–157.

Choi, B. H., and Lapham, L. W. (1978). Radial glia in the human fetal cerebrum: A combined Golgi, immunofluorescent and electron microscopic study. *Brain Res.* **148**, 295–311.

Cohen, J., Woodhams, P. L., and Balazs, R. (1979). Preparation of viable astrocytes from the developing cerebellum. *Brain Res.* **161**, 503–514.

Conley, F. K. (1979). The immunocytochemical localization of GFA protein in experimental murine CNS tumors. *Acta Neuropath.* **45**, 9–16.

Dahl, D. (1976). Glial fibrillary acidic protein from bovine and rat brain. Degradation in tissues and homogenates. *Biochem. Biophys. Acta* **420**, 142–154.

Dahl, D., and Bignami, A. (1973a). Glial fibrillary acidic protein from normal human brain. Purification and properties. *Brain Res.* **57**, 343–360.

Dahl, D., and Bignami, A. (1973b). Immunochemical and immunofluorescence studies of the glial fibrillary acidic protein in vertebrates. *Brain Res.* **61**, 279–293.

Dahl, D., and Bignami, A. (1974). Heterogeneity of the glial fibrillary acidic protein in gliosed human brains. *Neurol. Sci.* **23**, 551–563.

Dahl, D., and Bignami, A. (1975). Glial fibrillary acidic protein from normal and gliosed human brain. Demonstration of multiple related polypeptides. *Biochim. Biophys. Acta* **386**, 41–51.

Dahl, D., and Bignami, A. (1976). Immunogenic properties of the glial fibrillary acidic protein. *Brain Res.* **116**, 150–157.

Dahl, D., and Bignami, A. (1977a). Effect of sodium dodecyl sulfate on the immunogenic properties of the glial fibrillary acidic protein. *J. Immunol. Methods* **17**, 201–209.

Dahl, D., and Bignami, A. (1977b). Preparation of antisera to neurofilament protein from chicken brain and human sciatic nerve. *J. Comp. Neurol.* **176**, 645–658.

Dahl, D., and Bignami, A. (1978a). Immunochemical cross-reactivity of normal neurofibrils and aluminum-induced neurofibrillary tangles. Immunofluorescence study with antineurofilament serum. *Exp. Neurol.* **58**, 74–80.

Dahl, D., and Bignami, A. (1978b). Neurofilament protein in clonal lines of mouse neuroblastoma. *Dev. Neurosci.* **1**, 142–152.

Dahl, D., and Bignami, A. (1979). Astroglial and axonal proteins in isolated brain filaments. I. Isolation of the glial fibrillary acidic protein and of an immunologically active cyanogen bromide peptide from brain filament preparations of bovine white matter. *Biochim. Biophys. Acta* **578**, 305–316.

Das, G. D. (1976). Differentiation of Bergmann glia cells in the cerebellum: A Golgi study. *Brain Res.* **110**, 199–213.

Das, G. D., Lammert, G. L., and McAllister, J. P. (1974). Contact guidance and migratory cells in the developing cerebellum. *Brain Res.* **69**, 13–29.

Davison, P. F., and Hong, B.-S. (1977). Structural homologies in mammalian neurofilament proteins. *Brain Res.* **134,** 287–295.

Davison, P. F., Hong, B.-S., and Cooke, P. (1977). Classes of distinguishable 10 nm cytoplasmic filaments. *Exp. Cell Res.* **109,** 471–474.

Dawson, G., Sundarraj, N., and Pfeiffer, S. (1977). Synthesis of myelin glycosphingolipids [galactosylceramide and galactosyl (3-O-sulfate) ceramide (sulfatide)] by cloned cell lines derived from mouse neurotumors. *J. Biol. Chem.* **252,** 2777–2779.

Deck, J. H. N., Eng, L. F., Bigbee, J., and Woodcock, S. M. (1978). The role of glial fibrillary acidic protein in the diagnosis of central nervous system tumors. *Acta Neuropathol.* **42,** 183–190.

Del Cerro, M. P., and Knigge, C. M. (1979). The system of gap junctions and glial-neuronal contacts in the arcuate nucleus of the rat hypothalamus. *J. Comp. Neurol.* (in press).

Del Cerro, M. P., and Swarz, J. R. (1976). Prenatal development of Bergmann glial fibres in rodent cerebellum. *J. Neurocytol.* **5,** 669–679.

Delpech, A., Delpech, B., Girard, N., and Vidard, M-N. (1978a). Localization immunohistologique des 3 antigènes associés au tissu nerveux (GFA, ANS₂, brain glycoprotein). *Biol. Cell.* **32,** 207–214.

Delpech, B., Delpech, A., Vidard, M. N., Girard, N., Tayot, J., Clement, J. C., and Creissard, P. (1978b). Glial fibrillary acidic protein in tumours of the nervous system. *Br. J. Cancer* **37,** 33–40.

De Vries, G. H., Eng, L. F., Lewis, D. L., and Hadfield, M. G. (1976). The protein composition of bovine myelin-free axons. *Biochim. Biophys. Acta* **439,** 133–145.

Dittman, L., Axelsen, N. H., Norgaard-Pedersen, B., and Bock, E. (1977). Antigens in human glioblastomas and meningiomas. Search for tumour and onco-fetal antigens. Estimation of S-100 and GFA protein. *Br. J. Cancer* **35,** 135–141.

Duffy, P. E., Graf, L., and Rapport, M. M. (1977). Identification of glial fibrillary acidic protein by the immunoperoxidase method in human brain tumors. *J. Neuropathol. Exp. Neurol.* **36,** 645–652.

Eng, L. F., and Bigbee, J. W. (1978). Immunohistochemistry of nervous system specific antigens. *Adv. Neurochem.* **3,** 43–98.

Eng, L. F., and Rubinstein, L. J. (1978). Contribution of immunohistochemistry to diagnostic problems of human cerebral tumors. *J. Histochem. Cytochem.* **26,** 513–522.

Eng, L. F., Vanderhaeghen, J. J., Bignami, A., and Gerstl, B. (1971). An acidic protein isolated from fibrous astrocytes. *Brain Res.* **28,** 351–354.

Franke, W. W., Schmid, E., Osborn, M., and Weber, K. (1978). Different intermediate-sized filaments distinguished by immunofluorescence microscopy. *Proc. Natl. Acad. Sci. U.S.A.* **75,** 5034–5038.

Garber, B. B. (1978). Cell aggregation and recognition in the self assembly of brain tissues. *In* "Cell, Tissue and Organ Cultures in Neurobiology" (S. Fedoroff and L. Hertz, eds.), pp. 515–537. Academic Press, New York.

Goldman, J. E., Schaumburg, H. H., and Norton, W. T. (1978). Isolation and characterization of glial filaments from human brain. *J. Cell Biol.* **78,** 426–440.

Huston, J. S., and Bignami, A. (1977). Structural properties of the glial fibrillary acidic protein. Evidence for intermolecular disulfide bonds. *Biochim. Biophys. Acta* **493,** 97–103.

Jacobson, M. (1978). "Developmental Neurobiology." Plenum, New York.

Jacque, C. M., Jørgensen, O. S., and Bock, E. (1974). Quantitative studies of the brain specific antigens S-100, GFA, 14-3-2, D1, D2, D3 and C1 in quaking mouse. *FEBS Lett.* **49,** 264–266.

Jacque, C. M., Baumann, N. A., and Bock, E. (1976a). Quantitative studies of the brain specific antigens GFA, 14-3-2, synaptin C1, D1, D2, D3 and D5 in jimpy mouse. *Neurosci. Lett.* **3,** 41–44.

Jacque, C. M., Jørgensen, O. S., Baumann, N. A., and Bock, E. (1976b). Brain-specific antigens in the quaking mouse during ontogeny. *J. Neurochem.* **27**, 905-909.

Jacque, C. M., Vinner, C., Kujas, M., Raoul, M., Racadot, J., and Baumann, N. A. (1978). Determination of glial fibrillary acidic protein (GFAP) in human brain tumors. *J. Neurol. Sci.* **35**, 147-155.

Kozak, L. P., Eppig, J. J., Dahl, D., and Bignami, A. (1977). Ultrastructural and immunohistological characterization of a cell culture model for the study of neuronal-glial interactions. *Dev. Biol.* **59**, 206-227.

Kozak, L. P., Dahl, D., and Bignami, A. (1978a). Glial fibrillary acidic protein in reaggregating and monolayer cultures of fetal mouse cerebral hemispheres. *Brain Res.* **150**, 631-637.

Kozak, L. P., Eppig, J. J., Dahl, D., and Bignami, A. (1978b). Enhanced neuronal expression in reaggregating cells of mouse cerebellum cultured in the presence of poly-*L*-lysine. *Dev. Biol.* **64**, 252-264.

Latovit, N., Nilaver, G., Zimmerman, E. A., Johnson, W. G., Silverman, A.-J., Defendini, R., and Cote, L. (1979). Fibrillary astrocytes proliferate in response to brain injury. A study combining immunoperoxidase technique for glial fibrillary acidic protein and radioautography of tritiated thymidine. *Develop. Biol.* **72**, 381-385.

Lazarides, E., and Balzer, D. R., Jr. (1978). Specificity of desmin to avian and mammalian muscle cells. *Cell* **14**, 429-438.

Lazarides, E., and Hubbard, B. D. (1976). Immunological characterization of the subunit of the 100 Å filaments from muscle cells. *Proc. Natl. Acad. Sci. U.S.A.* **73**, 4344-4348.

Liao, C. L., Eng, L. F., Herman, M. M., and Bensch, K. G.(1978). Glial fibrillary acidic protein-solubility characteristics, relation to cell growth phases and cellular localization in rat C-6 glioma cells: An immunoradiometric and immunohistologic study. *J. Neurochem.* **30**, 1181-1186.

Liem, R. K. H., and Shelanski, M. L. (1978). Identity of the major protein in native, glial fibrillary acidic protein preparation with tubulin. *Brain Res.* **145**, 196-201.

Löwenthal, A., Noppe, M., Gheuens, J., and Karcher, D. (1978). a-Albumin (glial fibrillary acidic protein) in normal and pathological human brain and cerebrospinal fluid. *J. Neurol.* **219**, 87-92.

Ludwin, S. K., Kosek, J. C., and Eng, L. F. (1976). The topographical distribution of S-100 and GFA proteins in the adult rat brain: An immunohistochemical study using horseradish peroxidase-labeled antibodies. *J. Comp. Neurol.* **165**, 197-208.

Maunoury, R., Delpech, A., Delpech, B., Vidard, M. N., and Vedrenne, C. (1976). Presence of neurospecific antigen NSA 1 in fetal human astrocytes in long-term cultures. *Brain Res.* **112**, 383-387.

Maunoury, R., Courdi, A., Vedrenne, C., and Constans, J. P. (1978). Immunocytochemical localization of the GFAP in heterotransplanted human gliomas. *Neurochir.* **24**, 221-226.

Manthorpe, M., Adler, R., and Varon, S. (1979). Development, reactivity and GFA immunofluorescence of astroglia-containing monolayer cultures from rat cerebrum. *J. Neurocytol.* **8**, 605-622.

Møller, M., Ingild, A., and Bock, E. (1978). Immunohistochemical demonstration of S-100 protein and GFA protein in interstitial cells of rat pineal gland. *Brain Res.* **140**, 1-13.

Mori T., and Morimoto, K. (1975). Studies on the identity of astroprotein (Mori) and glial fibrillary acidic protein (Eng). *Igaku-no-ayumi* **92**, 16-17.

Paetau, A., Virtanen, I., Stenman, S., Kurki, P., Linder, E., Vaheri, A., Westermark, B., Dahl, D., and Haltia, M. (1979). GFA and intermediate filament protein in glioma cells. *Acta Neuropath.* **47**, 71-74.

Palfreyman, J. W., Thomas, D. G. T., Ratcliffe, J. G., and Graham, D. I. (1979). Glial fibrillary acidic protein (GFAP). Purification from human fibrillary astrocytoma, development and validation of a radioimmunoassay for GFAP-like immunoreactivity. *J. Neurol. Sci.* **41**, 101–113.

Pfeiffer, S. E., Betschart, B., Cook, J., Mancini, P., and Morris, R. (1978). Glial cell lines. *In* "Cell, Tissue and Organ Cultures in Neurobiology" (S. Fedoroff and L. Hertz, eds.), pp. 287–346. Academic Press, New York.

Privat, A.(1975). Postnatal gliogenesis in the mammalian brain. *Int. Rev. Cytol.* **40**, 281–323.

Raff, M. C., Fields, K. L., Hakomori, S.-I., Mirsky, R., Preuss, R. M., and Winter, J. (1979). Cell-type-specific markers for distinguishing and studying neurons and the major classes of glial cells in culture. *Brain Res.* **174**, 283–308.

Rakic, P. (1971). Guidance of neurons migrating to the fetal monkey neocortex. *Brain Res.* **33**, 471–476.

Rakic, P., and Sidman, R. L. (1973). Weaver mutant mouse cerebellum: Defective neuronal migration secondary to abnormality of Bergmann glia. *Proc. Natl. Acad. Sci. U.S.A.* **70**, 240–244.

Roback, H. N., and Scherer, H. J. (1935). Über die feinere Morphologie des frühkindlichen Hirnes unter besonderer Berücksichtigung der Gliaentwicklung. *Virchows Arch. Pathol. Anat. Physiol.* **294**, 365–413.

Royce, J. G. (1971). Morphology of neuroglia in the hypothalamus of the opossum *(Didelphis virginiana),* armadillo *(Dasypus novemcinctus mexicanus)* and cat *(Felis domestica). J. Morphol.* **134**, 141–180.

Rueger, D. C., Dahl, D., and Bignami, A. (1978a). Comparison of bovine glial fibrillary acidic protein with tubulin. *Brain Res.* **153**, 188–193.

Rueger, D. C., Dahl, D., and Bignami, A. (1978b). Purification of a brain specific astroglial protein by immunoaffinity chromatography. *Anal. Biochem.* **89**, 360–371.

Rueger, D. C., Huston, J. S., Dahl, D., and Bignami, A. (1979). Formation of 100 Å filaments from purified glial fibrillary acidic protein *in vitro. J. Mol. Biol.* **135**, 53–68.

Schachner, M., Ruberg, M. Z., and Carnow, T. B. (1976). Histological localization of nervous system antigens in the cerebellum by immunoperoxidase labeling. *Brain Res. Bull.* **1**, 367–377.

Schachner, M., Hedley-White, E. T., Hsu, D. W., Schoonmaker, G., and Bignami, A. (1977). Ultrastructural localization of glial fibrillary acidic protein in mouse cerebellum by immunoperoxidase labeling. *J. Cell Biol.* **75**, 67–73.

Sensenbrenner, M. (1978). Dissociated brain cells in primary cultures. *In* "Cell Tissue and Organ Cultures in Neurobiology" (S. Fedoroff and L. Hertz, eds.), pp. 191–214. Academic Press, New York.

Sipe, J. C., Rubinstein, L. J., Herman, M. M., and Bignami, A. (1974). Ethyl-nitrosourea-induced astrocytomas. Morphological observations on rat tumors maintained in tissue and organ culture systems. *Lab. Invest.* **31**, 571–579.

Sotelo, C., and Changeux, J.-P. (1974). Bergmann fibers and granular cell migration in the cerebellum of homozygous weaver mutant mouse. *Brain Res.* **77**, 484–491.

Sternberger, L. A., Hardy, P. H., Jr., Cuculis, F. F., and Meyer, H.G. (1970). The unlabeled antibody enzyme method of immunohistochemistry: Preparation and properties of soluble antigen-antibody complex (horse-radish peroxidase-antiperoxidase) and its use in identification of spirochetes. *J. Histochem. Cytochem.* **18**, 315–333.

Sternberger, N. H., Itoyama, Y., Kies, M. W., and Webster, H.D.(1978). Myelin basic protein demonstrated immunocytochemically in oligodendroglia prior to myelin sheath formation. *Proc. Natl. Acad. Sci. U.S.A.* **72**, 1927–1931.

Sun, T.-T., and Green, H. (1978). Keratin filaments of cultured human epidermal cells. Formation of intermolecular disulfide bonds during terminal differentiation. *J. Biol. Chem.* **253**, 2053–2060.

Sundarraj, N., Schachner, M., and Pfeiffer, S. E. (1975). Biochemically differentiated mouse glial lines carrying a nervous system specific cell surface antigen (NS-1). *Proc. Natl. Acad. Sci. U.S.A.* **72**, 1927–1931.

Uyeda, C. T., Eng, L. F., and Bignami, A. (1972). Immunological study of the glial fibrillary acidic protein. *Brain Res.* **37**, 81–89.

VandenBerg, S. R., Herman, M. M., Ludwin, S. K., and Bignami, A.(1975). An experimental mouse testicular teratoma as a model for neuroepithelial neoplasia and differentiation. *Am. J. Pathol.* **79**, 147–168.

VandenBerg, S. R., Ludwin, S. K., Herman, M. M., and Bignami, A. (1976). *In vitro* astrocytic differentiation from embryoid bodies of an experimental mouse testicular teratoma. *Am. J. Pathol.* **83**, 197–206.

van der Meulen, J. D. M., Houthoff, H. J., and Ebels, E. J.(1978). Glial fibrillary acidic protein in human gliomas. *Neuropathol. Appl. Neurobiol.* **4**, 177–190.

Velasco, M. E., Dahl, D., Rossmann, U., and Gambetti, P. L. (1980). Immunohistochemical localization of glial fibrillary acidic protein in human glial neoplasms. *Cancer* **45**, 484–494.

Vidard, M.-N., Girard, N., Chauzy, C., Delpech, B., Delpech, A., Maunoury, R., Laumonier, R., and Latarjet, M. R. (1978). Disparition de la protéine gliofibrillaire (GFA) au cours de la culture de cellules de glioblastomes. *C. R. Acad. Sc. Paris* **286**, 1837–1840.

Vraa-Jensen, J., Herman, M. M., Rubinstein, L. J., and Bignami, A. (1976). *In vitro* characteristics of a fourth ventricle ependymoma maintained in organ culture systems: Light and electron microscopy observations. *Neuropathol. Appl. Neurobiol.* **2**, 349–364.

Weigert, C. (1895). "Beiträge zur Kenntnis der Normalen Menschlichen Neuroglia." Moritz Diesterweg, Frankfurt a.M.

Yen, S.-H., Dahl, D., Schachner, M., and Shelanski, M. L. (1976). Biochemistry of the filaments of brain. *Proc. Natl. Acad. Sci. U.S.A.* **73**, 529–533.

Section 3

METHODOLOGIES

IN VITRO BEHAVIOR OF ISOLATED OLIGODENDROCYTES

SARA SZUCHET AND KARI STEFANSSON

Department of Neurology
Pritzker School of Medicine
University of Chicago
Chicago, Illinois

Those who have handled sciences have either been men of experiment or men of dogmas. The men of experiment are like the ant; they only collect and use; the reasoners resemble spiders, who make cob webs out of their own substance. But the bee takes a middle course; it gathers its material from the flowers of the garden and of the field, but transforms and digests it by a power of its own . . .

FRANCIS BACON
"Novum Organum"
Book I, XCV

I. Introduction

Oligodendrocytes of the central nervous system (CNS), like their cousins the Schwann cells of the peripheral nervous system (PNS), have as their specific established biological function the formation and maintenance of myelin. How this task is accomplished is incompletely understood and is a subject for challenging research. While much has been learned about the biochemistry and ultrastructure of the myelin membrane and of the macroscopic steps involved in myelin synthesis, the nature of the molecular and cellular interactions responsible for initiation of myelination and for myelin maintenance remain completely unknown. An orderly transition from ventricular cells, to glioblasts, to oligodendrocytes can be surmised on the basis of morphological and labeling studies *in vivo,* but the triggers for these transitions remain a total mystery.

The knowledge that under the rubric "oligodendrocytes" is harbored a family of differentiated cells with distinct morphological and ultrastructural characteristics, each prototype representing a stage in the development of the cell (Mori and Leblond, 1970; Imamoto and Leblond, 1977; Privat and Fulcrand, 1977), raises some fundamental questions. What is the relationship between the structure and function of these cells? Does each structural entity correspond to a specific functional role in the process from assembly to maturation of myelin? In other words, does the ripening of the mother cell involve a continuous functional adaptation to the demands and needs of its outgrowth: the myelin sheath? If so, is the "dark" oligodendrocyte—the truly mature cell—merely a caretaker? Is the synthesis of myelin the only duty of oligodendrocytes, or are they entrusted with other, as yet unknown, missions?

Several groups have begun to address these questions, and the central thesis of this article will be that purification of oligodendrocytes and *in vitro* study of their properties provide perhaps the most promising methodological approach to these problems.

The notion that myelinogenesis by oligodendrocytes may require input(s) from other cellular elements is gaining credence. This concept has evolved from experiments with explant cultures, which were highly suggestive of cooperative interaction among the various cellular components from within the explant (Bornstein and Murray, 1958; Allerand and Murray, 1968). More recently, this idea has been advanced a step further by implicating axons in the early stages of oligodendrocyte differentiation (Fulcrand and Privat, 1977; Bornstein, 1977; Diaz *et al.,* 1977).

Novel and interesting information is emerging from studies on Schwann cells in culture. These studies are further advanced than those on oligoden- drocytes (Bunge, 1968), and lessons learned from the examination of peripheral nerve myelination under controlled culture conditions offer tan- talizing hints as to what may be expected of oligodendrocytes as well. Wood and Bunge (1975; see also Wood, 1976) have shown that a mitogenic signal is provided to Schwann cells by axons and that the addition of ax- onal vesicles to cultures of Schwann cells leads to augmented thymidine in- corporation by the Schwann cells (Salzer *et al.,* 1977; Varon and Bunge, 1978). Similar results have also been obtained by McCarthy and Partlow (1976a,b). It appears that signals to Schwann cells differ depending on the nature of the axon with which contact is made. Weinberg and Spencer (1976) demonstrated that only certain types of axons were capable of in- ducing Schwann cells to envelop them with myelin. Recently Bunge *et al.* (1978) and Bunge and Bunge (1978) presented compelling evidence that myelination in the PNS required a concerted effort by Schwann cells, neurons, and fibroblasts. Demonstration of this fact required cultures of pure populations of each type of cell, which were then mixed under con- trolled conditions. Similar experiments with oligodendrocytes have not yet been performed but would obviously be of paramount interest.

The plasticity of the nervous system is receiving intensified attention as it becomes increasingly evident that individual glial cell classes are not im- mutable entities. Varon (1977) has put it neatly: "One needs to distinguish carefully between the intrinsic capabilities of a cell (differentiated program) and the extent to which they become expressed in a given cellular and humoral environment (extrinsic regulation)." As an example, Sundarraj *et al.* (1975) could not detect 2', 3-cAMP 3'-phosphodiesterase (CNP, EC 3.1.4.16) activity in a cloned glial cell line grown *in vitro,* but this activity appeared when the same line was grown as a tumor *in vivo.* Thus, proper- ties present *in vivo* may be lost *in vitro.* The reverse also holds. Synthesis of S-100 protein by the same clone was negligible *in vivo* but significant *in vitro.* Fedoroff (1978), studying cultures of dissociated cells obtained from embryos or newly born mice, has drawn attention to the influence of the immediate environment of a cell on cell differentiation and cell number.

Thus, *in vitro* studies of isolated oligodendrocytes may help to reveal new potentialities of these cells, including capabilities not expressed *in vivo.* Here the principle enunciated by Murray (1977) applies: " . . . anything a cell is seen to do in culture must be counted among its potentialities."

Investigation of isolated oligodendrocytes in culture may also prove im- portant in terms of diseases that selectively affect oligodendrocytes in humans. Multiple sclerosis is a relatively common disease of unknown etiology in which myelin is lost and oligodendrocytes die. Clearly, a better understanding of oligodendrocyte function obtained from *in vitro* study could profoundly impact on this presently incurable disease.

A review on culturing of isolated oligodendrocytes is possibly premature, since only recently have the problems involved in isolation of viable cells that survive in long-term culture been solved. It is like writing the biography of a newborn child. We therefore prefer to regard this essay as a forum for speculation rather than as a collection of established facts. The latter are scanty, indeed.

II. Isolation of Oligodendrocytes

Figure 1 portrays an oligodendrocyte and its multiple connections to myelinated axons. Inspection of this figure reveals some of the difficulties encountered when attempts are made to isolate *intact* and *viable* oligodendrocytes. During fractionation the processes connecting the cell to its myelin must be sheared from the cell body, and at the same time cell viability must be preserved. Several methods for isolation of oligodendrocytes have been published in recent years. It is beyond the scope of this chapter to catalog the merits and demerits of the various procedures, and we will limit ourselves to a description of the method developed in our laboratory (Szuchet *et al.,* 1978a, 1980b).

A. Cell Preparation

1. EQUIPMENT

   a. A refrigerated centrifuge with a rate controller and a swinging-bucket rotor: We use a Sorvall RC–5 refrigerated centrifuge equipped with a rate controller and an HS–4 rotor. The machine has been modified by the addition of a high-resolution slow-speed control which permits us to set and control any speed from 500 to 1500 rpm within better than 10 rpm.
   b. Laminar flow hood: Ours has been designed to include a well which when filled with ice and covered with a flat stainless steel plate provides a cold surface. This is not indispensable, since a cold surface can be improvised with a tray wrapped in aluminum foil and set on a bed of crushed ice.
   c. Orbit water bath shaker.

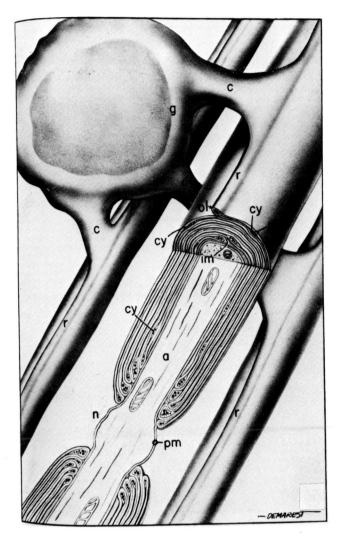

FIG. 1. Diagram showing an oligodendrocyte attached to three internodes of myelin. a, Axon; n, node; pm, plasma membrane; im, inner mesaxon; c, connection to oligodendroglial cell body (g); r, ridge formed by outer loop; ol, outer loop; cy, cytoplasm. Reproduced with kind permission from Bunge *et al.* (1961).

2. Solutions

Hanks' balanced salt solution (Hanks' BSS), without Ca^{2+}, Mg^{2+}, sodium bicarbonate, and phenol red, is used as solvent either at full (H) or at half strength ($\times/2$)H. The following solutions are made daily:

a. Hanks' BSS–2% (w/v) Ficoll (H–2%F)
b. 0.2% trypsin–0.04% EDTA in H–2%F
c. Trypsin inhibitor in H–2%F: The concentration of this solution is varied according to the manufacturer's specification of activity so as to inactivate all the trypsin used.
d. 0.9, 1.0, and 1.2 M sucrose in ($\times/2$)H

All solutions are adjusted to pH 6.00 ± 0.05, sterilized by filtration through 0.22-μm filters, and cooled to 4°C with the exception of the trypsin solution which is kept at room temperature.

3. Brains

Brains from 4- to 6-month-old lambs are obtained from a slaughterhouse immediately after death of the animals, immersed in chilled H–2%F, placed in a bucket of ice, and brought to the laboratory within 1 hr of death. They are then processed as described here.

4. Procedure

The entire operation is carried out in the laminar flow hood with the exception of the trypsinization and centrifugation steps. A diagrammatic representation of the technique is given in Fig. 2. The centra ovales and corpora callosa are dissected free of gray matter and placed in a preweighed beaker containing 80 ml of cold H–2%F. After all pieces have been cleaned, the net weight is determined. This usually amounts to 12–15 gm. The pieces are transferred onto a cold stainless steel plate, minced finely (< 1 mm), and put into a 500-ml Erlenmeyer flask. The trypsin solution (200 ml) is added, and the flask is positioned in a water bath at 37°C for 1.8 min/gm of white matter. The flask is cooled down for 2 min, and 200 ml of trypsin inhibitor is added, mixed, and centrifuged at 1000 rpm (192 g) for 4 min. The softened tissue is washed with 0.01% trypsin inhibitor and then with H–2%F, centrifuging at 192 g for 4 min after each washing. Final disruption of tissue is accomplished by screening. The base of a 50-ml plastic syringe is sawed off and replaced by a 7-cm² piece of 350-μm nylon screen secured with a 2.5-cm-wide Tygon sleeve and further fastened with a no. 20 stainless steel tubing clamp (Fig. 2). The tissue is suspended in 50 ml of 0.9 M sucrose in ($\times/2$)H, poured into the syringe, pushed (gently)

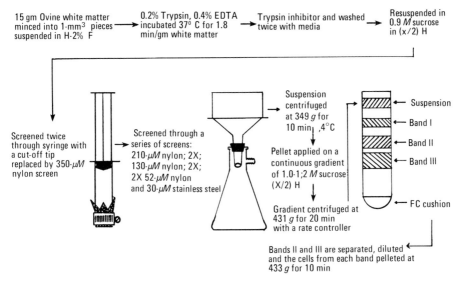

FIG. 2. Schematic representation of the procedure used for isolating oligodendrocytes. For details see text. FC, Fluorocarbon.

through the screen with the plunger, and collected in a beaker, care being taken to recover material adhering to the outer surface of the screen. The procedure is repeated using the same syringe but adding more solvent. A two-piece polypropylene funnel is mounted with a 210-μm screen, and the crude suspension is passed through it by spreading the material with the fingertips (surgical gloves are worn), squirting solvent (i.e., 0.9 M sucrose) with a wash bottle, and slightly tapping the funnel. No vacuum is used. Sequentially, the suspension is passed through double-thickness 130-μm nylon, double-thickness 52-μm nylon, and 30-μm stainless steel. A total of 300 ml of solvent/15 gm white matter is used during the screening. The crude suspension is centrifuged at 2100 rpm (850 g) for 10 min. During this step myelin floats to the top of the tube, and the cells pellet (P_1). Myelin and the suspension are aspirated off, leaving only P_1. Total removal of the suspension is critical, since failure to do so leads to streaming during gradient centrifugation.

P_1 is suspended in 2–3 ml of 0.9 M sucrose and applied on a linear sucrose gradient from 1.0 to 1.2 M. Gradients are usually made in 50-ml tubes containing 2 ml fluorocarbon (3M Company, Chicago, Ill.) plus 42 ml of sucrose gradient on which the cell suspension is layered. Centrifugation is performed at 1500 ± 15 rpm (431 g) for 20 min using the rate controller to start and stop the machine. No brakes are used. Three bands (I, II, and III) separate on this gradient. Band I contains predominantly red blood cells. The remaining suspension and band I are discarded. Bands II and III are removed carefully with a Pasteur pipet or a peristaltic pump, diluted fourfold by slow addition of Hanks' BSS (pH 6.00), and centrifuged at 1700 ± 15 rpm (554 g) for 10 min. As will become clear, the majority of cells in both bands II and III are oligodendrocytes. Further handling of cells depends on experimental requirements.

B. Criteria Used to Classify Isolated Oligodendrocytes

1. MORPHOLOGY AND ULTRASTRUCTURE

It is remarkable that, despite a substantial number of publications concerning procedures for isolating oligodendrocytes (for reviews, see Poduslo and Norton, 1975; Sellinger and Azcurra, 1974), ultrastructural characterization of the isolated cells has seldom been reported. Most authors have contented themselves with identifying their cells by phase-contrast microscopy. It should be pointed out that this technique does not always permit a distinction between bare nuclei and intact cells, a particularly vexing problem in the case of oligodendrocytes that have scanty cytoplasm. Fewster *et al.* (1973) characterized oligodendrocytes in thick sections stained with cresyl violet by "the size of the cells and chromatin distribution."

The first ultrastructural characterization of isolated oligodendrocytes was that of Raine *et al.* (1971). They identified oligodendrocytes as "round and dense cells containing microtubules and frequently associated with loops of myelin." Raine *et al.* (1971) stated that 90% of their cells were small and round when examined by light microscopy, implying that they were oligodendrocytes; it is, however, not clear whether the same percentage of cells could pass a scrutinous examination by electron microscopy.

There are intrinsic problems in identifying neural cells once they have been isolated. First, the cells have lost their cytoplasmic processes, and these can be important in their identification (Peters *et al.*, 1976); second, cells are often discriminated by the extent of osmiophilia relative to neighboring elements (e.g., axons) which are also absent; third, cells may change or be damaged during the process of isolation and fixation. In this regard, Raine *et al.* (1971) have observed swelling of cellular organelles, cytoplasmic vesiculation, and rarification. In our cell preparations, we too find a certain swelling of mitochondria, but the nuclear chromatin, endoplasmic reticulum, and Golgi apparatus are well preserved.

When pellets from cells obtained by the procedure described herein (see also Szuchet *et al.*, 1978b, 1980a) are embedded in Epon, stained with toluidine blue, and examined by light microscopy, they appear intact. There is a gradation in staining intensities which gives the pellet an appearance of heterogeneity. The number of darkly stained cells is more predominant in band III than in band II. Almost all cells possess cytoplasm; sometimes this is present only as a narrow rim surrounding a centrally located nucleus, and at other times as a relatively large mass at one pole of the cell with the nucleus at the other pole (Figs. 3a and b). Electron micrographs of these pellets (Figs. 3c and d, courtesy of R. L. Wollmann) confirm the quality of the cells and in particular show the overall preservation of the plasma membrane. However, small discontinuities in the plasma membrane are often encountered. These we interpret as sites where processes had been attached.

We require that certain minimal ultrastructural criteria be satisfied before a cell can be classified as an oligodendrocyte (Szuchet *et al.*, 1980a). Cells having myelin attached directly to their plasma membranes are accepted as oligodendrocytes. If no attached myelin is found in the plane of section, then two or more of the following elements must be present simultaneously for a cell to qualify as an oligodendrocyte: (1) abundant cytoplasmic microtubules; (2) fragments or loops of plasma membranes projecting from the cell surface, presumed to be sites where processes had been connected; (3) rough endoplasmic reticulum, single or stacked; (4) studs of ribosomes on the outer surface of the nuclear membrane; (5) a large number of ribosomes; and (6) intense osmiophilia.

Using the above criteria, Szuhet *et al.* (1980a) identified 65 ± 14% of the cells in band II and 91 ± 3% of the cells in band III as definite oligodendrocytes. The remaining cells were unidentified. It may be noted that 17% of cells in starting white matter could not be classified as to cell type on electron microscopy when the criteria listed above were employed for identification. Only an occasional astrocyte, recognized by the presence of 7- to 9-nm filaments, was found in band II and none in band III, despite the fact that on electron microscopy of starting material 13% of white matter cells were judged to be astrocytes. No microglial cells were encountered, but we concede that the distinction between microglia and oligodendrocytes is not always certain.

Two possible explanations were offered by Szuchet *et al.* (1980a) to account for the relatively large number of cells in band II that could not be identified: (1) Unidentified cells might include "Undifferentiated" cells or "transitional forms toward oligodendroglia" as described by Ramon-Moliner (1958). The latter cells are only found in young animals, but it will be recalled that Szuchet *et al.* (1980a) used lamb brains. (2) Unidentified cells might also include "light" oligodendrocytes (Mori and Leblond, 1970), but this explanation requires the further assumption that the cells

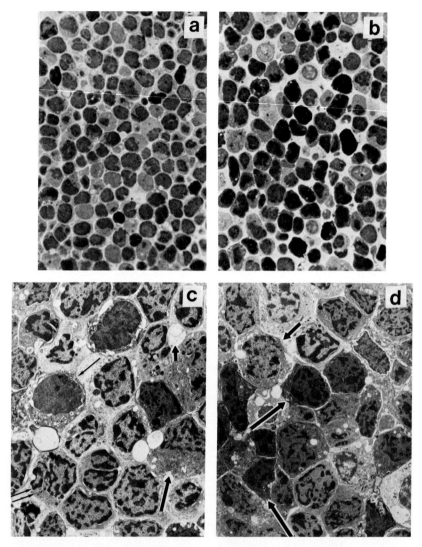

FIG. 3. Cross sections of pellets from cells fixed in a mixed aldehyde solution (2% gluteraldehyde–2% paraformaldehyde in 0.1 M cacodylate buffer, pH 7.2) for 24 hr, postfixed in 2% osmium tetroxide, pH 7.2, dehydrated in a graded series of alcohols, and embedded in Epon. Top: 0.5- to 1.0-μm sections stained with toluidine blue. Bottom: Thin sections stained with lead citrate and uranyl acetate. (a and b) Cells from bands II and III, respectively, as viewed by light microscopy of stained sections. Magnification: 60 × . (c) Electron micrograph of cells from band II. Almost all cells have clumped chromatin. Short, thick arrow indicates an oligodendrocyte with attached myelin; long, thin arrow points to an oligodendrocyte with an eccentrically located nucleus and abundant cytoplasm at the other end; thin double arrows indicate cells with rims of cytoplasm; and long, thin arrow points to an unidentified cell. Magnification: 4620 × . (d) Electron micrograph of cells from band III. Most cells are surrounded by a fair amount of cytoplasm. Long arrows indicate dark oligodendrocytes, some with eccentrically located nuclei, and short arrow indicates an oligodendrocyte with medium osmiophilia. Magnification: 4620 × .

lost their microtubules either during isolation or during fixation (Weber *et al.*, 1978). Light oligodendrocytes *in situ* contain large number of microtubules.

The results given above emphasize the limitations of ultrastructure as a sole criterion for cell classification and the need for additional criteria for identification of isolated oligodendrocytes.

2. BIOCHEMICAL CHARACTERISTICS

The chemical composition of isolated bovine oligodendrocytes has been analyzed by Fewster and Mead (1968), Fewster *et al.* (1973), and Poduslo and Norton (1972). Significant differences between the two sets of data exist (Table I), perhaps because of procedural differences in cell isolation or other technical considerations. Noteworthy in Table I is the high DNA/RNA ratio, a finding that was interpreted as reflecting the scanty cytoplasm of oligodendrocytes.

Early studies on tissue sections by Koenig (1964) led him to conclude that oligodendrocytes "possess substantial quantities of nucleolar, nucleoplasmic and cytoplasmic RNA." We have been investigating the DNA and RNA content of isolated oligodendrocytes with a cytofluorograph (Bio/Physics Systems, Inc., Mahopac, N.Y.), employing the technique of Malamed *et al.* (1977). Two-dimensional histograms of oligodendrocytes from bands II and III, respectively, reveal that both populations of cells are

TABLE I

CHEMICAL COMPOSITION OF ISOLATED BOVINE AND HUMAN OLIGODENDROCYTES

Component	Poduslo and Norton ± SD[a,b]	Fewster and co-workers ± SE[b,c]	Iqbal et al. ± SD[d,e]
DNA (pg/cell)	5.14 ± 0.75	6.51 ± 0.34	5.3 ± 0.6
RNA (pg/cell)	1.95 ± 0.43	1.52 ± 0.08	1.8 ± 0.1
DNA/RNA ratio	2.6	4.41 ± 0.34	3.2 ± 0.2
Protein (pg/cell)	—	38.7 ± 2.1	47 ± 3
Total lipid (% dry wt)	29.5	20.8 (16–26)	—
Total galactolipid[f]	9.9	14.8	—
Cerebroside	7.3	10.4	—
Sulfatide	1.5	4.3	—
Cerebroside/sulfatide	4.9	2.4	—
Total phospholipid[f]	62.2	42.2	—

[a] Poduslo and Norton (1972).
[b] Bovine.
[c] Fewster and Mead (1968); Fewster *et al.* (1973).
[d] Human.
[e] Iqbal *et al.* (1977).
[f] Expressed as percentage of total lipid.

fairly homogeneous with respect to their RNA and DNA contents (Figs. 4a and b). Presently, we only have qualitative data and cannot, therefore, compare our results with those of previous workers.

Poduslo and Norton (1972) have compared the chemical composition of isolated bovine oligodendrocytes with those of isolated rat neurons and rat astrocytes (see Table 3 in Poduslo and Norton, 1972). When possible species differences are allowed for, the chemical compositions of the three types of cells are not strikingly dissimilar. An exception is the exclusive presence of galactocerebroside and its sulfate derivative, sulfatide, in oligodendrocytes, providing an element of distinction between) oligodendrocytes on the one hand and astrocytes and neurons on the other. The

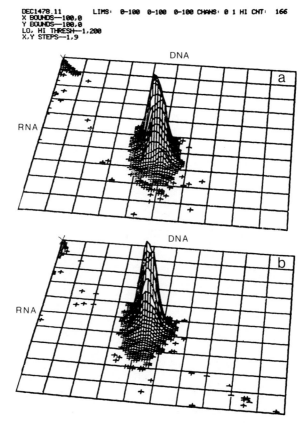

FIG. 4. Computer-drawn two-dimensional histograms of the frequency distribution of individual oligodendrocytes according to their green (DNA) and red (RNA) fluorescence (in arbitrary units) after staining with acridine orange. Each cell is represented by a single point. × marks the origin. (a) Cells from band II. (b) Cells from band III.

presence of CNP in isolated oligodendrocytes (Raine *et al.,* 1971; Poduslo, 1975; Pleasure *et al.,* 1977) might be another distinctive feature of these cells. CNP catalyzes the hydrolysis of 2',3'-cyclic monophosphates to their corresponding 2'-nucleotide monophosphates (Drummond and Perrott-Yee, 1961; Drummond *et al.,* 1962). The enzyme is a constituent of myelin (Kurihara and Tsukada, 1967, 1968; Norton, 1977; Benjamins and Smith, 1977), where it is found in relatively large amounts. The biological role of CNP has not been defined, nor has its natural substrate been recognized, but an increase in its specific activity with myelinogenesis has been well documented (see Section IV,D).

Significant levels of CNP have been found in isolated oligodendrocytes (Table II). At least part of this activity must originate in myelin which remains attached to the cells. CNP activity has also been reported in isolated astrocytes and neurons (Norton and Poduslo, 1971; Poduslo and Norton, 1972). In addition, it has been detected in C6 glioma cells (Zanetta *et al.,* 1972; Pfeiffer and Eagle, 1976; McMorris, 1977) which are generally considered to be astrocytic, RN-2 Schwannoma cells (Pfeiffer and Wechsler, 1972), and G26 oligodendroglioma cells (Sundarraj *et al.,* 1975). However, CNP activity seems to be highest in oligodendrocytes (Table II). Some CNP activity (0.03–0.1 mol/min/mg protein) is found in nonneural cells (Pfeiffer and Wechsler, 1972). Thus, a high activity of this enzyme may help in distinguishing oligodendrocytes from other cells but does not suffice as a sole criterion for their identification. Moreover, the use of CNP as a myelin marker (Mandel *et al.,* 1972; Norton, 1972) has recently been questioned, since nonmyelin brain membranes with a higher content of this enzyme than is found in myelin itself have been isolated (Waehneldt, 1977; Shapiro *et al.,* 1978).

Poduslo (1975, 1978) published a procedure in which two fractions labeled, respectively, "glial plasma membrane" and "glial myelin" are separated from isolated oligodendrocytes. The latter fraction appears to comprise chiefly myelin which has remained attached to the cells during the isolation procedure. Both fractions contain relatively high levels of enzyme markers traditionally associated with membranes, e.g., Na$^+$,K$^+$,-ATPase, 5'-nucleotidase, and CNP (see Table 1 in Poduslo, 1975). Poduslo compared the chemical compositions of the glial fractions with those of the intact cell and "mature" myelin and found them to be different (cf. Table 3 in Poduslo, 1975, and Table 5 in Poduslo, 1978). The glial plasma membrane fraction is enriched, relative to the oligodendrocyte itself, in galactocerebrosides and cholesterol, is impoverished in phospholipids, and has a similar content of gangliosides. It is of interest that the membrane fraction lacks basic protein but contains proteolipid protein.

The foregoing discussion merely points to the scarcity of current

TABLE II

SPECIFIC ACTIVITY OF CNP IN NEURAL CONSTITUENTS AND CELLS[a]

	Poduslo (1975)[b]	Poduslo and Norton, 1972	Pleasure et al., 1977 ±SD	Szuchet et al., 1980a ±SD	Pfeiffer and Wechsler, 1972[c]
White matter	—	—	6.8 ± 2.6	11.0 ± 3	—
Myelin	4.6	3.72	—	—	—
Neurons	—	0.42	—	—	—
Astrocytes	—	0.41	—	—	—
Oligodendrocytes	—	3.23	1.5 ± 0.7	2.0 ± 0.5	—
C6	—	—	—	—	0.9
C21	—	—	—	—	1.4
RN–2	—	—	—	—	1.2
Oligoplasma membrane	2.0	—	—	—	—

[a] Micromoles P_i per milligram protein per minute.
[b] Recalculated.
[c] Taken from Table 2 in Pfeiffer and Wechsler (1972). Specific activity in micromoles P_i per milligram per percentage dry weight.

knowledge of the biochemistry of isolated oligodendroctes. In contrast, the biochemical characterization of astrocytes is well advanced (Hertz, 1977).

3. Cell-Specific Markers

The application of chemical and immunocytochemical techniques to brain sections, and aggregate and monolayer cultures has been of great help in determining the cellular localization of individual brain components. These techniques have provided researchers with an invaluable tool for cell identification. It is outside the scope of this review to consider this topic; comprehensive reviews of this subject have recently appeared (Varon, 1977; Bock, 1977). From the viewpoint of oligodendrocytes these "molecular markers," as Varon (1978) calls them, can be classified as positive or negative. A specific positive marker is one whose presence identifies a cell as an oligodendrocyte; a specific negative marker, on the other hand, tells us what the cell *is not*.

Examination of the list of molecular markers discussed by Varon (1978) reveals that only a handful of markers with absolute specificity are known. Among these, cerebroside sulfotransferase (EC 2.8.2.1) is undoubtedly the most specific positive marker for oligodendrocytes thus far uncovered. Of less certain specificity are CNP and the hydrocortisone-induced increase in glycerolphosphate dehydrogenase (GPDH, EC 1.1.1.8), activity. The latter enzyme has been found, by immunofluorescence methods, exclusively in oligodendrocytes (de Vellis *et al.,* 1977). However, both enzymes are synthesized by C6 clonal cells supposedly of astrocytic origin. Whereas this fact need not invalidate use of these enzyme markers for the classification of freshly isolated cells, it should cause some concern when dealing with cells that might have dedifferentiated after a period in culture. This point is taken up in Section IV,D.

Glial fibrillary acidic (GFA) protein long considered an absolute marker for astrocytes (Eng *et al.,* 1971; Bignami *et al.,* 1972; Ludwin *et al.,* 1976; Kozak *et al.,* 1978; Bignami, this volume) should prove a valuable negative marker for oligodendrocytes, in the sense of proving the absence of astrocytic contamination.

Freshly isolated cells from bands II and III exhibit surface staining, as viewed by indirect immunofluorescence with antigalactocerebroside (a generous gift of M. Rapport) and anti-sheepmyelin basic protein antisera obtained in our laboratory by A. Noronha. The pattern of staining with both antisera is patchy, but the patches appear larger and more bulky with antimyelin basic protein than with antigalactocerebroside. This positive staining for myelin basic protein is consistent with the notion that components of myelin have remained attached to the cells during the process of isolation. The galactocerebroside on the surface of the cells need not be of myelin origin. Indeed, several reports have appeared (Raff *et al.*, 1978; Poduslo, 1978; Lisak *et al.*, 1979) suggesting that galactocerebroside is a normal constituent of the oligodendrocyte membrane.

4. ANTIOLIGODENDROCYTE ANTISERUM

Poduslo *et al.* (1977) were the first to raise antibodies against isolated lamb oligodendrocytes and show the antiserum to be specific to cell surface components. No cross-reaction with myelin was detected. Since then, other laboratories have also obtained sera against oligodendrocytes and found similar specificity (Abramsky *et al.*, 1977, 1978; Traugott *et al.*, 1978).

We, too, have raised antiserum against freshly isolated cells. Staining of tissue sections by indirect immunofluorescence and by the peroxidase–antiperoxidase method of Sternberger *et al.* (1970) proved the antiserum to be relatively specific for oligodendrocytes. Neither neurons nor astrocytes stained. However, the antiserum stained blood vessels to some extent, suggesting either that our cell preparation might have contained capillaries or that an antigen is shared by oligodendrocytes and brain endothelium. Even though the target antigen(s) against which the antibodies are directed remains unknown, the antiserum provides a powerful tool for cell identification and can be further exploited for cell isolation.

III. Oligodendrocyte Subpopulations

That oligodendrocytes comprise a heterogeneous population of cells was already recognized by light microscopists [Del Rio-Hortega and others; see Peters *et al.*, (1976), or Mori and Leblond (1970) for reviews], who observed a gradation in nuclear and cytoplasmic staining densities in cells with all the morphological characteristics of oligodendrocytes. In a detailed study of the cells from corpus callosum of young rats, Mori and Leblond (1970) confirmed earlier observations on the dispersity of oligodendrocytes and classified them into three groups: "Light," "medium," and "dark." It is to be understood that sharp boundaries between the three types of cells cannot be drawn and that, while the prototype cell of each class has distinct morphological and ultrastructural characteristics, for practical purposes one is essentially dealing with a spectrum. Thus, at one end of the spectrum is situated the light oligodendrocyte with its large, pale nucleus (7 μm), large, centrally located nucleolus, abundant cytoplasm, and numerous processes, and lacking clumped chromatin. Its ultrastructure reveals an abundance of organelles among which microtubules and the Golgi apparatus are prominent. At the other end lies the dark oligodendrocyte with a small nucleus often eccentrically located, clumped chromatin, and scanty cytoplasm. Only a few processes extend from the cells. The cisternae of the Golgi apparatus, stacks of rough endoplasmic reticulum, and lamellar bodies are the outstanding ultrastructural features of dark cells.

The concept has evolved that these cell types represent maturation stages, the dark cell being the most mature. All cell types are considered to be differentiated in the sense that they are believed to partake in myelinogenesis and/or maintenance of myelin (Privat and Fulcrand, 1977). It remains, however, an open question whether or not different cell types have specific roles.

As mentioned, the isolation technique outlined above separates oligodendrocytes into two fractions (bands II and III). Light and electron micrographs of cells from these bands are shown in Fig. 3. Figures 5a and b show computer-drawn scattergrams of the size distribution of individual cells in each band. It is clear from inspection of these figures that the majority of cells in band II are smaller than those in band III. Figure 5c shows

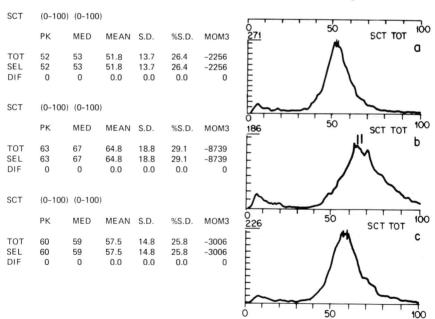

FIG. 5. Computer-drawn graphs of the frequency distribution of individual oligodendrocytes according to size. (a) Cells from band II. (b) Cells from band III. (c) Artificial mixture of an equal number of cells from bands II and III.

the size distribution of an artificial mixture of cells from both bands; as expected, and intermediate distribution is obtained (S. Szuchet and D. Richman, unpublished results). The difference in size between the cells of both fractions is consistent with their separation on a linear sucrose gradient. It should be mentioned that also Mori and Leblond (1970) found size differences among the three types of oligodendrocytes, the light oligodendrocyte being the largest.

The cells from bands II and III also differ in the amount of galactocerebroside and sulfatide they synthesize as measured by the incorporation of [^{35}S] sulfuric acid and [^{3}H]galactose. However, the ratio of sulfatide to total galactocerebroside is the same (Table III, Section IV,C).

While it is tempting to try to correlate the differences found between bands II and III with the subgroups of oligodendrocytes described by Mori and Leblond (1970), in our view it is premature to do so.

IV. Culture of Oligodendrocytes

Study of oligodendrocytes maintained and propagated *in vitro* should facilitate delineation of the molecular and cellular events that lead to myelin synthesis. It should also lead to an understanding of oligodendrocyte differentiation and dedifferentiation. In addition, it should permit a detailed biochemical characterization of oligodendrocytes; such knowledge is presently lacking (see Section II,B,2). Finally, oligodendrocytes propagated in culture could provide a self-renewing source of these cells. This would be especially desirable, since isolation procedures for these cells are both lengthy and expensive.

A. Procedures

The only report of successful long-term culture of isolated oligodendrocytes is that by Fewster and Blackstone (1975). These workers kept bovine cells isolated by the procedure of Fewster *et al.* (1973) in culture for up to 6 months. They investigated the effect of various parameters on cell appearance, growth, process development, survival, and overgrowth by other cells. Medium 199 with no added serum or a maximum of 2.5% serum proved to be best for cell attachment on glass or plastic surfaces. The addition of excess glucose (300 mg%) to this medium shortened the time required for attachment to 1 hr. Survival for an extended period and process formation were favored by these conditions. The development of processes was enhanced close to an air–liquid interphase. Cells isolated from fetal brains survived longer than cells from adult brains. On the other hand, the latter put out processes after a shorter period in culture than the former. Overgrowth by fibroblasts was a problem only when cells isolated from fetal brains were cultured in media containing high concentrations of serum.

The study of Fewster and Blackstone (1975) on culture conditions for oligodendrocytes appears to be the most detailed work in the field thus far published. However, our criticism of this work concerns the morphological criteria adopted by these authors for cell identification (Fewster *et al.* 1967, 1973).

Subsequent investigators have worked with short-term cultures maintained in various media. Poduslo *et al.* (1977, 1978) maintained oligodendrocytes isolated from calf and lamb brains in suspension cultures, using Dulbecco's enriched medium with 0.6% glucose, 5% fructose, 0.001% insulin, antibiotics, 5% fetal calf serum (FCS), and 0.3% 1,4-piperazine–diethanesulfonic acid buffer. Poduslo *et al.* (1977, 1978) considered gentle transfer of cells from isolation media to culture media critical for cell survival, since otherwise cell lysis occurred. They also emphasized the need for care in the selection of trypsin and FCS. If these precautions are taken, cells can be kept viable for 1–2 days as judged by their metabolic activity (see the following discussion). Pleasure *et al.* (1977) use Eagle's minimum essential medium with Earle's salts, 2 mM L-glutamine, and 10% (v/v) FCS to keep their isolated oligodendrocytes in culture. They obtain cells by the procedure of Poduslo and Norton (1972, 1975) and maintain them for 24–48 hr.

PROCEDURE USED IN OUR LABORATORY

When cells are destined for culture, 1.25 μg/ml of amphotericin B and 10 μg/ml of gentamycin are added to the medium in which brains are placed at the slaughterhouse. These concentrations of antibiotics are kept in the media throughout the whole isolation procedure, as well as during culture. This is necessary because yeasts remain permanent residents of slaughterhouses. Our experience in this respect accords with that of Poduslo *et al.* (1978).

Freshly isolated cells are suspended in Dulbecco's modified Eagle's medium (DMEM) supplemented with either 20% FCS (DMEM–FCS) or 20% horse serum (DMEM–HS). After being centrifuged at 1700 rpm (554 g) for 4 min, the cells are resuspended in the same medium and plated for culture. We grow cells on either plastic or glass. The plating concentration is 1.0×10^6 cells/ml, and the plating density is 2.5×10^5 cells/cm^2. The cells are kept at 37°C in an atmosphere of 5% carbon dioxide in air with the humidity at saturation.

DMEM–FCS seems to favor sticking more than DMEM–HS. On the other hand, DMEM–HS in conjunction with a plastic surface appears to cause process formation more than DMEM–HS on glass or DMEM–FCS on either glass or plastic. We have kept cells in culture for up to 60 days.

B. Morphology and Ultrastructure of Isolated Oligodendrocytes in Culture

Poduslo *et al.* (1977) describe their cells, maintained for 1–3 days and viewed by phase microscopy, as "small, glowing round cells with no nucleus." If cells become brown and granular and nuclei are visible, they are considered unhealthy.

Some indications of how isolated oligodendrocytes, maintained for an extended period in culture, might be expected to appear can perhaps be gleaned from descriptions of oligodendrocytes in explant cultures.

Wolfgram and Rose (1957) worked with culture of explants from corpus callosum of 6-week-old kittens and compared them with Golgi preparations from the same tissue. For each cell type *in vivo* they found a similar cell type *in vitro* and concluded that the cells retained their morphology in the culture system. This procedure allowed them to gain knowledge of the morphology of cells in culture plus a degree of certainty with regard to cell identification. These authors characterized the *living* oligodendrocyte, by phase-contrast microscopy, as highly refractile, "usually encompassed by a halo of refracted light," and having a spherical nucleus with a prominent nucleolus, a highly granular cytoplasm, and bipolar or multipolar processes, the first with a process coming from each pole, and the latter with, usually, four to six processes and beadlike swellings which they termed "gliosomes."

Because bipolar oligodendrocytes were rarely seen in tissue sections stained by the Golgi method and because they were also of infrequent occurrence in culture when special medium was used, Wolfgram and Rose (1957) considered these cells as either "unhappy" with their environment or immature, i.e., preparing for cell division. Interestingly, division was observed more frequently in bipolar than in multipolar cells. Other salient features of cultured oligodendrocytes observed by Wolfgram and Rose (1957) were the appearance of two distinct nucleoli prior to cell division or the apparent disappearance of the original nucleolus, "right-angle" branching of processes, veillike membranes particularly at the distal end of processes, and the presence of cells with very short or aborted processes without being true "adendroglia" (Andrew and Ashworth, 1945).

Korinkova and Lodin (1977), using explants from 12-day-old rats, also found undifferentiated bipolar cells either isolated or in groups. These cells possessed an ovoid nucleus, several nucleoli, and dense cytoplasm. Often binucleated cells were seen. When differentiated (e.g., by the addition of dibutyryl cyclic AMP), these cells acquired the morphological characteristics of mature oligodendrocytes.

Ramon-Moliner (1958) described two types of uncommon brain cells found only in young animals (rats): (1) an undifferentiated cell or "mossy" oligodendrocyte which gave off a few very thick processes ("trunks") which branched out at a distance from the cell body into semitransparent filaments or membranes, giving the appearance of a "halo," and (2) a "transitional form toward oligodendrocyte" which already showed signs of differentiation by having lost its mossy aspect and having acquired thin processes.

Our cells in culture exhibit almost all the distinctive features ascribed to oligodendrocytes by Wolfgram and Rose (1957), Ramon-Moliner (1958), and Korinkova and Lodin (1977).

We grow cells as monolayers. A large proportion of cells stick to the surface within 4 hr, particularly when placed on glass. Some cells begin to extend processes within 24 hr, but it usually takes 1 week before processes extend from the majority ($\sim 85\%$) of cells (Figs. 6a and b). Perhaps the most outstanding characteristics of these cultures is the variability in cell shape as well as in the number, length, branching, and thickness of processes. The ability of glial cells to change their shape and contents in response to different stimuli is well recognized (Brightman *et al.*, 1978). The majority of cells are bipolar (Figs. 6c and d), but multipolar cells (Fig. 6c) and cells with aborted processes (Fig. 8d) are also seen; veillike membranes (lamellipodia), especially at the foot of processes, are of frequent occurrence (Figs. 7a and b). Sometimes the cells extend a few short, thick processes which then branch out into very thin membranous processes (Figs. 7a and b). Such cells fit the description of mossy oligodendrocytes first given by Ramon-Moliner (1958). At other times globules staining dark blue with Giemsa are seen either close to the cell bodies (Fig. 7d) or along processes (Fig. 6c). These structures are reminiscent of the beadlike swellings (gliosomes) observed by Wolfgram and Rose (1957) in cultured oligodendrocytes. Some cells do not emit processes but surround themselves with extensive membranous sheets often covering surface areas 10–20 times that occupied by the perikarya (Figs. 7c and 8b and c). This phenomenon was only observed when cells were grown on plastic surfaces in the presence of horse serum. We have initiated the ultrastructural characterization of cells

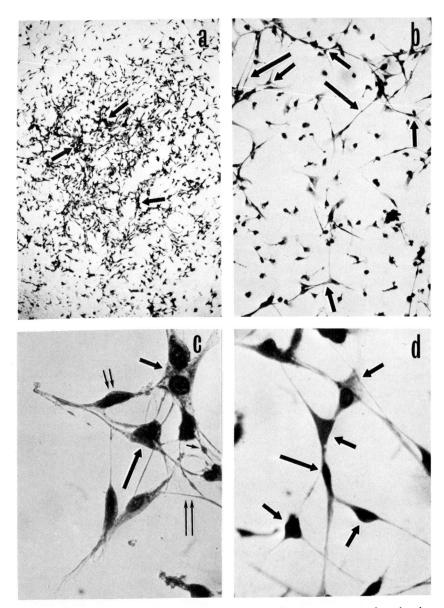

FIG. 6. Oligodendrocytes after 21 days in culture. Cells are grown on a glass chamber slide in DMEM supplemented with 20% horse serum. The medium is removed, and the cells are washed three times with phosphate-buffered saline (PBS), fixed for 2–4 min with methanol, and stained with Giemsa. (a) Magnification: × 75. Majority of cells have processes, mainly bipolar. Arrows point to aggregated cells. (b) Magnification: × 190. Short arrow indicates cells with three processes; long arrow points to contact between processes from

in culture as a function of time *in vitro* (R. L. Wollmann and S. Szuchet, unpublished results). In this manner, we hope to follow the fate of individual cells. The cells are fixed in the culture dishes after careful washing, thus, we only look at cells that remain attached to the surface. After 4 hr in culture the cells have the same ultrastructural characteristics as the cells that were originally plated. After 2 days in culture the cells still bear a great resemblance to the mother cells. However, from the fifth day on the cells take on the appearance of immature or "young" cells with a cytoplasm rich in organelles. These are still preliminary results.

C. Metabolic Studies on Maintained Cells

The incorporation of radioactive mevalonic acid into neutral lipids by a fraction enriched in glial cells was investigated by Jones *et al.* (1975). The duration of the experiment was 15 hr. A similar study was performed by Fewster *et al.* (1975) on isolated oligodendrocytes. These early studies were directed mostly to finding out the role of each of the neural cell types in lipid metabolism.

Recent metabolic studies on oligodendrocytes (Poduslo *et al.,* 1977, 1978; Poduslo and McKhann, 1977; Pleasure *et al.,* 1977; Poduslo, 1978) have been designed to (1) prove that isolated cells are oligodendrocytes and (2) demonstrate cell viability.

Poduslo *et al.* (1977, 1978) reported maintenance of oligodendrocytes isolated by the procedure of Poduslo and Norton (1972) in suspension culture for 1–2 days. During this period, cells remained metabolically active, as evidenced by their incorporation of radioactive precursors. Poduslo *et al.* (1977, 1978) found that, when cells were labeled with [³H]galactose, radioactivity was incorporated primarily into cerebrosides and that [³H]acetic acid was converted to cholesterol. However, Poduslo *et al.* (1978) reported difficulties in proving synthesis of sulfatides when sodium [³⁵S]sulfate was used as a precursor. These authors found no evidence of ganglioside synthesis and speculated that the reasons could be that no appropriate substrates had been fed to the cells. Based on the acrylamide gel electrophoresis profile shown by Poduslo *et al.* (1978), their cells incorporated [³H]proline into a variety of proteins with a continuous spectrum

---

different cells. (c) Magnification: × 740. Short, thick arrow points to an apparent syncytium which the cells sometimes form; long, thick arrow indicates cell with six processes; short, thin arrow points to a gliosome (?); short double arrows indicate bipolar cells, and long double arrows indicate a network of processes. (d) Magnification: × 740. Short arrow shows a variety of cell shapes, and long arrow points to a gliosome (?). Note how processes extend between cells.

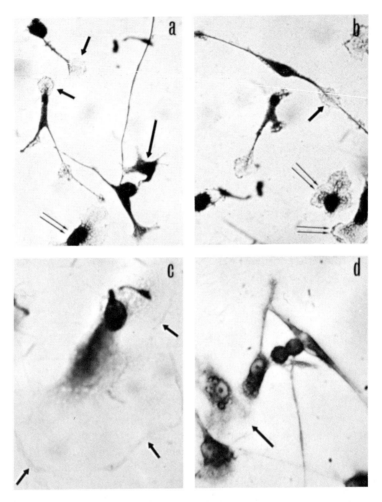

Fig 7a and b.   Cells after 21 days in culture under experimental conditions similar to those in Fig. 6. Stained with Giemsa. Magnification: × 650. Short, thick arrow indicates veillike processes at distal end of cells; long, thick arrow indicates cell with five processes; and thin double arrows point to mossy oligodendrocytes (see text).

Fig. 7c.   Note extensive membrane surrounding perikarya (arrows) Magnification: × 590.

Fig. 7d.   Cells grown on a plastic surface after 6 days in culture. Arrow indicates a group of cells with prominent centrally located nucleoli. Magnification: × 650.

of molecular masses, but assembly into myelin-specific proteins—proteolipid and basic proteins—was not detected.

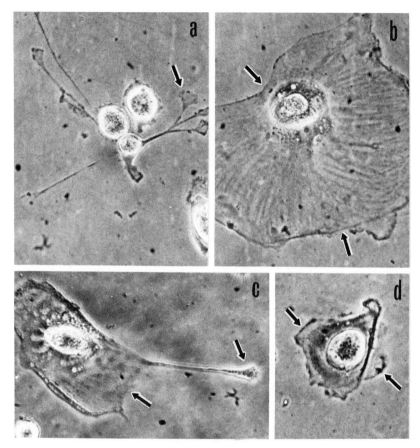

FIG. 8   Cells grown on a plastic surface in DMEM plus 20% HS for 14 days. Phase-contrast microscopy without fixation. Magnification: × 630. (a) Group of cells with branching processes (arrow) and veillike membranes. (b and c) Cells with extensive membranes (arrows). (d) Cell with aborted processes (arrows).

Pleasure *et al.* (1977) compared the incorporation of [^{14}C]acetate and [^{35}S]sulfate into unesterified sterols and sulfatides, respectively, by cells in culture for 0–24 hr and 24–48 hr. Their findings showed a threefold increase in the uptake of [^{14}C]acetate on the second day in culture, whereas there was an equivalent decrease in the incorporation of [^{35}S]sulfate. Measurements of specific activities of the enzymes that partake in these metabolic pathways, namely, 3-hydroxy-3-methylglutaryl coenzyme A reductase and cerebroside sulfotransferase (CST), corroborated the previous results (see Table III in Pleasure *et al.,* 1977).

Data obtained in our laboratory (Szuchet *et al.*, 1980a) are summarized in Table III. The measurements were performed after the cells had been in culture for 60–66 hr. In accord with previous workers (Poduslo *et al.*, 1977, 1978; Pleasure *et al.*, 1977), we too found evidence of galactocerebroside and sulfatide synthesis. However, several distinct features of our results deserve comment: first, we obtained a high incorporation of [^{35}S]sulfate into sulfatide; second, when cells were labeled with [^{3}H]galactose, the ratio of galactosylceramide to sulfogalactosylceramide was approximately 1:3, which was similar to that found in whole brain (O'Brien and Sampson, 1965) but different from the values reported by Poduslo *et al.* (1978); third, the ratio of long-chain fatty acids to α-hydroxy fatty acids in galactosylceramide was essentially 1, in agreement with the results of Poduslo *et al.* (1978). Finally, contrary to the experience of Poduslo *et al.* (1978), our cells incorporated [^{3}H]galactose into ganglioside. Interestingly enough, the highest label was found associated with the G_{D3}/G_{Dla} region; the former was reported by Poduslo (1975, 1978) to be a component of oligodendrocyte membrane.

TABLE III

BIOSYNTHESIS OF COMPLEX CARBOHYDRATES BY LAMB
OLIGODENDROGLIAL CELLS IN TISSUE CULTURE[a]

Compound	Band II (cpm/mg protein)		Band III (cpm/mg protein)	
	^{3}H	^{35}S	^{3}H	^{35}S
GalCer LCFA	2,200	—	3,200	—
GalCer αHOFA	1,660	—	3,000	—
LacCer	300	—	480	—
Sulfo-GalCer	1,350	1,750	2,500	3,400
Tetrahex-Cer	200	—	220	—
G_{M3}	800	—	1,000	—
G_{M2}	120	—	130	—
G_{M1}	130	—	140	—
G_{Dla}/G_{D3}	2,000	—	1,920	—
"G_{Dlb}"	1,400	—	1,440	—
Glycosaminoglycans				
Hyaluronic acid	47,500	—	47,900	—
Chondroitin-6-sulfate	—	104,500	—	110,100

[a] Cells were grown as monolayer cultures in modified Eagle's medium supplemented with 10% horse serum for 60 hr, during which time most of the cells extended processes. Cells were labeled with 5 μCi [^{3}H]galactose and 150 μCi of [^{35}S]sulfuric acid.

Abbreviations: Gal = galactose; Cer = ceramide; Lac = lactoside; Tetrahex = tetrahexoside; LCFA = long chain fatty acid; αHOFA = α-hydroxyl fatty acid; G = ganglioside.

D. Are the Cells that Survive in Long-Term Culture Oligodendrocytes?

It is apparent from the work of others as well as our own that isolated oligodendrocytes remain metabolically active in culture for several days. The question, however, still remains whether the cells that survive for a long period are oligodendrocytes. The question of cell identity after a period in culture is not trivial, for the possibility of cell dedifferentiation or overgrowth by contaminating species is real. This is more so, since no author has claimed more than 90–95% purity for isolated cells.

The most distinctive feature of oligodendrocytes is their ability to synthesize myelin. This synthesis can be considered at two different levels: the molecular level, i.e., the accumulation of necessary lipid and protein molecules, and the supramolecular level, i.e., the actual assembly of myelin sheaths. Evidence that cells in culture can accomplish either or both tasks proves their oligodendrocytic origin. However, there are good reasons to believe that olidogodendrocytes may be unable to accomplish these tasks in isolation and that help from other cells may be required (see Section I). Whether or not this help (signals?) is needed at all steps of the process has not been determined. Two sources can provide clues to enlighten on this point. These sources are (1) studies on developing brains and explant cultures, and (2) cultures of cell lines from brain tumors. Unfortunately, information gathered thus far, is *not* unequivocal.

A fair amount of knowledge on the course of myelinogenesis has accrued from studies on developing brains and explant cultures. Of relevance to this discussion are the observations of Davison (1971) and Davison and Sabri (1978) who state that myelin assembly is heralded by the accumulation of lipid droplets and dense rough endoplasmic reticulum in oligodendrocytes. Sternberger *et al.* (1978a,b), using immunocytochemical techniques, have presented evidence that myelin basic protein is present in oligodendrocytes prior to myelin sheath formation but *not* afterward. Coinciding with the onset of myelin synthesis there is also an increase in the activity of CNP and CST (Duch *et al.,* 1975; Tennekoon *et al.,* 1977; Latovitzki and Silberberg, 1977; Sheppard *et al.,* 1978; Sprinkle *et al.,* 1978a,b). Likewise, Pleasure and Kim (1976) reported an increase in the specific activity of the enzyme that catalyzes the synthesis of mevalonic acid. Recently, Sprinkle *et al.* (1978a) measured the specific activity of CNP and the amount of myelin basic protein in rat brains as a function of age and found a linear correspondence between CNP activity and myelin basic protein. It can be inferred from these and other observations (see Section I) that myelin synthesis from its onset to the point of maturation involves a highly ordered sequence of events. This implies strict control mechanisms with a turning on and off signals at each step of the process.

A perspective slightly different from that described above results when data on the culture of mouse clonal cell lines are analyzed. The ability of clonal cell lines of neural origin to express differentiated functions has been well documented (McMorris, 1971). Pertinent to our work are findings that clones from mouse glial tumors are capable of synthesizing myelin precursors. Dawson (1978) obtained incorporation of [^{35}S]sulfate into sulfogalactosylceramide by G26-20 and G26-24 clones from an oligodendroglioma. Synthesis of myelin basic protein, S-100 protein, and CNP was achieved with RN-2 cells from a schwannoma (Pfeiffer and Wechsler, 1972). The manifestation of these traits is highly dependent on culture conditions. For example, McDermott and Smith (1978) could not detect myelin basic protein in their cultures of RN-2 clonal cells. If extrapolation from tumor cells to normal cells is at all permissible, the aforementioned results suggest that oligodendrocytes and Schwann cells are programmed to accumulate myelin precursors and would do so in a favorable environment independent of the presence of other cellular elements. However, actual assembly of the myelin sheath may not be attainable in a culture of pure cells.

Whereas the issue of synthesis of myelin and/or its precursors by isolated oligodendrocytes has yet to be tested, other means of cell identification are becoming available. Several laboratories (Raff et al., 1978; Poduslo, 1978; Lisak et al., 1979) have recently reported that galactocerebroside is a specific surface marker for oligodendrocytes in culture. At variance with these results, however, is the work published by Sternberger et al. (1978b) who found no evidence of galactocerebrosides in oligodendrocytes or their processes but did obtain staining of myelin. This led them to conclude that galactocerebroside was a specific marker for myelin only. Sternberger et al. (1978b) used tissue sections for their staining, whereas the other investigators used cells in culture. The results, if confirmed, are of significance, since they suggest a change in the chemical composition of the plasma membrane of cells in culture. Interestingly enough, Schwann cells in culture do not appear to have galactocerebrosides in their plasma membranes (Raff et al., 1978).

We have shown that after 21 days in culture cells isolated by the procedure described in this review still possess most of the morphological characteristics ascribed to oligodendrocytes. Furthermore, cells kept in vitro for 12 days have galactocerebrosides in their plasma membranes, as evidenced by the fact that they stain with the corresponding antiserum by indirect immunofluorescence (Szuchet et al., 1980a). This again attests to their oligodendrocytic character. However, we could not detect CNP activity in these cells. It has been reported that CNP is highly dependent on cell density and is only expressed 2 weeks after cells have reached confluency. Thus, our failure to detect activity may be due to this factor. Alternatively, isolated oligodendrocytes may be incapable of expressing this function *in vitro,* as has been reported to be the case for the RN-2 clonal line. Cells kept in culture for 30 or more days lose their ability to incorporate [^{35}S]sulfate into sulfatide. Concomitantly they acquire a different morphology—they flatten—suggesting that dedifferentiation has occurred.

V. Conclusions

It is apparent from the data presented that methods for isolating viable oligodendrocytes that survive in long-term cultures are now available. There is still room for improvement in the present methodology; in particular, effort should be directed toward finding experimental conditions that would avoid subjecting the cells to osmotic trauma during isolation. This, in our experience, is not a trivial problem, since we have carefully explored without success the use of substances such as Ficoll and serum albumin for making gradients.

Our knowledge of the behavior of oligodendrocytes in culture is rudimentary, but it is clear that oligodendrocytes kept *in vitro* are active cells; they synthesize processes and membranes, assume various shapes, and interact with each other. Presently, we have no control over their behavior in culture. Thus, an urgent task will be to find experimental conditions that will allow us to direct and control the responses of the cells. A second task will be to determine which traits expressed by the cells *in vitro* are relevant to myelinogenesis. Finally, much of the biochemical machinery of oligodendrocytes remains to be explored.

We believe that the stage is now set for research into these and other aspects of oligodendrocyte function not mentioned explicitly. Thus, the availability of viable oligodendrocytes and their capacity to survive in culture opens up new and exciting avenues for research.

Acknowledgment

Dr. B. G. W. Arnason read this manuscript and offered valuable criticism and advice. We are indebted to him for this and for his continuous support and encouragement. We are grateful to Dr. R. L. Wollmann for the electron micrographs, to Dr. D. P. Richman for the cytofluorographic studies, to Dr. R. P. Roos for his help with the immunofluorescence studies, and to Dr. G. Dawson for performing the biochemical studies. Our thanks are also due to Mr. Paul Polak and Ms. Ann Speckman for their expert technical assistance and to Mr. David Barnhart for typing the manuscript. This research was supported by grant No. RG1223-A-2 from the National Multiple Sclerosis Society. One of us (K. S.) is grateful to the Icelandic Science Foundation for a fellowship.

References

Abramsky, O., Lisak, R. P., Silberberg, D. H., and Pleasure, D. E. (1977). Antibodies to oligodendroglia in patients with multiple sclerosis. *N. Engl. J. Med.* **297**, 1207-1211.

Abramsky, O., Lisak, R. P., Pleasure, D., Gilden, D. H., and Silberberg, D. H. (1978). Immunologic characterization of oligodendroglia. *Neurosci. Lett.* **8**, 311-316.

Allerand, C. D., and Murray, M. R. (1968). Myelin formation *in vitro* endogenous influence on cultures of newborn mouse cerebellum. *Arch. Neurol. (Chicago)* **19**, 292-301.

Andrew, W., and Ashworth, C. T. (1945). The adendroglia: A new concept of the morphology and reactions of the smaller neuroglial cells. *J. Comp. Neurol.* **82**, 101-110.

Benjamins, J. A., and Smith, M. E. (1977). Metabolism of myelin. *In* "Myelin" (P. Morrell, ed.), pp. 233-270. Plenum, New York.

Bignami, A., Eng, L. F., Dahl, D., and Uyeda, C. T. (1972). Localization of the glial fibrillary acidic protein in astrocytes by immunofluorescence. *Brain Res.* **43**, 429-435.

Bock, E. (1977). Immunochemical markers in primary cultures and in cell lines. *In* "Cell, Tissue and Organ Cultures in Neurobiology" (S. Fedoroff and L. Hertz, eds.), pp. 407-422. Academic Press, New York.

Bornstein, M. B. (1977). Differentiation of cells in primary cultures: Myelination. *In* "Cell, Tissue and Organ Cultures in Neurobiology" (S. Fedoroff and L. Hertz, eds.), pp. 141-146. Academic Press, New York.

Bornstein, M. D., and Murray, M. R. (1958). Serial observations on patterns of growth, myelin formation, maintenance and degeneration in cultures of newborn rat and kitten cerebellum. *J. Biophys. Biochem. Cytol.* **4**, 499-504.

Brightman, M. W., Anders, J. J., Schmechel, D., and Rosenstein, J. M. (1978). The lability of the shape and content of glial cells. *In* "Dynamic Properties of Glia Cells" (E. Schofeniels, G. Frank, L. Hertz, and D. B. Tower, eds.), pp. 21-44. Pergamon, Oxford.

Bunge, M. B., Bunge, R. P., and Ris, H. (1961). Ultrastructural study of remyelination in an experimental lesion in adult cat spinal cord. *J. Biophys. Biochem. Cytol.* **10**, 67.

Bunge, R. P. (1968). Glial cells and the central myelin sheath. *Physiol. Rev.* **48**, 197-251.

Bunge, R. P., and Bunge, M. B. (1978). Evidence that contact with connective tissue matrix is required for normal interaction between Schwann cells and nerve fibers. *J. Cell Biol.* **78**, 943-950.

Bunge, R. P., Bunge, M. B., and Cochran, M. (1978). Some factors influencing the proliferation and differentiation of myelin-forming cells. *Neurology* **28**, 59-67.

Davison, A. N. (1971). Myologenesis. *Neurosci. Res. Program, Bull.* **9**, 465-470.

Davison, A. N., and Sabri, M. I. (1978). Biosynthesis of myelin and neurotoxic factors in the serum of multiple sclerosis. *In* "Myelination and Demye-ination" (J. Palo, ed.), pp. 19-25. Plenum, New York.

Dawson, G. (1978). Induction of sulfogalactosylceramide (sulfatide) synthesis of hydrocortisone (cortisol) in mouse G-26 oligodendroglioma cell strains. *J. Neurochem.* **31**, 1091-1094.

de Vellis, J., McGinnis, J. F., Breen, G. A. M., Leveille, P. C., Bennett, K., and McCarthy, K. (1977). Hormonal effects on differentiation in neural cultures. *In* "Cell, Tissue and Organ Cultures in Neurobiology" (S. Fedoroff and L. Hertz, eds.), pp. 485–511. Academic Press, New York.

Diaz, M. B., Bornstein, M., and Raine, C. S. (1977). Dissociation of myelinogenesis in tissue culture by anti-CNS antiserum. *J. Neuropathol. Exp. Neurol.* **36**, 594 (abstr.)

Drummond, G. I., and Perrott-Yee, S. (1961). Enzymatic hydrolysis of adenosine 3′, 5′-phosphoric acid. *J. Biol. Chem.* **236**, 1126–1129.

Drummond, G. I., Iyer, N. T., and Keith, J. (1962). Hydrolysis of ribonucleoside 2′, 3′-cyclic phosphates by a diesterase from brain. *J. Biol. Chem.* **237**, 3535–3539.

Duch, D., Mandel, P., and Kaprowski, H. (1975). Demonstration of enzymes related to myelinogenesis in established human brain cell cultures. *J. Neurol. Sci.* **26**, 99–105.

Eng, L. F., Vanderhaeghen, J. J., Bignami, A., and Gersil, B. (1971). An acid protein isolated from fibrous astrocytes. *Brain Res.* **28**, 351–354.

Fedoroff, S. (1978). The development of glial cells in primary cultures. *In* "Dynamic Properties of Glia Cells" (E. Schoffeniels, G. Frank, L. Hertz, and D. B. Tower, eds.), pp. 83–92. Pergamon, Oxford.

Fewster, M. E., and Blackstone, S. (1975). *In vitro* study of bovine oligodendroglia. *Neurobiology* **5**, 316–328.

Fewster, M. E., and Mead, J. F. (1968). Lipid composition of glial cells isolated from bovine white matter. *J. Neurochem.* **15**, 1041–1052.

Fewster, M. E., Scheibel, A. B., and Mead, J. F. (1967). The preparation of isolated glial cells from rat and bovine white matter. *Brain Res.* **6**, 451–468.

Fewster, M. E., Blackstone, S. C., and Ihrig, T. J. (1973). The preparation and characterization of isolated oligodendroglia from bovine white matter. *Brain Res.* **63**, 263–271.

Fewster, M. E., Ihrig, T. J., and Mead, J. F. (1975). Biosynthesis of long chain fatty acids by oligodendroglia isolated from bovine white matter. *J. Neurochem.* **25**, 207–213.

Fulcrand, J., and Privat, A. (1977). Neuroglial reactions secondary to Wallerian degeneration in the optic nerve of the postnatal rat: Ultrastructural and quantitative study. *Comp. Neurol.* **176**, 189–224.

Hertz, L. (1977). Biochemistry of glial cells. *In* "Cell, Tissue and Organ Cultures in Neurobiology" (S. Fedoroff and L. Hertz, eds.), pp. 39–71. Academic Press, New York.

Imamoto, K., and Leblond, C. P. (1977). Presence of labelled monocytes, macrophages and microglia in a stab wound of the brain following an injection of bone marrow cells labelled with ^{3}H-thymidine into rats. *J. Comp. Neurol.* **174**, 255–279.

Iqbal, K., Grundke-Iqbal, I., and Wisniewski, H. M. (1977). Oligodendroglia from human autopsied brain: Bulk isolation and some chemical properties. *Neurochemistry* **28**, 707–716.

Jones, J. P., Nicholas, H. J., and Ramsey, R. B. (1975). Rate of sterol formation by rat brain glia and neurons *in vitro* and *in vivo*. *J. Neurochem.* **24**, 123–126.

Koenig, H. (1964). RNA metabolism in the nervous system. Some RNA dependent functions of neurons and glia. *In* "Morphological and Biochemical Correlates of Neural Activity" (M. M. Cohen and R. S. Snyder, eds.), pp. 36–59. Harper, New York.

Korinkova, P., and Lodin, Z. (1977). A transitional differentiation of glial cells of cultured corpus callosum caused by dibutyryl cyclic adenosine monophosphate. *Neuroscience* **2**, 1103–1114.

Kozak, L. P., Dahl, D., and Bignami, A. (1978). Glial fibrillary acidic protein in reaggregating and monolayer cultures in fetal mouse cerebral hemispheres. *Brain Res.* **150**, 631–637.

Kurihara, T., and Tsukada, Y. (1967). The regional and subcellular distribution of 2′,3′-

cyclic mononucleotide 3'-phosphohydrolase in the central nervous system. *J. Neurochem.* **11**, 1167–1174.

Kurihara, T., and Tsukada, Y. (1968). 2',3'-cyclic nucleotide 3'-phosphohydrolase in the developing chick brain and spinal cord. *J. Neurochem.* **15**, 827–832.

Latovitzki, N., and Silberberg, D. H. (1977). UDP-galactose:ceramide galactosyltransferase and 2',3'-cyclic nucleotide 3'-phosphohydrolase activities in cultured newborn rat cerebellum. Association with myelination and concurrent susceptibility to 5-bromodeoxyuridine. *J. Neurochem.* **29**, 611–614.

Lisak, R. P., Abramsky, O., Dorfman, S. H., George, J., Manning, G. C., Pleasure, D. E., Saida, T., and Silberberg, D. H. (1979). Antibodies to galactocerebroside bind to oligodendroglia in suspension culture. *J. Neurol. Sci.* **40**, 65–73.

Ludwin, S. K., Kosek, J. C., and Eng, L. F. (1976). The topographical distribution of S-100 and GFA proteins in the adult rat brain: An immuno-histochemical study using horseradish peroxidase-labelled antibodies. *J. Comp. Neurol.* **165**, 197–208.

McCarthy, K., and Partlow, L. (1976a). Preparation of pure neuronal and non-neuronal culture from embryonic chick sympathetic ganglia. A new method based on both differential cell adhesiveness and the formation of homotypic neuronal aggregates. *Brain Res.* **114**, 391–414.

McCarthy, K., and Partlow, L. (1976b). Neuronal stimulation of [^{3}H]thymidine incorporation by primary cultures of highly purified non-neuronal cells. *Brain Res.* **114**, 415–426.

McDermott, J. R., and Smith, A. R. (1978). Absence of myelin basic protein from glial cell lines and cultures. *J. Neurochem.* **30**, 1637–1639.

McMorris, F. A. (1971). Contributions of clonal systems to neurobiology. *Neuroscience Res. Program, Bull.* **11**, 411–536.

McMorris, F. A. (1977). Norpinephrine induces glial specific enzyme activity in cultured glioma cells. *Proc. Natl. Acad. Sci. U.S.A.* **74**, 4501–4504.

Malamed, M. R., Darzynkiewicz, Z., Traganos, F., and Sharpless, T. (1977). Cytology automation by flow cytometry. *Cancer Res.* **37**, 2806.

Mandel, P., Nussbaum, J. L., Nescovic, N. M., Sarlieve, L. L., and Kurihara, T. (1972). Regulation of myelinogenesis. *Adv. Enzyme Regul.* **10**, 101–118.

Mori, S., and Leblond, C. P. (1970). Electron microscopic identification of three classes of oligodendrocytes and a preliminary study of their proliferative activity in the corpus callosum of young rats. *J. Comp. Neurol.* **139**, 1–30.

Murray, M. R. (1977). Introduction. *In* "Cell, Tissue and Organ Cultures in Neurobiology" (S. Fedoroff and L. Hertz, eds.), pp. 1–8. Academic Press, New York.

Norton, W. T. (1972). Myelin. *In* "Basic Neurochemistry" (R. W. Albers, G. J. Siegel, R. Katzman, and B. W. Agranoff, eds.), pp. 365–386. Little, Brown, Boston, Massachusetts.

Norton, W. T. (1977). Isolation and characterization of myelin. *In* "Myelin" (P. Morrel, ed.), pp. 161–199. Plenum, New York.

Norton, W. T., and Poduslo, S. E. (1971). Neuronal perikarya and astroglia of rat brain. Chemical composition during myelination. *J. Lipid Res.* **12**, 84–90.

O'Brien, J. S., and Sampson, E. L. (1965). Fatty acid and fatty aldehyde composition of major brain lipids in normal human gray matter, white matter and myelin. *J. Lipid Res.* **6**, 537–543.

Peters, A., Palay, S. L., and Webster, H. F. (1976). In "The Fine Structure of the Nervous System," pp. 232–263. Saunders, Philadelphia, Pennsylvania.

Pfeiffer, S. E., and Eagle, H. (1976). Effect of ethanesulfonic acid buffers and pH on the accumulation of a nervous specific protein (S-100) and a glial-enriched enzyme in a clonal line of rat astrocytes (C6). *J. Biol. Chem.* **251**, 5112–5114.

Pfeiffer, S. E., and Wechsler, W. (1972). Biochemically differentiated neoplastic clone of Schwann cells. *Proc. Natl. Acad. Sci. U.S.A.* **69**, 2885-2889.

Pleasure, D., and Kim, S. W. (1976). Sterol synthesis by myelinating cultures of mouse spinal cord. *Brain Res.* **100**, 117-126.

Pleasure, D., Abramsky, O., Silberberg, D., Quinn, B., Parkis, J., and Saida, T. (1977). Lipid synthesis by an oligodendroglial fraction in suspension culture. *Brain Res.* **134**, 377-382.

Poduslo, S. E. (1975). The isolation and characterization of a plasma membrane and a myelin fraction derived from oligodendroglia of calf brain. *J. Neurochem.* **24**, 647-654.

Poduslo, S. E. (1978). Studies on isolated, maintained oligodendroglia: Biochemistry, metabolism and *in vitro* myelin synthesis. *In* "Myelination and Demyelination" (J. Palo, ed.), pp. 71-94. Plenum, New York.

Poduslo, S. E., and McKhann, G. M. (1977). Synthesis of cerebrosides by intact oligodendroglia maintained in culture. *Neurosci. Lett.* **5**, 159-163.

Poduslo, S. E., and Norton, W. T. (1972). Isolation and some chemical properties of oligodendroglia from calf brain. *J. Neurochem.* **19**, 727-736.

Poduslo, S. E., and Norton, W. T. (1975). Isolation of specific brain cells. *In* "Methods in Enzymology" (S. P. Colowick and N. Kaplan, eds.), Vol. 35, pp. 561-579. Academic Press, New York.

Poduslo, S. E., McFarland, H. F., and McKhann, G. M. (1977). Antiserum to neurons and to oligodendroglia from mammalian brain. *Science* **197**, 270-272.

Poduslo, S. E., Miller, K., and McKhann, G. M. (1978). Metabolic properties of maintained oligodendroglia purified from brain. *J. Biol. Chem.* **253**, 1592-1597.

Privat, A., and Fulcrand, J. (1977). Neuroglia—From the subventricular precursor to the mature cell. *In* "Cell, Tissue and Organ Cultures in Neurobiology" (S. Fedoroff and L. Hertz, eds.), pp. 11-37. Academic Press, New York.

Raff, M. C., Mirsky, R., Fields, K. L., Lisak, R. P., Dorfman, S. H., Silberberg, S., Greoson, N. A., Leibowitz, S., and Kennedy, M. C. (1978). Galactocerebroside is a specific cell-surface antigenic marker for oligodendrocytes in culture. *Nature (London)* **274**, 813-816.

Raine, C. S., Poduslo, S. E., and Norton, W. T. (1971). The ultrastructure of purified preparations of neurons and glial cells. *Brain Res.* **27**, 11-24.

Ramon-Moliner, E. (1958). A study on neuroglia. *J. Comp. Neurol.* **110**, 157-171.

Salzer, J. L., Glazer, L., and Bunge, R. P. (1977). Stimulation of Schwann cell proliferation by neurite membrane fraction. *J. Cell Biol.* **75**, 118a.

Sellinger, O. Z., and Azcurra, J. M. (1974). Bulk separation of neuronal cell bodies and glial cells in the absence of added digestive enzymes. *Res. Methods Neurochem.* **2**, 3-37.

Shapiro, R., Mobley, W. C., Thiele, S. B., Wilhelm, M. R., Wallace, A., and Kibler, R. F. (1978). Localization of 2′,3′-cyclic nucleotide 3′-phosphohydrolase of rabbit brain by sedimentation in a continuous sucrose gradient. *J. Neurochem.* **30**, 735-744.

Sheppard, J. R., Brus, D., and Weliner, J. M. (1978). Brain reaggregate cultures: Biochemical evidence for myelin membrane synthesis. *J. Neurobiol.* **9**, 309-315.

Sprinkle, T. J., Zaruba, M. E., and McKhann, G. M. (1978a). Activity of 2′,3′-cyclic-nucleotide 3′-phosphodiesterase in regions of rat brain during development: Quantitative relationship to myelin basic protein. *Neurochemistry* **31**, 309-314.

Sprinkle, T. J., Zaruba, M. E., and McKhann, G. M. (1978b). Radioactive measurement of 2′,3′-cyclic nucleotide 3′-phosphodiesterase activity in the central and peripheral nervous system and in extraneural tissue. *Anal. Biochem.* **88**, 449-456.

Sternberger, L. A., Hardy, P. H., Culculis, J. H., and Meyer, H. G. (1970). The unlabeled antibody-enzyme method of immunohistochemistry. Preparation and properties of soluble antigen-antibody complex (horseradish peroxidase-antihorse radish peroxidase) and its use in identification of spirochetes. *J. Histochem. Cytochem.* **18**, 315.

Sternberger, N. H., Itoyama, Y., Kies, M. W., and Webster, H. deF. (1978a). Immunocytochemical methods to identify basic protein in myelin-forming oligodendrocytes of newborn rat C.N.S. *J. Neurocytol.* **7**, 251–263.

Sternberger, N. H., Itoyama, Y., Kies, M. W., and Webster, H. deF. (1978b). Myelin basic protein demonstrated immunocytochemically in oligodendroglia prior to myelin sheath formation. *Proc. Natl. Acad. Sci. U.S.A.* **75**, 2521–2524.

Sundarraj, N., Schachner, M., and Pfeiffer, S. E. (1975). Biochemically differentiated mouse glial lines carrying a nervous system specific surface antigen (NS-1). *Proc. Natl. Acad. Sci. U.S.A.* **72**, 1927–1931.

Szuchet, S., Arnason, B. G. W., and Polak, P. E. (1978a). A new method for oligodendrocyte isolation. *Biophys. J.* **21**, 51a.

Szuchet, S., Wollmann, R. L., and Arnason, B. G. W. (1978b). Ultrastructural characterization of isolated oligodendrocytes. *J. Neuropathol. Exp. Neurol.* **37**, 349.

Szuchet, S., Stefansson, K., Dawson, G., and Arnason, B. G. W. (1979). Biochemical and immunological studies on isolated oligodendrocytes. *Abstr. 10th Annu. Meet., Am. Soc. Neurochem.* p. 22.

Szuchet, S., Stefansson, K., Wollmann, R. L., Dawson, G., and Arnason, B. G. W. (1980a). Maintenance of two subpopulations of oligodendrocytes in long-term culture, *Brain Res.* (in press).

Szuchet, S., Arnason, B. G. W., and Polak, P. E. (1980b). Separation of ovine oligodendrocytes into two distinct bands on a linear sucrose gradient. *J. Neurosci. Meth.* (in press).

Tennekoon, G. I., Cohen, S. R., Price, D., and McKhann, G. M. (1977). Myelinogenesis in optic nerve. *J. Cell Biol.* **72**, 604–616.

Traugott, U., Snyder, D. S., Norton, W. T., and Raine, C. S. (1978). Characterization of antioligodendrocyte serum. *Ann. Neurol.* **4**, 431–439.

Varon, S. (1977). Neural cell isolation and identification. *In* "Cell, Tissue and Organ Cultures in Neurobiology" (S. Fedoroff and L. Hertz, eds.), pp. 237–261. Academic Press, New York.

Varon, S. (1978). Macromolecular glial cell marker. *In* "Dynamic Properties of Glial Cells" (E. Schoffeniels, G. Frank, L. Hertz, and D. B. Tower, eds.), pp. 93–103. Pergamon, Oxford.

Varon, S. S., and Bunge, R. P. (1978). Trophic mechanisms in the peripheral nervous system. *Annu. Rev. Neurosci.* **1**, 327–361.

Waehneldt, T. V. (1977). Protein heterogeneity in rat CNS myelin subfractions. *In* "Myelination and Demyelination" (J. Palo, ed.), pp. 117–133. Plenum, New York.

Weber, K., Rathke, P. C., and Osborn, M. (1978). Cytoplasmic microtubular images in gluteraldehyde-fixed tissue culture cells by electron microscopy and by immunofluorescence microscopy. *Proc. Natl. Acad. Sci. U.S.A.* **75**, 1820–1824.

Weinberg, H. J., and Spencer, P. S. (1976). Studies on the control of myelinogenesis. II. Evidence for neuronal regulation of myelin production. *Brain Res.* **113**, 363–378.

Wolfgram, F., and Rose, A. (1957). The morphology of neuroglia in tissue culture with comparisons to histological preparations. *J. Neuropathol. Exp. Neurol.* **16**, 514–531.

Wood, P. M. (1976). Separation of functional Schwann cells and neurons from normal peripheral nerve tissue. *Brain Res.* **115**, 361–375.

Wood, P. M., and Bunge, R. P. (1975). Evidence that sensory axons are mitogenic for Schwann cells. *Nature (London)* **256**, 662–664.

Zanetta, J. P., Benda, P., Gombos, G., and Morgan, I. G. (1972). The presence of 2′,3′-cyclic AMP 3′-phosphohydrolase in glial cells in tissue culture. *J. Neurochem.* **19**, 831–883.

BIOCHEMICAL MAPPING OF SPECIFIC NEURONAL PATHWAYS

E. G. McGEER AND P. L. McGEER

Department of Psychiatry
University of British Columbia
Vancouver, British Columbia, Canada

I. Introduction

A fundamental challenge facing neuroscientists is to define the wiring diagram of the brain; clearly such knowledge is necessary for an understanding of the neurological basis of human behavior and of pathological processes in central nervous system (CNS) diseases. Traditionally, neuroanatomists have classified neurons according to such criteria as size, axonal length, degree of dendritic branching, and location; both neuroanatomists and neurophysiologists have attempted to trace neuronal paths between various brain nuclei. The relatively recent advent of axonal transport tracing techniques has led and is leading to the discovery of large numbers of previously unsuspected tracts, so that even classic neuroanatomical text books will need revision. An equally great challenge is raised by the recognition that the role and pharmacology of a neuron depends not only on its particular location in the brain but upon the particular chemical substance it uses as a transmitter and the biochemical nature of the neurons with which it interconnects. A major challenge to neuroscientists, therefore, is not only to describe the wiring diagram of the CNS but to "color-code" each neuron according to the particular neurotransmitter it uses.

The magnitude of this problem may be judged when it is realized that there are at least 15–20 different neurotransmitters used in the mammalian brain, that the neuroanatomy of only about three (dopamine, norepinephrine, and serotonin) of these biochemically distinctive systems has, as yet, been worked out in any detail, and that fewer than 10% of the neuronal systems in the brain (exclusive of the cerebellum) have been biochemically "fingerprinted." In this article we will discuss the techniques available to the biochemical neuroanatomist and give some examples of their use along with brief reference to some of the problems involved in each technique. These will be discussed under the general headings of histochemical and immunohistochemical procedures, lesion techniques, physiological methods, radioactive uptake and axonal transport, and selective neurotoxins. Labeling techniques useful in electron microscopic studies of the morphology and interconnections of a neuron of a particular biochemical type will also be briefly described.

II. Techniques for Biochemical Classification of Neurons

A. Histochemical and Immunohistochemical Procedures

These methods offer the best possibilities for precise anatomical localization of neurons of a given biochemical type. The development more than a decade ago of a satisfactory histofluorescence method for catecholamines and serotonin has allowed the neuroanatomy of these systems to be determined in great detail (Dahlstrom and Fuxe, 1964a,b; Lindvall and Björklund, 1978). There is no similarly useful histochemical method for any other biochemical type of neuron.

It must be emphasized that methods based on the metabolic enzymes for breakdown of neurotransmitters are not satisfactory, since such destructive enzymes are not specific to presynaptic neurons or even to neurons at all. In the absence of a simple, definitive histochemical method for cholinergic systems, for example, histochemical methods for acetylcholinesterase have been widely used and have yielded much valuable information (Lewis and Shute, 1978), but this hydrolytic enzyme exists in cholinoceptive as well as in cholinergic neurons. It may also exist in glia. For these reasons it cannot be used as a definitive marker for cholinergic systems at the cellular level. The difficulty is exemplified by the histochemical studies on acetylcholinesterase in the substantia nigra of normal and lesioned cats and monkeys, which led to a suggestion of the possible existence of a striatonigral cholinergic system as a direct feedback loop to the dopaminergic tract (Olivier *et al.*, 1970). Studies in a number of laboratories using biochemical methods for choline acetyltransferase (CAT), a specific marker of cholinergic neurons, have provided firm evidence against the existence of such a tract (McGeer and McGeer, 1979). It has now been established that there is considerable acetylcholinesterase activity in dopaminergic neurons, and it is undoubtedly this which renders acetylcholinesterase an unreliable marker for cholinergic systems in extrapyramidal nuclei (Butcher *et al.*, 1975; Lehmann and Fibiger, 1978).

The development of immunohistochemical procedures offers a broad new front for attack, and such techniques are already being applied with some success to GABA-ergic, cholinergic, dopaminergic, noradrenergic, serotonergic, and various peptidergic neuronal systems. Their sensitivity is made evident by the fact that immunohistochemical work with dopamine-β-hydroxylase has already yielded new refinements of the "map" of noradrenergic systems (Swanson and Hartman, 1975). There are, however, numerous technical difficulties in the development and application of an immunohistochemical method.

In the immunohistochemical technique antibodies are made to a protein marker characteristic of a cell type. In the case of nonpeptidergic neurotransmitters, specific synthetic enzymes have been used as the bases for immunochemistry. In the case of peptidergic neurotransmitters, such as substance P, the enkephalins, neurotensin, or somatostatin, the synthetic processes are not known and the peptides are of too low MW to be antigenic in themselves; they are, therefore, coupled with a protein such as bovine serum albumin, polylysine, or keyhole limpet hemocyanin in order to produce an antigenic molecule. A major problem is to ensure that the antibody produced by such an antigen is specific for the peptide neurotransmitter under investigation (McGeer *et al.*, 1978a, pp. 321–341).

The general techniques and some of the problems involved in immunohistochemical procedures will be illustrated here by the work with CAT used as a specific marker for cholinergic systems. CAT was purified from human neostriatum and injected into rabbits so that anti-CAT antibodies would develop in the rabbit serum. The enzyme must be absolutely pure, and the antibodies produced to it monospecific; otherwise nonspecific staining can be expected. Figure 1 illustrates the specificity for CAT and shows a double-diffusion plate on which a neostriatal extract containing many proteins was run against rabbit sera. The rabbit sera diffusing from the outer wells in the plate meet the brain proteins diffusing from the center well, and a precipitate is formed when the serum antibody meets the antigen. The fact that only one precipitin line forms is a strong indication that the antibodies are monospecific. Additional evidence that these are anti-CAT antibodies is provided by their ability to precipitate CAT activity from homogenates (Table I).

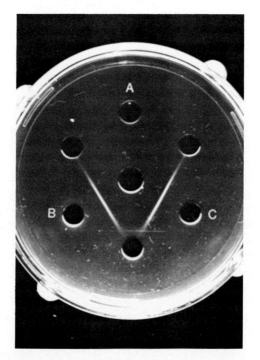

Fig. 1. Ouchterlony double-diffusion plate showing immunoprecipitin reaction of CAT with rabbit serum antibodies. Inner well contained CAT protein. Outer wells at lower left and lower right contained serum from two rabbits immunized against CAT. Outer well at top contained serum from a normal rabbit. Note the single precipitin line with each immune serum and the absence of reaction with normal serum.

TABLE I

PRECIPITATION OF CAT ACTIVITY BY ANTIBODIES[a]

	CAT activity	
	cmp/assay	Percentage
Pure CAT	2720	100
Pure CAT and rabbit anti-CAT	1423	52
Pure CAT plus rabbit anti-CAT and goat anti-rabbit IgG	260	9.5
Pure CAT and normal rabbit and goat anti-rabbit IgG	2609	96
Striatal homogenate plus rabbit anti-CAT plus goat anti-rabbit IgG	96	3
Striatal homogenate plus normal rabbit plus goat anti-rabbit IgG	3145	98

[a] Pure CAT or striatal homogenate (0.1 ml) and 0.1 ml anti-CAT (or normal rabbit) plus 0.1 ml goat anti-rabbit IgG plus 0.1 ml water kept at 4°C for 2 days, centrifuged at 29,000 g for 1 hr and CAT activity measured in supernate. Enzyme assays run on equivalent dilutions.

In essence, the immunohistochemical technique is a reproduction of the double-diffusion plate at the tissue level. Additional reactions are added to provide a marker that can be detected easily. One type of marker is the intensely fluorescent molecule fluorescein which can be chemically coupled to serum proteins. With this technique, the tissue antigen is reacted with rabbit antibody to form the first-stage tissue complex. The tissue complex is then treated with fluorescein-linked goat anti-rabbit serum (Fig. 2a) which attaches to the complex, forming a sandwich which is visible under a fluorescence microscope. Figure 2b shows anterior horn cells of beef spinal cord fluorescing after this treatment; it is part of classic neuroanatomy that these cells are cholinergic.

The most sensitive immunohistochemical technique is Sternberger's ingenious peroxidase–antiperoxidase (PAP) method (Fig. 3A–D) (Sternberger et al., 1970). In this technique horseradish peroxidase (HRP) is injected into an animal such as a rabbit to produce antiserum to the enzyme. This serum is then reacted in a test tube with HRP ("Perox" in Fig. 3) to produce an enzymatically active rabbit PAP complex. Brain tissue to be tested for CAT (or any other antigen) is first reacted with the specific rabbit antiserum (Fig. 3A). This tissue complex is treated with goat anti-rabbit serum as shown in Fig. 3B and then with the rabbit PAP complex (Fig. 3C). In the presence of hydrogen peroxide and diaminobenzidine, the peroxidase at the top of the sandwich catalyzes the splitting of the hydrogen peroxide which then oxidizes the diaminobenzidine to a brown reaction product which can be seen by both light and electron microscopy (Fig. 3D). The cleverness of the technique lies in its amplification. For each

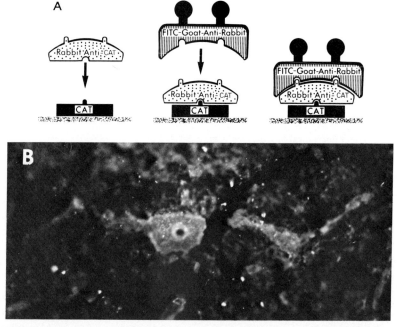

Fig. 2.   (A) Sandwich technique for immune fluorescent tagging of tissue proteins. Tissue
CAT acts as the antigen that couples with immune rabbit serum (left). This is reacted with
fluorescein-tagged goat anti-rabbit serum (center). The complete sandwich is shown on the
right. (B) Anterior horn cells of beef spinal cord fluorescing following the treatment described
in (A) (P. L. McGeer *et al.,* 1974).

tissue molecule of antigen, many thousands of diaminobenzidine molecules
are oxidized. A number of modifications of the original Sternberger procedure have been introduced in attempts to improve sensitivity and
specificity (e.g., de Olmos, 1977; Malmgren and Olsson, 1978).

The results using the PAP technique for striatal tissue are shown in Fig.
3E. One can see many striatal interneurons staining positively for CAT,
while others are clear, indicating that at least one population of noncholinergic striatal neurons exists.

It must be remembered that immunohistochemical techniques involve a
complex series of poorly understood chemical reactions, with obvious opportunities for artifacts to appear. Precise conditions must be found that
permit detection of a specific immunohistochemical reaction not masked
by the nonspecific background staining frequently present, particularly
with the PAP technique. The original antigen–antibody reaction must be
strong, dominating nonspecific reactions of serum globulin with brain pro-

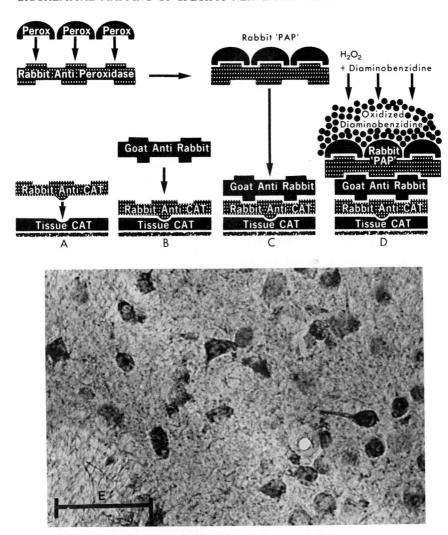

FIG. 3. (A–D) Schematic diagram of the PAP multiple-sandwich technique. (A) A reaction of tissue CAT with rabbit anti-CAT serum similar to that in Fig. 2A. (B) The complex is reacted with goat anti-rabbit serum to serve as a binding agent for the rabbit PAP complex. This is formed by reacting HRP with a rabbit serum made immune to the enzyme protein. (C) The multiple sandwich is completed when the rabbit PAP is reacted with the sandwich containing goat anti-rabbit serum. (D) The reaction of HRP with hydrogen peroxide and diaminobenzidine is shown. (E) Light microscopic immunohistochemical micrograph of guinea pig neostriatum stained by the PAP technique. Numerous positively staining cells of intermediate size are seen. Note the reticular network in the background, which probably represents staining of cellular processes. Calibration bar, 50 μm.

teolipids. The problem of achieving sera with sufficiently high antibody titer to be useful in immunohistochemical work has been a constant source of frustration in working with CAT and some peptidergic neurotransmitters.

Other possible difficulties must also be considered. Since the tissue must first be partially fixed in order to preserve subcellular structure, the catalytic activity, solubility, and antigenic properties of the substance being localized may be changed by the fixative. Because of its large size the serum antibody may not reach the antigen in adequate amounts, or the antigen–antibody complex may be bound to subcellular structures not reflecting the original *in vivo* antigenic sites. The PAP complex itself, which contains the peroxidase marker, has a very high MW which could well retard its ability to penetrate small structures even if the original antigen–antibody reaction were satisfactory. Finally, small amounts of the oxidized product of the peroxidase reaction may be generated from residual endogenous peroxidase in tissues or may diffuse from sites of the specific PAP reaction. These potential artifacts must be kept in mind in interpreting immunohistochemical results.

Lack of penetration has been blamed for the relatively poor staining of nerve endings seen in immunohistochemical work with both CAT (Hattori *et al.,* 1976b, 1977) and tyrosine hydroxylase (TH) (Pickel *et al.,* 1975). In general, the poorer the fixation, the more nerve endings were seen to be stained, which supports the interpretation that the problem is one of penetration.

On the other hand, some light staining of the nucleus has been seen in both CAT-containing cells and TH-containing cells. In subcellular biochemical studies, small amounts of CAT and TH can be found in the nuclear fraction (McGeer *et al.,* 1965), but thorough washing removes nearly all the activity, suggesting that this is weak, probably nonspecific, binding. Such a form of binding may also occur under the conditions used for immunohistochemistry.

Even when the technical problems in immunohistochemistry have been surmounted and a useful technique is available, the ubiquitous distribution of such systems as the cholinergic and GABA-ergic systems in brain makes them difficult to untangle and map. Many months and years of microscopic work were necessary before a map of the dopaminergic system was developed. Yet major dopaminergic activity is limited to relatively few regions of the brain, and almost every region is probably much richer in GABA-ergic and cholinergic structures than the striatum is in dopaminergic nerve endings. It can be anticipated that mapping such complicated systems will require great patience even when problems of penetration, fixation, and reliability have been overcome.

B. Lesion Techniques

These are probably the most generally used methods in biochemical neuroanatomy at present. They can be extremely valuable, as indicated by the fact that they have provided the initial clue to many of the tracts that have now been biochemically fingerprinted, for example, the habenulo-interpeduncular cholinergic system and the striatonigral GABA-ergic and substance-P systems. The basic reasoning underlying lesion techniques is simple: When a neuron is lesioned, there is always anterograde degeneration and there may also be retrograde degeneration if no sustaining collaterals remain (Fig. 4). Biochemical markers specific to that neuronal type should therefore decrease in the nerve ending area and may also decrease in the cell body area. The use and interpretation of lesion techniques are,

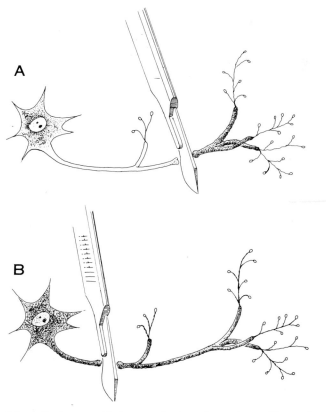

FIG. 4. Schematic diagram of lesions causing loss of neurotransmitters and synthetic enzymes. (A) Only anterograde degeneration-sustaining collateral. (B) Both anterograde and retrograde degeneration—no sustaining collateral.

however, more complex, because there is frequently no good specific marker available, it is difficult to determine the polarity and origin of the neuronal system under study, and the data may be confused by secondary changes that do not directly reflect degeneration of lesioned neurons.

1. Choice of Marker

Levels of neurotransmitter, of synthetic enzyme activity, of high-affinity uptake of transmitter or precursor, and of transmitter release have all been used as specific biochemical markers in lesion studies (Table II). The activity of a catabolizing enzyme, such as acetylcholinesterase or monoamine oxidase, is not a satisfactory index, since these compounds are not limited to specific neuronal types. Receptor binding assays likewise do not give information on the biochemical type of neuron destroyed in lesion experiments, since they are primarily postsynaptic rather than presynaptic markers.

The level of the neurotransmitter itself has often been used as an index of neuronal integrity. In some instances, most notably acetylcholine, the level of transmitter is very sensitive to postmortem changes (Table III), and its assay requires special techniques of sacrifice, thus reducing its reliability as an index. Typically also, the assays for neurotransmitters require much

TABLE II

Presynaptic Biochemical Markers Used in Lesion Studies Aimed
at Identifying Central Pathways Using Some Putative Transmitters[a]

Putative neurotransmitter	Neurotransmitter			Synthetic enzyme activity
	Level	Uptake	Release	
Dopamine	+	+	+	+
Norepinephrine	+	+		+
Serotonin	+	+	+	+
Acetylcholine	+	+[b]	+	+
GABA	+	+	+	+
Glutamate	(+)	+	+	
Aspartate	(+)	+	+	
Glycine	(+)	+		−
Histamine	−			+
Taurine	−			?
Substance P	+			
Enkephalins	+			
Neurotensin	+			
Carnosine	+			

[a] Technique: +, satisfactory; (+), partly satisfactory; −, unsatisfactory; ?, questionable
[b] Uptake of the precursor choline rather than the neurotransmitter.

TABLE III

CHOLINE AND ACETYLCHOLINE CONCENTRATIONS IN WHOLE RAT BRAIN
FOLLOWING VARYING METHODS OF SACRIFICE[a]

	Concentration (nmol/gm of brain)	
	Choline	Acetylcholine
Microwave, immediate homogenization	26.3	24.8
Microwave, homogenization after 5 min	25.4	26.1
Decapitation, homogenization after 5 min	156.7	17.2
Cervical dislocation, microwave after 5 min	148.4	14.9

[a] From Stavinoha and Weintraub, 1974.

more tissue than those for uptake or for synthetic enzymes, hence do not give as fine a localization. This is the only biochemical marker, however, that has so far proven of value in lesion experiments with putative peptide neurotransmitters where the synthetic process is not known and where high-affinity uptake of the transmitter does not appear to occur.

For many probable or putative transmitters (such as glutamate, aspartate, histamine, taurine, and glycine) the level of transmitter is a poor index, because these compounds occur in other large, nontransmitter pools. Glutamate, aspartate, and glycine are all used in many metabolic reactions in every cell in the body; histamine occurs in mast cells in the brain as well as in neurons; and taurine appears to occur in a large, slowly turning-over brain pool of uncertain origin. Whether taurine also occurs in nerve ending vesicles is still a matter of controversy. Despite this difficulty, measurements of changes in the levels of glutamate, aspartate, and glycine have been reported as presumptive evidence of specific neuronal systems using these materials as transmitters (for examples, see Table VI). Although the sensitivity is low, as indicated both by the data in Table IV and by the relatively little difference in levels between various brain regions (Table V), it is useful where other techniques such as uptake are not completely specific.

High-affinity uptake of the neurotransmitter (or a precursor in the case of cholinergic systems) into a synaptosomal fraction has been used as an index of neuronal integrity in lesion experiments. High-affinity uptake may, however, occur into glia as well as into synaptosomes and is limited in that it can only be used on fresh tissue. Freezing the tissue for microdissection, for example, will abolish high-affinity synaptosomal uptake, as will lengthy postmortem storage. High-affinity uptake may also not be specific to a given neurotransmitter. In the case of aspartate and glutamate, for example, the uptake mechanisms appeared to be indistinguishable (Table V).

TABLE IV

CHANGES IN GLUTAMATE UPTAKE AND GLUTAMATE LEVELS AFTER LESIONS
OF SOME GLUTAMATERGIC NEURONS[a]

	Lesioned as percentage of control	
	Glutamate uptake	Glutamate levels
Striatum after cortical lesions	55 ± 7	72 ± 2
Dentate gyrus of hippocampus after bilateral entorhinal lesions	78 ± 5	93 ± 5

[a] Data on hippocampus are from Nadler et al. (1978), on levels in striatum from Kim et al. (1977), and on uptake in striatum from McGeer et al. (1978c).

Despite this lack of specificity and other disadvantages inherent in uptake procedures, this has been the most widely used technique in investigating possible glutamate and aspartate pathways. The decreases in uptake typically found (Table IV) are greater than the changes in the levels of the corresponding neurotransmitters but much less than would be expected in a truly specific marker such as a synthetic enzyme. Thus, for example, lesions of the corticostriatal tract reduced glutamate uptake in the striatum by 40–50% but reduced glutamate levels by only 15–20%. Aspartate levels are not reduced, which is strong evidence that this corticostriatal system is glutamatergic and not aspartatergic, a distinction that could not be made on the basis of high-affinity uptake results alone.

For systems where a specific synthetic enzyme has been identified and is readily assayable, this probably provides the most convenient marker.

TABLE V

RELATIVE LEVELS OF GLUTAMATE AND ASPARTATE UPTAKE AND GLUTAMATE
CONCENTRATIONS IN VARIOUS BRAIN REGIONS AS PERCENTAGES
OF VALUES IN STRIATUM[a]

Area	Aspartate uptake	Glutamate uptake	Glutamate levels
Pons–medulla	9	6	62
Cerebellum	9	6	94
Midbrain	19	18	85
Thalamus–hypothalamus	61	68	91
Striatum	100	100	100
Cortex	119	121	103
Hippocampus–amygdala	136	124	103

[a] Note the very close parallelism between aspartate and glutamate uptakes and the relatively slight regional variation in glutamate levels. Glutamate levels are from Balcom et al. (1976) who also discuss postmortem decline; uptake data are from our laboratory.

Thus, most of the lesion experiments on cholinergic, GABA-ergic, catecholaminergic, or serotonergic systems have involved assays of CAT, glutamic acid decarboxylase (GAD), TH, or tryptophan hydroxylase. For many putative neurotransmitters, however, such as glutamate, aspartate, glycine, or the various peptides, no specific synthetic enzyme is known. In other cases, such as taurine and histamine, a synthetic enzyme may be tentatively identified but may, for one reason or another, be relatively unsatisfactory.

There is reasonably good evidence that the neuronal pool of histamine, unlike the mast cell pool, is dependent upon synthesis from histidine by the action of histidine decarboxylase, and this enzyme has been used as a marker in the relatively few lesion studies done on histaminergic systems (McGeer *et al.*, 1978a, pp. 344–345). So far, however, the activity of histidine decarboxylase measurable in brain homogenates has been very small, so that lesion-induced changes are difficult to detect. It seems probable that the total histidine decarboxylase activity is not being assayed with the present procedures and that technical developments in the future will render this a better marker for histaminergic systems.

Another difficulty is illustrated by the case of taurine. Its synthesis is believed to involve decarboxylation of cysteinesulfinic acid by a specific decarboxylase reported to have a synaptosomal localization (Agrawal *et al.*, 1971). Recent work on the purification of GAD and cysteinesulfinic acid decarboxylase (CSAD) activities from brain homogenates has indicated that there are indeed two separate enzymes but that, unfortunately, GAD will decarboxylate cysteinesulfinic acid (Wu *et al.*, 1978). Measurements of cysteinesulfinic acid decarboxylation on whole-brain homogenates or even on synaptosomal fractions can, therefore, not be taken as an index of the specific CSAD activity. In our hands, for example, we have found significant decreases in apparent CSAD activity in both the substantia nigra and striatum of animals with striatal kainic acid lesions (see Section II,B,3). The parallelism in the fall in apparent CSAD activity and GAD activity in these animals and in those with lesions in other areas of brain suggests to us, however, that we are not measuring primarily the activity of the specific CSAD, hence these results cannot be taken as an indication that taurine exists in specific neurons lesioned by these techniques.

Changes in the amount of neurotransmitter released in a given brain area after lesioning of an afferent tract has been used in a few instances and is believed by Nadler *et al.* (1978) to be the most specific and sensitive indicator available for the integrity of glutamatergic and aspartatergic neurons. This release is potassium-evoked and calcium-dependent. Release studies are technically more demanding than measurements of uptake, transmitter level or synthetic enzyme activity, and too few have yet been done to define the limitations and problems.

In general, as indicated in Table II, it seems clear that there are a number of neurotransmitters such as acetylcholine, GABA, and biogenic amines for which any one of a number of biochemical markers can be used; in such cases consistent results have been reported in studies using a variety of such markers. For many probable putative transmitters, however, there is no really reliable and satisfactory biochemical marker known, and identification of such markers is a major target for biochemical neuroanatomists.

2. POLARITY OF TRACTS

Since a lesioned neuron may undergo retrograde as well as anterograde degeneration, the polarity of the neuron cannot be deduced simply from a fall in synthetic enzyme or neurotransmitter concentration. In some cases a buildup proximal to the lesion can be found in animals examined shortly after the lesion is made; this buildup is the result of axonal transport of neurotransmitter synthetic enzyme from the still functioning cell body. Buildups of CAT have, for example, been detected proximal to lesions of the septohippocampal pathway (Lewis et al., 1967).

In other instances, a clue to the polarity of the tract can be obtained by studying the time course of degeneration. Thus, for example, after hemitransections at a midhypothalamic level, TH activity falls far more rapidly in the striatum than in the substantia nigra (Fig. 5); this is consistent with a more rapid anterograde than retrograde degeneration, and thus

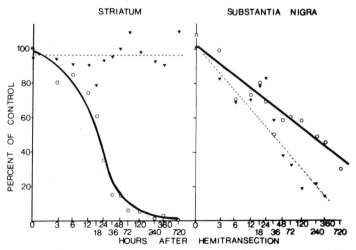

FIG. 5. The effect of hemitransection on amine-synthesizing enzyme activities in the corpus striatum and the substantia nigra. Hours after hemitransection are plotted on a log scale. Circles, TH; triangles, GAD.

with a nigrostriatal rather than a striatonigral dopaminergic tract. The marked fall in GAD in the substantia nigra in these rats (Fig. 5) is in accord with the existence of descending GABA-ergic paths; the lack of change in GAD in the striatum indicates that these fibers have sustaining collaterals ventral to the lesion, hence do not undergo retrograde degeneration.

3. SITE OF ORIGIN

Because the usual surgical or electrolytic lesioning techniques destroy neurons with their somata or axons of passage in the lesion area and may also cause degeneration of neurons afferent to that area (Fig. 6a), it is sometimes difficult to define the site of origin of a tract indicated by lesion experiments. Thus, for example, lesions of the habenula nucleus or knife cuts in the fasciculus retroflexus lead to an almost complete loss of cholinergic indices in the interpeduncular nucleus (IP) (McGeer and McGeer, 1979), and it has been tacitly assumed for several years that all the cholinergic innervation of the IP comes from neurons of the habenula. Recently, however, Gottesfeld and Jacobowitz (1978) have shown that lesions of the diagonal band of Broca (DBB) lead to drops of about 50% in CAT in the IP, and they suggest that the cholinergic innervation of the IP arises partly from habenular neurons and partly from neurons of the DBB which send their axons through the habenula on the way to the IP.

There are also many instances where it is known that a particular lesion produces a fall in some neuronal marker in a distant region, but the precise origin and localization of the tract have not yet been defined. For example, lesions of the septal area will produce falls in CAT in the amygdala as well as in the hippocampus; it is known that the hippocampal afferents come from the medial septal area, but it is suspected that the drops in amygdaloid CAT are a result of lesioning axons of passage. The polarity

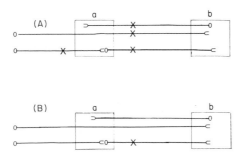

FIG. 6.   Presumed selectivity of the lesion caused by local injection (B) of kainic acid into area a as compared with an electrolytic or surgical lesion (A) of the same area. Neurons that would presumably be destroyed are indicated by an ×.

and origin or terminus of these axons are unknown. Similarly, there are several reports that undercutting the cortex leads to drops in cholinergic indices in the isolated cortical slab, but it is not known whether this loss reflects lesioning of cholinergic afferents or efferents or, indeed, if it indicates some secondary process (McGeer and McGeer, 1979).

The use of kainic acid or other neurotoxic amino acids as lesioning tools may help solve some of these problems. Kainic acid appears to destroy only neuronal somata in the injected area (Fig. 6b) and leave untouched axons of passage and afferent nerve endings. Although histological work must be done in each new region investigated to ensure that kainic acid has indeed destroyed the majority of neuronal somata in the injected area and left other structures untouched, it is already being widely applied as a useful tool (McGeer et al., 1978b). Thus, for example, intrastriatal injections of kainic acid lead to drops in both GAD and substance P in the substantia nigra; the existence of descending tracts with their cells of origin in the striatum is thus confirmed (Gale et al., 1977).

Injections of kainic acid into the striatum or nucleus accumbens, on the other hand, lead to very marked losses of CAT in these nuclei, which is consistent with the belief that cholinergic neurons are almost entirely intrinsic. This supposition was initially based on the failure to find any change in CAT in animals with electrolytic or surgical lesions of all known regions either projecting to these nuclei or receiving projections from them (McGeer et al., 1971; Fonnum et al., 1977). Such data, however, are entirely negative, and it is useful to have the positive support provided by the kainic acid technique.

4. POSSIBLE SECONDARY EFFECTS

Just because some change is noted in a neuronal marker at some distance from a lesion, it cannot be concluded that this change reflects a primary effect of neurons degenerating as a result of the lesion. There may be transsynaptic degeneration, or even a hypertrophic effect. Following administration of 6-hydroxydopamine (6-OHDA) under conditions that cause specific degeneration of catecholaminergic tracts, for example, there is a transient increase in acetylcholine levels in the striatum (Grewaal et al., 1974). Approximately 1 month after 6-OHDA administration, CAT is significantly increased in the head of the striatum but decreased in the tail. Changes in GAD in the striatum have also been found (Nagy et al., 1978). These all appear to be secondary effects of dopaminergic denervation.

In another instance, it has been reported that the decreases in acetylcholinesterase in isolated cortical slabs of cats can be prevented by electrical stimulation of the slabs (Chu *et al.,* 1971); this suggests the change is secondary rather than primary. The possibility of such secondary changes must be borne in mind in interpreting data from lesion experiments.

C. Physiological Methods

Pathways in the brain can be located by recording monosynaptic excitation or inhibition in one area of the brain when neuronal somata in another area are stimulated. For pathways served by a neurotransmitter for which a specific antagonist is available, the technique may be extended to color-coding by showing that the postsynaptic effect of stimulation is blocked by the specific antagonist. Spencer, for example, made the initial suggestion that the corticostriatal tract was glutamatergic on the basis of his finding that excitation of striatal neurons induced by either cortical stimulation or the application of glutamate was blocked by diethyl glutamate (Spencer, 1976). This evidence was, however, considered weak because of the probability that diethyl glutamate is not a good specific antagonist of glutamate and, unfortunately, good specific antagonists for many putative neurotransmitters have not yet been found.

Another technical difficulty is illustrated by the initial report suggesting the existence of a GABA-ergic habenuloraphe tract. This was made on the basis of findings that the inhibition of raphe neurons induced by habenular stimulation was blocked by the GABA antagonist picrotoxin (Wang and Aghajanian, 1977). Subsequent work, however, suggested that a polysynaptic pathway was involved and that the habenuloraphe tract probably excited GABA-ergic intrinsic neurons in the raphe area. Lesion studies in our own and other laboratories (Gottesfeld *et al.,* 1978) are in accord with the view that the habenuloraphe tract is not GABA-ergic.

At best, these physiological methods are slow and technically demanding, and their greatest use in the CNS may be to test the reality of neurotransmitter assignments made on the basis of other, less critical methods. There have been excellent examples of their use in this way. Some time after biochemical studies on the spinal cord had shown that glycine had the predicted distribution of the postsynaptic inhibitory transmitter (Aprison and Werman, 1965), for example, a rigorous neurophysiological comparison showed that glycine iontophoresed onto motor neurons duplicated the action of the inhibitory transmitter released on stimulation of these spinal interneurons. This was not the case with GABA, and strychnine, which had no effect on the action of GABA, blocked the inhibitory action of glycine on the motor neurons (Werman *et al.*, 1968; Curtis *et al.*, 1967). In another instance, iontophoretic application of GABA to single neurons in Deiters' nucleus was shown to mimic the effect of Purkinje cell stimulation, and both effects were blocked by picrotoxin and bicuculline but not by strychnine (McGeer *et al.*, 1978a, p. 203).

D. Radioactive Uptake and Axonal Transport

Uptake of radioactive neurotransmitter has been used in some instances to label neuronal structures for autoradiographic study at the light and electron microscopic levels. Uptake studies in the cerebellum using radioactive GABA, for example, clearly show heavy labeling in the basket, stellate, and Golgi cell regions, whereas granule cells (which are probably glutamatergic) do not accumulate GABA (Hökfelt and Ljungdahl, 1972). Preferential accumulation of radioactive glycine in synaptosomes of the ventral horn gray matter was used as one indicator that many spinal interneurons are glycinergic (Ljungdahl and Hökfelt, 1973). At the electron microscopic level, early studies on preferential labeling by tritiated GABA of nerve endings in the substantia nigra and of cell bodies in the globus pallidus provided some of the first morphological evidence consistent with the pallidonigral GABA-ergic tract indicated by lesion data (Hattori *et al.*, 1973). This technique has serious limitations, however, because of nonspecificity of uptake, possible metabolism and diffusion of the labeled material, and relatively poor structural preservation during the incubation procedure.

Axonal transport of the specific neurotransmitter or very closely related molecules has been demonstrated in a few systems (GABA-ergic, noradrenergic, and dopaminergic) and has been suggested as a possible method for fingerprinting systems. It is unlikely to have wide use, however, because of problems of metabolism, diffusion, and axonal pickup.

Axonal transport of transmitters has been extensively studied in the dopaminergic nigrostriatal tract (Fibiger *et al.*, 1973). In animals treated with a monoamine oxidase (MAO) inhibitor in order to inhibit metabolism of catecholamines, rapid transport of dopamine from the substantia nigra to the neostriatum in the rat occurred by a reserpine-sensitive process which depended upon the integrity of the dopaminergic neurons. Studies of the distribution of radioactivity in the brain following the injection of various labeled neurotransmitters, putative neurotransmitters, or related compounds into the substantia nigra suggested that only dopamine, norepinephrine, and octopamine were specifically accumulated in the ipsilateral striatum. It is noteworthy however, that many of the compounds caused diffuse labeling of the striatum along with other brain areas, particularly the hypothalamus and thalamus, and careful mathematical analysis of the data plus comparative experiments in animals where the nigrostriatal tract had been specifically destroyed with 6-OHDA were necessary in order to detect the specific transport process (E. G. McGeer *et al.,* 1974). Moreover, it has been shown that dopa can be picked up by axons in the medial forebrain bundle and thereafter converted to dopamine which is transported to the striatum (Fibiger and McGeer, 1974). The occurrence of such an axonal pickup poses an additional problem in the interpretation of data. Still another difficulty is indicated by the repeated attempts to demonstrate transport of serotonin in the raphe projection (Hattori *et al.,* 1976a); in this case considerable metabolism of the injected material occurred, and the labeling in areas remote from the injection site was clearly nonspecific.

Axonal transport of the neurotransmitter or related compound has been used in a few specific instances as an aid in identifying the morphology of the nerve terminals of a particular neuronal tract (E. G. McGeer *et al.,* 1975; P. L. McGeer *et al.,* 1975), but its potential for biochemical neuroanatomical investigations seems relatively small.

E. Selective Neurotoxins

Neurons can be labeled for both light and electron microscopic studies by degeneration (cf. Fig. 7A), and this can be a very useful technique for the biochemical neuroanatomist if a chemical lesioning tool specific to a given neuronal type can be found. The closest to such an agent is 6-OHDA which, under proper conditions of use, is specific for catecholamine neurons. 5,6-Dihydroxytryptamine and *p*-chloroamphetamine will cause degeneration of serotonergic neurons but are neither as powerful nor as specific as 6-OHDA (Hattori *et al.,* 1976a). There is a recent report suggesting that capsaicin may have some specificity for substance-P neurons (Jessell *et al.,* 1978). No toxin specific to any other type of neuron has yet been reported, and discovery of such toxins remains a worthwhile goal for biochemical neuroanatomists.

There are then a number of methods that can be used for the biochemical coding of neuronal tracts in the brain and, the more different tools that can be brought to bear on any given tract, the greater the chance of a firm identification of its neurotransmitter. Table VI gives a number of examples of CNS tracts that have been biochemically characterized and the methods that have been used.

TABLE VI
Some Biochemically Characterized Pathways and Methods Used to Study Them[a]

Transmitter	Pathways	Methods						
		Hst.	Imh.	Les.	Ax.T.	Upt.	Phy.	Ntx.
GABA	Purkinje cells		✓	✓			✓	
	Cerebellar Golgi cells		✓			✓	✓	
	Cerebellar basket cells		✓			✓	✓	
	Cerebellar stellate cells				✓	✓	✓	
	Hippocampal basket cells		✓	✓		✓	✓	
	Neostriatal interneurons			✓		✓		
	Striatonigral			✓				
	Pallidonigral			✓	✓			
	Spinal cord interneurons						✓	
	Cortical interneurons					✓	✓	
	Olfactory bulb interneurons		✓	✓		✓	✓	
	Retinal interneurons					✓	✓	
Acetylcholine	Septohippocampal			✓		✓		
	Septocingulate			✓		✓		
	Habenulointerpeduncular		✓	✓				
	Striatal interneurons		✓	✓				
	Interneurons in nucleus accumbens			✓				
	Mossy fibers of cerebellum		✓	✓				
	Tuberoinfundibular			✓			✓	
	DBB to interpeduncular			✓				
Glutamate and/ or aspartate	Corticostriatal			✓				
	Entorhinal-hippocampal			✓				

366

	Hst.	Imh.	Les.	Upt.	Ax.T.	Phy.	Ntx.
Retinotectal				✓			
Cerebellar granule cells				✓			
Spinal interneurons				✓			
Hippocampal ipsilateral				✓	✓		
Hippocampal/subiculoseptal/ mammillary				✓			
Olfactory bulb/olfactory cortex				✓			
Glycine							
Spinal interneurons	✓	✓	✓	✓			
Dopamine							
Nigrostriatal	✓	✓	✓	✓	✓	✓	✓
Mesolimbic	✓	✓	✓		✓	✓	✓
Mesocortical	✓	✓	✓				✓
Tuberoinfundibular	✓	✓					
Olfactory bulb interneurons	✓	✓					
Norepinephrine							
Locus coeruleus to hypothalamus, etc.	✓	✓	✓	✓	✓		
Epinephrine							
Brain stem to thalamus, hypothalamus, etc.	✓	✓					
Serotonin							
Raphe to hypothalamus, hippocampus, etc.	✓	✓	✓	✓	✓		
Substance P							
Striatonigral	✓	✓	✓		✓		
Habenulointerpeduncular	✓	✓			✓		
Primary sensory afferents	✓	✓	✓		✓		
Enkephalins							
Striatopallidal	✓	✓					
Carnosine							
Primary olfactory neurons	✓	✓			✓		
Histamine							
Brain stem to telencephalon	✓				✓		

[a] Methods are those discussed, i.e., Hst., histochemical; Imh., immunohistochemical; Les., lesion techniques; Ax.T., axonal transport of neurotransmitter; Upt., uptake of radioactive neurotransmitter; Phy., physiological; Ntx., specific chemically induced degeneration. List of reported pathways is not complete, particularly those served by aromatic amines and various pertides. For further information see McGeer et al. (1978a, pp. 186, 204, 226, 242, 287, 297, and 347) and McGeer and McGeer (1979).

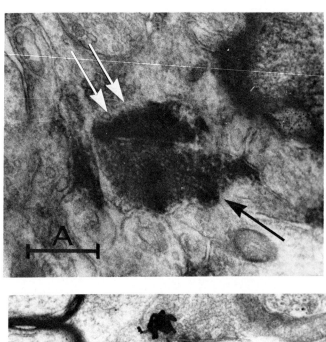

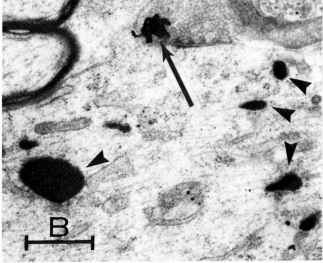

FIG. 7. Electron micrographs showing neuronal interconnections in double-labeling experiments. (A) Degenerating (dopaminergic) nerve ending (single arrow) in guinea pig neostriatum, following 6-OHDA administration, making asymmetrical contact with dendritic spine (double arrows) staining positively for CAT. Bar indicates 0.25 μm. (B) Nerve ending (arrow) of a habenular projection labeled by anterograde flow of radioactive protein synapsing on a dopaminergic dendrite in the A10 area labeled by retrograde transport of HRP from the nucleus accumbens (arrowheads indicate HRP reaction product). Bar indicates 50 μm.

III. Identification of Interconnections between Neurons Using Different Transmitters

It seems just as important to establish the interconnections between biochemically different types of neurons in various areas as to color-code the individual neuronal tracts themselves. There are two basic techniques used in an effort to deduce or establish interconnections. One is pharmacological. This can best be described by example. Dopamine agonists have been repeatedly shown to reduce acetylcholine release and increase acetylcholine levels in the striatum, whereas dopamine antagonists have the opposite effect (McGeer *et al.*, 1978a, p. 457). Only minor effects of these drugs on GABA release and GABA levels are noted (McGeer *et al.*, 1976). These data led to the supposition that there was prominent innervation of cholinergic neurons in the striatum by dopaminergic afferents. The lack of effect of dopaminergic drugs on cholinergic indices in the nucleus accumbens has suggested, on the other hand, that in that area these two systems are not directly interconnected (Consolo *et al.*, 1977).

A more definitive technique involves electron microscopic examination using double-labeling methods. In one example of such a technique, direct innervation of cholinergic dendrites (labeled by the immunohistochemical procedure for CAT) by dopaminergic nerve endings (labeled by degeneration with 6-OHDA) was demonstrated in the striatum of rats (Fig. 7A). In other examples, direct innervation of dopaminergic dendrites in the substantia nigra and A10 area by afferents from the globus pallidus and habenula, respectively, has also been shown. The dopaminergic dendrites were labeled in the first instance by degeneration with 6-OHDA and in the second instance by HRP transported in a retrograde fashion from the nucleus accumbens (Fig. 7B). The afferents were labeled in each instance by radioactive protein transported in anterograde fashion from the neuronal somata which were injected with radioactive leucine.

It can be expected that, as more and more neuronal tracts in the brain are color-coded for the neurotransmitter they use, increasing attention will be given to establishing the precise nature of the connections these tracts make with each other.

Acknowledgment

This work was supported by the Medical Research Council of Canada.

References

Agrawal, H. C., Davison, A. N., and Kaczmarek, L. K. (1971). Subcellular distribution of taurine and cysteinesulphinate decarboxylase in developing rat brain. *Biochem. J.* **122**, 759–763.

Aprison, M. H., and Werman, R. (1965). The distribution of glycine in cat spinal cord and roots. *Life Sci.* **4**, 2075–2083.

Balcom, G. J., Lenox, R. H., and Meyerhoff, J. L. (1976). Regional glutamate levels in rat brain determined after microwave fixation. *J. Neurochem.* **26**, 423–425.

Butcher, L. L., Talbot, K., and Bilezikjian, L. (1975). Localization of acetylcholine activity in the neostriatum. *Brain Res.* **71**, 167–171.

Chu, N. S., Rutledge, L. T., and Sellinger, O. Z. (1971). The effect of cortical undercutting and long-term electrical stimulation on synaptic acetylcholinesterase. *Brain Res.* **29**, 323.

Consolo, S., Ladinsky, H., Bianchi, S., and Ghezzi, D. (1977). Apparent lack of a dopaminergic-cholinergic link in the rat nucleus accumbens septituberculum olfactorium. *Brain Res.* **135**, 255–263.

Curtis, D. R., Hosli, L., and Johnston, G. A. R. (1967). Inhibition of spinal neurons by glycine. *Nature (London)* **215**, 1502–1503.

Dahlstrom, A., and Fuxe, K. (1964a). A method for the demonstration of monoamine-containing fibers in the central nervous system. *Acta Physiol. Scand.* **60**, 293–295.

Dahlstrom, A., and Fuxe, K. (1964b). Evidence for the existence of monoamine-containing neurons in the central nervous system. *Acta Physiol. Scand., Suppl.* **62**, 232.

de Olmos, J. S. (1977). An improved HRP method for the study of central connections. *Exp. Brain Res.* **29**, 541–552.

Fibiger, H. C., and McGeer, E. G. (1974). Accumulation and axoplasmic transport of dopamine but not of amino acids by axons of the nigro-striatal projection. *Brain Res.* **72**, 366–369.

Fibiger, H. C., McGeer, E. G., and Atmadja, S. (1973). Axoplasmic transport of dopamine in nigro-striatal fibers. *J. Neurochem.* **21**, 373–385.

Fonnum, F., Walaas, I., and Iversen, E. (1977). Localization of GABAergic, cholinergic and aminergic structures in the mesolimbic system. *J. Neurochem.* **29**, 221–230.

Gale, K., Hong, J. S., and Guidotti, A. (1977). Presence of substance P and GABA in separate striatonigral neurons. *Brain Res.* **136**, 371–375.

Gottesfeld, Z., and Jacobowitz, D. M. (1978). Cholinergic projection of the diagonal band to the interpeduncular nucleus of the rat brain. *Brain Res.* **156**, 329–342.

Gottesfeld, Z., Hoover, D. B., Muth, E. A., and Jacobowitz, D. M. (1978). Lack of biochemical evidence for a direct habenulo-raphe GABAergic pathway. *Brain Res.* **141**, 353–356.

Grewaal, D. S., Fibiger, H. C., and McGeer, E. G. (1974). 6-Hydroxydopamine and striatal acetylcholine levels. *Brain Res.* **73**, 372–375.

Hattori, T., McGeer, P. L., Fibiger, H. C., and McGeer, E. G. (1973). On the source of GABA-containing terminals in the substantia nigra. Electron microscopic, autoradiographic and biochemical studies. *Brain Res.* **54**, 103–114.

Hattori, T., McGeer, P. L., and McGeer, E. G. (1976a). Synaptic morphology in the neostriatum of the rat: Possible serotonergic synapse. *Neurochem. Res.* **1**, 451–467.

Hattori, T., Singh, V. K., McGeer, E. G., and McGeer, P. L. (1976b). Immunohistochemical localization of choline acetyltransferase containing neostriatal neurons and their relationship with dopaminergic synapses. *Brain Res.* **102**, 164–173.

Hattori, T., McGeer, E. G., Singh, V. K., and McGeer, P. L. (1977). Cholinergic synapse of the interpeduncular nucleus. *Exp. Neurol.* **55**, 102–111.

Hökfelt, T., and Ljungdahl, A. (1972). Autoradiographic identification of cerebral and cerebellar cortical neurons accumulating labelled gamma-aminobutyric acid (^{3}H-GABA). *Exp. Brain Res.* **14**, 331–353.

Jessell, T. M., Iversen, L. L., and Cuello, A. C. (1978). Capsaicin-induced depletion of substance P from primary sensory neurones. *Brain Res.* **152**, 183–188.

Kim, J. S., Hassler, R., Haug, P., and Paik, K. S. (1977). Effect of frontal cortex ablation on striatal glutamic acid level in rat. *Brain Res.* **132**, 370–374.

Lehmann, J., and Fibiger, H. C. (1978). Acetylcholinesterase in the substantia nigra and caudate-putamen of the rat: Properties and localization in dopaminergic neurons. *J. Neurochem.* **30**, 615–624.

Lewis, P. R., and Shute, C. C. D. (1978). Cholinergic pathways in CNS. *In* "Handbook of Psychopharmacology" (L. L. Iversen, S. D. Iverson, and S. H. Snyder, eds.), Vol. 9, pp. 315–356. Plenum, New York.

Lewis, P. R., Shute, C. C. D., and Silver, A. (1967). Confirmation from choline acetylase of a massive cholinergic innervation to the rat hippocampus. *J. Physiol. (London)* **191**, 215–224.

Lindvall, O., and Björklund, A. (1978). Organization of catecholamine neurons in the rat central nervous system. *In* "Handbook of Psychopharmacology" (L. L. Iversen, S. D. Iversen, and S. H. Snyder, eds.), Vol. 9, pp 139–231. Plenum New York.

Ljungdahl, A., and Hökfelt, T. (1973). Autoradiographic uptake patterns of (^{3}H)GABA and (^{3}H)glycine in central nervous tissues with special reference to the cat spinal cord. *Brain Res.* **62**, 587–595.

McGeer, E. G., Searl, K., and Fibiger, H. C. (1974). Chemical specificity of dopamine transport in the nigro-neostriatal projection. *J. Neurochem.* **24**, 283–288.

McGeer, E. G., Hattori, T., and McGeer, P. L. (1975). Electron microscopic localization of labeled norepinephrine transported in nigrostriatal neurons. *Brain Res.* **86**, 478–482.

McGeer, P. L., and McGeer, E. G. (1979). Central cholinergic pathways. *In* "The Uses of Choline and Lecithin in the Treatment of Neurologic and Psychiatric Diseases" (A. Barbeau, R. Wurtman, and J. Growdon, eds.), pp. 177–199. Raven Press, New York.

McGeer, P. L., Bagchi, S. P., and McGeer, E. G. (1965). Subcellular localization of tyrosine hydroxylase in beef caudate nucleus. *Life Sci.* **4**, 1859–1867.

McGeer, P. L., McGeer, E. G., Fibiger, H. C., and Wickson, V. (1971). Neostriatal choline acetylase and acetylcholinesterase following selective brain lesions. *Brain Res.* **35**, 308–314.

McGeer, P. L., McGeer, E. G., Singh, V. K., and Chase, W. H. (1974). Choline acetyltransferase localization in the central nervous system by immunohistochemistry. *Brain Res.* **81**, 373–379.

McGeer, P. L., Hattori, T., and McGeer, E. G. (1975). Chemical and autoradiographic analysis of γ-aminobutyric acid transport in Purkinje cells of the cerebellum. *Exp. Neurol.* **47**, 26–41.

McGeer, P. L., Grewaal, D. S., and McGeer, E. G. (1976). Effect on extra-pyramidal GABA levels of drugs which influence dopamine and acetylcholine metabolism. *In* "Advances in Parkinsonism" (W. Birkmayer and O. Hornykiewicz, eds.), pp. 132–140. F. Hoffmann-LaRoche & Co., Ltd., Basel.

McGeer, P. L., Eccles, J. C., and McGeer, E. G. (1978a). "Molecular Neurobiology of the Mammalian Brain." Plenum, New York.

McGeer, P. L., McGeer, E. G., and Hattori, T. (1978b). Kainic acid as a tool in Neurobiology. *In* "Kainic Acid as a Tool in Neurobiology" (E. G. McGeer, J. W. Olney, and P. L. McGeer, eds.), pp. 123–138. Raven Press, New York.

McGeer, P. L., McGeer, E. G., Scherer, U., and Singh, K. (1978c). A glutamatergic cortico-striatal path? *Brain Res.* **128**, 369–373.

Malmgren, L., and Olsson, V. (1978). A sensitive method for histochemical demonstration of horseradish peroxidase in neurons following retrograde axonal transport. *Brain Res.* **148**, 279-294.

Nadler, J. V., White, W. F., Vaca, K. W., Perry, P. W., and Cotman, C. W. (1978). Biochemical correlates of transmission mediated by glutamate and aspartate. *J. Neurochem.* **31**, 147-155.

Nagy, J. I., Vincent, S. R., and Fibiger, H. C. (1978). Altered neurotransmitter synthetic enzyme activity in some extrapyramidal nuclei after lesions of the nigro-striatal dopamine projection. *Life Sci.* **22**, 1777-1782.

Olivier, A., Parent, A., Simard, H., and Poirier, L. J. (1970). Cholinesterase striatopallidal and striatal nigral efferents in the cat and monkey. *Brain Res.* **18**, 273-282.

Pickel, V. M., Tong, H. J., and Reis, D. J. (1975). Ultrastructural localization of tyrosine hydroxylase in noradrenergic neurons of brain. *Proc. Natl. Acad. Sci. U.S.A.* **72**, 659-663.

Spencer, H. J. (1976). Antagonism of cortical excitation of striatal neurons by glutamic acid diethyl ester: Evidence for glutamic acid as an excitatory transmitter in the rat striatum. *Brain Res.* **102**, 91-101.

Stavinoha, W. B., and Weintraub, S. J. (1974). Choline content of rat brain. *Science* **183**, 964-965.

Sternberger, L. A., Hardy, P. H., Cuculis, J. J., and Myer, H. G. (1970). The unlabelled antibody enzyme method by immunohistochemistry: Preparation and properties of soluble antigen-antibody complex (horseradish peroxidase-antiperoxidase) and its use in identification of spirochetes. *J. Histochem. Cytochem.* **18**, 315-333.

Swanson, L. W., and Hartman, B. K. (1975). The central adrenergic system. An immunofluorescence study of the location of cell bodies and their efferent connections in the rat utilizing dopamine-beta-hydroxylase as a marker. *J. Comp. Neurol.* **163**, 467-505.

Wang, R. Y., and Aghajanian, G. K. (1977). Physiological evidence for habenula as major link between forebrain and midbrain raphe. *Science* **197**, 89-91.

Werman, R., Davidoff, R. A., and Aprison, M. H. (1968). Inhibitory effect of glycine in spinal neurons in the cat. *J. Neurophysiol.* **31**, 81-95.

Wu, J. Y., Chen, M. S., and Huang, W. M. (1978). Purification and immunochemical studies of cysteic acid decarboxylase and glutamate decarboxylase from bovine brain. *8th Annu. Meet. Soc. Neurosci. Abstr.* p. 454.

SEPARATION OF NEURONAL AND GLIAL CELLS AND SUBCELLULAR CONSTITUENTS

FRITZ A. HENN

Neurochemical Research Laboratories
University of Iowa
Iowa City, Iowa

I. Introduction

The analysis of any tissue usually proceeds at two distinct levels. One is an anatomical investigation in which the tissue structure is examined with ever-increasing resolution. The other is a biochemical examination in which the molecular components of the tissue are examined and their organization into units of structure is subsequently investigated. The ultimate marriage of these two approaches should result in an integrated picture of biological structure and function. This would include knowledge of molecular structure, the organization of subcellular organelles and their function, the properties and peculiarities of the cell types, and the interaction of various cells leading to the expression of organ function. The meeting place for these two approaches is in a study of the cell, and in no tissue is this more interesting or complex than in brain. Cellular neurobiology aims to understand the diverse cells of the central nervous system (CNS) and how they work together to process, store, and utilize information. To obtain an idea of the potentialities of various brain cells it would be desirable to isolate relatively pure samples of specific groups of neuronal and glial cells and study their function.

This isolation of homogeneous groups of neurons and glia is a long-term goal which has only been approached in the last 10–15 yr. Isolation of cells goes back over 40 yr (Rous and Beard, 1934), although such work on brain tissue began only 20 yr ago with Korey et al. (1958), who isolated glial cells from white matter. Previously, information on given CNS cell types was obtained using hand-dissected samples or the dissection of topographically defined brain regions. This generally resulted in small samples of mixed cell types. Large neurons could be hand-dissected and studied, but this was a tedious and limited approach in the study of neural cells. The more recent efforts began in 1965 when Rose described the isolation, with the aid of gradient centrifugation, of a neuronally enriched fraction and a neuropil fraction rich in glial elements. This technique led to renewed interest in the problem of isolating cells from the CNS, and several modifications of this method have been developed in laboratories throughout the world. Most of these methods separate a glial, and more specifically astrocytic (Norton and Poduslo, 1971; Hamberger et al., 1975), cell population from a neuronal cell population, and the two fractions are often referred to as bulk-prepared glial cells (astrocytes) and bulk-prepared neurons. Methods for preparing oligodendrocytes have been developed by Fewster and Mead (1968) and Poduslo and Norton (1971). While the general scheme of cell isolation by tissue dissociation and centrifugation is similar in all these techniques, a variety of separate procedures exist, unlike the situation in subcellular fractionation where methods are more standardized. This diversity of approaches points to the imperfections in current techniques and the difficulties inherent in studying CNS tissue. The brain, unlike other organs, contains a diverse set of cells whose principal features are long, entwined processes. Disaggregation of the CNS tissue inevitably leads to cell rupture, limiting both the net yield and functional integrity of the cells.

II. Methods of Cell Isolation

A. General Principles

The principal steps in cell isolation involve methods for disaggregating cells and then methods for separating them into relatively homogeneous fractions. The issues involved in tissue disaggregation include the media used in preparation of the tissue, incubation or lack of it, and the mechanical or enzymatic means for dissociating cells. The first question involves the media used to carry out cell isolation. Table I illustrates the variety of media used in the isolation of neurons and astrocytes by several workers. The early attempts utilized a sucrose medium by analogy with older subcellular fractionation techniques (Korey, 1957; Rappaport and Howze, 1966). Other preparations have substituted Ficoll for sucrose, for better cell preservation (Rose, 1965; Flangas and Bowman, 1968; Satake and Abe, 1966; Blomstrand and Hamberger, 1969). Polyvinylpyrrolidone (PVP) was introduced by Sellinger (Sellinger and Azcurra, 1974) in his procedure, by analogy with procedures widely used for sedimentation of blood cells. PVP serves as a stabilizing media for isolating CNS cells but inhibits certain membrane enzymes such as Na^+,K^+-ATPase (Sinha and Rose, 1972). Serum albumin is often used as a membrane-stabilizing system and together with increased levels of hexoses was used by Norton and Poduslo (1970) for incubating their tissue. The recent work from Norton's laboratory uses Ficoll in place of albumin (Farooq et al., 1977; Farooq and Norton, 1978). The ionic composition of isolation media is nearly as variable as the number of laboratories carrying out isolations. Most solutions are buffered with phosphate, Tris–HCl, or both buffers, but the pH used ranges from 6.0 to 7.8. The low pH is used in conjunction with proteolytic enzymes, whereas other procedures use pH values in the physiological range, with the exception of Sellinger's technique in which an unbuffered solution of about pH 4.7 is used (Sellinger and Azcurra, 1974). The use of divalent cations is variable, with some groups limiting Ca^{2+}, thinking that this ion is involved in cell–cell adhesion, and several workers (Satake and Abe, 1966; Flangas and Bowman, 1968; Sellinger and Azcurra, 1974) add exogenous Ca^{2+} for cell membrane stabilization. The role of potassium is also viewed differently by several groups. Some investigators use high concentrations (Rose, 1967; Satake and Abe, 1966) in an effort to retain high intracellular values of K^+, but others use physiological concentrations of K^+ (Blomstrand and Hamberger, 1969), and one group used

TABLE I

MEDIA FOR DISSOCIATION OF BRAIN TISSUE

Method	pH	Sucrose (mM)	Ficoll (%)	PVP (%)	BSA (%)	Glucose (%)	Fructose (%)	NaCl (mM)	KCl (mM)	CaCl₂ (mM)	MgCl₂ (mM)	Phosphate buffer (mM)	Tris–HCl (mM)	Trypsin (%)
Korey, 1957	—	0.25	—	—	—	—	—	—	—	—	—	3	—	—
Rappaport and Howze, 1966	7.8	0.50	—	—	—	—	—	140 TPB[a]	—	—	—	5	—	—
Rose, 1965	7.4	—	10	—	—	—	—	—	100	—	—	10	—	—
Satake and Abe, 1966	7.6	—	20	—	—	—	—	50	100	10	—	—	10	—
Flangas and Bowman, 1968	7.8	—	10	—	—	0.4	—	—	—	5	—	—	50	—
Blomstrand and Hamberger, 1969	7.4	—	2	—	—	—	—	120	5	—	2.5	5	35	—
Sellinger and Azcurra, 1974	~4.7	—	—	7.5	1	—	—	—	10	—	—	—	—	—
Norton and Poduslo, 1970	6.0	—	—	—	1	5	5	—	—	—	—	100	—	1
Farooq et al., 1977	6.0	—	2	—	—	8	5	—	—	—	—	100	—	0.1

[a] Tetraphenylboron.

tetraphenylboron, a K^+-complexing agent, to remove endogenous K^+ as an aid to tissue dissociation (Rappaport and Howze, 1966). However, tetraphenylboron has been shown to alter cell morphology and inhibit respiration (Mahaley and Wilfong, 1969) and thus is no longer used.

One of the principal differences in techniques for the isolation of cells from the CNS is whether or not enzymes are used to help disrupt the tissue. Norton and co-workers (Poduslo and Norton, 1975; Farooq et al., 1977; Farooq and Norton, 1978) have utilized trypsin or acetylated trypsin to carry out their cell separations. In general, the morphology and respiration of their cells appear equal to that of cells isolated without enzymes. In their more recently reported technique the morphological appearance of the cells is somewhat better than that observed with most other methods, and the dendrites and astrocytic filaments are better preserved. There are, however, serious reservations about the use of trypsin, especially in concentrations as high as 1%. These include the possibility of surface protein hydrolysis, especially in studies aimed at defining membrane transport systems or membrane receptors. The work of Hemminki et al. (1970) showed that incubation with trypsin decreased the DNA content of the tissue, suggesting increased leakage of intracellular contents during incubation. Hamberger et al. (1971) have also shown that incubation with trypsin at 1% markedly alters the size distribution of particles, suggesting increased tissue destruction when enzymes are used.

In a majority of the methods the CNS tissue is initially chopped into slices or cubes. The chopping is followed by an incubation step at 37°C in the methods developed by Hamberger et al. (1971) and Poduslo and Norton (1975), but not in those of Sellinger and Azcurra (1974) or Rose (1967), as shown in Table II. The next step is to dissociate tissue by passage through some type of mesh in all the isolation techniques except the recent one described by Farooq et al. (1977). In this procedure the use of various nozzles to aspirate tissue, followed by vortex mixing, was found to dissociate cells in a somewhat gentler fashion with more reproducible results. As shown in Table II, other workers either tease brain tissue directly through nylon mesh (Rose, 1965), using gentle positive or negative pressure, or use a metal tissue press with screening (Flangas and Bowman, 1968). In most procedures tissue disruption is followed by filtration through increasingly finer meshes which serve to trap large, undissociated tissue fragments and capillaries. Because normal neuronal perikarya are between 12 and 20 μm, the final filtration usually employs mesh of 35–100 μm. The smaller glia easily pass through, as do small blood capillaries up to 500 μm in length. A majority of these can be removed by filtration through multiple layers of mesh. An alternative procedure has been suggested by Rose and Sinha (1970) in which the filtered suspension is passed through a

TABLE II

TREATMENT OF BRAIN TISSUE TO OBTAIN CELL SUSPENSIONS

Method	Preparation	Dissociation	Filtration
Korey, 1957	Waring blender	—	250- and 100-μm mesh
Rappaport and Howze, 1966	Cutting	—	—
Rose, 1965	—	Teasing through 130-μm mesh	40-μm, glass-bead column
Satake and Abe, 1966	Chopping	Teasing through 740-μm mesh; teasing through 430-μm mesh; syringe passage, 340 μm 10 times	—
Blomstrand and Hamberger, 1969	Chopping	Syringe passage, 1000 μm 10 times	500-, 250-, 125-, and 50-μm mesh; triple-layered 50-μm mesh
Sellinger and Azcurra, 1974	Chopping	Syringe passage, 333 μm 3 times; syringe passage, 110 μm 3 times	73-μm mesh 3 times
Norton and Poduslo, 1970	Chopping	Vacuum suction, 105-μm mesh	74-μm mesh 3 times
Flangas and Bowman, 1968	—	Metal press, 75 and 37 μm	37-μm mesh
Farooq et al., 1977	Large slices	Aspiration through nozzles, 2.2 mm	420-μm mesh

column of glass beads having a diameter of 0.2 mm. This step can be added to any procedure having significant capillary contamination. After filtration or aspiration the cell suspensions are purified by centrifugation.

The centrifugation steps again display a wide variety of approaches. In our laboratory, following the work of Blomstrand and Hamberger (1969), a low-speed differential centrifugation is carried out to separate isolated cells from tissue debris. Other approaches proceed directly to density gradient centrifugation. These gradients usually utilize Ficoll or sucrose in a salt solution, often with 1% bovine serum albumin (BSA). Most preparations that go directly to density gradient centrifugation have two such steps. In the first centrifugation a purified neuronal fraction is obtained, and in a subsequent centrifugation a purified glial cell fraction is obtained (Sellinger and Azcurra, 1974; Poduslo and Norton, 1975). In general, the CNS cells appear to have densities related to the ratio of cell nucleus to perikaryal cytoplasm. The smaller nucleus of the astrocyte results in a lower density, whereas neurons with a large nucleus are considerably denser. Table III shows the approximate density of cortical cells. However, these numbers are only approximations and vary greatly depending on brain region, species, and age of the animal. We will detail three procedures for obtaining bulk-isolated cell fractions. These are techniques we use for obtaining relatively good yields of either caudate or cortical astroglia, a method we use for obtaining cerebellar Purkinje cells, and the procedure recently described by Farooq et al. (1977) using aspiration for dissociating tissue. Other nonenzymatic methodologies are well described by Sellinger and Azcurra (1974), and readers are referred to their article for detailed methods. The techniques using trypsin are well described by Poduslo and Norton (1975). Recently a method for obtaining astrocytes, oligodendrocytes, and neurons from the same brain tissue was described by Chao and Rumsby (1977), utilizing the fact that neurons and astrocytes lyse at pH 7.4 when incubated with trypsin for 1 hr, but oligodendrocytes remain. A method for obtaining oligodendrocytes from sheep brain is described in the chapter by Szuchet in this volume.

TABLE III

Isopynic Banding Densities of CNS Cells and Subcellular Particles

CNS preparation	Density	Reference
Neuronal soma	1.171–1.220	Hamberger et al., 1970
Astroglia	1.156–1.192	Hamberger et al., 1970
C6 glioma	1.156–1.187	Cotman et al., 1971
Synaptosomes	1.14–1.17	Whittaker, 1968
Gliosomes	1.14–1.17	Sieghart et al., 1978

B. *Specific Isolation Procedures*

1. ASTROGLIAL CELL PREPARATION FROM BOVINE CAUDATE NUCLEUS

Caudate nuclei are dissected from bovine brains at slaughter and immediately placed on ice. The caudate tissue is trimmed of all myelin and chopped into 600×400 μm rectangles on a McIlwain tissue chopper. The rectangles are placed in fresh incubation medium[1] sufficient to cover the tissue and incubated under oxygen at 37°C for 30 min. The incubated tissue is placed on ice and diluted fivefold with sucrose.[2] A 50-ml plastic syringe with the end cut off and 1-mm nylon mesh glued on is used gently to dissociate the tissue by drawing it up repeatedly. After dissociation, the solution of cells is gravity-filtered through nylon mesh of 149 μm pore size. The mesh can be washed with the sucrose solution, and the tissue further dissociated with the syringe and refiltered. After this filtration, the cells are filtered through a single layer of 47-μm nylon mesh. The filtration is carried out with gravity or slight positive pressure, and the debris is washed from the mesh and discarded. Finally, filtration through a double layer of 47-μm mesh is carried out to remove capillaries. The filtrate is collected and centrifuged at 100 g for 10 min at 4°C. The supernatant is poured off carefully, retaining the almost fluid pellet. This removes a great deal of the cell debris, membrane fragments, and synaptic particles. The supernatant is used to isolate synaptosomes if desired. The pellet is diluted with an equal volume of 40% w/w Ficoll.

A discontinuous Ficoll density gradient is then formed over the sample which is approximately 20% Ficoll. The gradient is formed in solutions containing 0.32 M sucrose, 10 mM Tris–HCl, 1 mM Na$_2$EDTA, and Ficoll to give a refractive index of 1.3714 (about 15% Ficoll). The final solution has a refractive index of 1.3633 (about 10% Ficoll), and the gradient is topped with sucrose solution. It is centrifuged at 100,000 g for 2 hr at 4°C. The top of the gradient consists of a myelin layer which can be removed with a spatula or by aspiration. Between the 10 and 15% Ficoll solutions is a cellular layer consisting of astroglia, an occasional capillary fragment, and some cell debris. Between the 15 and 20% Ficoll is a mixture of glia, neurons, capillaries, and debris. The pellet consists of red blood cells, free nuclei, and neurons. The astroglia cells are removed with a pasteur pipet, diluted with cold sucrose solution, and centrifuged at 800 g for 20 min. The cells are examined by phase microscopy and, if sufficiently pure, retained for experiments.

---

[1] Incubation medium: 35 mM Tris–HCl, pH 7.5, 5 mM NaPO$_4$ buffer, pH 7.5, 5 mM KCl, 120 mM NaCl, 2.5 mM MgCl$_2$, 1 mM EDTA, 5 mM dextrose, 2.5 mM ADP, and 2% (w/w) Ficoll. The ADP is added just prior to use.

[2] Sucrose solution: 10 mM Tris–HCl, pH 7.4, 1 mM Na$_2$ EDTA, 0.32 M sucrose, and 2% (w/v) Ficoll.

2. Preparation of Purkinje Perikarya

The cerebellum is removed from two to three beef brains (total volume about 100 ml) and cleaned of meninges, blood vessels, and pia. The cerebellum is cut into small pieces with scissors and diluted with an equal volume of sucrose solution (see the preceding discussion) with 5 mM MgCl$_2$ added and EDTA reduced to 0.5 mM. The tissue suspension is forced through a 1-mm nylon screen attached to the cut-off end of a plastic syringe. The resulting suspension is passed over a 150-μm nylon mesh filter. The filter is washed in sucrose, and the unfiltered tissue is forced through the 1-mm nylon mesh a second time and refiltered through the 150-μm mesh. This is repeated two to three times until the volume of the filtrate is about 500 ml. This suspension is then filtered through a 75-μm nylon mesh, and the solution is centrifuged at about 100 g for 5 min. The supernatant is discarded, retaining all of the fluid pellet. The pellet is mixed with a 40% buffered Ficoll solution (0.32 M sucrose, 10 mM Tris–HCl at pH 7.4, 0.5 mM EDTA) in the ratio of 2 parts pellet to 1 part Ficoll solution. This is placed on a gradient of 26% Ficoll and 21.5% Ficoll and covered with a 10% Ficoll solution, resulting in a four-step gradient with the sample in the second step. This gradient is centrifuged for 2 hr at 100,000 g, and the sample is removed from between 21.5 and 26% Ficoll. It is diluted in sucrose solution, centrifuged at 800 g for 10 min, and examined microscopically.

3. Preparation of an Astrocyte Fraction from Rat Brain Using Aspiration to Dissociate the Tissue

Recently Farooq and Norton (1978) described the isolation of morphologically well-preserved astrocytes from rat brain using a new technique for dissociating tissue. We have used their method and feel that it may represent an advance in bulk cell preparations. In this procedure density gradient solutions are prepared using the isolation media[3] with 2% Ficoll and, in addition, the amount of Ficoll indicated for each gradient solution (see the following discussion). For example, a 10% Ficoll solution contains 12% Ficoll plus hexoses plus buffer.

The brains of six rats are trimmed of cerebellum and brain stem. The meninges are removed, and the brains cut into 10–12 slices. The slices are placed in room-temperature medium containing 0.1% acetylated trypsin and incubated for 90 min at 37°C. After 90 min the medium is removed with a pasteur pipet and replaced with medium containing 0.1% soybean trypsin inhibitor. This medium is cooled and subsequently discarded. The slices are washed five times in ice-cold media. Disaggregation of the tissue

[3] Isolation media: 8% (w/v) glucose, 5% (w/v) fructose, 10 mM KH$_2$PO$_4$–NaOH buffer, pH 6.0, and 2% (w/v) Ficoll.

is achieved by repeated aspiration through a nozzle which is 2.3 mm in diameter at the tip and about 3.4 cm long. (A disposable automatic pipet tip with the nozzle trimmed to about 2.3 mm is used.) The tip is connected to a 250-ml two-necked bottle containing 30 ml of medium. The slices are aspirated into the bottle using a slight vacuum, and the suspension is poured over a 40-mesh nylon screen (420 μm). The filtrate is collected, and the tissue on the screen is again aspirated and refiltered. After four aspirations, the screen is aspirated a final time, and the filtrates are combined. This cell suspension is placed on ice for 15 min during which time large particles settle out, and the supernatant is decanted and saved. The precipitate is resuspended in 20 ml of medium and allowed to set for 15 min after gentle vortex stirring. The supernatant is collected, and the pellet is discarded. The supernatant is then centrifuged at 720 g for 10 min. The supernatant is discarded, and the pellet is suspended in 7% Ficoll and centrifuged at 280 g for 10 min to obtain a P_1 pellet. A P_2 pellet is obtained by spinning the P_1 supernatant at 720 g for 10 min. Finally, the P_2 supernatant is collected, diluted 1:1.125 with medium, and spun at 1120 g for 15 min for the P_3 pellet. The P_1, P_2, and P_3 pellets are suspended in 38 ml of isolation medium with an additional 7% Ficoll. The sample is layered on SW-27 tubes containing the following gradient from the bottom up: 32% Ficoll, 28% Ficoll, 22% Ficoll, and 10% Ficoll and sample. The tubes are centrifuged at 8500 g for 5 min, and the astrocytes are collected over the 22% Ficoll layer. They are diluted with medium and collected by centrifugation at 280 g for 10 min and examined microscopically.

These cell isolation procedures can be carried out in any biochemical laboratory with a preparative ultracentrifuge. Unfortunately, in our experience, none of the procedures is foolproof, and repeated attempts are often necessary for an investigator unfamiliar with cell isolation techniques to obtain suitable fractions. Even with experienced technicians we find that some preparations are heavily contaminated. For this reason each preparation should be assessed for cellular integrity and purity prior to use. The problem of assessing the purity of cellular fractions isolated from brain tissue is difficult. A lack of well-defined biochemical checks on purity results in reluctance to accept work done on these fractions. As with subcellular fractions, such as synaptosomes, there is no easy way to quantitate contamination.

C. Examinations of Fraction Purity

1. MICROSCOPY

Two basic approaches exist for assaying the purity of cell fractions, morphological and biochemical. For routine laboratory use, morphological investigations are much preferred. Examination of the fractions, either un-stained using phase-contrast microscopy or stained using light microscopy, should be carried out on all fractions. A careful examination of a slide of unstained sedimented cells viewed under phase-contrast is an excellent in-dicator of the number of contaminating particles relative to intact cells. Such an examination of the glial fractions described in the preceding discussion usually reveals 60–70% particle purity. Since the contaminating tissue fragments are usually considerably smaller than the cells, purity on a weight basis is usually better than 80%. There is a marked difference in the appearance of glia isolated by the methods outlined above. The caudate glia are isolated in considerably higher yield using our method, compared to that of Farooq and Norton (1978). We obtain about 2½–3 times more material, although the appearance of the cells is better using the aspiration method. When aspiration is used, the cells are most often found flattened out with their processes extended and appearing intact (Fig. 1), whereas, with the sieving method, we most frequently find clumps of two to five cells with processes intertwined (Fig. 2). These clumps rarely have their pro-cesses extended in one focal plane, but careful viewing can distinguish these cells from neurons by the size of the nucleus, structure of the processes, and ratio of cytoplasm to nucleus.

With most methods good neuronal fractions approach 95% particle

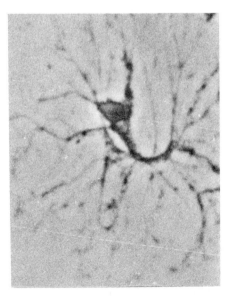

FIG. 1. An astroglial cell showing retained processes prepared by the method of Farooq and Norton (1978).

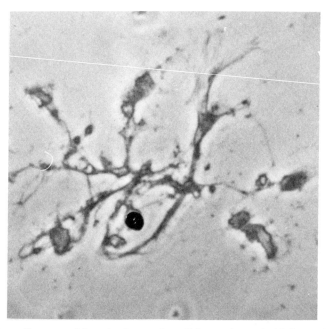

FIG. 2. Astroglia prepared from bovine caudate. Cells are clumped with intertwined processes.

purity (Fig. 3). The cells are often shorn of dendrites and only a portion of the axon preserved (Fig. 4). The method of Farooq *et al.* (1977) preserves more of the dendrites, providing the possibility of comparing perikaryal properties to dendritic properties. This of course depends on a demonstration that the use of trypsin does not significantly alter the properties being examined. Morphological examination of good neuronal fractions usually reveals only cell bodies with a large nucleus and characteristic shape, occasional free nuclei, or intact capillaries consisting of endothelial cells. If the tissue has not been perfused, red blood cells are also found in neuronal fractions. While morphological examination can easily distinguish a good preparation from a contaminated one, it is not sufficient to quantitate the amount and type of contaminating tissue.

2. BIOCHEMICAL AND IMMUNOLOGICAL CHECKS ON PURITY

Unequivocal demonstration of the composition of the cell fractions requires the use of biochemical markers. Ideally, they should be specific for a given cell type and localized by labeled antibodies using *in vivo* conditions, tissue slices, and primary cultures. The markers ought to be membrane-

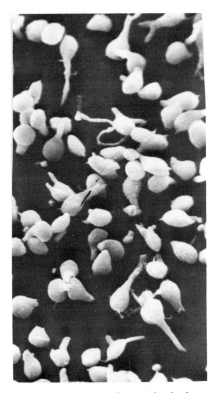

FIG. 3. Purkinje cells, a scanning micrograph of a low-power field.

bound and insoluble under ordinary tissue isolation conditions. One area that has offered promise in finding good biochemical markers is the study of CNS-specific proteins. Several CNS-specific proteins have become available as a result of the investigations of Moore and his collaborators (1968). Beginning with the description of S-100 (Moore, 1965), CNS-specific proteins have been studied. While there is some controversy over the localization of S-100, it is principally found in glia (Moore *et al.*, 1977). An exclusively glial protein, glia fibrillary acidic protein, was first isolated by Eng *et al.* (1971). This protein was highly immunogenic, and localization techniques revealed selective localization in astrocytes with more found in fibrous than in protoplasmic astrocytes (Bignami *et al.*, 1972; see also chapter by Bignami, Dahl, and Rueger, this volume). A soluble α_2 glycoprotein found principally in white matter has been described by Warecka and Bauer (1967). From tumor data (Warecka, 1975), it was concluded that this protein was present in astrocytes; its function is unknown. Another glycoprotein presumably specific to neurons has been described by

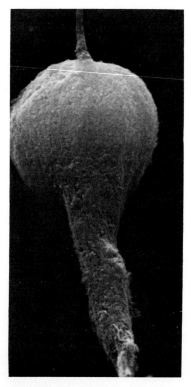

Fig. 4. Higher magnification of a Purkinje cell showing dendrites almost all shorn off and the axon fragment.

Van Nieuw Amerongen *et al.* (1972). The protein, called GP-350, is rich in regions with ganglia (Van Nieuw Amerongen and Roukema, 1973), and immunofluorescence studies suggest that it is restricted to neuronal cells (Van Nieuw Amerongen *et al.,* 1974).

Unfortunately, all these proteins are of unknown function and have marked solubility under fractionation conditions, making them unsuitable markers for quantifying the contamination of fractions. Membrane-specific antigens are potentially of more value in estimating contamination, but to date no membrane proteins of known function have been found to be suitable as antigens. However, recently some progress has been made in identifying the function of two cell-specific proteins. The acidic protein, 14-3-2, was first identified by Moore and Perez (1968). Bock and Dissing (1975), using immunoprecipitation, were able to show that 14-3-2 demonstrated enolase activity. Subsequently, Marangos *et al.* (1976) demonstrated that two of the brain-specific isoenzymes of enolase were associated with

14-3-2. Subsequent studies using immunocytochemical techniques by Schmachel *et al.* (1978) have clearly shown that the enolase activity in brain has a neuron-specific isoenzyme and a glia-specific form. These proteins are immunologically distinguishable and provide cell-specific markers of known function. Studies of the 14-3-2 form of enolase reveal that not all neurons contain this protein. In fact, three types of neurons have been demonstrated: those that contain this form of enolase, those that contain it during a developmental stage and later lose it, and those that never contain it (Cimino *et al.,* 1977). These enzymes are therefore not ideal neuronal or glial markers, since their distribution is not uniform in a given cell type. Also, since enolase activity is widely distributed and soluble, these enzymes, while helpful in culture studies, are not useful in bulk cell preparations in which a major proportion of cells are broken during tissue disruption, allowing redistribution of soluble enzymes.

Recently, three different groups of neuroimmunologists have attempted to define surface-specific antigens on CNS cells (Bock, 1978; Fields, 1976; Schachner, 1974). This work offers great promise of finding a specific marker for a variety of cell types that would be useful under the conditions of bulk isolation procedures. Thus far, the results of Bock and her coworkers are the most promising for proteins specific to the synaptic plasma membrane. Jorgensen and Bock (1974) have described a series of antibodies to synaptic plasma membrane proteins. The synaptic membrane proteins are called D1, D2, and D3, and all are enriched in synaptic membrane preparations three- to fourfold over brain homogenates. These proteins are absent in primary astroglial cultures (Bock *et al.,* 1975), and they appear in all CNS regions. Thus, they have the characteristics of good neuronal markers. Another protein identified by Bock *et al.* (1974) is specific to the nervous system and found on synaptic vesicle membranes and plasma membranes. It is designated synaptin. The use of tumor material has also led to some cell-specific antibodies. Schachner (1974), using a glioblastoma membrane preparation, was able to prepare an antibody specific to the nervous system. The antigenic protein is found predominantly in white matter. It is called nervous system antigen 1 (NS1) and appears to be specific to glial cells. With use of the cytotoxicity test the protein was shown to be on the cell membrane and, because of its distribution and the limited amount found in myelin-deficient mice, it may be principally on oligodendrocytes. In the chapter by Szuchet in this volume this specific cell type is discussed in more detail.

This brief review suggests that it may now be feasible to quantitate contamination in bulk cell fractions immunologically. Such a demonstration would provide the necessary documentation for assessing specific cell function critically using these preparations. To date, work with these fractions has been suspect because of the difficulty of precisely defining cell contamination. We felt that our own work, which suggested that astroglia have high-affinity transport systems for amino acids that may act as neurotransmitters (see Henn, 1976), required a demonstration in *both* bulk cell and cell culture systems for verification. The reason is that culture systems derived from tumor lines are subject to dedifferentiation, whereas primary cultures from young animals lack fully differentiated expression. The bulk-isolated cells are subject to problems of both viability and contamination. Since the bulk cells clearly are fully differentiated if derived from adult animals, and the cloned culture lines are pure and viable, demonstration that both contain a given property is strong support that this reflects a real *in vivo* function of the cell. This is still, in our opinion, a satisfactory way to determine general cell properties, but with a realization that both neuronal and glial cells contain many subtypes and many have distinct regional specialization, this approach may not have the sensitivity necessary for future problems in cellular neurobiology.

III. Subcellular CNS Fractions

The problem of contamination is present not only in cellular CNS fractions but also in subcellular fractions derived from brain. All the general cell organelles—nuclei, mitochondria, microsomes, ribosomes, lysosomes, and plasma membrane preparations—if derived from whole brain, are bound to be a heterogeneous collection from very different types of cells. Thus, to obtain a better picture of the varieties of cell organelle function, a bulk cell preparation should be used as the starting material for comparative studies. Figures 5 and 6 illustrate preparative schemes that provide crude preparations of cell organelles starting from bulk-isolated cell fractions.

Synaptic Particle Isolation and Fraction Purity

One subcellular fraction unique to nervous tissue is enriched in neuronal tissue. This, the synaptic ending particle, or synaptosome preparation, has been of enormous importance in neurobiology. It provides a preparation enriched in the area of neurons involved in chemical transmission. Problems of contamination of this fraction tend to be minimized because of the great interest in the biochemical activities of the synapse. The same problems concerning marker enzymes discussed in relation to whole cells apply

SEPARATION OF NEURONAL AND GLIAL CELLS389

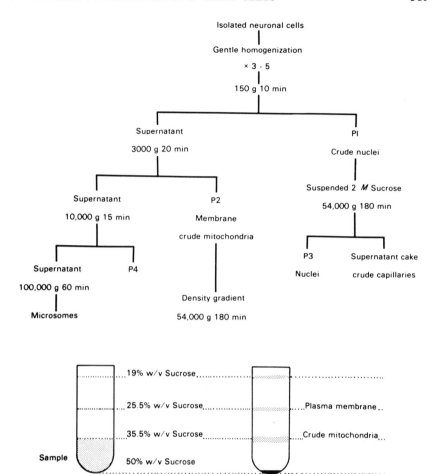

FIG. 5. Diagram outlining the steps in the purification of neuronal plasma membrane.

to synaptic particles, with the exception of some enzymes specific to neurotransmitter synthesis. Dopamine decarboxylase, for example, is only found in norepinepherine-containing neurons, and γ-aminobutyric (GABA) neurons are certainly enriched in glutamic acid decarboxylase. Even though these enzymes are specific for a given neurotransmitter, they are not general markers of nerve endings. The standard procedure for nerve ending preparations was described by Gray and Whittaker (1962). Briefly, it usually involves homogenization, centrifugation at approximately 1000 g for 10 min to sediment nuclei and debris, and centrifugation at 10,000 g for 20 min to sediment a crude mitochondrial pellet. This is then fractionated on a gradient containing 1.2 M sucrose at the bottom, followed by 0.8 M

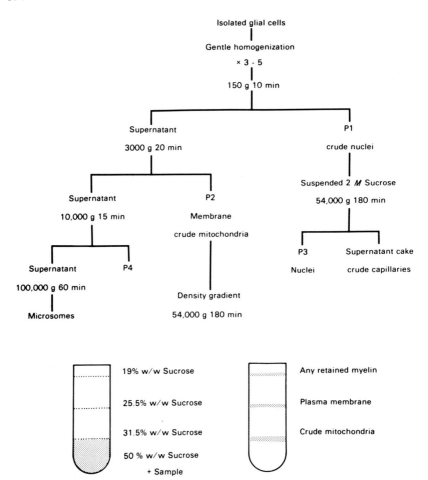

FIG. 6. Diagram outlining the steps in the purification of glial plasma membrane.

sucrose with the sample in 0.32 M sucrose applied on the top. The gradient is centrifuged to equilibrium at about 100,000 g for 2 hr, and the synaptosome fraction is found at the 0.8–1.2 M interface. An alternative gradient used Ficoll and was initially described by Autilio *et al.* (1968). This procedure works well when the final gradient consists of Ficoll solutions made in buffered 0.32 M sucrose. The bands are 20, 13, 7.5, and 5% Ficoll. Synaptosomes are found at the 7.5–13.0% Ficoll interface. Now, studies looking at the isopynic banding densities of synaptic particles repeatedly find them in the range of 1.14–1.18 gm/ml (Whittaker, 1968; Cotman *et al.*, 1968), and glial cells have usually been reported between densities of 1.15 and 1.19 gm/ml (Cotman *et al.*, 1971; Hamberger *et al.*, 1970). Since the preparation of synaptosomes involves complete homogenization, it may not be a problem that glial cells and synaptic particles have such similar densities. This problem was first investigated by Cotman et al. (1971) when they fractionated a clonal line of glia. They showed that a vesicular fraction from this glial homogenate sedimented with a density similar to that of synaptosomes and that this fraction constituted about one-fifth of the total protein of the glia. The study of Cotman et al. suggested that a Ficoll-sucrose gradient was capable of separating glial membranes from synaptic particles with greater resolution, with only about 7% of the glial protein appearing at the interface where synaptosomes were isolated. This question was reinvestigated by Henn et al. (1976) using a cloned glial culture grown in ^{14}C-leucine. These labeled cells were then added to brain tissue and homogenized. Synaptic particles were isolated on the standard sucrose gradient and Ficoll gradient and examined morphologically and counted for ^{14}C. The results are shown in Table IV. This suggests heavy glial vesicle contamination. When we examined glutamate uptake by glial vesicles, they retained this transport system (Henn et al., 1974), suggesting this contaminant could cause serious problems in interpreting data obtained with synaptic particles. To further examine this question Sieghart et al. (1978) compared the sedimentation of three glial preparations with the commonly used P_2 fractions on sucrose density gradients. The results showed first that the different glial systems have different densities depending on their origin, and secondly that bulk isolated glia are essentially indistinguishable from P_2 particles (Fig. 7). This result illustrates the serious problem of contamination in synaptosomal fractions and the heterogeneous nature of vesicles found in these fractions. Data obtained on P_2 fractions do not suggest localization to nerve endings in and of themselves. Further studies involving autoradiography, lesion studies, and culture

TABLE IV

PROPORTION OF LABELED GLIAL HOMOGENATE IN
SYNAPTOSOMAL PREPARATIONS[a]

Fraction	Counts per minute per milligram	Percentage
Combined C6 glia and rabbit brain homogenates	2318	100
Ficoll gradient synaptosomes	1165	50
Sucrose gradient synaptosomes	1095	47
Sucrose followed by a Ficoll gradient	1018	44

[a] Results of one experiment. Analogous results have been triplicated with various amounts of label in the original homogenate.

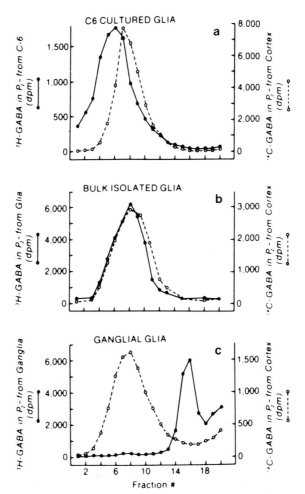

Fɪɢ. 7. Subcellular distribution of glial and synaptosomal fractions. C6 cells, bulk-isolated glial cells, and dorsal root ganglia were incubated with [³H]GABA, and slices from cerebral cortex with [¹⁴C]GABA. Glial and cortical tissue was mixed and cohomogenized. P_2 fractions derived from the combined tissues were centrifuged for 2 hr at 27,000 rpm in a SW-27 rotor in a linear density gradient. (a) Subcellular distribution of P_2 from C6 and rat cerebral cortex. (b) Subcellular distribution of P_2 from bulk-isolated glia and P_2 from rabbit cortex. (c) Subcellular distribution of P_2 from dorsal root ganglia and P_2 from rabbit cortex. (From Sieghart et al., 1978.)

systems are necessary to define the localization of biochemical findings made using P_2 fractions. The development of antibodies specific to synaptic plasma membrane will help define the contamination in these fractions and possibly lead to the development of purer preparations.

IV. New Isolation Techniques

The development of cell- or organelle-specific antibodies appears not only to offer the best hope for quantitative analysis of contamination in bulk-isolated fractions but also suggests several new approaches to the isolation of these fractions. Three methods appear to offer promise of finding more specific cell isolation procedures. These include affinity chromatography (see the chapter by Varon, this volume), the use of an automated cytofluorograph to sort cells, and the use of magnetic microspheres to separate cells. All three methods utilize the specificity of defined antibodies to identify the cell or particle in question. The isolation of the particle takes place directly by means of the binding in affinity systems. In cell sorter systems a fluorescent tag is used to label the antibody and identify the particle, whereas magnetic microspheres may be coupled to antibodies and then the particles or cells moved in a magnetic field.

The automatic cell sorter has a proven capacity to separate fractions of similar cells with defined surface antigens in fractions having 95% purity. We have carried out preliminary experiments on the fluorescence-activated cell sorter (FACS), and they suggest that this system can be applied to some neural tissues. The machine originally designed by immunologists at Stanford University was described in detail by Julius *et al.* (1973).

In brief, the FACS takes a cell suspension and forms a coaxial liquid sheath about it, resulting in a 50-μm diameter stream with single cells in the center. This stream passes through a laser beam which is scattered by the cell, establishing particle size by low-angle light detection. The analyzer also establishes whether the cell is labeled by fluorescent markers. Machines with the capability of detecting two fluorescent markers can sort viable cells containing a specific cell surface marker. One application of the FACS that was recognized during early analytical studies was in detecting and analyzing viable cells. Treatment of cell populations with the fluorogenic compound fluorescein diacetate (FDA) renders only viable cells fluorescent (Rotman and Papermaster, 1966). The dead cells do not become fluorescent, so that analysis of the scatter signals derived from the fluorescent cells only (i.e., fluorescence-gated scatter, FGS) in a population treated with FDA, generates the scatter profile of viable cells in that population. The dead cells (nonfluorescent) constitute a subpopulation that can be clearly differentiated on the basis of scatter alone. Appropriate threshold settings allow a "gating out" of these dead cells, so that any subsequent separation or analysis of the population can be done on the basis of viable cell content. Viable cells in an appropriate size range and with the desired fluorescence-labeled antibody binding can be sorted by charging the stream when a desired cell is in the incipient drop just prior to breaking off the stream. The stream is broken into approximately 40,000 droplets per second at a precisely determined point by means of an acoustically excited nozzle. The desired cells in the charged droplets are isolated by passage through an electric field and subsequent collection of the deflected droplets. Using a fluorescent antibody to label a specific surface antigen allows the isolation of cell fractions of greater than 95% purity in the FACS (Tyrer et al., 1974).

Another new approach that may be useful in the isolation of functionally specific cell populations from the CNS involves magnetic microspheres. These are small particles of magnetite coated with a hydrophylic polymer capable of covalent linkage to a cell surface ligand. A recent report by Kronick et al. (1978) reports on a process that gives high yields of magnetic hydrogel microspheres of relatively homogeneous size distribution. These particles were utilized to bind cholera enterotoxin, and the product was found useful in separating neuroblastoma cells. The microsphere separation was effected under conditions in which sterility was maintained and the cells remained viable. The technique was capable of concentrating cells containing the ganglioside G_{MI} in better than 99% purity in about 6 min. The cells are reacted with the microspheres and then run through polyvinyl chloride tubing wound around the edges of the pole pieces of an 8-in. divergent magnet. The cells that have reacted with the ligand are retained in the tubing with the magnetic field on. This technique appears capable of separating cells with very few bound microspheres; the usefulness of the technique depends on the availability of ligands specific for cell surface receptors. The great advantage this technique offers is speed and gentle conditions. Normal flow characteristics in cell sorters may be severe enough to destroy fragile cells such as astroglia. However, the forces involved in magnetic microsphere separations appear controllable and small and offer some hope of isolating well-defined cell fractions that are essentially homogeneous and viable.

The utilization of cell sorters or magnetic microspheres in neurobiology depends upon the ability to dissociate cells and to find parameters that will allow their identification and separation while retaining viability. To date, the most successful approaches in analyzing cellular functions in the CNS have involved the use of tissue culture of pathological material (Sensenbrenner *et al.*, 1970; Gilman and Nirenberg, 1971; Benda *et al.*, 1968) or bulk-isolated separation of CNS fractions (Gray and Whittaker, 1962; Rose, 1967; Norton and Poduslo, 1970; Blomstrand and Hamberger, 1969; Sellinger and Azcurra, 1974). Both methods have been utilized in this laboratory (Henn *et al.*, 1974; Henn, 1975), and both have significant limitations. The use of cultured material requires pathological material for neuronal models, and problems of interpretation due to dedifferentiation are inevitable. These systems do not aid in attempting to understand regional differences between cells involved in neurotransmission. Bulk-isolated cell systems are not subject to problems of dedifferentiation, and with some difficulty it is possible to look at regional variations within populations (Henn, 1975). Problems of contamination and cell viability are, however, of great concern when working with these systems. The use of the FACS system or magnetic microspheres may provide the technology necessary to isolate pure fractions of functionally homogeneous cells in a viable state amenable to maintenance in a tissue culture system. In addition, it may be possible to isolate pure fractions of synaptic particles, possibly even functionally homogeneous particles using specific antibodies.

V. The Use of Cell Fractions

The impetus for developing cell isolation techniques is to analyze the unique potentialities of a given cell type and specify how that cell may interact with others to form a functional area of brain tissue. Initial studies on bulk-isolated fractions defined the biochemical composition of the different fractions. Following such efforts at defining the chemical characteristics, studies began appearing aimed at understanding functional differences between cell types. In the area of neuron–glia interactions, bulk cell studies have led the way in defining possible mechanisms of cell interactions.

One of the earlier areas investigated was protein synthesis. Differences between neuronal and glial capacities for protein synthesis were examined. *In vivo* experiments done by Satake and Abe (1966) and Blomstrand and Hamberger (1969) showed considerably more protein-bound radioactivity in neuronal perikarya than in glial cells. Johnson and Sellinger (1971, 1973a,b), in a series of papers looking at protein synthesis *in vivo,* also showed higher total neuronal incorporation of labeled amino acid into protein for neurons using rats under 18 days of age. In mature rats the reverse appeared to be true. However, using pulse-labeling techniques, they found 10-day-old rats had higher rates of protein synthesis in glia and that this decreased by day 18 when neuronal and glial fractions had similar rates of incorporation of labeled amino acids. This conclusion depends critically upon where in the time course one looks. The *in vivo* approach, in which labeled amino acid is injected into the animal prior to cell isolation, offers no control of precursor pool sizes. Use of *in vitro* labeling in which isolated cells are incubated with known pools of labeled amino acids also suggests greater protein synthesis in neuronal fractions (Tiplady and Rose, 1971; Hamberger *et al.*, 1971). This is of some interest because of the suggestion that glia may synthesize macromolecules and transfer them to neurons (Lasek *et al.*, 1974). The bulk-prepared cell data offer little support for this idea, but the experiments done do not provide a critical test of the question. This remains an area of potential investigation.

Another area of cellular interaction is in the control of extracellular ion concentrations. Astrocytes are critically situated around unmyelinated axons and synapses. This location allows them to react quickly to extracellular K^+ buildup after neuronal firing. Bulk cell fractions have been used to examine the characteristics of Na^+, K^+-ATPase, the enzyme that controls active K^+ transport. Medzihradsky *et al.* (1971) have shown that this enzyme is concentrated 7-fold in glia, compared to a neuronal fraction. Since astrocytes contain much more membrane per cell than isolated neuronal perikarya, it appears that a more accurate assessment of the Na^+, K^+-ATPase would compare neuronal and glial plasma membranes. Such a study was carried out (Henn *et al.*, 1972) and showed the enzyme was 3.5- to 4-fold more active in glial membranes. More interesting is the finding that the enzyme is stimulated by just the range of K^+ expected extracellularly after neuronal firing (Fig. 8).

Studies of amino acid transport using bulk-isolated cell fractions led us to the surprising finding that astroglia had "high-affinity" transport systems. Initially, we examined GABA (Henn and Hamberger, 1971), but, subsequently, amino acids that were electrically active were shown to be transported in this fashion by glia. We have completed uptake studies on GABA, glutamate (Henn *et al.*, 1974), glycine, taurine, alanine, and valine (Henn, 1976). The first four amino acids have apparent K_a values in the range of $10^{-5}M$, which is consistent with high-affinity transport systems. Alanine and valine, on the other hand, showed K_a values in the range of $10^{-3} M$, which is consistent with cellular uptake of amino acids in most tissue. That glia possess the ability to accumulate these amino acids has been verified with autoradiographic studies and tissue culture experiments, substantiating the results of bulk-isolated glial cell fraction studies.

A final area in which cell fractions may prove useful is receptor studies. Studies using neurotransmitters or their analogs to label specific receptors in the CNS have contributed greatly to our understanding of the regional

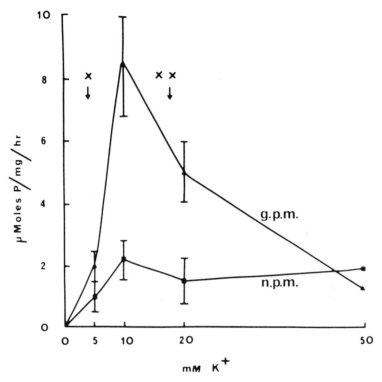

FIG. 8. Na⁺-, K⁺-ATPase activity of neuronal (n.p.m.) and glial (g.p.m.) plasma membrane fractions as a function of K⁺ ion concentration in the incubation medium. (From Henn *et al.*, 1972.)

distribution and function of neurotransmitter pathways over the last 10 yr. In general, these studies employ a low concentration of a ligand having a high specific radioactivity. This is incubated with two tissue samples, one with an excess of a displacing ligand and one without. The difference between the binding under the two conditions is taken as the specific binding, provided certain conditions are met. The rationale is that the binding in the presence of an excess of unlabeled ligand is nonspecific, and this is subtracted from the binding of the ligand that is not displaced, leaving an estimate of specific receptor binding. The problem is to demonstrate specificity, and this is done by attempting to demonstrate strict structural requirements for binding including stereochemical specifity, and correlations of binding with other measures of the effect of stimulating the receptor. These studies are particularly useful in defining drug receptor interactions, and results on opiates and neuroleptic drugs are now well known. We have found that nonenzymatically prepared cell fractions can be used in receptor studies. These fractions suggest that certain receptors may have a nonsynaptic localization. Before giving examples of such studies it may be useful to illustrate the problems trypsin can cause when cell surface enzymes or receptors are studied. In carrying out receptor binding studies we utilized the astroglial preparation just described. In addition, because of the excellent morphological appearance of the cells and the purity of the fraction we were able to achieve using the aspiration technique, we examined fractions prepared using trypsin and aspiration. The results of binding studies done on homogenates and isolated cell fractions for the two ligands whose receptors appear to be on glia are shown in Table V. These results show that the binding is almost eliminated when cell homogenates are exposed to trypsin under the isolation conditions. Furthermore, the isolated cells have considerably lower binding when isolated with trypsin than without. Note that the receptors are enriched in our glial fraction compared to the homogenate, suggesting that contamination of the fractions does not account for the binding. Contamination has also been estimated using studies of enzymatic markers such as GAD.

The two ligands used in the comparison both define drug receptors of considerable importance. Spiroperidol is a neuroleptic drug of the butyrophenone class, which appears to bind to dopamine receptors and to a lesser extent to serotonin receptors. The excess ligand is (+)-butaclamol, the stereochemically active form of this neuroleptic.

This assay correlates with the clinical potency of a wide variety of neuroleptic drugs. There are clearly several dopamine receptors, some coupled to adenylate cyclase and some not (Kebabian and Caine, 1979). What role glial receptors may play in antipsychotic drug action is a question that deserves more study and one in which the use of bulk-isolated fractions is vital. These receptors are regionally specific and are not found in areas without dopaminergic innervation. Nor are they found in cultured cell lines having a tumor origin.

TABLE V

SPECIFIC BINDING

CNS Fraction	[³H]Spiroperidol (fmol/mg)[a]	[³H]Diazepam (pmol/mg)[b]
Homogenate without trypsin	76	0.88
Homogenate with trypsin	11	0.19
Astroglia without trypsin	142	1.31
Astroglia with trypsin	15	0.49

[a]Binding on fractions from beef caudate.
[b]Binding on fractions from rabbit frontal cortex.

The second ligand investigated, diazepam (Valium), is one of the most commonly used drugs in the world. It apparently has natural receptors, and recently reports of a natural ligand for this receptor have appeared (Colello *et al.*, 1978; Skolnick *et al.*, 1978; Asano and Spector, 1979; Möhler *et al.*, 1979). Our studies clearly indicate that this receptor is principally on glia in the frontal cortex (Henn and Henke, 1978). This is supported by lesion studies of R. Chang and S. Snyder (Chang *et al.*, 1978). The exploration of these data promises to lead to understanding of fundamental pharmacological mechanisms at the cellular level. Such studies will benefit from the use of isolated cell fractions.

One final example of the usefulness of cell isolation techniques comes from recent studies on GABA receptors in the cerebellum. Chan-Palay (1978) and Chan-Palay and Palay (1978) studied the cellular localization of GABA receptors in cerebellum. They used autoradiographic localization of [³H]muscimol and concluded that there was a high density of receptors on the membrane of the Purkinje cell soma and the initial portion of the axon (Fig. 9). We were able to isolate a fraction of Purkinje cells having just this portion of the cell body. GABA binding studies and muscimol binding on this fraction clearly substantiate the presence of receptors on Purkinje cell plasma membranes and provide support for the concept of neuronal perikaryal receptors not associated with synaptic regions.

This brief overview of the use of bulk-isolated cells does not begin to cover all the current areas in which these fractions are being used but does suggest several potentially significant areas of cellular neurobiology where

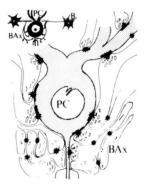

FIG. 9. Summary of the location and relative density of GABA receptor sites detected on surfaces of the Purkinje cell soma, dendrites, and initial axonal segment by [³H]M autoradiography. The basket cell (B)–Purkinje cell (PC) relationships are diagrammed in the inset. GABA receptor sites (*) are found on the membranes of the Purkinje cell as well as on the membranes between basket axons (BAx). The diagram is intended to show only the relative density of receptors, not the absolute number. (From Chan-Palay and Palay, 1978.)

such fractions can make contributions. The isolation procedures remain difficult and somewhat of an art. However, with the development of specific cell surface markers there is hope of greatly improved cell isolation techniques.

Acknowledgments

This article was made possible by the technical assistance of David Anderson, Rick Venema, and James Deering; the suggestions and comments of Suella Weiland Henn and the support of the Scottish Rite Schizophrenia Research Program, N.M.J.

References

Asano, T., and Spector, S. (1979). Identification of inosine and hypoxanthine as endogenous ligands for the brain benzodiazepine-binding sites. *Proc. Natl. Acad. Sci. U.S.A.* **76**, 977–981.

Autilio, L. A., Appel, S. H., Pettis, P., and Gambetti, P. L. (1968). Biochemical studies of synapses *in vitro*. I. Protein synthesis. *Biochemistry* **7**, 2615–2622.

Benda, P., Lightbody, J., Sato, G., Levine, L., and Sweet, W. (1968). Differentiated rat glial cell strain in tissue culture. *Science* **161**, 370–372.

Bignami, A., Eng, L. F., Dahl, D., and Uyeda, C. T. (1972). Localization of the glial fibrillary acidic protein in astrocytes by immunofluorescence. *Brain Res.* **43**, 429–435.

Blomstrand, C., and Hamberger, A. (1969). Protein turnover in cell-enriched fractions from rabbit brain. *J. Neurochem.* **16**, 1401–1407.

Bock, E. (1978). Nervous system specific proteins. *J. Neurochem.* **30**, 7–14.

Bock, E., and Dissing, J. (1975). Demonstration of enolase activity connected to the brain-specific protein 14-3-2. *Scand. J. Immunol.* **4**, Supple. 2, 31–36.

Bock, E., Jorgensen, O. S., and Morris, S. J. (1974). Antigen-antibody crossed electrophoreses of rat brain synaptosomes and synaptic vesicles: Correlation to water-soluble antigens from rat brain. *J. Neurochem.* **22**, 1013–1017.

Bock, E., and Jorgensen, O. S., Dittmann, L., and Eng, L. F. (1975). Determination of brain-specific antigens in short term cultivated rat astroglial cells and in rat synaptosomes. *J. Neurochem.* **25**, 867–870.

Chang, R. S. L., Tran, V. T., Poduslo, S. E., and Snyder, S. H. (1978). Glial localization of benzodiazepine receptors in the mammalian brain. *Soc. Neurosci.* **4**, 511 (abstr.).

Chan-Palay, V. (1978). Autoradiographic localization of γ-aminobutyric acid receptors in the rat central nervous system by using [³H]muscimol. *Proc. Natl. Acad. Sci. U.S.A.* **75**, 1024–1028.

Chan-Palay, V., and Palay, S. L. (1978). Ultrastructural localization of γ-aminobutyric acid receptors in the mammalian central nervous system by means of [³H]muscimol binding. *Proc. Natl. Acad. Sci. U.S.A.* **75**, 2977–2980.

Chao, S. W., and Rumsby, M. (1977). Preparation of astrocytes, neurons and oligodendrocytes from the same rat brain. *Brain Res.* **124**, 347–351.

Cimino, M., Hartman, B. K., and Moore, B. W. (1977). *Proc. Int. Soc. Neurochem.* **6**, 304.

Colello, G. D., Hockenbery, D. M., Bosmann, H. P., Fuchs, S., and Folkers, L. (1978). Competitive inhibition of benzodiazepine binding by fractions from porcine brain. *Proc. Natl. Acad. Sci. U.S.A.* **75**, 6319–6323.

Cotman, C., Mahler, H. R., and Anderson, N. G. (1968). Isolation of a membrane fraction enriched in nerve end membranes from rat brain by zonal centrifugation. *Biochim. Biophy. Acta* **163**, 272–275.

Cotman, C., Herschman, H., and Taylor, D. (1971). Subcellular fractionation of cultured glial cells. *J. Neurobiol.* **2**, 169–180.

Eng, L. F., Vanderhaeghe, J. J., Bignami, A., and Gerstl, B. (1971). An acidic protein isolated from fibrous astrocytes. *Brain Res.* **28**, 351–354.

Farooq, M., and Norton, W. T. (1978). A modified procedure for isolation of astrocyte- and neuron-enriched fractions from rat brain. *J. Neurochem.* **31**, 887–894.

Farooq, M., Ferszt, R., Moore, C. L., and Norton, W. T. (1977). The isolation of cerebral neurons with partial retention of processes. *Brain Res.* **124**, 69–81.

Fewster, M. E., and Mead, J. F. (1968). Lipid composition of glial cells isolated from bovine white matter. *J. Neurochem.* **15**, 1041–1052.

Fields, K. L. (1976). Brain-specific cell-surface antigens. *In* "Membranes and Disease" (L. Bolis, J. Hoffman, and A. Leaf, eds.), pp. 369–379. Raven Press, New York.

Flangas, A. L., and Bowman, R. E. (1968). Neuronal perikarya of rat brain isolated by zonal centrifugation. *Science* **161**, 1025–1027.

Gilman, A., and Nirenberg, M. (1971). Effect of catecholamines on the adenosine 3′:5′-cyclic monophosphate concentrations of clonal satellite cells of neurons. *Proc. Natl. Acad. Sci. U.S.A.* **68**, 2165–2168.

Gray, E. G., and Whittaker, V. P. (1962). The isolation of nerve endings from brain: An electron-microscopic study of cell fragments derived by homogenization and centrifugation. *J. Anat.* **96**, 79–89.

Hamberger, A., Blomstrand, C., and Lehninger, A. L. (1970). Comparative studies on mitochondria isolated from neuron-enriched and glia-enriched fractions of rabbit and beef brain. *J. Cell Biol.* **45**, 221–234.

Hamberger, A., Eriksson, O., and Norrby, K. (1971). Cell size distribution in brain suspensions and in fractions enriched with neuronal and glial cells. *Exp. Cell Res.* **67**, 380–388.

Hamberger, A., Hansson, H. A., and Sellström, Å. (1975). Scanning and transmission electron microscopy on bulk-prepared neuronal and glial cells. *Exp. Cell Res.* **92**, 1–10.

Hemminki, K., Huttunen, M. O., and Järnefelt, J. (1970). Some properties of brain cell suspensions prepared by a mechanical-enzymic method. *Brain Res.* **23**, 23–24.

Henn, F. A. (1975). The structure and function of glial cell membranes. *5th Annu. Meet. Int. Soc. Neurochem.* No. S4–4.

Henn, F. A. (1976). Neurotransmission and glial cells: A functional relationship? *J. Neurosci. Res.* **2**, 271–282.

Henn, F. A., and Hamberger, A. (1971). Glial cell function: Uptake of transmitter substances. *Proc. Natl. Acad. Sci. U.S.A.* **68**, 2686–2690.

Henn, F. A., and Henke, D. (1978). Cellular localization of [³H]diazepam receptors. *Neuropharmacology* **17**, 985–988.

Henn, F. A., Haljamäe, H., and Hamberger, A. (1972). Glial cell function: Active control of extracellular K⁺ concentration. *Brain Res.* **43**, 437–443.

Henn, F. A., Goldstein, M., and Hamberger, A. (1974). Uptake of the neurotransmitter candidate glutamate by glia. *Nature (London)* **249**, 663–664.

Henn, F. A., Anderson, D., and Rustad, D. (1976). Glial contamination of synaptosomal fractions. *Brain Res.* **101**, 341–344.

Johnson, D. E., and Sellinger, O. Z. (1971). Protein synthesis in neurons and glial cells of the developing rat brain: An *in vivo* study. *J. Neurochem.* **18**, 1445-1460.

Johnson, D. E., and Sellinger, O. Z. (1973a). Age-dependent utilization of phenylalanine for the synthesis of neuronal and glial proteins. *Neurobiology* **3**, 113-124.

Johnson, D. E., and Sellinger, O. Z. (1973b). Synthesis of soluble neuronal proteins *in vivo*. Age-dependent differences in the incorporation of leucine and phenylalanine. *Brain Res.* **54**, 129-142.

Jorgensen, O. S., and Bock, E. (1974). Brain specific synaptosomal membrane proteins demonstrated by crossed immunoelectrophoresis. *J. Neurochem.* **23**, 879-880.

Julius, M., Sweet, R., Fathman, C., and Herzenberg, L. (1973). Fluorescence-activated cell sorting and its applications. *Annu. Life Sci. Symp.* pp. 1-53.

Kebabian, J. W., and Caine, D. B. (1979). Multiple receptors for dopamine. *Nature (London)* **277**, 93-96.

Korey, S. R. (1957). Concentration of neuroglia cells. *In* "Biology of Neuroglia" (W. Windle, ed.), Chapter 13, pp. 203-218. Thomas, Springfield, Illinois.

Korey, S. R., Orchen, M., and Brotz, M. (1958). Studies of white matter. I. Chemical constitution and respiration of neuroglial and myelin-enriched fractions of white matter. *J. Neuropathol. Exp. Neurol.* **17**, 430-438.

Kronick, P., Campbell, G., and Joseph, K. (1978). Magnetic microspheres prepared by redox polymerization used in cell separation based on gangliosides. *Science* **200**, 1074-1076.

Lasek, R., and Gainer, H., and Przbylski, R. J. (1974). Transfer of newly synthesized proteins from Schwann cells to the squid giant axon. *Proc. Natl. Acad. Sci. U.S.A.* **71**, 1188-1192.

Mahaley, M. S., Jr., and Wilfong, R. F. (1969). The nature of tetraphenylboron-dissociated brain and brain tumor cells. *J. Neurosurg.* **30**, 250-259.

Marangos, P. J., Zomzely-Neurath, C., and York, C. (1976). Determination and characterization of neuron specific protein (NSP)-associated enolase activity. *Biochem. Biophys. Res. Commun.* **68**, 1309-1316.

Medzihradsky, F., Nandhasri, P., Idoyaga-Vargas, V., and Sellinger, O. Z. (1971). A comparison of the ATPase activity of the glial cell fraction and the neuronal perikaryal fraction isolated in bulk from rat cortex. *J. Neurochem.* **18**, 1599-1603.

Möhler, H., Polc, P., Cumin, R., Pieri, L., and Kettler, R. (1979). Nicotinamide is a brain constituent with benzodiazepine-like actions. *Nature (London)* **278**, 563-565.

Moore, B. W. (1965). A soluble protein characteristic of the nervous system. *Biochem. Biophys. Res. Commun.* **19**, 739-744.

Moore, B. W., and Perez, V. J. (1968). Specific acidic proteins of the nervous system. *In* "Physiological and Biochemical Aspects of Nervous Integration" (F. D. Carlson, ed.), pp. 343-360. Prentice-Hall, Englewood Cliffs, New Jersey.

Moore, B. W., Perez, V. J., and Gehring, M. (1968). Assay and regional distribution of a soluble protein characteristic of the nervous system. *J. Neurochem.* **15**, 265-272.

Moore, B. W., Cimino, M., and Hartman, B. K. (1977). *Proc. Int. Soc. Neurochem.* **6**, 31.

Norton, W. T., and Poduslo, S. E. (1970). Neuronal soma and whole neuroglia of rat brain: A new isolation technique. *Science* **167**, 1144-1146.

Norton, W. T. and Poduslo, S. E. (1971). Neuronal perikarya and astroglia of rat brain: Chemical composition during myelination. *J. Lipid Res.* **12**, 84-90.

Poduslo, S. E., and Norton, W. T. (1971). Isolation and some chemical properties of oligodendroglia from calf brain. *J. Neurochem.* **19**, 727-736.

Poduslo, S. E., and Norton, W. T. (1975). Isolation of specific brain cells. *In* "Methods in Enzymology" (J. M. Lowenstein, ed.), Vol. 35, pp. 561-579. Academic Press, New York.

Rappaport, C., and Howze, G. G. (1966). Further studies on the dissociation of adult mouse tissue. *Proc. Soc. Exp. Biol. Med.* **121**, 1016-1021.

Rose, S. P. R. (1965). Preparation of enriched fractions from cerebral cortex containing isolated, metabolically active neuronal cells. *Nature (London)* **206**, 621–622.

Rose, S. P. R. (1967). Preparations of enriched fractions from cerebral cortex containing isolated, metabolically active neuronal and glial cells. *Biochem. J.* **102**, 33–43.

Rose, S. P. R., and Sinha, A. K. (1970). Separation of neuronal and neuropil cell fractions: A modified procedure. *Life Sci.* **9**, 907–915.

Rotman, B., and Papermaster, B. W. (1966). Membrane properties of living mammalian cells as studied by enzymatic hydrolysis of fluorogenic esters. *Proc. Natl. Acad. Sci. U.S.A.* **55**, 134–141.

Rous, P., and Beard, J. W. (1934). Selection with the magnet and cultivation of reticulo-endothelial cells (Kupffer cells). *J. Exp. Med.* **59**, 577–592.

Satake, M., and Abe, S. (1966). Preparation of enriched fractions from cerebral cortex containing isolated, metabolically active neuronal and glial cells. *J. Biochem. (Tokyo)* **59**, 72–75.

Schachner, M. (1974). NS-1 (nervous system antigen-1), a glial cell-specific antigenic component of the surface membrane. *Proc. Natl. Acad. Sci. U.S.A.* **71**, 1795–1799.

Schmachel, D., Marangos, P., Zis, A., Brightman, M., and Goodwin, F. (1978). Brain enolases as specific markers of neuronal and glial cells. *Science* **199**, 313–315.

Sellinger, O. Z., and Azcurra, J. M. (1974). Bulk separation of neuronal cell bodies and glial cells in the absence of added digestive enzymes. *In* "Research Methods in Neurochemistry" (N. Marks and R. Rodnight, eds.), pp. 3–38. Plenum, New York.

Sieghart, W., Sellström, A., and Henn, F. (1978). Sedimentation characteristics of subcellular vesicles derived from three glial systems. *J. Neurochem.* **30**, 1587–1589.

Sensenbrenner, M., Treska-Ciesielski, J., Lodin, Z., and Mandel, P. (1970). Autoradiographic study of RNA synthesis in isolated cells in culture from chick embryo spinal culture. *Z. Zellforsch. Mikrosk. Anat.* **106**, 615–626.

Sinha, A. K., and Rose, S. P. R. (1972). Compartmentation of lysosomes in neurons and neuropil and a new neuronal marker. *Brain Res.* **39**, 181–196.

Skolnick, P., Marangos, P., Goodwin, F., Edwards, M., and Paul, S., Identification of ionsine and hypanthine as endogenous inhibitors of [^{3}H]diazepan binding in the central nervous system. *Life Sci.* **23**, 1473–1480.

Tiplady, B., and Rose, S. P. R. (1971). Amino acid incorporation into protein in neuronal cell body and neuropil fractions *in vitro*. *J. Neurochem.* **18**, 549–558.

Tyrer, H., Sharrow, S., Ryan, J., and Wunderlich, J. (1974). *Proc. 27th Annu. Congr. Eng. Med. Biol.* Vol. 16, p. 52.

Van Nieuw Amerongen, A., and Roukema, P. A. (1973). Physico-chemical characteristics and regional distribution studies of GP-350, a soluble sialoglycoprotein from brain. *J. Neurochem.* **21**, 125–136.

Van Nieuw Amerongen, A., Van Den Eijnden, D. H., Heijlman, J., and Roukema, P. A. (1972). Isolation and characterization of a soluble glucose-containing sialoglycoprotein from the cortical grey matter of calf brain. *J. Neurochem.* **19**, 2195–2205.

Van Nieuw Amerongen, A., Roukema, P. A., and Van Rossum, A. S. (1974). Immunofluorescence study of the cellular localization of GP-350, a sialoglycoprotein from brain. *Brain Res.* **81**, 1–19.

Warecka, K. (1975). Immunological differential diagnosis of human brain tumors. *J. Neurol. Sci.* **26**, 511–516.

Warecka, K., and Bauer, H. (1967). Studies of "brain-specific" proteins in aqueous extracts of brain tissue. *J. Neurochem.* **14**, 783–787.

Whittaker, V. P. (1968). The morphology of fractions of rat forebrain synaptosomes separated on continuous sucrose density gradients. *Biochem. J.* **106**, 412–417.

ADVANCES IN CELLULAR NEUROBIOLOGY, VOLUME 1

SEPARATION OF NEURONS AND GLIAL CELLS BY AFFINITY METHODS

SILVIO VARON AND MARSTON MANTHORPE

Department of Biology and
School of Medicine
University of California at San Diego
La Jolla, California

I. Introduction

The nervous system is a biological machinery whose apparent stability is the expression of a dynamic equilibrium between the inherent properties of its component cells and the humoral and cellular environments in which they operate. The properties of neural cells and the regulatory mechanisms by which they adjust to the environment are the subjects of a new and rapidly expanding discipline, cellular neurobiology. Its main impetus derives from the addition of the study of neural systems *in vitro* to the traditional *in vivo* approaches. Increasing use of *in vitro* approaches

405

depends fundamentally on the progress to be made in techniques that will provide separate populations of viable and functional glial and nerve cells. Attempts to separate neural cells into homotypic populations have been numerous and have used a variety of approaches, most of which start with cell suspensions obtained from the source tissue by several dissociation procedures (Varon and Saier, 1975; Varon, 1977a). Two important approaches are bulk fractionation (Henn, this volume) and the use of differential culture environments (Section III, Table I).

An ideal way to separate cells from a mixed cell suspension is to take advantage of cell-specific surface constituents. Potential techniques directed to this approach require three critical components. The first is the availability of ligands that can selectively recognize surface constituents of the cells to be isolated. Ligands can be known molecules binding with high affinity to cell receptors, or antibody molecules directed against known or unknown cell surface antigens. The second component is a test system for cell recognition to determine whether a given ligand will in fact recognize the desired cells in the mixed cell suspension and at the same time ignore accompanying cells of a different category. The last component is a *separation system* that will achieve sequestration, and subsequent release, of the desired cells from corresponding ligands previously immobilized on a solid surface of semisolid matrix. This is the basis of the affinity methods with which this review will be concerned.

In the review, we shall address separately each of the above three components, although for greater clarity the last component will be taken up first. The thrust of the review will be on examining potential rather than actual approaches to affinity separation of neural cells because, thus far, only three reports have dealt directly with glial or neuronal cells (Venter *et al.*, 1976; Dvorak *et al.*, 1978; Au and Varon, 1979). Affinity methods have been mainly described for lymphoid cell populations, but such studies provide concepts and technical approaches that may be equally valuable when adapted to neural cells.

II. Techniques of Separation by Affinity Systems

In this section, we shall assume that (1) cell-selective ligands have already become available (see Section III,A), and (2) they have been shown to discriminate, within the cell suspension under study, between "positive" and "negative" cells (i.e., cells carrying or, respectively, lacking surface acceptors for the ligand) (see Section III,B). One approach, column cell chromatography, will be examined first and in considerable detail. Other affinity separation approaches will then be discussed, which share many technical features with the first one but also present different problems and/or potential advantages.

A. Column Cell Chromatography

The principles of this approach are schematically represented in Fig. 1. A mixed cell suspension, comprising positive (+) and negative (−) cells, is loaded on a column whose matrix has been derivatized with a cell-selective surface ligand. The (−) cells will be washed through and collected, resulting in their *purification* from (+) cell contaminants. In a second step, the (+) cells that have become sequestered on the immobilized ligand of the column may be retrieved by the use of appropriate "releasing" treatments or eluants, completing a true *separation* process with both cell classes recovered. A detailed analysis of the sequestration step in such an approach has been provided recently by Au and Varon (1979), using spinal cord cell suspensions as a (+) cell population and antibody generated against them as the surface-recognizing ligand. The main components of a

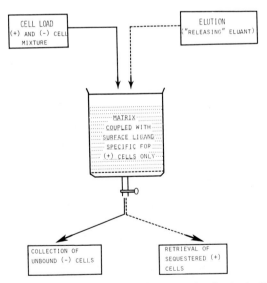

FIG. 1. Principle of column cell affinity chromatography. A mixed cell suspension of (+) and (−) cells is loaded on a column whose matrix has been derivatized with a surface ligand selective for (+) cells. The (−) cells elute unattached, and the column is washed with a "releasing" agent to elute (+) cells. The collection of both (−) and (+) cells constitutes a true separation.

column cell chromatography system are described in the following discussion.

1. THE CHROMATOGRAPHIC MATRIX

An efficient matrix should meet three sets of criteria: (1) It should be available in a stable bead form, provide good flow rates, and be amenable to use under sterile conditions; (2) it should allow flow-through of all the cells loaded, without physical entrapment or damage to their viability; and (3) it should nermit highly efficient coupling of the selected ligands with minimal loss of their binding competence, gain no tendency to retain cells except through the ligand, and maintain its stability after the coupling procedure. Several matrices have been reported, mainly for use with lymphoid cells which pose minimal flow-through problems because of their small diameter. Among them are the following.

a. Glass and Plastic Beads. Kondorosi *et al.* (1977) have provided a detailed analysis of noncovalent protein binding to glass or plastic beads (130–160 μm) for use with lymphoid cells. Since protein binding was not specific, highly purified protein ligands were needed to attain selective cell sequestration. Glass or plastic beads have also been used for lymphoid cells after covalent ligand coupling (e.g., Wigzell *et al.*, 1972; Binz *et al.*, 1974; Isturiz *et al.*, 1975).

b. G-200 Sephadex Beads. Lymphoid cells have been separated on derivatized G-200 Sephadex columns (Schlossman and Hudson, 1973; Crum and McGregor, 1976; Scott, 1976). The gel was activated with cyanogen bromide (CNBr) before coupling of an appropriate immunoglobulin and could be stored after coupling. For use, the G-200 gel was packed over a bottom layer of G-25 Sephadex and sterilized by sterile washes and irradiation.

c. Polyacrylamide Beads. BioGel P6 (BioRad) consists of autoclavable beads (150–300 μm hydrated diameter). This matrix has been used for lymphoid cell separation (e.g., Truffa-Bachi and Wofsy, 1970; Godfrey and Gell, 1976). In addition, Au and Varon (1979) investigated the use of P6 columns for cell suspensions from neural and muscle chick embryo tissues and found them to allow 90–100% flow-through recoveries of the cells loaded.

d. Agarose Beads. Blood cells have been effectively chromatographed on Pharmacia's Sepharose 4B (e.g., Weinstein *et al.*, 1973; Shearer *et al.*, 1977; Ljungsted-Påhlman *et al.*, 1977; Light and Tanner, 1978) and 6B

(Edelman *et al.,* 1971) columns. Sepharose 4B (60- to 140-μm beads), derivatized with insulin or lectin, has been used to fractionate much larger cells such as rat adipocytes (Soderman *et al.,* 1973). More recently, Sepharose 6MB (200- to 300-μm beads) has been successfully used for perinatal chick sympathetic ganglion cells (Dvorak *et al.,* 1978) and for cells from various embryonic chick tissues (Au and Varon, 1979).

2. DERIVATIZATION OF THE MATRIX WITH CELL-SELECTIVE LIGANDS

Covalent coupling of different ligands to different matrices will require different derivatization procedures reflecting the chemical nature of both components.

a. Nonprotein Ligands. The need of developing diverse procedures may be particularly serious with small ligands of a nonpolypeptidic nature. Approaches to their covalent coupling can draw from the extensive information available on affinity systems directed to the separation of molecules (cf. Cuatrecasas and Anfinsen, 1971) and membrane fragments (e.g., Cuatrecasas, 1972). Particular attention ought to be given to the interposition of a "spacer" or "arm" (e.g., hydrocarbon chains) between the ligand and the matrix. Examples of interest for neural cell separation are isoproterenol and triiodothyronine for glass beads (Venter *et al.,* 1976), serotonin-related compounds for Sepharose 4B (Mehl and Weber, 1974), and histamine for Sepharose 4B (Weinstein *et al.,* 1973; Shearer *et al.,* 1977).

b. Polypeptide Ligands. In the case of proteins, the coupling procedures differ mainly in accordance with the matrix to be derivatized. Glass can be derivatized with protein by the use of silane coupling agents (Weetall, 1969). However, thus far, the two major matrices for neural cell chromatography are polyacrylamide (BioGel P6) and agarose (Sepharose of various grades).

With *P6,* the derivatization procedure of Inman and Dintzis (1969), using hydrazine hydrate for gel activation, has been used to couple albumin (Godfrey and Gell, 1976) and immunoglobulin (Au and Varon, 1979), among other ligands. A second procedure (Weston and Avrameas, 1971) involves activation with glutaraldehyde prior to coupling of the ligand (Godfrey and Gell, 1976; Au and Varon, 1979).

Sepharose is used as a CNBr derivative (March *et al.,* 1974), which is commercially available and becomes activated directly on suspension in hydrochloric acid. The CNBr procedure has been used for insulin (Cuatracasas, 1972; Soderman *et al.,* 1973), concanavalin A (Con A) (Edelman *et al.,* 1971), histamine–albumin conjugates (Weinstein *et al.,* 1973), transferrin (Light and Tanner, 1978), and immunoglobulin (Scott, 1976; Au and Varon, 1979), among others. Agarose–CNBr can also be used to attach amino alkyl spacer arms of desired length (Cuatrecasas and Anfinsen, 1971; Cuatrecasas, 1972; Cuatrecasas and Parikh, 1972), as has been done for insulin (Cuatrecasas, 1972; Ljungstedt-Pohlman *et al.,* 1977), α-bungarotoxin (α-BTx) (Dvorak *et al.,* 1978), and smaller ligands (Venter *et al.,* 1976).

With either matrix, two features of the derivatization procedure should be carefully examined. One is the coupling efficiency of the procedure, i.e., the ratio between the amount of ligand that becomes coupled and the amount (or the concentration) of it used in the coupling reaction. A low coupling efficiency is a disadvantage with protein ligands available in very small amounts (e.g., specific antibodies and factors). Au and Varon (1979) have compared immunoglobulin-coupling efficiencies of the hydrazine and glutaraldehyde procedures for P6 and the CNBr procedure for Sepharose and found them to be about 5, 15, and 95%, respectively. The second important feature is whether the activated beads acquire binding sites for molecules and/or complex structures (membrane, cells) beyond those occupied by the coupled ligand. Such a ligand-independent binding ability has led to high levels of nonspecific cell sequestration (e.g., Au and Varon, 1979). With Sepharose–CNBr, the gel sites remaining available after coupling of the protein ligand can be blocked with excess ethanolamine, glycine, or lysine (Cuatrecasas, 1972; Ljungstedt-Pohlman *et al.,* 1977; Au and Varon, 1979).

3. Chromatographic Setup and Sequestration Step

Chromatographic columns of the desired capacity can be readily assembled using glass tubes (Edelman *et al.,* 1971; Godfrey and Gell, 1976; Dvorak *et al.,* 1978) or glass or plastic syringes (Scott, 1976; Ljungstedt-Pohlman *et al.,* 1977; Au and Varon, 1979) fitted with sintered or nylon mesh disks. Before addition of the matrix beads, assembled glass columns can be autoclaved, and presterile plastic syringes are commercially available. The temperature can be controlled by the use of water-jacketed columns (Dvorak *et al.,* 1978) or temperature-controlled rooms. Flow rates can be controlled by regulating column outlet, head pressure of the eluant, or appropriate pumping systems. Sterile media to be used for packing and eluting the matrix gel should be chosen to suit the needs of the cells under study and the duration of the operation.

Two modes of sequestration can be tested (Au and Varon, 1979):

1. In a *flow-through* mode, the cells are allowed to move at once through the column at a selected slow rate sustained by a corresponding introduction of medium. The emerging fluid can be collected fractionally, and cell contents determined by counting (Coulter counter, hemocytometer) or by radioactivity measurement of prelabeled cells. In subsequent experiments, collection can be restricted to the volume of eluate previously found to contain cells. This mode is particularly useful in evaluating the suitability of matrix packing, cell load, and other characteristics of the setup with regard to nonspecific retention of cells. However, it may offer insufficient opportunity for the cells to establish a stable binding to the matrix-carried ligand and thus become effectively sequestered.

2. In a *stationary-stage* mode, the loaded cells are first allowed to sink into the gel bed (by stopping outflow before cell emergence). They can then be kept within the matrix for a long enough time to ensure the most stable sequestration, and the unbound cells washed through with medium to be collected and measured as in the other mode.

4. The Cell Retrieval Step

Retrieval methodologies will depend on the nature of the immobilized ligand and the connections between (1) matrix and immobilized ligand, (2) ligand and ligand-recognizing "acceptor" on the cell surface, and (3) acceptor and cell membrane. The principal aim will be to preserve integrity and viability of the cells to be released. A lesser priority should be to preserve the affinity competence of the matrix, or at least its amenability to subsequent regeneration.

a. Matrix–Ligand Connection. It appears unlikely that covalent connections along a spacer arm can be broken without risking cell damage as well. However, the matrix itself or the first link between it and the spacer might be amenable to selective enzymatic attack. An example of this approach is the use of dextranase to recover lymphocytes by digestion of the sequestering Sephadex matrix (Schlossman and Hudson, 1973; Crum and McGregor, 1976). Alternatively, it might be possible to "digest" the ligand itself (e.g., by proteolytic attack of a polypeptidic ligand). Such a disruption may have been involved in the release by trypsin of sympathetic neurons from α-BTx–Sepharose (Dvorak *et al.,* 1978).

b. Ligand–Acceptor Connection. Resolution of the noncovalent association between immobilized ligand and cell-carried acceptor may permit reuse of the column as well. Depending on the nature of the ligand–acceptor bond, it may be possible to resolve it by altering the aqueous environment in the column (cf. Edelman *et al.*, 1971). However, the main way to attempt this is to provide the system with large excesses of either free ligand to compete with the immobilized one for the cell-carried acceptor, or free acceptor (if available) to compete with the cell-carried one for the immobilized ligand. Several examples illustrate this approach:

1. Insulin was found to release adipocytes from insulin–Sepharose columns (Soderman *et al.*, 1973).
2. Specific analogs of the ligand may be employed, as demonstrated with serotonin-binding membrane (Mehl and Weber, 1974).
3. Specific sugar haptens can release cells bound to corresponding lectins by competing with the reactive (carbohydrate) sites of the cell acceptors. Such was the case with lymphoid cells on Con-A–Sepharose (Edelman *et al.*, 1971) and on Con-A–nylon (Killion and Kollmorgen, 1976).
4. Rat IgG-bearing lymphocytes have been sequestered on matrix-coupled rabbit anti-rat IgG F (ab)' fragments and released with excess free rat IgG (Crum and McGregor, 1976). The use of free IgG as a competitor of cell surface IgG acceptors need not be restricted to cells naturally endowed with IgG surface constituents (such as B lymphocytes). For example, neural cells can be converted to IgG-bearing cells by pretreatment with rabbit IgG directed against specific surface antigens, then sequestered on matrix-coupled goat anti-rabbit IgG antibody, and finally released with excess free rabbit globulin. This would be but one case of a more general concept to be discussed further, namely, the supply of an exogenous acceptor to the cell surface to increase the versatility of the affinity system. A variation of the same concept is illustrated by the next example.
5. Antigen-binding lymphocytes have been provided with fluorescein-labeled antigens, sequestered on columns derivatized with antifluorescein antibody, and released with an excess of heterologous fluorescein conjugates (Scott, 1976). Such an approach can be readily extended to neural cells capable of binding fluorescein-conjugated molecules.

c. Acceptor–Membrane Connection. Cell surface constituents are amenable to selective enzymatic attack and so, therefore, should be their interaction with the immobilized ligand. To what extent such enzymatic treatments affect the integrity and viability of the cells themselves remains, of course, a question to be carefully addressed in each particular case. Proteolytic enzymes are often used for neural tissue dissociation and are undoubtedly compatible with viability in many cases. The release of viable sympathetic neurons from α-BTx–Sepharose by brief trypsin incubation (Dvorak *et al.*, 1978) provides a convincing demonstration of the feasibility of such an approach. Also to be considered is the use of neuraminidase

(Ljungstedt-Pohlman *et al.*, 1977) or other glycosidases. A much greater flexibility would be provided if an exogenous acceptor were used. One could (1) attack it directly with less concern for cell damage, (2) dissociate it from its binding site on the cell surface, or (3) compete for the binding site with excess free exogenous acceptor.

5. PAST USES OF CELL SEPARATION COLUMNS

Application of the preceding principles to affinity separation by cell chromatography is illustrated in the following examples.

Ljungstedt-Pohlman *et al.* (1977) sequestered lymphoid cells on Sepharose 4B derivatized with either insulin or Con A. The cell surfaces were prelabeled with ^{125}I, and sequestration was evaluated by radioactivity measurements. A stationary-stage mode was used (30 min at room temperature). Cell retention was 5% of the load on control columns and rose to 25–40% on insulin- and Con-A-derivatized gels. The study contains instructive evaluations on how nonspecific and specific cell sequestration may be affected by the use of spacers, cell loads, pH and temperature, and height of the packed gel bed.

Dvorak (1978) used α-BTx-derivatized Sepharose 6MB to sequester and subsequently retrieve neurons from chick embryo sympathetic ganglionic dissociates. The medium used for loading and sequestration contained albumin (to stabilize the cells), deoxyribonuclease (to minimize cell aggregation), and nerve growth factor (required for survival of sympathetic neurons). The flow-through mode was employed (1–2 ml/hr at 0°–4°C). After the eluate became practically cell-free, the column was washed with medium containing 0.1% trypsin, clamped off, incubated for 2–3 min at 37°C, supplied with fetal calf serum (to block further trypsin action), and eluted with serum-containing culture medium. The first eluate contained essentially only nonneuronal cells, whereas the cell population of the final eluate was made up almost exclusively of neurons (>95%, as compared with the 10% neuronal presence in the initial cell load). The retrieved neuronal fraction was viable and remained free of nonneuronal elements after 6 days in monolayer culture. No data were supplied on load size, final neuronal yields, or comparative survival in culture.

Au and Varon (1979) analyzed sequestration of chick embryo spinal cord cells on P6 and Sepharose 6MB gels before and after derivatization with normal rabbit globulin or immunoglobulin against the same cells. Underivatized columns allowed 90–100% cell recovery (all in the first 1 ml of effluent) by the flow-through mode, and 80–85% by the stationary-stage mode (60 min at room temperature). P6 gels derivatized with normal globulin by the glutaraldehyde procedure retained 50–70% of the cell load by either modality (presumably because of residual active sites), thereby greatly reducing the suitability of this material for specific cell separation. In contrast, Sepharose 6MB derivatized with normal globulin by the CNBr procedure caused the same nonspecific cell losses as before derivatization (15–20%). The same matrix, derivatized with two different batches of antispinal cord immunoglobulin, sequestered 50% and 80–85% of the cell load, respectively, when the stationary-stage modality was used. In addition, Sepharose 6MB derivatized with immunoglobulin against αBTx sequestered 60–70% of a load of embryonic chick muscle cells previously treated with αBTx, demonstrating that artificially supplied (exogenous) acceptors may render the cells susceptible to immunoaffinity sequestration.

6. DIRECTIONS FOR FUTURE PROGRESS

Many of the difficulties encountered in past work on cell separation by affinity chromatography have arisen from having to confront simultaneously several uncertainties. Principal among them is the fact that the cell subset to be segregated usually occurs in undefined proportions in the initial, heterogeneous cell suspension and is endowed with undefined (and, possibly, variable) affinity for the immobilized ligand. In turn, the uncertain availability of sequestrable cells makes it difficult to define rigorously the optimal features of the sequestering system and of the sequestration procedure. The ideal way to circumvent these problems is to concentrate the initial efforts on a population that is as near homogeneity as possible, at least with regard to its ability to react with the chosen ligand and to define with it all the necessary parameters of (1) cell sequestration and (2) cell release. A first step in this direction has been taken by studying sequestration of a spinal cord cell suspension on an immobilized IgG that recognized all cells within the suspension (Au and Varon, 1979). The use of a homogeneous clonal cell line of neural origin and of an antibody reactive to its surface membrane—a combination that has not yet been investigated—may prove even more advantageous. Further work should be directed to several important problems.

a. Nonspecific Cell Retention. Nonspecific entrapment of intact cells represents a serious source of contamination for specifically sequestered cells, and efforts should be made either to reduce the nonspecific retention (e.g., by additional washes or the use of more suitable media) or to eliminate the contaminant cells after retrieval of the sequestered population (e.g., by recycling).

b. Efficiency of the Sequestration Step. What controls the sequestration efficiency of a system needs to be better understood, as do the ways in which such efficiency can be optimized. Among the features to be determined are (1) ability of unsequestered cells to become sequestered when rerun through a fresh column; (2) maximal capacity and variations in efficiency of a given setup under different, or repeated, cell loads; (3) influence of medium, time and temperature, and geometric parameters of the packed gel on the efficiency; (4) advantages and disadvantages of varying the spacing between matrix-anchored and cell-carried reactive sites, whether in terms of spacer arm length or by interposition of an exogenous acceptor; and (5) density and distribution of reactive sites on the derivatized matrix, e.g., by varying the amount of ligand coupled per unit matrix, derivatizing in the presence of nonrelevant ligands, or packing the column with mixtures of derivatized and underivatized beads.

c. Versatility of the Separation System. Figure 2 summarizes several situations in which an immobilized cell selector (the cell-specific ligand) could be expected to sequester the relevant (positive) cells.

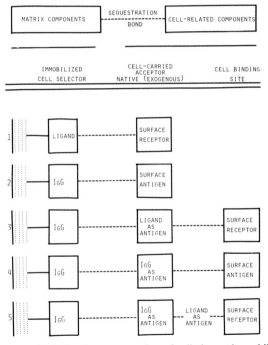

FIG. 2. Various methods for the sequestration of cells by an immobilized matrix. The matrix can be derivatized with a cell selector such as a surface-binding ligand (1), a cell surface specific IgG (2), or IgG directed against exogenous acceptors carried by the cell (3–5). The cells are presented to the immobilized selector, either with no prior treatment (1 and 2) or after exposure to exogenous acceptors (3–5).

1. The first situation is the typical one, in which the cell selector is a nonimmune ligand—such as a transmitter, a hormone, or a certain toxin or lectin (see Section III)—that will bind in a noncovalent manner to corresponding cell surface receptors. The various features of derivatization, sequestration, and retrieval will have to be tailored to the nature of the ligand. Thus, each case will require its own procedural investigation and at least substantial adjustments before it can be extended to a different ligand.

2. The second situation may be viewed as merely a special case of the first one, with immunoglobulin (IgG) directed against a cell surface antigen serving as the immobilized ligand and the antigen acting as a receptor for it. One important difference, however, is that, once an optimal procedure for IgG coupling to the matrix is established, the same procedure can be applied to any desired cell-selective immunoglobulin. Sequestration and retrieval steps should also vary only moderately, depending on the degree of antibody–antigen affinity and the extent to which the antigen is represented on the cells.

3. Given the advantage of IgG as the cell selector, it may be possible to extend this approach to cell-carried antigens that are exogenous, rather than native components of the cell surface. Antibody could be obtained against a known (and available) molecule with high and specific affinity for the desired cells, the cell surface receptors could be saturated with this molecule, and the cell-carried molecule could serve as the antigen that would bind to the immobilized IgG. This is the third situation depicted in Fig. 2. As in the second situation, the sequestration bond is an antibody-antigen one, but advantage is being taken of the cell-selective ligands involved in the first situation. Examples are the already mentioned use of anti-αBTx-Sepharose for αBTx-coated muscle cells (Au and Varon, 1979) and of antifluorescein-Sepharose for fluorescein-conjugated ligands bound to lymphoid cells (Scott, 1976).

4. The fourth situation is like the one just described but it takes advantage of cell-specific antibodies (against native surface antigens) rather than of nonimmune exogenous ligand. The IgG to be immobilized is obtained (in a different species) against IgG from the animal species in which the antineural antibody was raised. If antibodies against different neural cell surface antigens are raised in the same animal species (e.g., rabbit), we now have a situation in which all the steps of the sequestration and retrieval procedures remain identical regardless of which cells and which surface antigens are being addressed (see also Section II,A,4). The same principle has been used very successfully in the technique of indirect immunofluorescence and immunoperoxidase labeling of cytological preparations (see Section 3,B).

5. The last situation combines the principles and the advantages of the previous two. A known, available ligand is bound to the cell surface, IgG against the ligand is then bound to it, and this IgG enters into a sequestration bond with immobilized IgG against normal IgG of the same source. Besides covering all types of cell-specific molecules considered in this review (see Section III) and providing a single generalized technique for all of them, such an approach offers a multiplicity, and thus a choice, of points of attack for the cell retrieval steps.

B. Other Affinity Systems

1. AFFINITY BATCH SEPARATION

Like the fractionation of protein molecules, column cell chromatography may be replaced by batch cell separation procedures. The cell suspension is mixed with a suspension of beads, derivatized with the appropriate cell-selective ligand, and the free cells are mechanically separated from the bead-bound ones. The cell-carrying beads are then subjected to a releasing treatment, and the released cells are again separated from the beads by a mechanical procedure. Thus, the main distinction of this approach is that separation of unsequestered or subsequently released cells, from the sequestering matrix, occurs by means other than an elution process. Herein, too, lie the potential advantages and disadvantages of the batch approach. Possible advantages are: (1) matrix options are extended to smaller beads, since there is no flow-through requirement; (2) the procedure may be scaled up with no need for special equipment; and (3) the greater access to the cell-carrying beads may facilitate cell retrieval. Possible disadvantages lie essentially in difficulties to be encountered in the separation step.

a. Glass Beads. Venter *et al.* (1976) have coupled norepinephrine and isoproterenol, corticotropin, thyroxine, and diptheria toxin to glass beads by diazotization procedures using hexanediamine as a spacer. The derivatized beads were mixed with radiolabeled clonal cell suspensions (glial C6 cells and nonneural Y, GH_3, and HeLa cells) in serum-free medium and allowed to react for various times at 37°C under shaking. Separation was achieved by unit gravity sedimentation of the beads in isotonic sucrose, followed by filtration of the free cells through a 200-μm silk mesh. Free and bound cells were determined by measuring radioactivity in the corresponding fractions. Maximal binding time was about 15 min at 37°C. Nonspecific binding amounted to a variable and substantial 15–50% of the total cell binding. Specific binding occurred between all cell lines and each ligand but was highest between ligands and the cell type known to have high numbers of surface receptors for them. Even so, 15–20% of the C6 cells loaded remained unbound to isoproterenol–beads even at high bead/cell ratios. Specific binding was prevented by pretreating the cells with free ligand or an appropriate analog, or by carrying the sequestration in its continuous presence. Free ligand or analogs were not tested for ability to release already bound cells. Similarly derivatized Sepharose–beads bound cells with 15–30 times lower efficiency.

b. Agarose Beads. Kinzel *et al.* (1977) have described binding and retrieval of HeLa cells and transformed mouse fibroblasts (SV3T3) by use of Sepharose 2B. Larger beads (100–200 μm) were selected and derivatized with lectin (LCA) by a CNBr procedure (50–85% coupling efficiency). A cell suspension was mixed with the lectin–beads for 30 min at room temperature in a petri dish or in a flask with two compartments separated by a 100-mesh gauze. The unbound cells were filtered through the 100-μm gauze and counted. Cell binding by the lectin–beads was rapid (75% in 10 min), nearly exhaustive (up to 90% by 20–30 min), and specific (suppressed only by the correct sugar haptens, and in a dose-dependent manner). Bound cells could be retrieved by shaking the cell–bead complexes for 60 min in the presence of specific sugar inhibitors. The retrieved cells were viable (albumin was included in the sequestration medium to "stabilize" the cells) and grew at rates comparable to those of untreated cells. The lectin–beads could be reused through several cycles (sugar displacement of the bound cells, followed by very extensive washes with water) with no apparent loss of competence. The amount of coupled lectin per bead was important for both binding and retrieval of the cells, with lower levels reducing cell binding and higher levels reducing cell retrieval. Larger-sized agarose beads (Sepharose 6B), derivatized with equal amounts of lectin, worked as well as Sepharose 2B with HeLa cells but performed more poorly with SV3T3 fibroblasts.

c. Magnetic Beads. An interesting approach to the separation of free and bead-bound cells has been described by Molday *et al.* (1977). Magnetic polymeric microspheres (MPMs) were synthesized from ferric oxide particles and a variety of methacrylates, selected for their 40-nm size, and derivatized with lectins or antibodies using glutaraldehyde. The unique feature of these beads is their susceptibility to magnetic fields (e.g., a horseshoe magnet placed around the separation vessel). For example, MPMs derivatized with goat anti-mouse immunoglobin were mixed with mouse spleen lymphocytes, layered over 5% bovine serum albumin in buffered saline, and placed within the magnet for 2 hr at 4°C. Cells that had become labeled with MPMs (antibody-carrying B lymphocytes) were attracted to the glass walls of the tube with 97–99% efficiency, whereas unattracted free cells could be readily decanted.

d. Blood Cells. In immunological studies, a "rosette-forming" approach has often been used in which red blood cells (RBCs) coated with a suitable ligand are allowed to cluster on the surface of ligand-recognizing cells and thus permit visualization of the latter. Schachner and Hammerling (1974) used such a method to separate mouse brain cell subpopulations bearing the surface antigen Thy-1. After the rosette-forming step, the cell preparation was centrifuged first through a discontinuous density gradient to remove unreacted RBC and then on a 25% albumin layer to remove the unreacted cells. The rosette-bearing cells (the pellet) could be subsequently freed of their RBCs. An alternate method for rosette formation (Ljungstedt-Pohlman *et al.,* 1977) has been to couple to RBCs, by glutaraldehyde treatment, a staphylococcal cell wall protein called protein A, which is characterized by the ability to bind the Fc region of immunoglobulins specifically (Forsgren and Sjoquist, 1966). The protein-A–RBCs, thus, will form rosettes on IgG-carrying lymphocytes—or, conceivably, any nonlymphoid cell bearing an exogenous surface-specific IgG.

2. AFFINITY PLATING SEPARATION

When primary cell suspensions are seeded for monolayer cultures, cell attachment depends on cell–substratum adhesion, among other features (cf. Varon, 1979). Differential cell attachment, based on unidentified differences in cell surface adhesiveness, has provided ways to separate neurons from nonneuronal cells (Varon and Raiborn, 1969, 1972; Varon *et al.,* 1973; McCarthy and Partlow, 1976). A much more specific cell separation should be possible if the substratum provided attachment exclusively for cells endowed with unique surface constituents, that is, if an otherwise nonadhesive substratum were derivatized with cell-specific ligands. The main advantages of such an "affinity plating" approach would be simplicity and rapidity of the sequestration step, availability of the sequestered cells to direct examination, and either no need for their retrieval (if the bound cells are to be studied in culture) or easier and more versatile approaches to the retrieval step.

The main obstacle to affinity plating separation lies in the tendency of most cells to attach nonspecifically to culture substrata. This can be avoided by (1) selecting a substratum that is nonadhesive before the derivatization treatment, and (2) finding derivatization conditions that will not introduce nonspecific adhesiveness. Plastic petri dishes, unlike the sulfonated tissue culture plastic, usually permit no cell attachment in serum-free media for several hours. Other nonadhesive surfaces to be explored may be agar, selected lipids (cf. Margolis *et al.*, 1978), or polyanionic surfaces or coatings. Covalent coupling of protein ligands to plastic surfaces can be readily obtained in the presence of water-soluble carbodiimide (cf. Edelman and Rutishauser, 1974). However, dishes derivatized by this method with nonspecific protein (e.g., normal IgG) have been found to acquire adhesiveness for various cell suspensions, thereby interfering with selective attachment to cell-specific ligands (A. M. J. Au and S. Varon, unpublished). Possible solutions to this problem may be sought in the future by, for example: (1) reducing the amount of specific ligand to be coupled; (2) minimizing nonspecific cell attachment by media manipulation (cf. Varon and Saier, 1975), positioning the dish vertically to avoid gravity deposition of the cells (cf. Edelman and Rutishauser, 1974), or interfering with nonspecific attachment by mechanical vibrations (cf. McCarthy and Partlow, 1976); or (3) exploring procedures that will differentially release the nonspecifically or the specifically attached cells.

Affinity plating separation on dishes has been reported, thus far, only with lymphoid cells, which have little propensity to attach nonspecifically to plastic surfaces. Edelman's group has explored the extension to plastic dishes of the techniques it has developed for affinity separation on nylon fibers (for review, see Edelman and Rutishauser, 1974), which others have used to select subsets of intestinal cells (Podolsky and Weiser, 1973) and of pituitary thyrotropic cells (Tal *et al.*, 1978). Barker *et al.* (1975) have derivatized dishes with human IgG, bound rabbit anti-human IgG to them, and used them to capture selectively B lympocytes under centrifugal force. Nash (1976) has described (for B-lymphocyte separation) an interesting variation which consists of coating a dish sequentially with (1) any convenient IgG (covalent coupling), (2) staphylococcal protein A, and (3) an antiallotype IgG antibody (i.e., directed against the species from which IgG-bearing lymphocytes are to be fractionated)—the overall effect being to amplify the number of reactive sites presented by the matrix-bound ligand (anti-IgG) to the desired cell acceptors (IgG on the lymphocyte surface). Lymphoid cells have also been separated on a matrix coated with derivatized gelatin (cf. Edelman and Rutishauser, 1974) or collagen (Maoz and Shellam, 1976), from which cell retrieval can be readily effected by melting or by collagenase treatment, respectively.

III. Ligands for Neural Cell Surfaces

Ultimately, the critical element of the overall separation process lies in the availability of ligands specific for the desired neural cell surfaces.

Several ligand classes are already either known or reasonably presumed to fit this definition. The fact that they have been so rarely applied as yet to neural cell separation underlines the importance of a separate quest for efficient separation technologies (Section II). One should note that selective ligands, particularly surface-specific antibodies, may also be used for cell separation by other than affinity procedures (cf. Varon, 1977a), such as selective cell destruction (Varon *et al.*, 1979a) and electronic cell sorting (Campbell *et al.*, 1977). In this section, we shall consider first several classes of putative ligands and next how to ascertain their suitability for the cell populations under actual study.

A. Categories of Neural Cell Ligands

Affinity separation depends on ligands that are selective for *different subsets* of neural cells rather than for the nervous system in general. Much of the available information, however, concerns ligands that are nervous system-specific, and efforts to define their restriction to one or another neural cell subclass have often been hampered by inadequate visualization techniques (see Section III,B). Thus, the following sections will review potential as well as actual availability of neural cell ligands with restricted specificity.

1. HORMONES, TRANSMITTERS, TOXINS

a. Hormones. Hormone-binding cells have been sequestered, for example, on immobilized insulin (Soderman *et al.*, 1973), corticotropin and triiodothyronine (Venter *et al.*, 1976), and thyrotropin-releasing hormone (Tal *et al.*, 1978). One difficulty in selecting hormones for neural cell separation is the current paucity of information as to which neural cells respond to which hormones (cf. Balazs, 1976). Among pertinent observations are the effects of thyroid hormones on myelinogenesis (e.g., Hamburgh, 1966), ACTH on serotonin metabolism (Halgren and Varon, 1972), insulin on ganglionic RNA and protein labeling (Partlow and Larrabee, 1971; Burnham *et al.*, 1974), and gonadal hormones on brain development (e.g., Toran-Allerand, 1978), as well as the inclusion of insulin, progesterone, and transferrin in a serum-free defined medium recently shown to sustain neuronal but not nonneuronal survival in culture (Bottenstein *et al.*, 1980; Skaper *et al.*, 1979). A special category of hormones is that of trophic factors for survival, growth, and functional maturation, and specifying factors for selective expression of certain cell programs (Varon, 1977b; Varon and Bunge, 1978). Neuron-specific factors are, for example, nerve growth factor (cf. Levi-Montalcini and Angeletti, 1968; Varon, 1975; Varon and

Bunge, 1978) and, more recently, a ciliary neuronotrophic factor (Varon *et al.*, 1979b; Adler *et al.*, 1979b; Landa *et al.*, 1980; Manthorpe *et al.*, 1980) and a cholinergic-specifying factor (cf. Patterson, 1978). Glial cells also respond to several protein agents, such as epidermal growth factor (Westermark, 1976), fibroblast growth factor (Westermark and Wasteson, 1975), a surface constituent of neuritic membrane (Salzer *et al.*, 1977), and the glial maturation factor (Lim *et al.*, 1977). All such agents are known or presumed to affect their target cells via interaction with surface receptors.

b. Transmitters. Postsynaptic neurons carry surface receptors for their own neurotransmitters (and related drugs), although in many cases the transmitter involved is yet to be ascertained. In addition, some transmitters also appear to interact with glial cell surfaces (cf. Varon and Somjen, 1979; van Calker and Hamprecht, this volume). Cells, or cell membranes, have been separated on matrices derivatized with transmitter or their analogs, such as catecholamines (Venter *et al.*, 1976), serotonin (Mehl and Weber, 1974), and histamine (Weinstein *et al.*, 1973; Shearer *et al.*, 1977). A special class of ligand, not yet considered for affinity separation, is that of endorphins and related compounds. Opiate receptors have been demonstrated in neural tissues of various species, as well as in neural cultures (cf. Hiller *et al.*, 1978).

c. Neurotoxins. Several toxins, derived from a variety of sources, are known to affect nerve cells via surface receptors (cf. Narahashi, 1974):

1. αBTx binds specifically and nearly irreversibly to nicotinic acetylcholine receptors of neurons in the avian mammalian central nervous system (CNS) (Moore and Brady, 1976; McQuarrie *et al.*, 1976; Kouvelas and Greene, 1976; Vogel *et al.*, 1977; among others) and peripheral nervous system (Fumagalli *et al.*, 1976, 1978; Greene, 1976; Kouvelas *et al.*, 1978; Carbonetto *et al.*, 1978; Chiappinelli and Giacobini, 1978). The occurrence of αBTx receptors is developmentally regulated. The binding is inhibited by nicotinic agonists and antagonists. Thus, αBTx is a good example of a neuron-restricted ligand and has indeed been used to separate viable chick embryo sympathetic neurons on affinity columns (Dvorak *et al.*, 1978; see Section II,A,5). We have also already mentioned the possibility of using anti-αBTx immunoglobulin to sequester αBTx-bearing cells (Au and Varon, 1979).

2. *Tetanus toxin* (TeTx) has also emerged from recent *in vitro* work as a reliable marker for neuronal surfaces, albeit with no restriction to particular neuronal subsets. Selective binding to neurons has been demon-

strated by autoradiography (Dimpfel and Habermann, 1977) or indirect immunofluorescence (Mirsky *et al.,* 1978; Fields *et al.,* 1978) in primary cell cultures from mouse, rat, or chick central and peripheral nervous tissues (see Section III,B,2). TeTx has no adverse effect on survival and neurite extension, at least with chick neurons (Mirsky *et al.,* 1978), which justifies expectations for future uses of this toxin for sequestration and retrieval of neuronal cells.

3. *Tetrodotoxin* (TTX) is known to bind to the outside of voltage-dependent sodium channels in electrically excitable cells (neurons, muscle), thereby blocking action potential responses. Although there have been reports of TTX binding to Schwann cells in the squid (Villegas *et al.,* 1976), TTX may provide another reliable marker for neuronal surfaces (Guillary *et al.,* 1977).

2. LECTINS AND OTHER ADHESION-PROMOTING AGENTS

a. Plant Lectins. Plant lectins have become important probes for study of the structure and function of cell surfaces (e.g., Lis and Sharon, 1973; Poste and Nicholson, 1977; Brown and Hunt, 1978). Lectins bind to selective glycoconjugates on surface membranes and cause cell agglutination because of their multivalent nature. Binding and agglutination are blocked by specific carbohydrate haptens, presumably competing with the glycoconjugate surface receptors. Vertebrate cells, including neural cells, possess surface receptors for plant lectins. In general, a given lectin should not be expected to address uniquely a particular class of neural cells. However, neural cell receptors for different plant lectins have been shown to vary independently of one another with regard to developmental age, regional distribution, and sensitivity to enzymatic treatments (Kleinschuster and Moscona, 1972; McDonough and Lilien, 1975; Hatten and Sidman, 1977; Isenberg *et al.,* 1978; among others). Lectin receptors have been visualized (see Section III,B) on neuronal membranes in histological sections (e.g., Kelly *et al.,* 1976; Reeber *et al.,* 1977) and in neural cultures (e.g., Koda and Partlow, 1976; Gonatas *et al.,* 1977). Two recent studies are particularly instructive:

1. Sieber-Blum and Cohen (1978) showed, in cell cultures from early quail neural crest, that Con A, wheat germ agglutinin (WGA), and soybean agglutinin (SBA) did not label pigment cells and that SBA selectively labeled adrenergic cells in parallel with their adrenergic development.

2. Denis-Donini *et al.* (1978) reported that C1300 neuroblastoma cells had receptors on their neurites only for SBA, even though their somata bound Con A and WGA as well. Similarly, cultured chick dorsal root

ganglionic neurons displayed differential binding for these three lectins both with regard to developmental age and subcellular localization.

   b. *Animal Lectins.* Sugar-specific lectins have also been reported to occur in several animal tissues (cf. Barondes, 1976; Simpson *et al.,* 1978), including extracts from C1300 neuroblastoma (Teichberg *et al.,* 1975), rat brain (Margolis *et al.,* 1976; Simpson *et al.,* 1977), chick brain (Barondes *et al.,* 1974; Kobiler and Barondes, 1977; Kobiler *et al.,* 1978), chick spinal cord and retina (Eisenbarth *et al.,* 1978), and chick optic lobe (Gremo *et al.,* 1978). Most of these animal lectins are not tissue-specific. In neural tissues, their presence is developmentally and regionally regulated, but no clear definition has yet been provided as to which neural cells are either the source or the recipients of such lectins. A similar lack of definition characterizes, thus far, the study of aggregation-promoting agents, despite their known regional specificity and developmental regulation (cf. Hausman and Moscona, 1975; Thiery *et al.,* 1977). It is worth noting, for affinity separation purposes, that the LETS protein (cf. Yamada *et al.,* 1975; Hynes, 1976) appears to be present on mesenchymal cell surfaces but not on neuronal or glial elements (Schachner *et al.,* 1978; Fields *et al.,* 1978; however, see Vaheri *et al.,* 1976).

3. IMMUNE LIGANDS

   A major class of potential ligands for neural cell separation, as already discussed in Section II, is that of antibody molecules directed against neural cell surface antigens, even when the identity of the latter has not yet been established. It should also be possible to exploit in a similar manner antibodies directed against known cell-specific ligands, available in a pure form, and immobilize the antibody rather than the ligand for the selective affinity sequestration of "ligand-bearing" neural cells. For sequestration purposes, the immune reagent used should be purified immunoglobulin, IgG (e.g., Kondorosi *et al.,* 1977; Au and Varon, 1979), or its active fragment, F(ab)′ (e.g., Brockes *et al.,* 1977; Thiery *et al.,* 1977).

   Many immune sera directed against neural surface antigens have been obtained by several research groups, based on the use of neural materials such as immunogens and absorption of the crude antisera with extraneural tissues. These studies were mainly aimed at defining nervous system-specific surface antigens and investigating their regional and developmental distributions. However, for affinity separation of different neural cell classes from one another, the critical concern is with immune reagents restricted to specific cell subsets, rather than with possible cross-reactivities with extraneural cells. To this effect, one must either start with a homogeneous material like immunogen or have available homogeneous neural cell populations as immunoabsorbants of the unwanted antibodies (shared with, or exclusively directed against, the other cell subsets). The following brief survey of different sources of neural immunogens will show that class-specific antibodies against neural surface antigens have already been successfully obtained in a few cases but also that a vast potential for them remains to be tapped.

a. *Neural Tumor Cells.* This source of neural surface immunogens has been attractive because of ready availability in substantial amounts, a reasonable presumption of defined cell types, and the known retention of at least some properties characteristic of their normal counterparts. Cell suspensions from neural tumors grown *in vivo* have been used, for example, to define sets of neural surface antigens predominantly found on glial cells, such as NS-1 (Schachner, 1974; SundarRaj *et al.,* 1975) and NS-2 (Schachner and Carnow, 1975; Schachner *et al.,* 1976; Yuan *et al.,* 1977). Clonal cell lines developed from neuronal and glial tumors have been used even more extensively (e.g., Pfeiffer *et al.,* 1971; Herschman *et al.,* 1972; Martin, 1974; Akeson and Herschman, 1974; Akeson and Seeger, 1977; Akeson and Hsu, 1978). Two such studies deserve special mention here:

1. Fields *et al.* (1975), using several rat clonal lines, have defined a "common" antigen present in nearly all the neural lines tested and in adult rat central and peripheral neural tissues. In later studies by the same group (see Section III,B,2), antiserum specific for the common antigen was shown to label a surface antigen (Ran-1) on Schwann cells but not on other cells in monolayer cultures of rat peripheral nerve or ganglia.
2. Stallcup and Cohn (1976) examined surface antigens in a large number of rat clonal lines (Schubert *et al.,* 1974) previously classified as either neuronal (by electrophysiological and ionic tests) or glial (by several glial markers). They were able to define, and purify antisera against, three neuronal antigens (N1, N2, and N3) present in various combinations only in the six neuronal lines examined, and two glial antigens (G1 and G2) absent from all the neuronal lines and present (either or both) in all the glial ones.

b. *Primary Neural Cells.* The use of heterogeneous primary neural cell populations has yielded promising results for future development, but no defined, class-restricted antibodies as yet. Varon *et al.* (1979a) report that antisera against whole spinal cord suspensions can be diluted to levels where they would still be toxic to spinal neurons but no longer so to spinal glial elements (see Section III,B,2). Seeds (1974, 1975), using reaggregate cultures of cerebellar cells, recognized surface antigens restricted to a small subset of cerebellar cells. Rutishauser and co-workers (1976; Brackenbury *et al.,* 1977; Thiery *et al.,* 1977) obtained antisera against chick neuroretinal cells, which interfere with their aggregation and react with molecules released by them in the culture medium. These cell adhesion molecules (CAMs) appear restricted to neural cell surfaces, but not to individual subsets of neuroretinal cells, and may play a role in neurite-neurite interactions (Rutishauser *et al.,* 1978a,b). Considerable help can be expected in the near future from the use, both as immunogens and as immunoabsorbants, of several purified neuronal and glial cultures that have become available in recent years by manipulation of culture conditions. A list of such preparations is provided in Table I. A potential limitation in the use of purified neural cell cultures may be that the purified cells are available in relatively modest amounts.

TABLE I

PURIFIED NEURAL CULTURES

Species and organ	Cell class	References
Chick cerebrum	Neurons	Varon and Raiborn, 1969
	Glioblasts	Sensenbrenner, 1977
Chick sympathetic ganglia	Neurons	Varon and Raiborn, 1972
	Nonneurons	McCarthy and Partlow, 1976
Chick dorsal root ganglia	Neurons	Varon, 1979
Mouse dorsal root ganglia	Neurons, nonneurons	Varon *et al.,* 1973
Rat dorsal root ganglia	Neurons, Schwann cells	Wood, 1976
Rat superior cervical ganglia	Neurons, nonneurons	Mains and Patterson, 1973
Chick spinal cord	Neurons	Popiela *et al.,* 1978
Chick optic lobe	Neurons, nonneurons	Adler *et al.,* 1979a
Rodent cerebrum	Astroglia	Sensenbrenner, 1977
		Hertz, 1977
		Manthorpe *et al.,* 1979
N1-defined medium		
Chick dorsal root ganglia	Neurons	Bottenstein *et al.,* 1980
Chick spinal cord, chick neural retina, chick optic lobe, chick telencephalon, rat telencephalon, rat mesencephalon, mouse telencephalon	Neuron-like cells	Skaper *et al.,* 1979

No such limitation applies to bulk-purified neural cells. Poduslo *et al.* (1977) injected into rabbits neuronal bulk fractions from rat brain. The antisera were active on the original rat neurons and similarly prepared mouse brain neurons, and the antineuron activity was not absorbed by bulk-isolated astrocytes or oligodendrocytes, rat myelin, or rat nonneural tissues. The same investigators also obtained rabbit antisera against oligodendroglial bulk fractions from lamb white matter. The antisera reacted with the lamb oligodendrocytes and similar bulk preparations from calf or human brain, and the activity was not absorbed by rat neurons, lamb brain gray matter, lamb myelin, or lamb liver or red blood cells. Bulk calf oligodendrocytes were used by another group of investigators (Abramsky *et al.*, 1978). Their antiserum was shown (by immunofluorescence) to react not only with the original oligodendrocyte preparations but also with the surface of perineuronal and interfascicular oligodendroglia in both bovine and human brain sections. The activity was absorbed by oligodendroglial cells and white matter but not by purified myelin, neuroblastoma cells, liver.

c. Subcellular Neural Fractions. Whole homogenates of neural tissue would be expected to contain immunogens as heterogeneous as unfractionated primary cell suspensions (cf. Schachner *et al.*, 1975; Zimmermann and Schachner, 1976). Nevertheless, antisera against corpus callosum homogenates have been shown to react prevalently with glial surface antigens (Schachner *et al.*, 1977; Campbell *et al.*, 1977). More specific results may be obtained by using selected subfractions of neural membrane. Synaptosomes have been employed to elicit antisera (e.g., Herschman *et al.*, 1972; Livett *et al.*, 1974; Matus *et al.*, 1976), and antibodies against two neuronal surface antigens, D1 and D2, have been obtained by Bock and her co-workers (cf. Bock, 1977) by use of purified synaptic plasma membrane. Central white matter, injected into various animals, causes the development of experimental allergic encephalomyelitis (EAE), which can also be elicited by injection of purified myelin basic protein (e.g., Kies, 1965; Eylar *et al.*, 1970). EAE sera contain antibodies that will specifically recognize oligodendrocyte surface antigens (e.g., Bornstein and Raine, 1970; Latovitzky and Silberberg, 1975; Dorfman *et al.*, 1976; Bornstein, 1977). Corresponding sera directed against Schwann cells may be obtained by inducing experimental allergic neuritis (EAN) through injection of peripheral myelin or basic proteins from it (e.g., Waksman and Adams, 1956; Arnason, 1969; Brostoff *et al.*, 1972). EAE and EAN sera have not yet been exploited for the separation of oligodendrocytes or Schwann cells, respectively.

d. Isolated Neural Cell Surface Antigens. A vast reservoir of immunogens, nearly untapped as yet, is provided by free antigens related to specific neural cell surfaces. A successful illustration of this approach is the use of purified cerebrosides to obtain specific antioligodendrocyte antibodies (Fry *et al.*, 1974; Dorfman *et al.*, 1976; Raff *et al.*, 1978; see also Section III,B,2). Other potential immunogens are, for example, gangliosides (Rapport and Karpiack, 1977), myelin glycoproteins (Everly *et al.*, 1977; Quarles and Everly, 1977; Roomi *et al.*, 1977; Gould, 1977), purified surface receptors for αBTx (Lindstrom, 1976), TTX (cf. Narahashi, 1974), or NGF (Banerjee *et al.*, 1976; Andres *et al.*, 1977), ectoenzymes such as the glial 5'-nucleotidase (Bosman and Pike, 1971; Kreutzberg *et al.*, 1978), and the alloantigen Thy-1 (Reif and Allen, 1964, 1966; Schachner and Hammerling, 1974; Zwerner *et al.*, 1977; Williams *et al.*, 1977; Acton and Pfeiffer, 1978; see also Section III,B,2). Finally, it appears worthwhile to reiterate the opportunity to impose exogenous antigens on neural cell surfaces, such as subset-specific ligands, antibody to which could be readily raised and used for affinity cell separation (see Section II,A,6).

B. Applicability of a Given Ligand to the Cells under Study

Affinity approaches always address a cell suspension. Therefore, the question of appropriate dissociation techniques is always critical in affinity cell separation studies. It is by no means to be assumed that any dissociation procedure will yield viable cells of all the classes represented in the source tissue. Dissociation procedures greatly influence composition, yield and viability, and also tendency to reaggregate, amount of debris, and other interfering features (cf. Varon and Saier, 1975; Varon, 1977a). In turn, the effectiveness of a given procedure is considerably dependent on the region and age of the source tissue, particularly in the CNS where different treatments may be demanded by increasing neuropil complexity and synapse numbers. A first task, therefore, is to verify that the suspension obtained contains the desired cell subsets. The general question of identification of neural cells outside their tissue context has often been a hazardous one. Past studies on neural cell separation were frequently limited to an operational definition of cell subpopulations based on their very susceptibility to the separation procedure. More recently, batteries of neural cell markers have become increasingly available (cf. Varon, 1978; Bock, 1978; Varon and Somjen, 1979), even though controversies persist as to whether some markers are truly restricted to neurons, glia, or subsets thereof.

Dissociation procedures, as well as species, region, and age of the source tissue, also affect the occurrence, abundance, distribution and/or reactivity of neural cell surface acceptors (receptors, antigens). A second task, therefore, is to verify whether surface acceptors for the chosen ligand (1) are present in the initial cell suspension, and (2) are restricted to the cell subset to be sequestered. The two questions are addressed with different effectiveness in cell suspensions and in cell cultures.

1. CELL SUSPENSIONS

To ascertain the presence of reactive acceptors, ligand and cells (or membrane derived from them) are brought together under conditions conducive to the binding of one to the other. Free ligand and cell materials are then separated, and binding is verified by demonstrating either a depletion of free ligand in the medium or an acquisition of bound ligand by the cell membrane. Depletion measurements may encounter problems, such as low proportions of removed versus residual ligand, nonspecific loss of free ligand on noncellular and cellular components of the binding system (cf. Almon and Varon, 1978), or interference by cell- or medium-derived soluble materials with the assay procedure. In addition, they obviously offer no opportunity to recognize which cells within the suspension may have been selectively responsible for the depletion. Binding measurements also may encounter difficulties due to low signal-to-noise ratios (i.e., low specific binding versus high nonspecific background), particularly when the cells used are in small amounts or contain a relative paucity of acceptors. Moreover, since most cells tend to round up in suspension, different subsets may be distinguished at best only by features like differences in size.

Both depletion and binding evaluations rely on detection of the ligand in its free or, respectively, bound form. Even though some ligands may be detected and measured by their intrinsic properties, detection is most often effected via radioactive, fluorescent, or histochemical tags with which the ligand has been previously labeled. With cell suspensions, two approaches that allow visualization of the positive cells are complement-dependent immunocytolysis (detected by changes in the number of dye-excluding cells) and the already mentioned rosette formation. Where possible, specificity of the binding is tested by use of appropriate competitive inhibitors, including excess unlabeled ligand itself.

2. CELL CULTURES

The ideal approach both to recognize the presence and to verify the restriction of a given acceptor to a subset of the cells to be fractionated is to visualize the entire cell population under conditions where (1) each subset is distinguishable from the others, and (2) the bound ligand is manifest. Explant cultures may be inspected directly, but only in the cellular and neuritic outgrowth zone. Monolayer cell cultures provide the most effective inspection opportunities, particularly as more refined culture procedures become available for neural cells (see Table I). However, one may have to distinguish between ligand bound to the cell surface and ligand internalized by endocytotic activities (e.g., Gonatas *et al.*, 1977). Despite their advantages for visual inspection, monolayer cultures of normal neural cells have only recently begun to be used for recognition of ligand-specific cells.

a. Immunocytolysis. Varon *et al.* (1979a) have described the use of monolayer cultures of chick embryo spinal cord cells for both recognition and quantitative titration of anti-spinal cord sera and IgG. Cells are seeded into collagen- or polyornithine-coated culture wells, incubated for 24 hr, washed, and exposed for 45 min at 37°C to medium containing a nontoxic complement and serially diluted antibody. Controls receive only antibody, only complement, or neither one. After another wash, the wells are filled with 2% glutaraldehyde and inspected under phase microscopy for residual healthy cells. Figure 3 illustrates morphological differences between control and treated cells. Quantitative analysis of older cultures, where non-neuronal flat cells (presumptive glial elements) have become more numerous, have also shown that nonneurons are not affected by these antisera at concentrations that are still fully damaging to the neurons, demonstrating the possibility of distinguishing cell subpopulations with

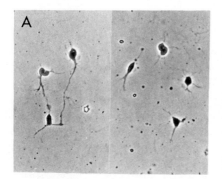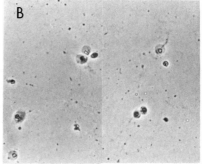

FIG. 3. Complement-dependent immunocytolysis of chick spinal cord neurons treated with anti-spinal cord cell antisera. Eleven-day chick embryo spinal cord was dissociated and grown in culture for 24 hr at 37°C, washed, and exposed for 45 min (37°C) to medium containing diluted anti-chick spinal cord antisera with or without complement. After another wash the cells were fixed with glutaraldehyde. (A) Antibody alone. (B) Antibody plus complement.

qualitatively or quantitatively different surface antigens. Last, titration of immune activity by use of these monolayers was 10–20 times more sensitive than more traditional procedures applied to spinal cord cells in suspension.

b. Indirect Immunofluorescence. Mirsky and Thompson (1975) applied fluorescent antibody against the Thy-1 alloantigen to visualize positive cells in mouse brain monolayer cultures. Brockes *et al.* (1977) used dissociated rat sciatic nerve to demonstrate the presence of Thy-1 antigen on the surface of fibroblasts but not of Schwann cells, and the converse presence of another surface antigen Ran-1 on Schwann cells but not on fibroblasts. The procedure consists in culturing the cells on collagen-coated glass coverslips and presenting them with mouse anti-Thy-1 and rabbit anti-Ran-1 antisera (in the absence of complement). Washed coverslips are then incubated with rhodamine-conjugated goat anti-mouse IgG antibody, fluorescein-conjugated goat anti-rabbit IgG antibody, or both. The cultures are washed, fixed, mounted on a slide, and examined with a fluorescence microscope. In this way, rhodamine labeled the Thy-1-bearing fibroblasts and fluorescein the Ran-1-bearing Schwann cells.

These elegant studies have been extended by Fields *et al.* (1978) to monolayer cultures of rat dorsal root ganglia which comprise three main cell classes: neurons, Schwann cells, and fibroblasts. Figure 4A shows that Ran-1 was displayed only on Schwann cells. Thy-1 was expressed on fibroblasts and also on neurons, but not on Schwann cells. In addition, neurons but not the other two cell classes bound TTX which could be visualized by indirect immunofluorescence using rabbit antitoxin serum and rhodamine-tagged goat anti-rabbit IgG antibody. The technique could also be combined with autoradiography to demonstrate [^{3}H]thymidine incorporation in Schwann and fibroblast elements.

Raff *et al.* (1978) have cultured dissociated cells from several rat neural tissues and examined the cultures by use of antiserum against purified galactocerebrosides from bovine or human brain. Some process-bearing cells were stained with the anticerebroside serum (Figure 4B) but were negative with anti-glial fibrillary acidic (GFA) protein serum, indicating their oligodendroglial rather than astroglial identity (GFA protein is an intracellular antigen believed to be specific for astroglial elements). No positive cells were seen in sciatic nerve cell cultures (although they occur in sections of this tissue), suggesting that Schwann cells also can be identified with anticerebroside serum but only under conditions favoring cerebroside synthesis.

c. Indirect Immunoperoxidase. Schachner *et al.* (1976) have described a method in which the anti-IgG antibody to be used is conjugated with

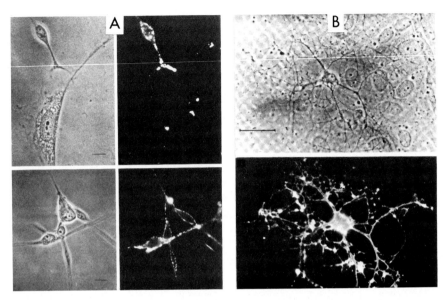

FIG. 4. Immunofluorescence demonstration of cell-specific antisera. (A) Mouse dorsal root ganglion cells in monolayer culture stained with rabbit anti-RAN-1 antisera and fluorescein-conjugated goat anti-rabbit IgG antibody. Top left: Phase-contrast micrograph of bipolar Schwann cell and flat fibroblast. Top right: Same field showing specific staining of Schwann cell. Bottom left: Phase-contrast micrograph of large neuron surrounded by several bipolar Schwann cells. Bottom right: Immunofluorescent staining of only the Schwann cells. (From Fields *et al.*, 1978). (B) Rat optic nerve monolayer cultures stained with rabbit antigalactocerebroside antisera and rhodamine-conjugated goat anti-rabbit IgG antibody. Top: Phase-contrast micrograph of a monolayer of flat cells overlaid by one process-bearing putative oligodendrocyte. Bottom: Immunofluorescent staining of oligodendrocyte. (From Raff *et al.*, 1978).

horseradish peroxidase rather than a fluorescent compound and visualization is effected via the product of the enzymatic tag. This procedure has been recently used (Schachner *et al.*, 1978) on histological sections of various mouse and chick neural tissues in conjunction with antiserum against the LETS protein (see Section III,A,2). The method has not yet been applied to monolayer cell cultures but promises to be an even more powerful tool than indirect immunofluorescence. Peroxidase-conjugated lectins have already been used to follow binding and internalization by mouse dorsal root ganglionic neurons in explant cultures (e.g., Gonatas *et al.*, 1977).

IV. Concluding Remarks

From our own direct experience and from the survey and analysis presented here, we find it difficult to escape the conclusion that we are entering a time when neural cell separation by affinity methods will become successfully and widely available to the neuroscientist. The technical questions have been defined and are well on the way to being resolved. Visualization in neural cell cultures of cells specifically addressed by different ligands is being reported at a greatly accelerated pace. A vast number of ligands directed to neural cells and specific subsets of them are waiting to be exploited, as well as to be further refined. The immune approaches provide ways to bring together separation techniques, visualization procedures, and ligands of various natures. It is hoped that this perception will be shared by the readers of this review and that the review may be at least in a small part instrumental in stimulating their interest and experimental efforts to make neural cell separation an accomplished reality in the near future.

Acknowledgments

This work was supported by USPHS grant NS-12893 from the National Institute of Neurological and Communicative Disorders and Stroke. The authors wish to thank Dr. Ruben Adler for helpful editorial comments.

References

Abramsky, O., Lisak, R. P., Pleasure, D., Gilden, D. H., and Silberberg, D. H. (1978). Immunologic characterization of oligodendroglia. *Neurosci. Lett.* **8,** 311–316.

Acton, R. T., and Pfeiffer, S. E. (1978). Distribution of Thy-1 differentiation alloantigen in the rat nervous system and in a cell line derived from a rat peripheral neurinoma. *Dev. Neurosci.* **1,** 110–117.

Adler, R., Manthorpe, M., and Varon, S. (1979a). Separation of neuronal and nonneuronal cells in monolayer cultures from chick optic lobe. *Dev. Biol.* **69,** 424–435.

Adler, R., Landa, K. B., Manthorpe, M., and Varon, S. (1979b). Cholinergic neuronotrophic factors. II. Selective intraocular distribution of soluble trophic activity for ciliary ganglionic neurons. *Science* **204,** 1434–1436.

Akeson, R., and Herschman, H. R. (1974). Neural antigens of morphologically differentiated neuroblastoma cells. *Nature (London)* **243,** 620–623.

Akeson, R., and Hsu, W.-C. (1978). Identification of a high molecular weight nervous system specific cell surface glycoprotein on murine neuroblastoma cells. *Exp. Cell Res.* **115,** 367–377.

Akeson, R., and Seeger, R. C. (1977). Interspecies neural membrane antigens on cultured human and murine neuroblastoma cells. *J. Immunol.* **118,** 1995–2003.

Almon, R. R., and Varon, S. (1978). Associations of beta nerve growth factor with bovine serum albumin as well as with the alpha and gamma subunits of the 7S macromolecule. *J. Neurochem.* **30,** 1559–1567.

Andres, R. Y., Jeng, I., and Bradshaw, R. A. (1977). Nerve growth factor receptors: Identification of distinct classes in plasma membrane and nuclei of embryonic dorsal root neurons. *Proc. Natl. Acad. Sci. U.S.A.* **74**, 2785-2789.

Arnason, B. G. W. (1969). Idiopathic polyneuritis and experimental allergic encephalomyelitis. A comparison. *In* "Immunological Disorders of the Nervous System" (L. P. Rowland, ed.), pp. 156-177. Williams & Wilkins, Baltimore, Maryland.

Au, A. M.-J., and Varon, S. (1979). Neural cell sequestration on immunoaffinity columns. *Exp. Cell Res.* **120**, 269-276.

Balazs, R. (1976). Hormones and brain development. *Prog. Brain Res.* **45**, 139-159.

Banerjee, S. P., Cuatrecasas, P., and Snyder, S. H. (1976). Solubilization of NGF receptors of rabbit SCG. *J. Biol. Chem.* **251**, 5680-5685.

Barker, C. R., Worman, C. P., and Smith, J. L. (1975). Purification and quantification of T and B lymphocytes by an affinity method. *Immunology* **29**, 765-777.

Barondes, S. H., ed. (1976). "Neuronal Recognition." Plenum, New York.

Barondes, S. H., Rosen, S. D., Simpson, D. L., and Kafka, J. A. (1974). Agglutinins of formalinized erythrocytes: Changes in activity with development of *Dictyostelium discoideum* and embryonic chick brain. *In* "Dynamics of Degeneration and Growth in Neurons" (K. Fluxe, L. Olson, and Y. Zotterman, eds.), pp. 449-454. Pergamon, Oxford.

Binz, H., Lindenmann, J., and Wigzell, H. (1974). Cell-bound receptors for alloantigens on normal lymphocytes. *J. Exp. Med.* **139**, 877-887.

Bock, E. (1977). Immunochemical markers in primary cultures and in cell lines. *In* "Cell, Tissue and Organ Cultures in Neurobiology" (S. Fedoroff and L. Hertz, eds.), pp. 407-422. Academic Press, New York.

Bock, E. (1978). Nervous system specific proteins. *J. Neurochem.* **30**, 7-14.

Bornstein, M. B. (1977). Differentiation of cells in primary cultures: Myelination. *In* "Cell, Tissue and Organ Cultures in Neurobiology" (S. Fedoroff and L. Hertz, eds.), pp. 141-146. Academic Press, New York.

Bornstein, M. B., and Raine, C. A. (1970). Experimental allergic encephalomyelitis antiserum. Inhibition of myelination *in vitro*. *Lab. Invest.* **23**, 536-542.

Bosman, H. B., and Pike, G. L. (1971). Membrane marker enzymes: Isolation, purification, and properties of 5'-nucleotidase from rat cerebellum. *Biochim. Biophys. Acta* **227**, 402-412.

Bottenstein, J. E., Skaper, S. D., Varon, S., and Sato, G. (1980). Selective survival of neurons in chick embryo sensory ganglionic cultures by use of a defined, serum-free medium. *Exp. Cell Res.* (in press).

Brackenbury, R., Thiery, J.-P., Rutishauser, U., and Edelman, G. M. (1977). Adhesion among neural cells of the chick embryo. I. An immunological assay for molecules involved in cell-cell binding. *J. Biol. Chem.* **252**, 6835-6840.

Brockes, J., Fields, K. L., and Raff, M. C. (1977). A surface antigenic marker for rat Schwann cells. *Nature (London)* **266**, 364-366.

Brostoff, S., Burnett, P., Lampert, P., and Eylar, E. H. (1972). Isolation and characterization of a protein from sciatic nerve myelin responsible for experimental allergic neuritis. *Nature (London), New Biol.* **235**, 210-212.

Brown, J. C., and Hunt, R. C. (1978). Lectins. *Int. Rev. Cytol.* **52**, 277-349.

Burnham, P. A., Silva, J., and Varon, S. (1974). Anabolic responses of embryonic dorsal root ganglia to nerve growth factor, insulin, concanavalin A or serum *in vitro*. *J. Neurochem.* **23**, 689-697.

Campbell, G. L., Schachner, M., and Sharrow, S. (1977). Isolation of glial cell-enriched and -depleted populations from mouse cerebellum by density gradient centrifugation and electronic cell sorting. *Brain Res.* **127**, 69-86.

Carbonetto, S. T., Fambrough, D. M., and Muller, K. J. (1978). Nonequivalence of α-bungarotoxin receptors and acetylcholine receptors in chick sympathetic neurons. *Proc. Natl. Acad. Sci. U.S.A.* **75**, 1016–1020.

Chiappinelli, V. A., and Giacobini, E. (1978). Time course of appearance of α-bungarotoxin binding sites during development of chick ciliary ganglion and iris. *Neurochem. Res.* **3**, 465–478.

Crum, E. D., and McGregor, D. D. (1976). Functional properties of T and B cells isolated by affinity chromatography from rat thoracic duct lymph. *Cell. Immunol.* **23**, 211–222.

Cuatrecasas, P. (1972). Affinity chromatography and purification of the insulin receptor of liver cell membranes. *Proc. Natl. Acad. Sci. U.S.A.* **69**, 1277–1281.

Cuatrecasas, P., and Anfinsen, C. B. (1971). Affinity chromatography. *Annu. Rev. Biochem.* **40**, 259–278.

Cuatrecasas, P., and Parikh, I. (1972). Adsorbents for affinity chromatography. Use of N-hydroxysuccinimide esters for agarose. *Biochemistry* **11**, 2291–2299.

Denis-Donini, S., Estenoz, M., and Augusti-Tocco, G. (1978). Cell surface modification in neuronal maturation. *Cell Differ.* **7**, 193–202.

Dimpfel, W., and Habermann, E. (1977). Binding characteristics of ¹²⁵I-labelled tetanus toxin to primary tissue cultures from mouse embryonic CNS. *J. Neurochem.* **29**, 1111–1120.

Dorfman, S., Holtzer, H., and Silberberg, D. (1976). Effect of 5-bromo-2-deoxyuridine or cytosine-D-arabinofuranoside hydrochloride on myelination in newborn rat cerebellum cultures following removal of myelination-inhibiting antiserum to whole cord or cerebroside. *Brain Res.* **104**, 283–294.

Dvorak, D. J., Gipps, E., and Kidson, C. (1978). Isolation of specific neurons by affinity methods. *Nature (London)* **271**, 564–566.

Edelman, G. M., and Rutishauser, U. (1974). Specific fractionation and manipulation of cells with chemically derivatized fibers and surfaces. *In* "Methods in Enzymology" (W. B. Jakoby and M. Wilchek, eds.), Vol. 34, pp. 195–225. Academic Press, New York.

Edelman, G. M., Rutishauser, U., and Millette, C. F. (1971). Cell fractionation and arrangements on fibers, beads, and surfaces. *Proc. Natl. Acad. Sci. U.S.A.* **68**, 2153–2157.

Eisenbarth, G. S., Ruffolo, R. R., Walsh, F. S., and Nirenberg, M. (1978). Lactose sensitive lectin of chick retina and spinal cord. *Biochem. Biophys. Res. Commun.* **83**, 1246–1252.

Everly, J., Quarles, R., and Brady, R. (1977). Proteins and glycoproteins in myelin purified from the developing bovine and human CNS's. *J. Neurochem.* **28**, 95–101.

Eylar, E. H., Caccam, J., Jackson, J. J., Westall, F., and Robinson, A. B. (1970). Experimental allergic encephalomyelitis: Synthesis of disease-inducing site of basic protein. *Science* **168**, 1220–1223.

Fields, K. L., Gosling, C., Megson, M., and Stern, P. L. (1975). New cell surface antigens in rat defined by tumors of the nervous system. *Proc. Natl. Acad. Sci. U.S.A.* **72**, 1296–1300.

Fields, K. L., Brockes, J. P., Mirsky, R., and Wendon, L. M. B. (1978). Cell surface markers for distinguishing different types of rat dorsal root ganglion cells in culture. *Cell* **14**, 43–52.

Forsgren, A., and Sjoquist, J. (1966). "Protein A" from *S. aurens*. I. Pseudoimmune reaction with human γ-globulin. *J. Immunol.* **97**, 822–827.

Fry, J., Weissbarth, S., Lehrer, G. M., and Bornstein, M. B. (1974). Cerebroside antibody inhibits sulfatide synthesis and myelination and demyelinates in cord tissue culture. *Science* **183**, 540–542.

Fumagalli, L., DeRenzis, G., and Miani, N. (1976). Acetylcholine receptors: Number and distribution in intact and deafferented superior cervical ganglion of the rat. *J. Neurochem.* **27**, 47–52.

Fumagalli, L., DeRenzis, G., and Miani, N. (1978). α-Bungarotoxin-acetylcholine receptors in the chick ciliary ganglion: Effects of deafferentation and axotomy. *Brain Res.* **153**, 87–98.

Godfrey, H. P., and Gell, P. G. H. (1976). Separation by column chromatography of cells active in delayed-onset hypersensitivities. *Immunology* **30**, 695–703.

Gonatas, N., Kim, S., Stieber, A., and Avrameas, S. (1977). Internalization of lectins in neuronal GERL. *J. Cell Biol.* **73**, 1–13.

Gould, R. M. (1977). Incorporation of glycoproteins into peripheral nerve myelin. *Trans. Am. Soc. Neurochem.* **8**, 156.

Greene, L. A., (1976). Binding of α-bungarotoxin to chick sympathetic ganglia: Properties of the receptor and its rate of appearance during development. *Brain Res.* **111**, 135–145.

Gremo, F., Kobiler, D., and Barondes, S. H. (1978). Distribution of an endogenous lectin in the developing chick optic tectum. *J. Cell Biol.* **79**, 491–499.

Guillary, R. J., Rayner, M. D., and D'Arrigo, J. S. (1977). Covalent labeling of the tetrodotoxin receptor in excitable membranes. *Science* **196**, 883–885.

Halgren, E., and Varon, S. (1972). Serotonin turnover in cultured newborn rat raphe nuclei: *In vitro* development and drug effects. *Brain Res.* **48**, 438–442.

Hamburgh, M. (1966). Evidence for a direct effect of temperature and thyroid hormone on myelinogenesis *in vitro*. *Dev. Biol.* **13**, 15–30.

Hatten, M. E., and Sidman, R. L. (1977). Plant lectins detect age and region specific differences in cell surface carbohydrates and cell reassociation behavior of embryonic mouse cerebellar cells. *J. Supramol. Struct.* **7**, 267–275.

Hausman, R., and Moscona, A. (1975). Purification and characterization of the retina-specific cell-aggregating factor. *Proc. Natl. Acad. Sci. U.S.A.* **72**, 916–920.

Herschman, H., Breeding, J., and Nedrud, J. (1972). Sialic acid-masked membrane antigens of clonal functional glial cells. *J. Cell. Physiol.* **79**, 249–258.

Hertz, L. (1977). Biochemistry of glial cells. *In* "Cell, Tissue and Organ Cultures in Neurobiology" (S. Fedoroff and L. Hertz, eds.), pp. 39–72. Academic Press, New York.

Hiller, J. M., Simon, E.J., Crain, S. M., and Peterson, E. R. (1978). Opiate receptors in cultures of fetal mouse dorsal root ganglia (DRG) and spinal cord: Predominance in DRG neurites. *Brain Res.* **145**, 396–400.

Hynes, R. O. (1976). Cell surface proteins and malignant transformation. *Biochim. Biophys. Acta* **458**, 73–107.

Inman, J. K., and Dintzis, H. M. (1969). The derivatization of cross-linked polyacrylamide beads. Controlled introduction of functional groups for the preparation of special-purpose, biochemical adsorbents. *Biochemistry* **8**, 4074–4082.

Isenberg, G., Czolonkowska, A., and Rieske, E. (1978). Spontaneous and concanavalin A-induced agglutination of chick-embryo neuronal cells. *Neurosci. Symp.* **3**, 633–640.

Isturiz, M. A. de E., de Bracco, M. M., and Manni, J. A. (1975). Reaction of human lymphocytes with aggregated IgG columns. Effect on antibody-dependent cytotoxicity and EAC rosette formation. *Cell. Immunol.* **16**, 82–91.

Kelly, P., Cotman, C. W., Gentry, C., and Nicholson, G. L. (1976). Distribution and mobility of lectin receptors on synaptic membranes of identified neurons in the central nervous system. *J. Cell Biol.* **71**, 487–496.

Kies, M. W. (1965). Chemical studies on an encephalitogenic protein from guinea pig brain. *Ann. N.Y. Acad. Sci.* **122**, 161–170.

Killion, J. J., and Kollmorgen, G. M. (1976). Isolation of immunogenic tumor cells by cell-affinity chromatography. *Nature (London)* **259**, 674–676.

Kinzel, V., Richards, J., and Kubler, D. (1977). Lectin receptor sites at the cell surface employed for affinity separation of tissue culture cells. Basic requirements as realized by lens culinarislectin (LCL) immobilized on 2tb-Sepharose. *Exp. Cell Res.* **105**, 389–400.

Kleinschuster, S. J., and Moscona, A. A. (1972). Interactions of embryonic and fetal neural retina cells with carbohydrate-binding phytoagglutinins: Cell surface changes with differentiation. *Exp. Cell Res.* **70,** 397–410.

Kobiler, D., and Barondes, S. H. (1977). Lectin activity from embryonic chick brain, heart and liver: Changes with development. *Dev. Biol.* **60,** 326–330.

Kobiler, D., Beyer, E. C., and Barondes, S. H. (1978). Developmentally regulated lectins from chick muscle, brain and liver have similar chemical and immunological properties. *Dev. Biol.* **64,** 265–272.

Koda, L. Y., and Partlow, L. M. (1976). Membrane marker movement on sympathetic axons in tissue culture. *J. Neurobiol.* **7,** 157–172.

Kondorosi, E., Nagy, J., and D'enes, G. (1977). Optimal conditions for the separation of rat T lymphocytes on anti-immunoglobulin-immunoglobulin affinity columns. *J. Immunol. Methods* **16,** 1–13.

Kouvelas, E. D., and Greene, L. A. (1976). The binding and regional ontogeny of receptors for α-bungarotoxin in chick brain. *Brain Res.* **113,** 111–126.

Kouvelas, E. D., Dichter, M. A., and Greene, L. A. (1978). Chick sympathetic neurons develop receptors for α-bungarotoxin *in vitro,* but the toxin does not block nicotinic receptors. *Brain Res.* **154,** 83–93.

Kreutzberg, W., Barron, K. D., and Schubert, P. (1978). Cytochemical localization of 5'-nucleotidase in glial plasma membranes. *Brain Res.* **158,** 247–257.

Landa, K. B., Adler, R., Manthorpe, M., and Varon, S. (1980). Cholinergic neuronotrophic factors. III. Developmental increase of trophic activity for chick embryo ciliary ganglionic neurons in their intraocular target tissues *Dev. Biol.* **74,** 401–408.

Latovitzki, N., and Silberberg, D. H. (1975). Ceramide glycosyltransferases in cultured rat cerebellum: Change with age, with demyelination, and with inhibition of myelination by 5-bromo-2'-deoxyuridine or experimental allergic encephalomyelitis serum. *J. Neurochem.* **24,** 1017–1022.

Levi-Montalcini, R., and Angeletti, P. U. (1968). Nerve growth factor. *Physiol. Rev.* **48,** 534–569.

Light, N. D., and Tanner, M. J. (1978). An affinity chromatography method for the preparation of human reticulocytes. *Anal. Biochem.* **87,** 263–266.

Lim, R., Troy, S. S., and Turrif, D. E. (1977). Fine structure of cultured glioblasts before and after stimulation by a glia maturation factor. *Exp. Cell Res.* **106,** 357–372.

Lindstrom, J. (1976). Immunological studies of acetylcholine receptors. *J. Supramol. Struct.* **4,** 389–404.

Lis, H., and Sharon, N. (1973). The biochemistry of plant lectins (phytohemagglutinins). *Annu. Rev. Biochem.* **42,** 541–574.

Livett, B. G., Rostas, J. A., Jeffrey, P. L., and Austin, L. (1974). Antigenicity of isolated synaptosomal membranes. *Exp. Neurol.* **43,** 330–338.

Ljungstedt-Pohlman, I., Seiving, B., and Oholm, I. (1977). Heterogeneous insulin- and concanavalin A-binding among spleen lymphocytes established by affinity chromatography. *Exp. Cell Res.* **110,** 191–200.

McCarthy, K., and Partlow, L. (1976). Preparation of pure neuronal and non-neuronal cultures from embryonic chick sympathetic ganglia: A new method based on both differential cell adhesiveness and the formation of homotypic neuronal aggregates. *Brain Res.* **114,** 391–414.

McDonough, J., and Lilien, J. (1975). Spontaneous and lectin-induced redistribution of cell surface receptors on embryonic chick neural retina cells. *J. Cell Sci.* **19,** 357–368.

McQuarrie, C., Salvaterra, P. M., DeBlas, A., Routes, J., and Mahler, H. R. (1976). Studies on nicotinic acetylcholine receptors in mammalian brain. *J. Biol. Chem.* **251,** 6335–6339.

Mains, R. E., and Patterson, P. H. (1973). Primary cultures of dissociated sympathetic neurons. I. Establishment of long term growth in culture and studies of differential properties. II. Initial studies on catecholamine metabolism. III. Changes in metabolism with age in culture. *J. Cell Biol.* **59,** 329–366.

Manthorpe, M., Adler, R., and Varon, S. (1979). Development, reactivity and GFA immunofluorescence of astroglia-containing monolayer cultures from rat cerebrum. *J. Neurocytol.* **8,** 605–621.

Manthorpe, M., Skaper, S., Adler, R., Landa, K., and Varon, S. (1980). Cholinergic neuronotrophic factors. IV. Fractionation properties of an extract from selected chick embryonic eye tissues. *J. Neurochem.* **34,** 69–75.

Maoz, A., and Shellam, G. R. (1976). Fractionation of cytotoxic cells from tumor-immune rats on derivatized collagen gels. *J. Immunol. Methods* **12,** 125–130.

March, S. C., Parikh, I., and Cuatrecasas, P. (1974). A simplified method for cyanogen bromide activation of agarose for affinity chromatography. *Anal. Biochem.* **60,** 149–152.

Margolis, L. B., Dyathlovitskaya, E. V., and Bergelson, L. D. (1978). Cell-lipid interactions. Cell attachment to lipid substrates. *Exp. Cell Res.* **111,** 454.

Margolis, R. J., Lalley, K., Kiang, W.-L., Crockett, C., and Margolis, R. K. (1976). Isolation and properties of a soluble chondroitin sulfate proteoglycan from brain. *Biochem. Biophys. Res. Commun.* **73,** 1018–1024.

Martin, E. E. (1974). Mouse brain antigen detected by rat anti-C1300 antiserum. *Nature (London)* **249,** 71–73.

Matus, A. I., Jones, D. H., and Mughal, S. (1976). Restricted distribution of synaptic antigens in the neuronal membrane. *Brain Res.* **103,** 171–175.

Mehl, E., and Weber, L. (1974). Affinity chromatography for subfractionation of 5-hydroxytryptamine-, LSD-binding proteins from cerebral and nerve-ending membranes. *Adv. Biochem. Psychopharmacol.* **11,** 105–108.

Mirsky, R., and Thompson, E. J. (1975). Thy-1 (theta) antigen on the surface of morphologically distinct brain cell types. *Cell* **4,** 95–101.

Mirsky, R., Wendon, L. M. B., Black, P., Stolkin, C., and Bray, D. (1978). Tetanus toxin: Cell surface marker for neurons in culture. *Brain Res.* **148,** 251–259.

Molday, R. S., Yen, S. P. S., and Rembaum, A. (1977). Application of magnetic microspheres in labelling and separation of cells. *Nature (London)* **268,** 437–438.

Moore, W. M., and Brady, R. N. (1976). Studies of nicotinic acetylcholine receptor protein from rat brain. *Biochim. Biophys. Acta* **444,** 252–260.

Narahashi, T. (1974). Chemicals as tools in the study of excitable membranes. *Physiol. Rev.* **54,** 813–889.

Nash, A. A. (1976). Separation of lymphocyte sub-populations using antibodies attached to staphylococcal protein A-coated surfaces. *J. Immunol. Methods* **12,** 149–161.

Partlow, L., and Larrabee, M. (1971). Effects of a NGF, embryo age and metabolic inhibitors on growth of fibers and on synthesis of RNA and protein in embryonic sympathetic ganglia. *J. Neurochem.* **18,** 2101–2118.

Patterson, P. (1978). Environmental determination of autonomic neurotransmitter functions. *Annu. Rev. Neurosci.* **1,** 1–19.

Pfeiffer, S. E., Herschman, H. R., Lightbody, J. E., Sato, G., and Levine, L. (1971). Modification of cell surface antigenicity as a function of culture conditions. *J. Cell. Physiol.* **78,** 145–152.

Podolsky, D. K., and Weiser, M. M. (1973). Specific selection of mitotically active intestinal cells by concanavalin-A-derivatized fibers. *J. Cell Biol.* **58,** 497–500.

Poduslo, S., McFarland, H. F., and McKhann, G. M. (1977). Antiserum to neurons and to oligodendroglia from mammalian brain. *Science* **197,** 270–272.

Popiela, H., Manthorpe, M., Adler, R., and Varon, S. (1978). Choline acetyl transferase-promoting activity of medium exposed to skeletal muscle cell cultures from chick embryo. *Trans. Am. Soc. Neurochem.* **9**, 49.

Poste, G., and Nicholson, G. L., eds. (1977). "Dynamic Aspects of Cell Surface Organization," Vol. 3. Elsevier, Amsterdam.

Quarles, R. H., and Everly, J. L. (1977). Glycopeptide fractions prepared from purified central and peripheral rat myelin. *Biochim. Biophys. Acta* **466**, 176–186.

Raff, M. C., Mirsky, R., Fields, K. L., Lisak, R. P., Dorfman, S. H., Silberberg, D. H., Gregson, N. A., Leibowitz, S., and Kennedy, M. C. (1978). Galactocerebroside is a specific cell-surface antigenic marker for oligodendrocytes in culture. *Nature (London)* **274**, 813–816.

Rapport, M. M., and Karpiak, S. E. (1977). *In vivo* activities of anti-ganglioside antibodies. *Trans. Am. Soc. Neurochem.* **8**, 200.

Reeber, A., Gandhour, M. S., Roussel, G., Vincendon, G., Gombos, G., and Zanetta, J. P. (1977). Postnatal development of rat cerebellum: Massive and transient accumulation of concanavalin-A-binding glycoproteins in parallel fiber axolemma. *Proc. Int. Soc. Neurochem.* **6**, 224.

Reif, A. E., and Allen, J. M. V. (1964). The AKR thymic antigen and its distribution in leukemias and nervous tissue. *J. Exp. Med.* **120**, 413–433.

Reif, A. E., and Allen, J. M. V. (1966). Mouse nervous tissue iso-antigens. *Nature (London)* **209**, 523–524.

Roomi, M., Ishaque, A., Breckenridge, W., Khan, N., and Eylar, E. H. (1977). The PO protein: A glycoprotein of PNS myelin. *Trans. Am. Soc. Neurochem.* **8**, 158.

Rutishauser, U., Thiery, J.-P., Brackenbury, R., Sela, B. A., and Edelman, G. M. (1976). Mechanisms of adhesion among cells from neural tissues of the chick embryo. *Proc. Natl. Acad. Sci. U.S.A.* **73**, 577–581.

Rutishauser, U., Thiery, J.-P., Brackenbury, R., and Edelman, G. M. (1978a). Adhesion among neural cells of the chick embryo. III. Relationship of the surface molecule CAM to cell adhesion and the development of histotypic patterns. *J. Cell Biol.* **79**, 371–381.

Rutishauser, U., Gall, W. E., and Edelman, G. M. (1978b). Adhesion among neural cells of the chick embryo. IV. Role of the cell surface molecule CAM in the formation of neurite bundles in cultures of spinal ganglia. *J. Cell Biol.* **79**, 382–393.

Salzer, J. L., Glaser, L., and Bunge, R. P. (1977). Stimulation of Schwann cell proliferation by a neurite membrane fraction. *J. Cell Biol.* **75**, 118a (abstr.).

Schachner, M. (1974). NS-1 (nervous system antigen-1), a glial-cell-specific antigenic component of the surface membrane. *Proc. Natl. Acad. Sci. U.S.A.* **71**, 1795–1799.

Schachner, M., and Carnow, T. B. (1975). Nervous system antigen-2 (NS-2), an antigenic cell surface component expressed on a murine glioblastoma. *Brain Res.* **88**, 394–402.

Schachner, M., and Hammerling, U. (1974). The postnatal development of antigens on mouse brain cell surfaces. *Brain Res.* **73**, 362–371.

Schachner, M., Wortham, K., Carter, L., and Chaffee, J. (1975). NS-4 (nervous system antigen-4) a cell surface antigen of developing and adult mouse brain and mature sperm. *Dev. Biol.* **44**, 313–325.

Schachner, M., Ruberg, M., and Carnow, T. (1976). Histological localization of nervous system antigens in the cerebellum by immunoperoxidase labelling. *Brain Res. Bull.* **1**, 367–377.

Schachner, M., Wortham, K. A., Ruberg, M. Z., Dorfman, S., and Campbell, G. LeM. (1977). Brain cell surface antigens detected by anti-corpus callosum antiserum. *Brain Res.* **127**, 87–97.

Schachner, M., Schoonmaker, G., and Hynes, R. O. (1978). Cellular and subcellular localization of LETS protein in the nervous system. *Brain Res.* **158**, 149–158.

Schlossman, S. F., and Hudson, L. (1973). Specific purification of lymphocyte populations on a digestible immunoabsorbant. *J. Immunol.* **110**, 313–315.

Schubert, D., Heinemann, S., Carlisle, W., Tarikas, H., Kimes, B., Patrick, J., Steinbach, J. H., Culp, W., and Brandt, B. L. (1974). Clonal cell lines from the rat central nervous system. *Nature (London)* **249**, 224–227.

Scott, D. W. (1976). Antifluorescein affinity columns. Isolation and immunocompetence of lymphocytes that bind fluoresceinated antigens *in vivo* or *in vitro*. *J. Exp. Med.* **144**, 69–78.

Seeds, N. W. (1974). Cerebellar specific cell surface antigens. *J. Cell Biol.* **63**, 307.

Seeds, N. W. (1975). Cerebellar cell surface antigens of mouse brain. *Proc. Natl. Acad. Sci. U.S.A.* **72**, 4110–4414.

Sensenbrenner, M. (1977). Dissociated brain cells in primary cultures. *In* "Cell, Tissue and Organ Cultures in Neurobiology" (S. Fedoroff and L. Hertz, eds.), pp. 191–214. Academic Press, New York.

Shearer, G. M., Simpson, E., Weinstein, Y., and Melmon, K. L. (1977). Fractionation of lymphocytes involved in the generation of cell-mediated cytotoxicity over insolubilized conjugated histamine columns. *J. Immunol.* **118**, 756–761.

Sieber-Blum, M., and Cohen, A. M. (1978). Lectin binding to neural crest cells. Changes of the cell surface during differentiation *in vitro*. *J. Cell Biol.* **76**, 628–638.

Simpson, D. L., Thorne, D. R., and Loh, H. H. (1977). Developmentally regulated lectin in neonatal rat brain. *Nature (London)* **266**, 367–369.

Simpson, D. L., Thorne, D. R., and Loh, H. H. (1978). Lectins: Endogenous carbohydrate-binding proteins from vertebrate tissues. Functional role in recognition processes? *Life Sci.* **22**, 727–748.

Skaper, S. D., Adler, R., and Varon, S. (1979). A procedure for purifying neuron-like cells in cultures from central nervous tissues with a defined medium. *Dev. Neurosci.* **2**, 233–237.

Soderman, D. D., Gemershausen, J., and Katzen, H. M. (1973). Affinity binding of intact fat cells and their ghosts to immobilized insulin. *Proc. Natl. Acad. Sci. U.S.A.* **70**, 792–796.

Stallcup, W., and Cohn, M. (1976). Correlation of surface antigens and cell type in cloned cell lines from the rat CNS. *Exp. Cell Res.* **98**, 285–297.

SundarRaj, N., Schachner, M., and Pfeiffer, S. (1975). Biochemically differentiated mouse glial lines carrying a nervous system specific cell surface antigen (NS-1). *Proc. Natl. Acad. Sci. U.S.A.* **72**, 1927–1931.

Tal, E., Savion, S., Hanna, N., and Abraham, M. (1978). Separation of rat pituitary thyrotrophic cells. *J. Endocrinol.* **78**, 141–146.

Teichberg, V. I., Silman, I., Beitsch, D. D., and Resheff, G. (1975). A β-D-galactoside binding protein from electric organ tissue of *Electrophorus electricus*. *Proc. Natl. Acad. Sci. U.S.A.* **72**, 1383–1387.

Thiery, J. P., Brackenbury, R., Rutishauser, U., and Edelman, G. M. (1977). Adhesion among neural cells of the chick embryo. II. Purification and characterization of a cell adhesion molecule from neural retina. *J. Biol. Chem.* **252**, 6841–6845.

Toran-Allarand, C. D. (1978). Gonadal hormones and brain development: Cellular aspects of sexual differentiation. *Am. Zool.* **18**, 553–565.

Truffa-Bachi, P., and Wofsy, L. (1970). Specific separation of cells on affinity columns. *Proc. Natl. Acad. Sci. U.S.A.* **66**, 685–692.

Vaheri, A., Ruaslahti, E., Westermark, B., and Pontén, J. (1976). A common cell-type specific surface antigen in cultured human glial cells and fibroblasts: Loss in malignant cells. *J. Exp. Med.* **143**, 64–72.

Varon, S. (1975). Nerve growth factor and its mode of action. *Exp. Neurol.* **48**, 75–92.

Varon, S. (1977a). Neural cell isolation and identification. *In* "Cell, Tissue and Organ

Cultures in Neurobiology" (S. Fedoroff and L. Hertz, eds.), pp. 237–261. Academic Press, New York.

Varon, S. (1977b). Neural growth and regeneration: A cellular perspective. *Exp. Neurol.* **54**, 1–6.

Varon, S. (1978). Macromolecular glial markers. *In* "Dynamic Properties of Glial Cells" (E. Schoffeniels, G. Franck, L. Hertz, and D. Tower, eds.), pp. 93–103. Pergamon, Oxford.

Varon, S. (1979). Culture of chick embryo dorsal root ganglion cells on polylysine-coated plastic *Neurochem. Res.* **4**, 155–173.

Varon, S., and Bunge, R. (1978). Trophic mechanisms in the peripheral nervous system. *Annu. Rev. Neurosci.* **1**, 327–362.

Varon, S., and Raiborn, C. W. (1969). Dissociation, fractionation and culture of embryonic brain cells. *Brain Res.* **12**, 180–199.

Varon, S., and Raiborn, C. (1972). Dissociation, fractionation and culture of chick embryo sympathetic ganglionic cells. *J. Neurocytol.* **1**, 211–221.

Varon, S., and Saier, M. (1975). Culture techniques and glial-neuronal interrelationships *in vitro*. *Exp. Neurol.* **48**, 135–162.

Varon, S., and Somjen, G. (1979). Neuron-glia interactions. *Neurosci. Res. Program, Bull.* **17**, 1–239.

Varon, S., Raiborn, C., and Tyszka, E. (1973). *In vitro* studies of dissociated cells from newborn mouse dorsal root ganglia. *Brain Res.* **54**, 51–63.

Varon, S., Au, A. M.-J., Hewitt, E., and Adler, R. (1979a). An immunocytotoxicity assay for neural cells in monolayer cultures. *Exp. Cell Res.* **120**, 257–267.

Varon, S., Manthorpe, M., and Adler, R. (1979b). Cholinergic neuronotrophic factors. I. Survival, neurite outgrowth and choline acetyltransferese activity in monolayer cultures from chick embryo ciliary ganglia. *Brain Res.* **173**, 29–45.

Venter, B. R., Venter, J. C., and Kaplan, N. O. (1976). Affinity isolation of cultured tumor cells by means of drugs and hormones covalently bound to glass and Sepharose beads. *Proc. Natl. Acad. Sci. U.S.A.* **73**, 2013–2017.

Villegas, J., Sevcik, C., Barnola, F. V., and Villegas, R. (1976). Grayanotoxin-, veratrine-, and tetrodotoxin-sensitive sodium pathways in the Schwann cell membrane of squid nerve fiber. *J. Gen. Physiol.* **67**, 369–380.

Vogel, Z., Maloney, J., Ling, A., and Daniels, M. P. (1977). Identification of synaptic acetylcholine receptor sites in retina with peroxidase-labelled α-bungarotoxin. *Proc. Natl. Acad. Sci. U.S.A.* **74**, 3268–3272.

Waksman, B. H., and Adams, R. D. (1956). A comparative study of experimental allergic neuritis in the rabbit, guinea pig, and mouse. *J. Neuropathol. Exp. Neurol.* **15**, 293–314.

Weetall, H. H. (1969). Trypsin and papain covalently coupled to porous glass: Preparation and characterization. *Science* **166**, 615–617.

Weinstein, Y., Melmon, K. L., Bourne, H. R., and Sela, M. (1973). Specific leukocyte receptors for small endogenous hormones: Detection by cell binding to insolubilized hormone preparations. *J. Clin. Invest.* **52**, 1349–1361.

Westermark, B. (1976). Density dependent proliferation of human glia cells stimulated by epidermal growth factor. *Biochem. Biophys. Res. Commun.* **69**, 304–310.

Westermark, B., and Wasteson, A. (1975). The response of cultured human normal glial cells to growth factors. *Adv. Metab. Disord.* **8**, 85–100.

Weston, P. D., and Avrameas, S. (1971). Proteins coupled to polyacrylamide beads using glutaraldehyde. *Biochem. Biophys. Res. Commun.* **45**, 1574–1580.

Wigzell, H., Sundquist, K. G., and Yoshida, T. O. (1972). Separation of cells according to surface antigens by the use of antibody-coated columns. Fractionation of cells carrying immunoglobulins and blood group antigens. *Scand. J. Immunol.* **1**, 75–87.

Williams, A., Barklay, A. N., Letarte-Muirhead, M., and Morris, R. J. (1977). Rat Thy-1 antigens from thymus and brain: Their tissue distribution, purification and chemical composition. *Cold Spring Harbor Symp. Quant. Biol.* **41**, 51–61.

Wood, P. (1976). Separation of functional Schwann cells and neurons from normal peripheral nerve tissue. *Brain Res.* **115**, 361–375.

Yamada, K. M., Yamada, S. S., and Pastan, I. (1975). The major cell surface glycoprotein of chick embryo fibroblasts is an agglutinin. *Proc. Natl. Acad. Sci. U.S.A.* **72**, 3158–3162.

Yuan, D., Vitetta, E., and Schachner, M. (1977). Partial characterization of nervous system-specific cell surface antigen(s) NS-2. *J. Immunol.* **118**, 551–557.

Zimmermann, A., and Schachner, M. (1976). Nervous system antigen-5, an antigenic cell surface component of neuroectodermal origin. *Brain Res.* **115**, 297–310.

Zwerner, R. K., Acton, R. T., and Seeds, N. W. (1977). The developmental appearance of Thy-1 in mouse reaggregating brain cell cultures. *Dev. Biol.* **60**, 331–335.

SUBJECT INDEX

A

α-Adrenergic antagonist, 36
α-Adrenergic receptor, 36, 41, 42, 47, 50–51
α-Ketoglutarate, 158, 159
α-receptor, *see* α-Adrenergic receptor
A1 adenosine receptor, 41, 52
A2 adenosine receptor, 41, 52
100–A filament, 290
Acetylcholine, 39, 40, 41, 50, 52, 53, 121, 153
    160, 248, 249, 356, 360, 362, 366, 369
  in early development, 249
  as growth-regulating factor, 248
  synthesis in invertebrate neurons, 155
Acetylcholine receptor, 4, 42, 422
Acetylcholinesterase, 72, 76, 86, 87, 97, 153,
    155, 230, 239–249, 349, 363
  activity in dopaminergic neurons, 349
  in aging, 240, 243, 244
  in cerebellum, 244, 245
  in cerebral hemispheres, 244
  in clonal cell lines, 246
  in cultured neurons, 247
  dendritic secretion of, 87
  development, 242, 244
  in diencephalon, 244
  in dissociated brain cell cultures, 247
  effect of hatching, 243
  effect of light stimulation, 243
  effect of transmitters, 241
  isoenzymes, 245
  in mammalian brain, 240
  as marker, 239
  maturation, 240
  in midbrain, 244
  in neuroblastoma cells, 248
  nonspecific, *see* butyrylcholinesterase
  in optic lobes, 243
  in reaggregated cultures, 246
  specific, 239
Acetyltransferase, 230
ACh, *see* Acetylcholine
AChE, *see* Acetylcholinesterase
Acid phosphatase, 88, 152, 157
  axonal transport of, 88
Actin, 128, 129, 287
Actinomycin D, 43, 46, 94, 133, 135, 145
Action potential, 124, 135
Adenine, 131, 133, 135
Adenosine, 40, 41, 50, 269, *see also* A1 and
    A2 receptors
  effect on cyclic AMP, 40–41
Adenosine diphosphate, 159
Adenosine receptor, 41
Adenosine triphosphatase, 81, 152, 265–267
Adenosine triphosphate, 82, 121, 152, 157–
    159
Adenylate cyclase, 35, 37, 41, 43, 46, 48, 51,
    52, 268–269, 270
  catylytic component, 268
  development of, 269
  in glial cells, 270
  hormone sensitivity, 269
  inhibition, 48
  in neurons, 270
  regulatory component, 268
Adhering junction, 4, 6
ADP ribosyltransferase, 35
ADP, *see* Adenosine diphosphate
Adrenergic projections, 54
  lack of synapses, 54
Adrenergic receptors, *see* α-Adrenergic
    receptor and β-Adrenergic receptor
Affinity chromatography, 393

443